VOLUME ONE HUNDRED AND FORTY NINE

ADVANCES IN CANCER RESEARCH

Mechanisms and Therapy of Liver Cancer

VOLUME ONE HUNDRED AND FORTY NINE

ADVANCES IN CANCER RESEARCH

Mechanisms and Therapy of Liver Cancer

Edited by

DEVANAND SARKAR

Department of Human and Molecular Genetics; VCU Institute of Molecular Medicine, Virginia Commonwealth University, School of Medicine; VCU Massey Cancer Center, Virginia Commonwealth University, Richmond, VA, United States

PAUL B. FISHER

Department of Human and Molecular Genetics, VCU Institute of Molecular Medicine; VCU Massey Cancer Center, Virginia Commonwealth University, School of Medicine, Richmond, VA, United States

ELSEVIER

ACADEMIC PRESS

An imprint of Elsevier

Academic Press is an imprint of Elsevier
50 Hampshire Street, 5th Floor, Cambridge, MA 02139, United States
525 B Street, Suite 1650, San Diego, CA 92101, United States
The Boulevard, Langford Lane, Kidlington, Oxford OX5 1GB, United Kingdom
125 London Wall, London, EC2Y 5AS, United Kingdom

First edition 2021

ISBN: 978-0-12-824030-4
ISSN: 0065-230X

For information on all Academic Press publications
visit our website at https://www.elsevier.com/books-and-journals

Publisher: Zoe Kruze
Editorial Project Manager: Leticia Lima
Production Project Manager: James Selvam
Cover Designer: Alan Studholme

Typeset by SPi Global, India

Contents

Contributors ix
Preface xi

1. Hepatocellular carcinoma (HCC): Epidemiology, etiology and molecular classification **1**

Saranya Chidambaranathan-Reghupaty, Paul B. Fisher, and Devanand Sarkar

1. Introduction 6
2. Etiology and risk factors 6
3. Liver cancer staging 31
4. Molecular classification of HCC 34
5. Discussion 47

Acknowledgments 48
References 48

2. Signaling pathways in hepatocellular carcinoma **63**

Teresa Garcia-Lezana, Juan Luis Lopez-Canovas, and Augusto Villanueva

1. Introduction 64
2. Pathways involving growth factors 65
3. Pathways involved in cell differentiation and development 72
4. Nuclear signaling pathways 79
5. Emerging pathways 82
6. Conclusions 87

Acknowledgment 87
Conflict of interest 87
References 87

3. Hepatitis C virus associated hepatocellular carcinoma **103**

Mousumi Khatun, Ranjit Ray, and Ratna B. Ray

1. Introduction 104
2. HCV infection and immune responses 106
3. Risk factors for development of HCC from chronic infection 108
4. HCV regulation on metabolic pathways 109
5. Mechanisms of HCV associated HCC 111

6. Biomarkers for detection of HCV associated HCC 122
7. Therapeutic strategies for HCV associated HCC 123
8. Perspectives 125
Acknowledgments 127
References 127

4. NAFLD-related HCC 143

Bubu A. Banini and Arun J. Sanyal

1. The problem of NAFLD-related HCC 144
2. Metabolic syndrome and NAFLD in driving HCC 146
3. NAFLD-related HCC 149
4. Molecular mechanisms of NAFLD-related HCC 150
5. Prevention and surveillance of NAFLD-related HCC 156
6. Management of NAFLD-related HCC 157
7. Conclusions/future directions 158
Acknowledgment 159
Disclosures 159
References 160

5. Atypical immunometabolism and metabolic reprogramming in liver cancer: Deciphering the role of gut microbiome 171

Rachel M. Golonka and Matam Vijay-Kumar

1. Hepatocellular carcinoma (HCC): A heterogenous and complex liver disease 174
2. Gut microbiota dysbiosis as an identifier of HCC 177
3. HCC microenvironment favors immunosuppression: Is gut microbiota a driver of this immunologic signature? 179
4. Hijacking energy sources: Does gut microbiota have an influence on metabolic reprogramming in HCC? 201
5. Immunological and gut microbiota therapeutics for HCC 209
6. Summary and future perspectives 217
Acknowledgment 218
References 218

6. Systemic therapy of liver cancer 257

Tarik Demir, Sunyoung S. Lee, and Ahmed O. Kaseb

1. Introduction 258
2. Staging 259
3. Liver function 260
4. Current systemic treatment 262

5. Histological variants 283
6. Conclusion 285
Acknowledgment 286
References 286

7. Immunotherapy in hepatocellular cancer 295

Christos Fountzilas, Rachel Evans, Sabah Alaklabi, and Renuka Iyer

1. Introduction 295
2. Immune checkpoint inhibitors in hepatocellular carcinoma 298
3. Future directions 307
4. Conclusion 312
References 312

8. Intrahepatic cholangiocarcinoma: Morpho-molecular pathology, tumor reactive microenvironment, and malignant progression 321

Alphonse E. Sirica, Mario Strazzabosco, and Massimiliano Cadamuro

1. Introduction 322
2. iCCA classification, cell origins, and molecular subtypes 324
3. Fibroinflammatory cholangiopathies, mechanisms, and cholangiocarcinogenesis 339
4. The desmoplastic reaction and iCCA progression 344
5. Immune milieu in iCCA 355
6. Challenges and clinical implications 359
Grant Support 367
Conflict of interest statement 367
References 367

Contributors

Sabah Alaklabi
Roswell Park Comprehensive Cancer Center, Buffalo, NY, United States

Bubu A. Banini
Section of Digestive Diseases, Yale University, New Haven, CT, United States

Massimiliano Cadamuro
Department of Molecular Medicine, University of Padova, Padova, Italy

Saranya Chidambaranathan-Reghupaty
C. Kenneth and Dianne Wright Center for Clinical and Translational Research, Virginia Commonwealth University, Richmond, VA, United States

Tarik Demir
Department of Gastrointestinal Medical Oncology, The University of Texas M.D. Anderson Cancer Center, Houston, TX, United States

Rachel Evans
Roswell Park Comprehensive Cancer Center, Buffalo, NY, United States

Paul B. Fisher
Department of Human and Molecular Genetics; VCU Institute of Molecular Medicine, Virginia Commonwealth University, School of Medicine; VCU Massey Cancer Center, Virginia Commonwealth University, Richmond, VA, United States

Christos Fountzilas
Roswell Park Comprehensive Cancer Center, Buffalo, NY, United States

Teresa Garcia-Lezana
Division of Liver Diseases, Liver Cancer Program, Department of Medicine, Tisch Cancer Institute, Icahn School of Medicine at Mount Sinai, New York, NY, United States

Rachel M. Golonka
Department of Physiology and Pharmacology, University of Toledo College of Medicine and Life Sciences, Toledo, OH, United States

Renuka Iyer
Roswell Park Comprehensive Cancer Center, Buffalo, NY, United States

Ahmed O. Kaseb
Department of Gastrointestinal Medical Oncology, The University of Texas M.D. Anderson Cancer Center, Houston, TX, United States

Mousumi Khatun
Department of Pathology, Saint Louis University, St. Louis, MO, United States

Sunyoung S. Lee
Department of Gastrointestinal Medical Oncology, The University of Texas M.D. Anderson Cancer Center, Houston, TX, United States

Juan Luis Lopez-Canovas
Department of Cell Biology, Physiology and Immunology, Maimonides Institute of Biomedical Research of Córdoba (IMIBIC), University of Córdoba, Córdoba, Spain

Ranjit Ray
Department of Internal Medicine, Saint Louis University, St. Louis, MO, United States

Ratna B. Ray
Department of Pathology, Saint Louis University, St. Louis, MO, United States

Arun J. Sanyal
Division of Gastroenterology, Hepatology and Nutrition, Virginia Commonwealth University, Richmond, VA, United States

Devanand Sarkar
Department of Human and Molecular Genetics; VCU Institute of Molecular Medicine, Virginia Commonwealth University, School of Medicine; VCU Massey Cancer Center, Virginia Commonwealth University, Richmond, VA, United States

Alphonse E. Sirica
Department of Pathology, Virginia Commonwealth University School of Medicine, Richmond, VA, United States

Mario Strazzabosco
Liver Center and Section of Digestive Diseases, Department of Internal Medicine, Section of Digestive Diseases, Yale University School of Medicine, New Haven, CT, United States

Matam Vijay-Kumar
Department of Physiology and Pharmacology, University of Toledo College of Medicine and Life Sciences, Toledo, OH, United States

Augusto Villanueva
Division of Liver Diseases, Liver Cancer Program, Department of Medicine, Tisch Cancer Institute; Division of Hematology and Medical Oncology, Department of Medicine, Icahn School of Medicine at Mount Sinai, New York, NY, United States

Preface☆

Globally, primary liver cancer is the fifth most common cancer and the second most common cause of cancer-related deaths in males. Primary liver cancer shows male predominance with male to female ratio of 3–4:1 and as such in females the global incidence is not significant although it is the sixth most common cause of cancer-related deaths. In the United States in 2020, the estimated new cases of liver cancer in both sexes are 42,810 of which 30,160 are expected to die. These statistics demonstrate that the primary liver cancer is a fatal disease and this grim scenario is further complicated by the fact that the projected incidence and mortality of liver cancer will continue to show steep increases at least until 2030. The disease arises from a variety of etiologies all causing chronic inflammation. Viral hepatitis from HBV and HCV is the most common cause of liver cancer globally. However, because of availability of HBV vaccine and a cocktail of drugs that can efficiently mitigate HCV, viral hepatitis is gradually becoming a less important cause of liver cancer. Nevertheless, liver cancer incidence is still projected to disconcertingly increase because of a direct causative contribution of fatty liver disease or nonalcoholic steatohepatitis (NASH) developing as a consequence of obesity, a worldwide pandemic. Because of this connection, liver cancer is getting a significant focus from researchers and clinicians from a variety of backgrounds centered on understanding the disease process and developing effective therapies.

Liver cancer is inherently resistant to conventional chemo- and radiotherapy. Liver is the primary metabolic organ in the body and the preceding events leading to chronic inflammation, fibrosis, and eventually cirrhosis greatly compromise liver function, and against this backdrop liver cancer develops. As such, the same drug that works well in other cancers does not work well in liver cancer patients because of the inability of liver to metabolize the drug resulting in significant toxicity which reduces a patient's compliance in taking the drug. Liver transplantation is indicated in early stage of the disease. However, in most cases the disease is diagnosed at later stages, which is managed by strategies to reduce the size of the primary tumor by radiofrequency ablation or transarterial chemoembolization and a variety of tyrosine kinase

☆This volume is dedicated to a close colleague and friend, Dr. Zhao-zhong Su, who sadly died of HCC. It is our hope that continued research in this area will lead to improved methods of prevention and ultimately a cure for this aggressive and invariably fatal cancer.

inhibitors (TKIs), the most common of which is sorafenib. In advanced stages of the disease, these TKIs provide a survival benefit of only 2–3 months. Recent studies on immunotherapy in liver cancer are showing promise and many combinatorial immunotherapy trials are currently ongoing.

In this thematic issue, we endeavored to assemble chapters that comprehensively address different aspects of liver cancer, which will be of interest to both basic researchers and practicing clinicians. In Chapter 1 (Chidambaranathan-Reghupaty, Fisher, and Sarkar), the epidemiology and etiology of hepatocellular carcinoma (HCC), which arises from hepatocytes and comprises more than 80% of primary liver cancer cases, are described in detail. Efforts are focused on generating comprehensive profiles of molecular abnormalities in HCC patients through in-depth next-generation sequencing worldwide. The classification of HCC based on molecular, metabolic, and immunologic abnormalities is highlighted in this chapter which will facilitate stratifying patients and matching them with appropriate therapies. This chapter also discusses in detail the mechanism by which HCC develops from HBV infection, which remains quite common in developing countries. The oncogenic signaling pathways that are perturbed in HCC are discussed in Chapter 2 (Garcia-Lezana, Lopez-Canovas, and Villanueva) with subsequent discussion of the ways these vulnerabilities are exploited to develop targeted therapies. Even though effective drugs are now available, HCV is still a major cause of HCC in the Western world. As such a whole chapter, Chapter 3 (Khatun, Ray, and Ray) is written on the mechanism of HCV infection resulting in HCC and the biomarkers used to diagnose the disease and the currently available therapeutic approaches.

As discussed earlier, obesity-induced NASH is becoming the most important cause of HCC globally. HCC is usually associated with advanced ages. However, with a rising shift in preteen and teen overweight and obese individuals HCC is progressively being diagnosed in younger patients. Understanding the metabolic consequences of NASH that contribute to HCC is essential and identifying molecular changes in this disease process is required to develop therapies that can target both the metabolic problems as well as cancer. NASH-HCC is a growing field requiring appropriate mouse models for studies the results of which can be correlated to the human disease process. In Chapter 4 (Banini and Sanyal), a comprehensive picture of the current scenario, mechanistic and translational, of NASH-HCC is provided. High fat and sugar containing diet changes the gut microbiome, damages the barrier function of the gut facilitating seepage of bacterial endotoxins that contribute to chronic inflammation in the liver. In this context, the gut microbiome plays a very important role in regulating HCC development and progression and indeed probiotic treatment has shown significant efficacy in harnessing NASH-HCC progression. The mechanisms by which the gut microbiome regulate the metabolic and immune functions to cause HCC and the ways these findings are leveraged to develop appropriate therapies are delineated in Chapter 5 (Golonka and Vijay-Kumar).

Chapters 6 (Demir, Lee, and Kaseb) and 7 (Fountzilas, Evans, Alkalabi, and Iyer) are devoted to system therapy and immunotherapy of liver cancer. In Chapter 6, the first and second lines of treatments of HCC are discussed. Additionally, management of rarer histological variants of primary liver cancer, such as mixed HCC and cholangiocarcinoma and fibrolamellar HCC, a disease predominantly affecting young individuals with a specific mutational signature, is described. With recent FDA approval of immunotherapy for advanced HCC, the entire Chapter 7 is dedicated to immunotherapy of HCC. This is still a developing field with a multitude of ongoing clinical trials and this chapter discusses not only the current status, but also the future directions with comprehensive detail. Chapter 8 (Sirica, Strazzabosco, and Cadamuro) discusses intrahepatic cholangiocarcinoma (iCCA) which is the second most common primary liver cancer. iCCA is characterized by a desmoplastic milieu and this chapter deals with the challenges of treating this cancer and the ways the signaling pathways in fibroblasts are targeted to overcome these challenges.

We considered developing a comprehensive thematic volume on primary liver cancer would be timely and important because of the alarming increase in its incidence and mortality. With the development of more appropriate mouse models, which closely resemble human disease, and generation of mouse models with humanized liver along with utilization of cutting-edge NGS technology at single cell level, exciting new findings are coming out every day from a large volume of research in HCC and iCCA. These mechanistic findings coupled with clinical trials using combinatorial therapy involving immune modulators bode well for the future with the hope that this fatal disease might be more efficiently managed prolonging and saving lives of scores of patients.

Devanand Sarkar and Paul B. Fisher
Virginia Commonwealth University, Richmond, VA

CHAPTER ONE

Hepatocellular carcinoma (HCC): Epidemiology, etiology and molecular classification

Saranya Chidambaranathan-Reghupaty[a], Paul B. Fisher[b,c,d], and Devanand Sarkar[b,c,d,]*

[a]C. Kenneth and Dianne Wright Center for Clinical and Translational Research, Virginia Commonwealth University, Richmond, VA, United States
[b]Department of Human and Molecular Genetics, Virginia Commonwealth University, School of Medicine, Richmond, VA, United States
[c]VCU Institute of Molecular Medicine, Virginia Commonwealth University, School of Medicine, Richmond, VA, United States
[d]VCU Massey Cancer Center, Virginia Commonwealth University, Richmond, VA, United States
*Corresponding author: e-mail address: devanand.sarkar@vcuhealth.org

Contents

1. Introduction 6
2. Etiology and risk factors 6
   2.1 Viral infections 7
   2.2 Alcohol 19
   2.3 Non-alcoholic fatty liver disease (NAFLD) 21
   2.4 Aflatoxins 22
   2.5 Predisposing genetic factors 23
3. Liver cancer staging 31
   3.1 Child-Pugh (CP) score 31
   3.2 Model for end-stage liver disease (MELD) 32
   3.3 Tumor node metastasis (TNM) system 32
   3.4 The Okuda system 33
   3.5 Cancer of the liver Italian program (CLIP) system 33
   3.6 Barcelona clinic liver cancer (BCLC) system 34
4. Molecular classification of HCC 34
   4.1 Gene signature-based classification of HCC 38
   4.2 Metabolic classification of HCC 43
   4.3 Immunological classification of HCC 44
   4.4 Chromosomal classification of HCC 46
5. Discussion 47
Acknowledgments 48
References 48

Advances in Cancer Research, Volume 149
ISSN 0065-230X
https://doi.org/10.1016/bs.acr.2020.10.001

Abstract

Hepatocellular carcinoma (HCC), the primary malignancy of hepatocytes, is a diagnosis with bleak outcome. According to National Cancer Institute's SEER database, the average five-year survival rate of HCC patients in the US is 19.6% but can be as low as 2.5% for advanced, metastatic disease. When diagnosed at early stages, it is treatable with locoregional treatments including surgical resection, Radio-Frequency Ablation, Trans-Arterial Chemoembolization or liver transplantation. However, HCC is usually diagnosed at advanced stages when the tumor is unresectable, making these treatments ineffective. In such instances, systemic therapy with tyrosine kinase inhibitors (TKIs) becomes the only viable option, even though it benefits only 30% of patients, provides only a modest (~3 months) increase in overall survival and causes drug resistance within 6 months. HCC, like many other cancers, is highly heterogeneous making a one-size fits all option problematic. The selection of liver transplantation, locoregional treatment, TKIs or immune checkpoint inhibitors as a treatment strategy depends on the disease stage and underlying condition(s). Additionally, patients with similar disease phenotype can have different molecular etiology making treatment responses different. Stratification of patients at the molecular level would facilitate development of the most effective treatment option. With the increase in efficiency and affordability of "omics"-level analysis, considerable effort has been expended in classifying HCC at the molecular, metabolic and immunologic levels. This review examines the results of these efforts and the ways they can be leveraged to develop targeted treatment options for HCC.

List of abbreviations

A1AT	alpha 1-antitrypsin
A1ATD	alpha 1-antitrypsin deficiency
ABCB1	ATP binding cassette subfamily B member 1
ADAM17	ADAM metallopeptidase domain 17
ADAR1	RNA-specific adenosine deaminase
ADH	alcohol dehydrogenase
ADPRM	ADP-ribose/CDP-alcohol diphosphatase
AEG-1	Astrocyte elevated gene-1
AFB1	aflatoxin B
AFL	alcohol-associated fatty liver
AFP	alpha fetoprotein
AH	alcoholic hepatitis
AHP	acute hepatic porphyria
AHRR	aryl hydrocarbon receptor repressor
AIH	autoimmune hepatitis
AIP	acute intermittent porphyria
AJCC	American Joint Committee on Cancer
ALAD	aminolevulinic acid dehydratase deficiency
ALBI	albumin-bilirubin
ALD	alcohol-related liver disease
ALDH	aldehyde dehydrogenase
anti-LKM1	anti-liver/kidney microsomal type 1

AP	acute porphyria
APC	adenomatous polyposis coli
API5	apoptosis Inhibitor-5
ARHGEF12	Rho guanine nucleotide exchange factor 12
ARID	AT-rich interaction domain
AXIN1	axis inhibition protein 1
BCL2L10	BCL2-Like 10
BCL2L2	BCL2-like protein 2
BCLC	Barcelona clinic liver cancer
BubR1	benzimidazol related protein 1
CALD1	caldesmon 1
CBX7	chromobox 7
cccDNA	covalently closed circular DNA
CCNE1	cyclin E1
CD	cluster of differentiation
CDH1	cadherin 1
CDK6	cyclin-dependent Kinase 6
CDKN2A	cyclin dependent kinase inhibitor 2A
CEP	congenital erythropoietic porphyria
CLDN15	claudin 15
CLIP	cancer of the liver Italian program
COBL	cordon-Bleu WH2 repeat protein
COX2	cyclooxygenase-2
CP	Child-Pugh
CTLA4	cytotoxic T lymphocyte associated protein 4
CTNNB1	catenin beta 1
DAMPs	damage-associated molecular patterns
DEGs	differentially expressed genes
DNL	*de novo* lipogenesis
DRB1	DR beta 1
dslDNA	double stranded linear DNA
ECM	extracellular matrix
ECOG	eastern cooperative oncology group
EGF	epidermal growth factor
Enh1	enhancer 1
EpCAM	epithelial cellular adhesion molecule
EPHA1	ephrin type-A receptor 1
EPP	erythropoietic protoporphyria
ER	endoplasmic reticulum
ETC	electron transport chain
EZH2	enhancer of Zeste 2 polycomb repressive complex 2 subunit
FAA	4-fumarylacetoacetate
FAO	fatty acid β-oxidation
FAS	Fas cell surface death receptor
FASN	fatty acid synthase
FFA	free fatty acids
FGF	fibroblast growth factor

FN1	fibronectin 1
Foxa2	forkhead box A2
FOXO3a	forkhead box O3a
FZD7	frizzled class receptor 7
GAB1	GRB2-associated binding protein 1
GSD	glycogen storage disease
HBcAg	HBV core antigen
HBeAg	HBV pre-core antigen
HBV	hepatitis B virus
HBx	HBV regulatory protein
HBXIP	hepatitis B virus X-interacting protein
HCC	hepatocellular carcinoma
HCP	hereditary coproporphyria
HCV	hepatitis C virus
HDAC2	histone deacetylase 2
HH	hereditary hemochromatosis
HLA	human leukocyte antigen
HNF4A	hepatocyte nuclear factor 4 alpha
HOTTIP	HOXA Transcript at the distal tip
HSC	hepatic stellate cells
HT1	hereditary tyrosinemia type 1
hTERT	human telomerase reverse transcriptase
ICC	intrahepatic cholangiocarcinoma
IGF	insulin like growth factor
IL	interleukin
INR	international normalized ratio
IRES	internal ribosome entry site
IRF2	interferon regulatory factor 2
IRS1	insulin receptor substrate 1
ISGs	interferon stimulated genes
JAG1	jagged 1
JNK	c-Jun N-terminal kinase
KLF5	Kruppel-like factor 5
KRT32	keratin 32
LEF	lymphoid enhancer factor
LEF1	lymphoid enhancer binding factor 1
lncRNAs	long non-coding RNAs
LPS	lipopolysaccharide
MACC1	metastasis-associated gene in colon cancer 1
MAD1L1	mitotic arrest deficient 1 like 1
MCAM	melanoma cell adhesion molecule
MD	myeloid differentiation factor
MELD	model for end-stage liver disease
miRNA	microRNA
MLL4	mixed lineage leukemia gene homolog 4
MYCN	N-myelocytomatosis
MYOD1	myogenic differentiation 1

NASH	non-alcoholic steatohepatitis
NDRG3	N-myc downstream-regulated gene 3
NE	neutrophil elastase
NFE2L2	nuclear factor, erythroid 2 Like 2
NF-κB	nuclear factor kappa-light-chain-enhancer of activated B cells
NK	natural killer
NOD	nucleotide-binding oligomerization domain
NRG3	neuregulin 3
NTBC	2-(2 nitro-4-3 trifluoro-methylbenzoyl)-1, 3-cyclohexanedione
NTPase	nucleoside-triphosphatase
OAS1	5′-oligoadenylate synthetase 1
ORF	open reading frame
PAMPs	pathogen-associated molecular patterns
PCT	porphyria cutanea tarda
PD1	programmed cell death protein 1
PDL1	programmed death ligand 1
PHF2	plant homeodomain finger protein 2
PIK3CA	phosphatidylinositol-4,5-bisphosphate 3-kinase catalytic subunit alpha
PKR	protein kinase R
PLG	plasminogen
POLD2	DNA polymerase delta 2
PPAR	peroxisome proliferator-activated receptor
PPR	pattern recognition receptors
PS	performance status
PTEN	phosphatase and tensin homolog
PTPRD	protein tyrosine phosphatase receptor type-D
RBFOX1	RNA binding fox-1 homolog 1
rcDNA	relaxed circular DNA
RIG-I	retinoic acid-inducible gene I
ROCK1	Rho associated coiled coil containing protein kinase 1
RPS6	ribosomal protein S6
RPS6KA3	ribosomal protein S6 kinase A3
SERPIN	SERine proteinase INhibitor
SOX9	SRY-box transcription factor 9
SREBP-1	sterol regulatory element-binding protein 1
STAT1	signal transducer and activator of transcription 1
TACE	trans-arterial chemoembolization
TCF1	T cell factor 1
TGF	transforming growth factor
TKI	tyrosine kinase inhibitor
TLR	toll-like receptor
TNFSF14	tumor necrosis factor superfamily member 14
TNM	tumor node metastasis
Treg	regulatory T-cells
TXNRD1	thioredoxin reductase 1
UGT1A6	UDP glucuronosyltransferase family 1 member A6
UICC	Union for International Cancer Control

UNC5D	UNC-5 netrin receptor D
UTR	untranslated region
VEGF	vascular endothelial growth factor
VP	variegate porphyria
WHO	World Health Organization
XLDPP	X-linked dominant protoporphyria
ZEB2	zinc finger E-box binding homeobox 2

1. Introduction

Liver cancer is the fifth most common cancer and the fourth leading cause of cancer-related deaths worldwide. Among men, it is the fourth most frequent cancer and the second leading cause of cancer-related deaths. Men are at a higher risk of developing liver cancer compared to women. The global HCC incidence ratio of men to women is 2.8:1 (Kulik & El-Serag, 2019). It is worth noting that, despite being only the ninth most common cancer that affects women, it is the fourth leading cause of cancer-related deaths in women, emphasizing its high mortality rate (Bray et al., 2018). There are two main types of primary liver cancers—hepatocellular carcinoma (HCC) and intrahepatic cholangiocarcinoma (ICC), with less frequent cancers like angiosarcoma, hemangiosarcoma and hepatoblastoma. While ICC originates from bile ducts, HCC arises from hepatocytes, the main parenchymal cells of the liver. HCC accounts for more than 80% of primary liver cancer cases worldwide (El-Serag & Rudolph, 2007). Secondary liver cancer occurs when a tumor from another part of the body metastasizes to the liver. Although breast, esophageal, stomach, pancreatic, lung, kidney and several other cancers can metastasize to the liver, most secondary liver cancers originate from colorectal cancers. Approximately 70% of colorectal cancer patients develop secondary liver cancer (Valderrama-Trevino, Barrera-Mera, Ceballos-Villalva, & Montalvo-Jave, 2017).

2. Etiology and risk factors

HCC pathogenesis is complex and involves multiple molecular failures such as cell cycle deregulation, DNA methylation alteration, chromosomal instability, immunomodulation, Epithelial-to-Mesenchymal transition, increase in HCC stem cells and microRNA (miRNA) dysregulation (Hui, Makuuchi, & Li, 1998; Ogunwobi et al., 2019). Although the disease

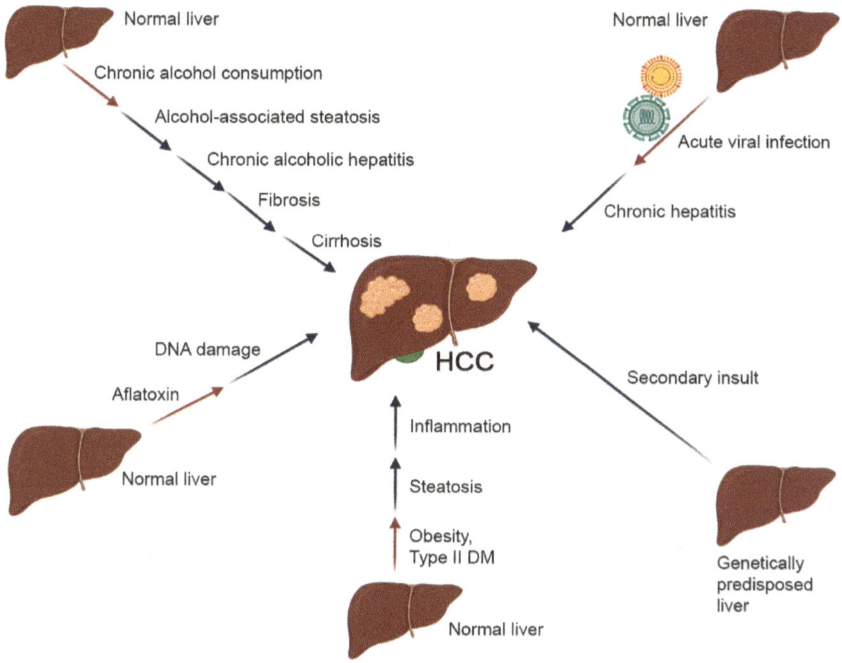

Fig. 1 An overview of different HCC etiologies and the potential mechanism causing HCC. *Created with Biorender.com.*

mechanism varies depending on underlying etiology, the usual sequence is liver injury, chronic inflammation, fibrosis, cirrhosis and HCC (Fig. 1). The release of molecular mediators like Damage-Associated Molecular Patterns (DAMPS) from damaged or necrotic cells, and Pathogen-Associated Molecular Patterns (PAMPS) by viruses elicits an innate immune response by activation of Pattern Recognition Receptors (PRR) like Toll-like receptor (TLR), C-type lectin Receptors, Nucleotide-binding Oligomerization Domain (NOD)-like receptors and retinoic acid-inducible gene I (RIG-I)-like receptors present on various immune cells causing an inflammatory response. An acute inflammatory response is resolved once the external stimulus is eliminated, whereas unresolved chronic inflammation can cause fibrosis and eventually cirrhosis (Villanueva & Luedde, 2016). Most HCCs (80–90%) are preceded by a cirrhotic event (Leong & Leong, 2005).

2.1 Viral infections

Chronic hepatitis caused by Hepatitis B Virus (HBV) and Hepatitis C Virus (HCV) infections is an important risk factor for HCC. HBV and HCV are both blood-borne pathogens and HCV can also be transmitted through other body fluids. Needle sharing, perinatal transmission and other routes of blood contact from an infected individual are major routes of transmission. In regions with high prevalence of these infections, HBV/HCV coinfection can occur, further increasing risk of HCC development (Donato, Boffetta, & Puoti, 1998). The World Health Organization (WHO) estimates that 257 million people or 3.3% of the global population lives with chronic HBV infection whereas 71 Million live with chronic HCV infection. In 2018, 54.5% and 21.2% of new global HCC cases were attributable to HBV and HCV infections, respectively (Bray et al., 2018). There is a difference in global prevalence of HBV and HCV that is based on geography.

WHO estimates global HBV prevalence through detection of Hepatitis B Surface Antigen (HBsAg). To facilitate reporting, the world is divided into six WHO regions and the most recent HBV prevalence data from these regions are as follows—Western Pacific Region (6.2%), African Region (6.1%), Eastern Mediterranean Region (3.3%), South-East Asian Region (2%), European Region (1.6%) and the Americas Region (0.7%). Although there is no specific treatment for HBV infection, it is preventable by a safe and effective vaccine that provides 98–100% protection against the virus and antiviral treatments can slow the progression of liver damage. The probability of an acute infection becoming chronic is inversely related to age. If infection occurs before the age of 1, there is an 80–90% chance of chronic infection and the probability drops to 30–50% at 1–6 years of age and less than 5% in adults.

HCV prevalence, on the other hand, is estimated based on the presence of antibodies to HCV in a person's blood. It is estimated to be the highest (>3.5%) in Central Asia and Central Africa, moderate (1.5–3.5%) in South, East and Southeast Asia, and North, East and West Africa, the Middle East, Southern and Tropical Latin America, the Caribbean (Cuba, Dominican Republic and Puerto Rico), Australasia (Australia and New Zealand), and Eastern Europe and low (<1.5%) in Western and Central Europe, North America, Southern Africa, Andean and Central Latin America and Pacific Asia (Japan and South Korea) (Petruzziello, Marigliano, Loquercio, Cozzolino, & Cacciapuoti, 2016). Roughly 30% of people infected with HCV clear the virus spontaneously within 6 months of infection. The remaining 70% of HCV-infected individuals develop chronic HCV infection. More than 95% of HCV-infected people can be cured with antiviral drugs.

2.1.1 Hepatitis B virus

HBV is a hepatotropic virus with a relaxed circular and partially double stranded, 3kb DNA genome encoding four overlapping Open Reading Frames (ORF)—S, C, P and X (Liang, 2009). ORF S encodes the surface antigen HBsAg. ORF C encodes the core protein (HBcAg) that forms the viral nucleocapsid or the precore protein (HBeAg) that directs the translation product to the Endoplasmic Reticulum (ER). ORF P is translated to a multidomain pol protein with terminal, reverse transcriptase and ribonuclease H domains. ORF X produces a regulatory protein that enhances viral gene expression and replication (Liang, 2009). Once HBV is imported into the nucleus of hepatocytes, the viral genome is repaired to form a covalently closed circular DNA (cccDNA). This repaired DNA is then transcribed to form a pregenomic RNA. The pregenomic RNA is encapsulated and reverse transcribed by DNA polymerase to single-stranded linear DNA, which can reproduce the partially double-stranded relaxed circular DNA (rcDNA) in the cytoplasm or form a double stranded linear DNA (dslDNA). The rcDNA is recycled to the nucleus for synthesis of cccDNA or enveloped and released from the hepatocytes. The dslDNA can be integrated into the host genome and this phenomenon occurs in 85–90% of HBV-associated HCC cases (Furuta et al., 2018; Liang, 2009; Tu, Budzinska, Shackel, & Urban, 2017).

Only 5–10% of adults progress to chronic infection, which is defined as persistence of HBsAg for >6months in a patient's body (Hyams, 1995). A cure for HBV is difficult because of the persistence of the highly stable viral replication intermediate, cccDNA, as a non-integrated episome in the nucleus of infected patients (Zoulim, 2005). Both innate and adaptive immune responses are triggered in response to HBV infection. HBV remains quiescent for a few weeks after infection causing a lack of robust early innate immune responses. Adaptive immunity involving HBV-specific antigens and T cells is the major player in clearing HBV infection (Tan, Koh, & Bertoletti, 2015). Improper T-cell memory maturation causes a lack of HBV antigen-specific T cells and a failure to mount an immune response (Lumley, McNaughton, Klenerman, Lythgoe, & Matthews, 2018). This can cause chronic infection, recurrent liver inflammation, fibrosis, cirrhosis and HCC. HBV integration into the host genome can cause oncogenesis when integration sites are close to genes like human Telomerase Reverse Transcriptase (hTERT) or other genes involved in cell cycle, proliferation and metabolism (Tu et al., 2017). The viral regulatory protein is suspected to have a role in oncogenesis in addition to its role in viral replication. It is involved in modulating oncogenic pathways, chromosomal instability, oxidative stress, DNA methylation, angiogenesis and migration (Tarocchi, Polvani, Marroncini, & Galli, 2014). Cirrhosis develops in 1.3–2.4% of HBV infected patients (Liaw, Tai, Chu, & Chen, 1988) and 3.9% of chronic HBV patients develop HCC (Takano, Yokosuka, Imazeki, Tagawa, & Omata, 1995).

2.1.1.1 Oncogenic properties of HBx

The viral regulatory protein HBx has a role in oncogenesis in addition to its role in viral replication and chromosomal instability. It is involved in modulating oncogenic pathways, oxidative stress, DNA methylation, angiogenesis and migration (Tarocchi et al., 2014). HBx disrupts p53-mediated apoptosis by forming a p53-HBx complex that prevents nuclear localization of p53 (Elmore et al., 1997; Ueda et al., 1995; Wang et al., 1995) and p53 DNA binding (Wang et al., 1994). Cyclooxygenase-2 (COX-2) overexpression by HBx also prevents p53-mediated apoptosis (Cheng, Yu, Lai, Chan, & Sung, 2008). Other anti-apoptotic mechanisms include caspase-3 inactivation *via* the HBx-PI3K-Akt-Bad pathway (Lee, Kang-Park, Do, & Lee, 2001), survivin accumulation *via* the p38/MAPK pathway (Kuo & Chao, 2010), and activation of the WNT/β-catenin pathway (Shen et al., 2013). In addition to promoting cellular proliferation by inhibiting apoptosis, HBx causes liver injury by sensitizing cells to apoptosis. HBx promotes oxidative stress-related apoptosis *via* caspase-3 cascade and loss of anti-apoptotic protein MCL1 (Hu et al., 2011), and can be p53- (Chirillo et al., 1997) or TNFα-mediated (Su & Schneider, 1997). HBx is negatively correlated to E-cadherin, positively correlated to vimentin and increases migration and invasion *in vitro* (Jin, Wu, et al., 2017). HBx increases transcription of Vascular Endothelial Growth Factor (VEGF) (Lee et al., 2000), angiopoietin (Sanz-Cameno et al., 2006) and increases transcriptional activity of HIF-1α by protein stabilization (Moon et al., 2004) to promote angiogenesis. HBx is also involved in aberrant cell cycle (Lee et al., 2002), increased chromosomal aberration and micronuclei formation (Livezey, Negorev, & Simon, 2002) and abnormal mitotic spindle formation (Forgues et al., 2003; Fujii et al., 2006; Kim et al., 2008; Wen, Golubkov, Strongin, Jiang, & Reed, 2008; Yun et al., 2004). HBx binds to different cellular proteins like Hepatitis B virus X-interacting protein (HBXIP) (Fujii et al., 2006; Wen et al., 2008) and the mitotic checkpoint protein, budding uninhibited by benzimidazol related protein 1 (BubR1),

(Kim, Park, et al., 2008) to disrupt spindle formation and chromosome segregation. The RAS/MEK/MAPK pathway is involved in amplification of centrosomes by HBx (Yun et al., 2004).

2.1.1.2 HBV integration and genomic instability

HBV integration into the host genome is random and causes oncogenesis when integration sites are close to genes involved in cell cycle, proliferation and metabolism (Tu et al., 2017). Comparison of paired non-tumor and tumor liver samples have revealed the presence of distinct patterns of increased HBV integration in tumors compared to matched non-tumors (Sung et al., 2012; Zhao et al., 2016). Several frequently altered genes and pathways have been described of which the most recurrent genes are TERT and Mixed Lineage Leukemia Gene Homolog 4 (MLL4) (Ding et al., 2012; Fujimoto et al., 2012; Jin et al., 2019; Shiraishi et al., 2014; Sung et al., 2012; Zhao et al., 2016). Earlier studies suggested random integration of HBV in the host genome and attributed increased expression of an altered gene to positive selection and clonal propagation (Furuta et al., 2018). However, preferred HBV integration sites have also been identified (Zhao et al., 2016). Telomerase, made up of an RNA and a protein (TERT) subunit, prevents telomere shortening in immortal cells and is overexpressed in HCC (Amisaki, Tsuchiya, Sakabe, Fujiwara, & Shiota, 2019). Integration of HBV Enhancer 1 (Enh1) at the TERT promoter enhances telomerase activation (Sze et al., 2020). HBV integration also increases TERT expression by making the promoter responsive to the androgen pathway (Li et al., 2019). HBV HBx integrates into intron 3 of MLL4, a histone methyltransferase, to form HBx/MLL4 fusion protein that downregulates a specific set of genes to promote HCC (Saigo et al., 2008). Integration into exons 3, 5 and 6 and exon-intron boundaries 3 and 5 were later reported (Dong et al., 2015; Sung et al., 2012) and MLL4 expression pattern was found to vary depending on the integration site. In addition to TERT and MLL4, several other altered genes like cyclin E1 (CCNE1), fibronectin 1 (FN1), protein tyrosine phosphatase receptor type D (PTPRD), unc-5 netrin receptor D (UNC5D), neuregulin 3 (NRG3), aryl hydrocarbon receptor repressor (AHRR), Rho guanine nucleotide exchange factor 12 (ARHGEF12), RNA binding Fox-1 homolog 1 (RBFOX1), SMAD5, ADP-ribose/CDP-alcohol diphosphatase (ADPRM), keratin 32 (KRT32) and CDK15 have been reported (Ding et al., 2012; Jin et al., 2019; Shiraishi et al., 2014; Sung et al., 2012; Zhao et al., 2016).

2.1.1.3 miRNAs and HBV-HCC

microRNAs (miR) are small non-coding RNAs that prevent gene expression by RNA silencing. They can act as oncogenes or tumor suppressor genes depending on their target genes. Deregulation of miRNAs occurs in every step of HCC development after HBV infection (Table 1). The miRNA profile of late-stage HCC differs from that of early infection, inflammatory phase or fibrosis. Additionally, miRNAs involved in major pathways like pI3K/MAPK, WNT/β-Catenin, TP53 and JAK/STAT pathways are deregulated during HBV-HCC development (Sartorius et al., 2019). Some of the most frequently upregulated miRNAs in HBV-HCC include miR-18a, miR-21, miR-221, miR-222, and miR-224 and the most commonly downregulated miRNAs include miR-26a, miR-101, miR-122, miR-125b, miR-145, miR-199a, miR-199b, miR-200a, and miR-223 (Xu, An, Winkler, & Yu, 2020).

Expression levels of miR-18a positively correlate with AFP levels, tumor size, TNM stage and intrahepatic vascular invasion. miR-18a over expression *in vitro* promotes whereas knock down inhibits proliferation, migration, and invasion of HCC cells. Interferon regulatory factor 2 (IRF2) and chromobox 7 (CBX7) are direct targets of miR-18a and BCL2-Like 10 (BCL2L10) is a downstream mediator of miR-18a (Wang et al., 2018; Yongyu et al., 2018). miR-21 is involved in tumor proliferation, migration and invasion and directly targets phosphatase and tensin homolog (PTEN) and Kruppel-like factor 5 (KLF5) (Meng et al., 2007; Wang et al., 2019). miR-221 promotes HCC cell proliferation, invasion and migration *in vitro* (Fu et al., 2019; Liu et al., 2016). miR-221 targets plant homeodomain finger protein 2 (PHF2) and miR-221/Astrocyte elevated gene-1 (AEG-1) axis regulates PTEN/PI3K/AKT signaling in HCC (Fu et al., 2019; Kannan et al., 2019). miR-222 regulates cell migration through increased AKT signaling (Wong et al., 2010). miR-224 is involved in cell proliferation, migration, invasion, targets apoptosis inhibitor-5 (API5) to promote apoptosis and activates AKT signaling (Li et al., 2010; Ma et al., 2012; Wang et al., 2008). miR-26a is a complex miRNA since it acts as an oncogene in certain cancers and as a tumor suppressor in others. In HCC, it targets cyclins D2 and E2 to promote cell cycle arrest and Maternally Expressed Gene 3 (MEG3) to inhibit proliferation and metastasis (Kota et al., 2009; Li et al., 2017). miRNA-26a also increases migration and invasion through targeting PTEN (Li et al., 2017; Zhao et al., 2019). This necessitates further exploration of this miRNA. miRNA-101 is downregulated by HBx and promotes apoptosis, inhibits autophagy and targets enhancer of zeste 2 polycomb repressive

Table 1 A list of miRNAs deregulated in HBV-related HCC.

miRNA	Deregulation	Genes/pathways regulated	HCC properties regulated	References
miR-18a	Upregulated	IRF2, CBX7 and BCL2L10	Proliferation, migration, and invasion	Wang et al. (2018), Yongyu, Lewei, Jian, and Yuqin (2018)
miR-21	Upregulated	PTEN and KLF5	Proliferation, migration and invasion	Meng et al. (2007), Wang et al. (2019)
miRNA-221	Upregulated	PHF2, PTEN/PI3K/AKT signaling	Proliferation, invasion and migration	Fu et al. (2019), Kannan et al. (2019), Liu et al. (2016)
miR-222	Upregulated	AKT signaling	Migration	(Wong et al. (2010)
miR-224	Upregulated	API5, AKT signaling	Proliferation, migration, invasion and apoptosis	Li et al. (2010), Ma, Tao, Gao, Fan, and Wu (2012), Wang et al. (2008)
miR-26a	Downregulated	Cyclin D2 and E2, MEG3, PTEN	Cell cycle arrest, proliferation and metastasis, migration and invasion	Kota et al. (2009), Li et al. (2017), Zhao et al. (2019)
miR-101	Downregulated	EZH2, DNA methyltransferase 3A, SOX9 and Girdin	Apoptosis and autophagy	Cao et al. (2016), Su et al. (2009), Wei et al. (2013), Xu et al. (2014), Xu et al. (2013), Zhang et al. (2012)
miR-122	Downregulated	BCL2L2, NDRG3, snail1 and snail2	Cell survival, cell proliferation and EMT	Fan et al. (2011), Jin, Wang, et al. (2017), Lin, Gong, Tseng, Wang, and Wu (2008)
miR-125b	Downregulated	sirtunin7, HOTTIP, TXNRD1, Sirtuin 6	Proliferation and metastasis	Hua et al. (2019), Kim et al. (2013), Song et al. (2018), Tsang et al. (2015)

Continued

Table 1 A list of miRNAs deregulated in HBV-related HCC.—cont'd

miRNA	Deregulation	Genes/pathways regulated	HCC properties regulated	References
miR-145	Downregulated	ROCK1, IRS1, HDAC2, ADAM17	Proliferation and motility	Ding et al. (2016), Liu et al. (2014), Noh et al. (2013), Wang et al. (2014)
miR-199a	Downregulated	FZD7	Cell proliferation	Song et al. (2014)
miR-199b	Downregulated	JAG1	proliferation, invasion and migration	Li, Yuan, Zhuang, and Wu (2018)
miR-200a	Downregulated	MACC1, ZEB2, Foxa2, CDK6, GAB1	cell growth, metastasis and migration	Chen et al. (2017), Feng et al. (2015), Wang et al. (2017), Xiao et al. (2013), Yang et al. (2015)
miR-223	Downregulated	stathmin1, ABCB1, FOXO3a	Autophagy, chemoresistance	Wong et al. (2008), Yang et al. (2013), Zhou et al. (2019)

For each miRNA type of deregulation (upregulation/downregulation), genes or pathways regulated, and HCC properties affected are given.

complex 2 subunit (EZH2), DNA methyltransferase 3A, SRY-box transcription factor 9 (SOX9), and Girdin in HCC (Cao et al., 2016; Su et al., 2009; Wei et al., 2013; Xu et al., 2013, 2014; Zhang et al., 2012). miR-122 targets the anti-apoptotic gene, BCL2-like protein 2 (BCL2L2), N-myc downstream-regulated gene 3 (NDRG3), snail1 and snail2 to suppress tumor progression (Fan et al., 2011; Jin, Wang, et al., 2017; Lin et al., 2008). miR-125b regulates sirtunin7, long non-coding RNA HOXA transcript at the distal tip (HOTTIP), Thioredoxin Reductase 1 (TXNRD1) and Sirtuin 6 (Hua et al., 2019; Kim et al., 2013; S. Song et al., 2018; Tsang et al., 2015). miR-145 inhibits Rho associated coiled-coil containing protein kinase 1 (ROCK1), insulin receptor substrate 1 (IRS1), histone deacetylase 2 (HDAC2) and ADAM metallopeptidase domain 17 (ADAM17) (Ding et al., 2016; Liu et al., 2014; Noh et al., 2013; Wang et al., 2014). miR-199a inhibits cell proliferation and targets frizzled class receptor 7 (FZD7) (Song et al., 2014). miR-199b promotes its tumor suppressive activity by directly targeting jagged 1 (JAG1) (Li et al., 2018). miR-200a is a prognostic marker that suppresses cell growth, metastasis and migration by targeting metastasis-associated gene in colon cancer 1 (MACC1), Zinc finger *E*-box binding homeobox 2 (ZEB2), forkhead box A2 (Foxa2), cyclin-dependent kinase 6 (CDK6) and GRB2-associated-binding protein 1 (GAB1) (Chen et al., 2017; Feng et al., 2015; Wang et al., 2017; Xiao et al., 2013; Yang et al., 2015). miR-223 downregulates stathmin1, ATP binding cassette subfamily B member 1 (ABCB1) and forkhead box O3a (FOXO3a) (Wong et al., 2008; Yang et al., 2013; Zhou et al., 2019).

2.1.1.4 Signaling pathways involved in HBV-HCC

Multiple signaling pathways are known to be deregulated in HBV-related HCC (Fig. 2). Binding of WNT to frizzled family receptor stabilizes β-catenin, which translocates to the nucleus, interacts with T-cell factor/lymphoid enhancer factor (TCF/LEF) and promotes transcription. β-catenin is also bound to E-cadherins in the cytoplasm. E-cadherin loss during EMT is associated with nuclear translocation of β-catenin. The WNT/β-catenin pathway is commonly deregulated in HCC. Interaction between the ligand Wnt3 and receptor Frizzled-7 receptor is observed in HBV-related HCC (Kim, Lee, et al., 2008). The HBx protein is required for activating WNT/β-catenin signaling in HCC cells. HBx stabilizes cytoplasmic β-catenin by src kinase activation and activates WNT/β-catenin signaling *in vitro* (Cha, Kim, Park, & Ryu, 2004). HBx also activates WNT/β-catenin pathway by binding to the tumor suppressor, adenomatous

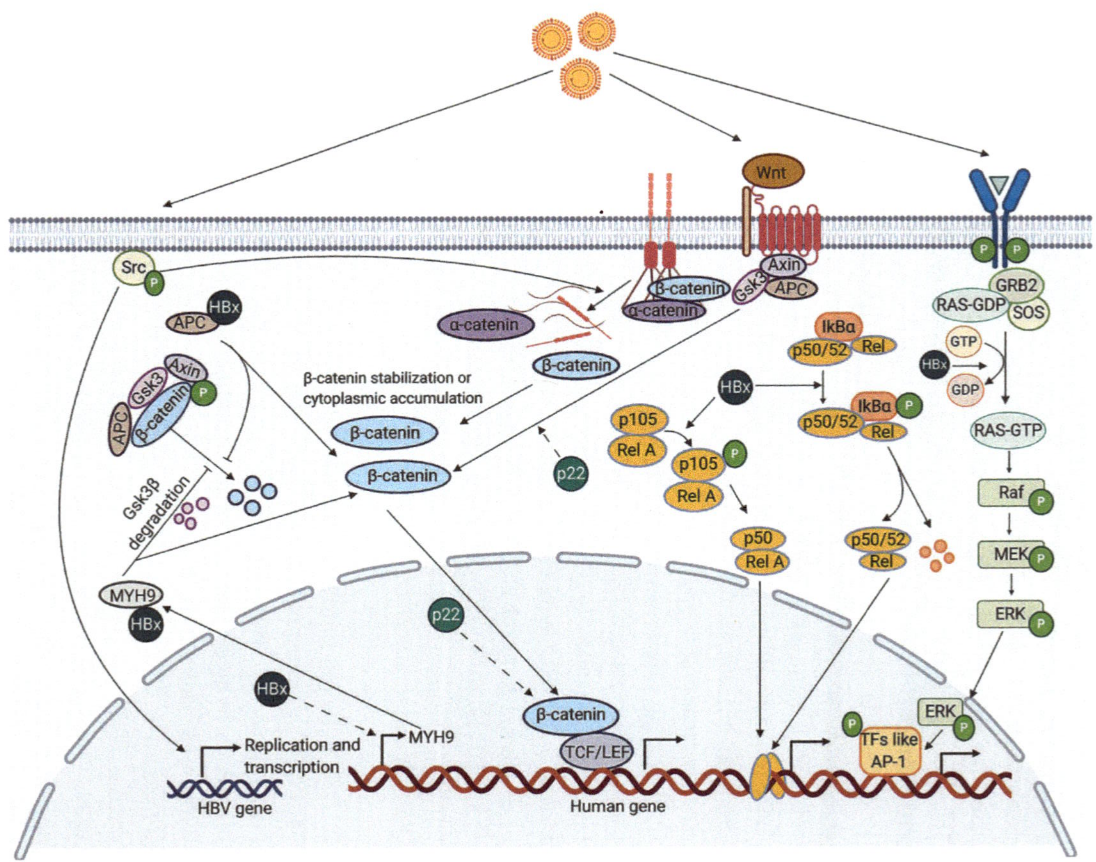

Fig. 2 See figure legend on opposite page.

polyposis coli (APC), displacing β-catenin leading to its upregulation in the nucleus (Hsieh et al., 2011). The precore protein, p22, is also an activator of TCF/β-catenin transcription (Tran et al., 2020). The RAS/RAF/MEK/ERK pathway is a cascade of phosphorylation reactions that leads to nuclear localization of ERK and transcription of target genes involved in cellular functions like proliferation, adhesion, angiogenesis and mobility. HBx transactivates the transcription factor AP1 through the MAPK pathway. An increase in levels of activated Raf and ERK was observed due to stimulation of Ras-GTP cycling by HBx (Benn & Schneider, 1994). This pathway increases expression of transcription factors like Activator Protein 1 (AP1) (Cross et al., 1993). HBx also activates NF-κB *via* the Ras pathway (Su & Schneider, 1996). Association of β-catenin with the scaffolding protein APC is required for its degradation by the APC/Axin/GSK3β complex. HBx stabilizes β-catenin by preventing its degradation through inhibiting APC binding or increasing expression of Non-muscle myosin heavy chain IIA (MYH9), a protein that degrades GSK3-β (Hsieh et al., 2011; Lin et al., 2020). The precore protein p22 also activates Wnt signaling (Tran et al., 2020).

Fig. 2 An overview of multiple signaling pathways deregulated by HBV and their potential mechanisms. Dotted lines indicate an unknown mechanism. Binding of Wnt3 to Frizzled-7 receptor activates Wnt/β-catenin signaling through β-catenin accumulation and enhanced TCF-mediated transcription (Kim et al., 2008). HBV infection increases phosphorylation of c-Src kinase, causing an increase in HBV transcription and replication and disassembly of adherens junctions. The disassembly of adherens junctions, made of *E*-Cadherin bound to β-catenin, results in increased levels of intracellular E-Cadherin and β-catenin leading to EMT and increased TCF/LEF dependent β-catenin signaling (von Olshausen et al., 2018). Association of β-catenin with the scaffolding protein APC is required for its degradation by the APC/Axin/GSK3β complex. HBx protein competitively binds to APC, preventing its association with β-catenin, leading to increased β-catenin stabilization (Hsieh, Kim, Lim, Yu, & Jung, 2011). Increased stabilization of β-catenin leads to an increase in levels of its downstream effectors like c-MYC and a decrease in E-cadherin expression (Not shown in illustration). HBx also transcriptionally regulates MYH9 through β-catenin signaling and interacts with Non-muscle myosin heavy chain IIA (MYH9) to facilitate ubiquitin mediated GSK3β degradation, thus stabilizing β-catenin (Lin et al., 2020). HBx also enhances MAPK/ERK pathway by activating Ras-GTP complex formation (Benn & Schneider, 1994). The transcription factor AP1 is activated through this pathway (Cross, Wen, & Rutter, 1993). HBx induces phosphorylation and degradation of the NF-κB inhibitor, IκBα and a decrease in levels of the p52 precursor, p105 (Su & Schneider, 1996). The precore protein p22 also activates Wnt signaling potentially both in the nucleus and the cytoplasm (Tran et al., 2020). *Created with Biorender.com.*

2.1.2 Hepatitis C virus

HCV is a single-stranded RNA virus with a 9.6 kb long genome composed of 5′ and 3′ noncoding regions and a single ORF that encodes a polyprotein cleaved by cellular and viral proteases in a specific order co- and post-translationally to produce structural (C, E1 and E2) and nonstructural (NS2, NS3, NS4A, NS4B, NS5A, NS5B and P7) viral proteins (Bukh, 2016). The structural protein C forms the nucleocapsid and E1 and E2 are envelope glycoproteins. The nonstructural protein NS2 is a cysteine protease and NS3 a serine protease with nucleoside-triphosphatase (NTPase) /RNA helicase activity. N4A forms a non-covalent complex with and is a cofactor of NS3, and NS5A is involved in modulation of RNA replication and viral assembly. NS4B forms an intracellular membrane structure on which viral replication occurs. NS5B is an RNA-dependent RNA polymerase and p7 is a viroporin involved in viral secretion (Ashfaq, Javed, Rehman, Nawaz, & Riazuddin, 2011).

HCV enters host cells through interaction of the two viral surface glycoproteins E1 and E2 with heparan sulfate proteoglycans on the surface of hepatocytes followed by endocytosis and fusion of cell membranes. The 5'-Untranslated Region (UTR) has an internal ribosome entry site (IRES) element, which recruits the translational machinery to the viral RNA in the cytosol and translation occurs. Encoded viral proteins involved in viral replication and other cellular factors associate with a double-membraned vesicle formed from the ER, where replication takes place. Lipid droplets containing triacylglycerides and cholesteryl esters and surrounded by a phospholipid monolayer are also a part of this structure and have an important role in HCV life cycle. Polymerase NS5B uses the positive stranded RNA as a template and replication occurs through a negative stranded intermediate. The RNA generated can act as a template to further produce RNA or undergo translation to produce new viral proteins that are assembled to form viral particles. Once assembled, HCV particles associated with distinct lipid and apolipoprotein profiles are egressed from the infected cell through the secretory pathway (Kim & Chang, 2013).

The initial immune response to HCV infection is triggered by activation of PRRs like TLR3 and RIG-I through PAMPS released by HCV. Activation of these receptors initiates a cascade of reactions that results in transcription of Interferon-β gene and the production of Interferon stimulated genes (ISGs), that can mutate RNA-specific adenosine deaminase 1 (ADAR1), degrade 2′-5′-oligoadenylate synthetase 1/RNAse L (OAS1/ RNAse L) system and inhibit translation (P56 and PKR) of the viral genome. HCV attenuates innate immune response through various mechanisms. HCV NS3/NS4A complex can block TLR3 and RIG-I signaling, and NS5A inhibits OAS. HCV core inhibits signal transducer and activator of transcription 1 (STAT1) activation and NS5A induces interleukin (IL)-8 both of which inhibits ISG expression downstream. HCV E2 and NS5A can inhibit the ISG protein kinase R (PKR) (Rehermann, 2009). There is also evidence of T cell-meditated immune response at 5–9 weeks post-infection. Specific cluster of differentiation (CD) 4+ and CD8+ T cells can act against HCV-derived antigens and eliminate the virus. When this process to clear HCV infection is unsuccessful, acute HCV infection can progress to chronic HCV infection. Chronic HCV infection progresses to HCC through fibrosis and cirrhosis. Fibrosis is accelerated by fat deposition in the liver or hepatic steatosis, specifically in HCV genotype 3 (Asselah, Rubbia-Brandt, Marcellin, & Negro, 2006). Several miRNAs and long non-coding RNAs (lncRNAs) are also involved in HCV-related HCC which will be discussed in detail in Chapter 3 (Plissonnier, Herzog, Levrero, & Zeisel, 2018).

2.2 Alcohol

Alcohol consumption increases the risk of not just liver cancer, but also oropharyngeal, colorectal, breast and stomach cancers. The risk of developing cancer is proportional to the amount of alcohol intake. Liver cancer risk increases 46% for consumption of 50 g of ethanol per day and 66% for 100 g per day compared to non-drinkers (Turati et al., 2014). Alcohol-related liver disease (ALD) accounts for about 30% of HCC cases, including HCC occurrences where other risk factors, like obesity, diabetes, hepatitis infections might co-exist with ALD (Ganne-Carrie & Nahon, 2019).

The first sign of liver damage due to alcohol abuse is steatosis or alcohol-associated fatty liver (AFL), which is usually asymptomatic. Although >90% of heavy drinkers develop AFL, it is a reversible condition and only about 30% of them progress to severe ALD. The accumulation of fat in hepatocytes happens primarily through an activation of sterol regulatory element-binding protein 1 (SREBP-1) causing increased lipogenesis or down-regulation of peroxisome proliferator-activated receptor (PPAR)-α causing decreased fatty acid β-oxidation (FAO) (Gao & Bataller, 2011). In hepatocytes, alcohol is converted to toxic acetaldehyde by the enzyme alcohol dehydrogenase (ADH). Acetaldehyde is then converted to non-toxic acetate by aldehyde dehydrogenase (ALDH) in the mitochondria and subsequently broken down to H_20 and CO_2. The first two steps involve reduction of NAD^+ to NADH. As more molecules of ethanol are metabolized, intracellular $NADH/NAD^+$ ratio increases disrupting mitochondrial FAO. Increased $NADH/NAD^+$ can also increase the production of reactive oxygen species (ROS) through the electron transport chain (ETC) (Donohue., 2007). Certain factors like polymorphisms in ADH and ALDH can cause accumulation of acetaldehyde molecules that form adducts with proteins, DNA or both, inhibiting their endogenous function and/or eliciting an intense immune reaction (Brooks & Theruvathu, 2005; Setshedi, Wands, & Monte, 2010).

A steatotic liver is vulnerable to secondary insults that can lead to alcoholic hepatitis (AH) or liver inflammation. It is characterized by activation of liver resident macrophages and recruitment of monocytes, macrophages and neutrophil leukocytes into the liver producing cytokines. The intestinal lumen is surrounded by a mucosal membrane that acts as a barrier between the gut and the liver. A small amount of lipopolysaccharide (LPS) derived from Gram negative bacteria can enter the liver through the portal vein, but it is readily cleared by the liver. Metabolites derived from alcohol can disrupt the tight junction proteins, adherens junction proteins, microtubule cytoskeleton and epithelial cell-cell adhesion, making the gut-liver barrier highly permeable. As a result, high levels of LPS enter the liver and are recognized by receptors like complement receptor 3 (in macrophages), CD14/TLR4 (in monocytes, macrophages, B cells, liver parenchymal cells, and some fibroblast cells) and myeloid differentiation factor (MD)-2/TLR4 (in myeloid and endothelial lineage cells). These receptors present in multiple cell types including liver-resident macrophages are activated to produce cytokines like TNFα, IL-6, IL-1β and IL-8 that cause inflammation. AH is also a reversible stage in alcohol-related HCC, with the solution being cessation of alcohol intake (Szabo & Bala, 2010).

Chronic Alcoholic hepatitis can cause fibrosis by replacement of normal liver parenchymal tissues by connective tissues as a result of an imbalance between synthesis and degradation of extracellular matrix (ECM) components. Activation of Kupffer cells, release of DAMPs from dying hepatocytes and PAMPS from a leaky gut, and activation of hedgehog-responsive genes can cause activation of hepatic stellate cells (HSC). Activated HSCs transdifferentiate to myofibroblasts that produce ECM components collagen types I, III and IV, fibronectin, hyaluronan, and proteoglycans. Natural killer (NK) cells can kill activated HSCs by inducing apoptosis and cell cycle arrest through different mechanisms. Alcohol-related fibrosis is linked to a decrease in the number and efficiency of NK cells, increasing HSC activity.

The histology of alcohol-related fibrosis is also different from non-alcoholic steatohepatitis (NASH) or hepatitis-related fibrosis (Lackner & Tiniakos, 2019; Schuppan, Ashfaq-Khan, Yang, & Kim, 2018). Regression of early fibrosis is possible through cessation of intake of alcohol and research on anti-fibrotic drugs is promising. However, prolonged fibrosis leads to cirrhosis and HCC.

2.3 Non-alcoholic fatty liver disease (NAFLD)

NAFLD refers to a spectrum of liver conditions ranging from steatosis to its more aggressive manifestation NASH. It is the most common liver disorder with a global prevalence of ~25% (Maurice & Manousou, 2018). Twenty percent of patients with early NAFLD or steatosis progress to NASH-cirrhosis, from which 2.6% undergo further progression to HCC (Ascha et al., 2010). Even though the number is lower than for other etiologies, the fact that NAFLD cases are forecasted to increase by 21% and NASH-related HCC by 137% in the US between 2015 and 2030 is of significant concern (Estes, Razavi, Loomba, Younossi, & Sanyal, 2018). Obesity and diabetes are two major risk factors for NAFLD, and the increase in NAFLD cases in the past decade parallels an increase in the worldwide incidence of obesity and diabetes (Younossi, 2019). As the name suggests, the pathology of NAFLD is similar to alcohol-related fatty liver disease, except that the etiology cannot be ascribed to chronic alcohol consumption.

The two-hit hypothesis for NAFLD progression to NASH and cirrhosis is based on steatosis being the first hit, which sensitizes hepatocytes to inflammation, which is the second hit, ultimately leading to fibrosis and cirrhosis. However, with a disease as complicated as NASH, support for the multi-hit hypothesis, which includes genetic and environmental factors, is gaining traction (Buzzetti, Pinzani, & Tsochatzis, 2016). The first stage of NAFLD is simple fatty liver, which is characterized by an excessive accumulation of triglycerides in hepatocytes. This can occur through increased lipolysis of visceral adipose tissue, which is transported to the liver, excess *de novo* lipogenesis (DNL) in the liver or consumption of high fat diet. Excess Free Fatty Acids (FFA) disrupt the lipid balance within cells, damage cellular organelles like mitochondria and ER, and cause cell injury and cell death. ER stress can cause chronic inflammation by increasing ROS levels and activation of NF-κB and c-Jun N-terminal kinase (JNK) pathways. Saturated fatty acids can also activate TLR4 directly in hepatocytes causing increased TNFα production and inflammation (Pierantonelli & Svegliati-Baroni, 2019).

In response to hepatocyte injury, lymphocytes and natural killer T cells are recruited to the liver and produce IFN-γ and tumor necrosis factor superfamily member 14 (TNFSF14). Signals from the gut can also activate an innate immune response through TLRs, causing inflammation. This activates liver-resident macrophages that release inflammatory cytokines and the resultant chronic inflammation culminates in HCC.

2.4 Aflatoxins

Aflatoxins are mycotoxins produced by fungal species like *Aspergillus flavus* and *A. parasiticus*. Chemically, they are furanocoumarin compounds, composed of a furan ring fused with a coumarin and are synthesized by secondary metabolic pathways in fungi. Most contaminations occur in grains and nuts during their harvest, transport or storage. Upon consumption, aflotoxins can lead to complications like hepatotoxicity, teratogenicity and immunotoxicity in plants and animals. There are more than 20 known aflatoxins, but the four important ones are aflatoxin B1, B2, G1 and G2, with aflatoxin B1 (AFB1) being the most potent hepatocarcinogen (Kumar, Mahato, Kamle, Mohanta, & Kang, 2016).

Aflatoxins are hepatotoxins that cause acute or chronic poisoning. Several outbreaks related to acute aflatoxin poisoning have occurred in countries like India, Nepal, Bangladesh, Kenya and Tanzania causing hundreds of deaths (Sarma, Bhetaria, Devi, & Varma, 2017). Acute poisoning results in liver failure that can present with jaundice, nausea and abdominal pain or can even be fatal. Chronic aflatoxin poisoning is carcinogenic, specifically related to HCC. AFB1-related HCC is most prevalent in Southeast Asia and Sub-Saharan Africa, where the dry and humid climates are favorable for fungal proliferation. The most common mutation seen in aflatoxin-related HCC patients is an AGGarg→AGTser missense mutation in codon 249 of the p53 gene (Aguilar, Hussain, & Cerutti, 1993). This mutation is known to promote HCC by affecting differentiation and metastasis (Peng, Peng, & Yao, 1998). Aflatoxin can also be metabolized by cytochrome P450 to form reactive genotoxic intermediates like aflatoxin B1–8, 9-oxide (AFBO) which can form DNA adducts in hepatocytes causing DNA damage (Hamid, Tesfamariam, Zhang, & Zhang, 2013). Aflatoxin is capable of inducing chromosomal aberrations and chromosomal strand breaks in human cells (Turkez & Sisman, 2012). In addition to purely mutagenic effects, a meta-analysis of 5 studies found that aflatoxin exposure is associated with a higher risk of liver cirrhosis, which could also lead to HCC (Mekuria,

Routledge, Gong, & Sisay, 2020). Aflatoxin exposure and HBV infection can synergistically increase HCC risk up to 70% compared to aflatoxin (0.3–17.4%) or HBV (4.8–17.4%) alone (Kew, 2003).

2.5 Predisposing genetic factors

In addition to environmental factors, certain genetic factors also play a role in increasing HCC risk (Table 2). In most conditions, HCC is a secondary outcome of the underlying genetic condition in adulthood. These genetic changes are also predisposing and generally, interaction with a secondary risk factor is essential for HCC development.

2.5.1 Alpha 1-antitrypsin deficiency (A1ATD)

Alpha 1-antitrypsin (A1AT) belongs to the protein super family SERPIN (SERine Proteinase INhibitor). Members of this family are characterized by three beta sheets (A, B and C) and a mobile reactive loop that inhibits the target proteinase by presenting it a peptide sequence as a pseudo-substrates that are >30% homologous in their structure to each other (Topic, Ljujic, & Radojkovic, 2012). The primary role of A1AT is to inhibit the serine protease neutrophil elastase (NE), a major protease that cleaves a number of connective tissue substrates. It is an acute phase response, anti-inflammatory protein that blocks TNFα effects and is produced as a host response to tissue injury and inflammation (Stockley, 2015).

A1ATD is characterized by low levels of serum A1AT and affects 1:1800 to 1:2000 live births. It is the most common genetic cause of liver disease in children and increases risk for cirrhosis and HCC (Eriksson et al., 1986; Perlmutter, 2006). Lung disorders associated with A1AT occur due to a decrease in circulating A1AT and are attributed to a loss-of-function mechanism whereas liver abnormalities occur through a gain-of-toxic function mechanism (Ghouse, Chu, Wang, & Perlmutter, 2014). AT1ATD is associated with accumulation and polymerization of mutant A1AT in hepatocytes, mitochondrial dysfunction (Perlmutter, 2002; Teckman, An, Blomenkamp, Schmidt, & Perlmutter, 2004), delayed ER protein degradation (Wu et al., 1994) and altered regulation of cyclin D1 and melanoma cell adhesion molecule (MCAM) (Marcus et al., 2010).

2.5.2 Autoimmune hepatitis (AIH)

AIH is a chronic liver disease in which the body's immune system attacks liver cells in genetically predisposed individuals. It is a polygenic, multifactorial disease that does not follow a typical inheritance pattern. AIH is linked

Table 2 A list of predisposing genetic conditions for HCC.

Genetic condition	Gene(s) mutated	Type of study	Cirrhosis frequency	HCC frequency	References
Alpha 1-antitrypsin deficiency	A1AT	Retrospective study of 17 autopsied patients with A1ATD	5 of 17	8 of 17	Eriksson, Carlson, and Velez (1986)
		Retrospective analysis of clinical data from 246 A1ATD patients	19% of patients with age over 50	12 of 91 deceased patients had liver disease	Larsson (1978)
		Retrospective study of 31 autopsied patients with A1ATD	13 of 31	5 of 31	Elzouki and Eriksson (1996)
		Retrospective study of 675 patients with end-stage liver disease	All patients	4 of 47 A1ATD patients	Antoury, Lopez, Zein, Stoller, and Alkhouri (2015)
Autoimmune hepatitis	Multiple genes including DRB1, FAS, TNFA and CTLA4	Meta-analysis of 25 studies with 6528 AIH patients	92 of 93 patients with known cirrhosis status at diagnosis	118 of 6528	Lleo, de Boer, Liberal, and Colombo (2019)
		Meta-analysis of 11 studies with 8460 AIH patients	18.7% to 83.3% in Japan and 12% to 50.2% in other areas	0–12.3%	Valean et al. (2019)

Hereditary hemochromatosis	Type 1 HH—HFE	Retrospective study of 163 patients with HH	112 of 163	16 of 53 deceased patients had liver cancer	Strohmeyer, Niederau, and Stremmel (1988)
		Prospective study of 208 HH patients	80 of 208	16 of 208	Bradbear et al. (1985)
		Prospective study of 230 HH patients	134 of 230	49 of 230	Fracanzani et al. (2001)
		population-based cohort study of 1847 HH patients	20–25%	49 of 1847	Elmberg et al. (2003)
Glycogen Storage Disease	Type 1 GSD—glucose-6-phosphatase	Retrospective analysis of 72 GSD patients	Data not available	4 of 72	Jang et al. (2020)
	Type III GSD—glycogen debranching enzymes	Systemic review of 45 GSD type III patients	2 of 45	2 of 45	Demo et al. (2007)
	Type VI GSD—glycogen phosphorylase	Case report of 2 patients with GSD-associated HCC	0 of 2	2 of 2	Manzia et al. (2011)

Continued

Table 2 A list of predisposing genetic conditions for HCC.—cont'd

Genetic condition	Gene(s) mutated	Type of study	Cirrhosis frequency	HCC frequency	References
Porphyria	Enzymes involved in heme biosynthesis	Retrospective population-based mortality study including 33 AIP	4 of 33	9 of 33	Andersson, Bjersing, and Lithner (1996)
		Prospective study of 62 AIP gene carriers	4 of 22 HCC patients	22 of 62	Innala and Andersson (2011)
		A nationwide cohort study including 612 PCT patients (of which 23 were diagnosed with cancer prior to PCT)	2 of 612	6 of 589	Baravelli, Sandberg, Aarsand, and Tollanes (2019)
		Prospective study of 83 PCT patients	13 of 13 HCC patients	13 of 83	Salata et al. (1985)
		Retrospective analysis of 245 AHP patients	2 of 7 HCC patients	7 of 206	Kauppinen and Mustajoki (1988)
		Prospective analysis of 650 AHP patients	2 of 7 HCC patients	7 of 605	Andant et al. (2000)
		Case report of 2 AP-related HCC patients	1 of 2	2	Schneider-Yin et al. (2010)
Tyrosinemia type 1	Fumarylacetoacetate hydrolase	Retrospective study of 16 HT patients	10 of 10 (with known cirrhosis status)	4 of 16	Ozcan et al. (2019)
		Retrospective study of 16 HT patients	16 of 16	12 of 16	Seda Neto et al. (2014)

For each condition, gene(s) mutated, type of study done to understand its association with HCC and frequencies of cirrhosis and HCC observed in these studies are given.

to mutation in a human leukocyte antigen (HLA) class II gene—DR beta 1 (DRB1), which is involved in antigen presentation to CD4+ T cells. Additionally, mutations in Fas cell surface death receptor (FAS), TNFA and cytotoxic T lymphocyte associated protein 4 (CTLA4) genes are also linked to AIH (Liberal, Longhi, Mieli-Vergani, & Vergani, 2011). There are two main types of AH depending on what autoantigens are mediators of this disease. Type 1 AIH is seen in children and adults and characterized by the presence of anti-smooth muscle and antinuclear antibodies. Type 2 AIH is predominantly observed in children and characterized by anti-liver cytosol type 1 and/or anti-liver/kidney microsomal type 1 (anti-LKM1) antibodies (Sucher et al., 2019). The production of autoantibodies can be explained by a phenomenon termed "molecular mimicry." For example, anti-LKM1 recognizes the autoantigen cytochrome P450IID6, which is molecularly similar to HCV antigens (Manns, Johnson, Griffin, Tan, & Sullivan, 1989; Zanger, Hauri, Loeper, Homberg, & Meyer, 1988). Prevalence of AH is 11–25/100,000 per year, with more women than men being affected. A meta-analysis of 11 studies indicate that the percentage of cirrhosis in AIH patients varies from 12% to 83% and 5–6% of AH patients develop HCC (Valean et al., 2019).

2.5.3 Hereditary hemochromatosis (HH)

Hemochromatosis is an iron overload disorder characterized by increased absorption of iron from the diet and accumulation of iron in organs like the liver, heart, pancreas and joints. There are two types of Hemochromatosis—hereditary/primary (which is further classified into types 1, 2, 3 and 4) and non-hereditary/secondary. Type 1 hereditary hemochromatosis is the most common (90%) hemochromatosis and is a result of mutations in the hemostatic iron regulator (HFE) gene (Radford-Smith, Powell, & Powell, 2018).

HFE encodes cell surface protein with a role in maintaining iron homeostasis by indirectly regulating the function of two proteins—ferroportin and transferrin. Ferroportin is an iron exporter found mainly in the intestinal epithelium where it transports absorbed dietary iron out of the cell and into the blood stream, increasing plasma iron concentration. Hepcidin is a protein that decreases iron delivery to plasma by binding to and destroying ferroportin. HFE regulates hepcidin expression through an unknown mechanism. In HH, HFE is mutated causing low levels of hepcidin, high ferroportin expression, increased iron uptake and systemic iron overload (Pietrangelo, 2006). Transferrin is the iron bound ligand of transferrin receptor 1 involved in cellular uptake of iron. HFE also binds to transferrin receptor 1 blocking its interaction with transferrin, preventing iron uptake by cells. In HH, there is increased uptake of iron by cells as a consequence of lack of functional HFE protein (Roy, Penny, Feder, & Enns, 1999). An increase in circulating iron levels and cellular uptake of iron leads to its deposition in various organs including the liver.

Iron overload can promote tumor growth by increasing cellular proliferation, cause DNA damage by increasing ROS levels and damage cell membranes of subcellular organelles through peroxidative damage (Bacon & Britton, 1989; Kowdley, 2004). Iron overload can also inhibit lymphocyte proliferation (Kowdley, 2004). Incidence of cirrhosis in HH patients is 10–25% and HCC incidence in HH patients is 8–10% (Kew, 2014) and the increase in risk for HCC development has been reported to be ~20-fold compared to healthy individuals (Elmberg et al., 2003).

2.5.4 Glycogen storage diseases (GSD)

Glycogen, a form of energy reserve for the body, is a branched polymer of linear chains of 8–12 glucose units linked by alpha ($1 \rightarrow 4$) glycosidic bonds. GSDs are rare metabolic disorders involving glycogen synthesis or breakdown pathways. There are over 12 types of GSDs each due to deficiency of a different enzyme. The liver is the major site for glycogen metabolism and hence most affected by any error in the pathways. The most common types of GSDs are types I, III, and IV (Burda & Hochuli, 2015).

Type I or von Gierke's disease is the most common form of GSD. This disease involves an autosomal recessive transmission with an incidence of 1 in 100,000 individuals (Ozen, 2007). There are four subtypes of Type I GSD—Type 1a, 1b, 1c and 1d, of which type 1a is the most common. People with type I GSD have a deficiency of the enzyme glucose-6-phosphatase, which is involved in breaking down glycogen to glucose (glycogenolysis) and synthesizing glucose from non-carbohydrates sources when glycogen reserves are depleted (gluconeogenesis). This enzyme is primarily present in the liver, kidney, and intestine and is involved in the terminal step, the conversion of glucose-6-phosphate to glucose, in both these pathways. Deficiency of this enzyme leads to severe hypoglycemia, due to low availability of glucose. Also, glucose-6-phosphate allosterically inhibits glycogen phosphorylase, the first enzyme in glycogenolysis, causing an accumulation of glycogen in the liver. Other biochemical abnormalities are lactic acidosis, hyperlipidemia and hyperuricemia. 57% and 8% of GSD type 1 patients develop hepatocellular adenoma and HCC, respectively (Jang et al., 2020). Autophagy impairment, mitochondrial dysfunction, metabolic reprogramming, oncogene activation and downregulation of tumor suppressors are potential mechanisms causing type 1GSD-related HCC (Bianchi, 1993; Cho, Lee, Bae, & Chou, 2020). Animal models of this disease did not develop cirrhosis (Resaz et al., 2014).

Type III, Cori disease, or Forbes disease is an autosomal recessive disease with an incidence of 1 in 100,000 individuals, initiated by a deficiency of the glycogen debranching enzymes—glucosyltransferase and glucosidase (Ozen, 2007). There are two subtypes—type IIIa (~80% of cases) where enzyme deficiency occurs in liver and muscle and type IIIb where only the liver is involved. Glycogen is a branched molecule with α-1,6 glycosidic bonds at branch points. Since the enzyme glycogen phosphorylase can only act on α-1,4 bonds, during glycogen breakdown a debranching enzyme removes the α-1,6 glycosidic bond leaving a linear chain for glycogen phosphorylase to continue the break down process. Deficiency of this enzyme leads to accumulation of short chain length branched glycogen molecules in the liver and muscles. Liver involvement in this condition is limited and the major cause of morbidity is muscle diseases in early adulthood (Kishnani et al., 2010). There have been isolated reports of liver cirrhosis in type III GSD patients, with less than 10 cases of HCC reported overall (Demo et al., 2007).

Type VI or Hers disease is an extremely rare condition with unknown prevalence, except in older Mennonite population where the disorder has a prevalence of 1:1000 (Ozen, 2007). It is an autosomal recessive disorder caused by a defect in the glycogenolysis enzyme glycogen phosphorylase. Symptoms of this disorder include hepatomegaly, poor growth, ketonic hypoglycemia, elevated hepatic transaminases, hyperlipidemia and low prealbumin level. There are limited reports of HCC development in type VI GSD patients. However, there is evidence that while rare, these patients can develop HCC. Also, fibrosis is more common than cirrhosis in these patients (Manzia et al., 2011).

2.5.5 Porphyria

Porphyrin is a compound formed of four modified pyrrole subunits interconnected at their alpha carbon atoms by methine bridges. Heme is chemically a protoporphyrin, a type of porphyrin, linked to an iron molecule. Porphyria represents a group of rare mostly genetic disorders caused by a deficiency of enzymes involved in heme biosynthesis, causing an accumulation of porphyrins and porphyrin precursors in the body. It can be broadly classified into acute and cutaneous porphyria. Acute porphyria (AP) causes sudden attacks of severe abdominal pain that can last for several days and primarily affects the nervous system. The subtypes of APs are acute intermittent porphyria (AIP), variegate porphyria (VP), aminolevulinic acid dehydratase deficiency porphyria (ALAD) and hereditary coproporphyria (HCP), each uniquely characterized by deficiency of a different enzyme. Non-acute or cutaneous porphyrias primarily cause blistering, swelling, reddening, itching and scarring of the skin as a result of exposure to sunlight. Their subtypes are X-linked dominant protoporphyria (XLDPP), congenital erythropoietic porphyria (CEP), porphyria cutanea tarda (PCT) and erythropoietic protoporphyria (EPP). Porphyria can also be classified into acute hepatic porphyria (AHP) and erythropoietic porphyria, based on the origin of excess precursors. AIP, PCT, HCP and VP are hepatic, with precursors originating in the liver and CEP and EPP are erythropoietic, where precursors originate in the bone marrow (Ramanujam & Anderson, 2015).

AHP are specifically associated with increased risk for HCC. They are autosomal dominant traits and one mutant allele is sufficient to reduce enzyme levels to 50%. However, other susceptibility factors like viral infections, alcohol abuse, HFE mutations or smoking are required to display the disease phenotype since more than a 50% decrease in enzyme activity is required to manifest this phenotype. Since these additional susceptibility factors in themselves can cause HCC, proper controlled studies are required to estimate the actual incidence and increased risk associated strictly with AHP (Gisbert et al., 2004). AHP patients have a 35-fold increased risk of developing HCC (Lang, Schafer, Schwender, Neumann, & Frank, 2015) and the incidence rate for AHP-related HCC is 0.16–0.35% (Peoc'h et al., 2019). PCT patients have abnormal liver enzymes and histological abnormalities. A 20-fold increase in risk for HCC (Baravelli et al., 2019) has been reported and a study with 39 patients reported 0.26% incidence per year. HCC was found in 4–27% of AIP patients (Andersson et al., 1996) and 1.3–5.4% of VP patients.

2.5.6 *Tyrosinemia type1*

Hereditary Tyrosinemia type 1 (HT1) is an autosomal recessive genetic disorder resulting from a deficiency in the enzyme fumarylacetoacetate hydrolase, which breaks down 4-fumarylacetoacetate (FAA) and water into acetoacetate, fumarate, and H+ in the ultimate step of tyrosine breakdown (Bergeron, D'Astous, Timm, & Tanguay, 2001). FAA and associated toxic metabolites accumulate in the liver and cause severe liver dysfunction. Treatment for this condition involves 2-(2 nitro-4-3 trifluoro-methylbenzoyl)-1, 3-cyclohexanedione

(NTBC), which is a compound that blocks the tyrosine catabolic pathway at an earlier step preventing the accumulation of most toxic metabolites. However, accumulation of tyrosine and phenylalanine (a precursor for tyrosine) still occurs which needs to be regulated through diet. Reports of HCC development have been observed in untreated patients or patients treated with NTBC after symptoms develop. If treated asymptomatically, HCC development in tyrosinemia type 1 patients can be prevented. However, the lack of newborn screening and NTBC availability still causes HCC due to untreated HT1 (van Ginkel, Pennings, & van Spronsen, 2017).

Twenty-five to 75% of HT1 positive individuals developed HCC (Ozcan et al., 2019; Seda Neto et al., 2014). In untreated patients, oxidative stress, inhibition of DNA repair enzymes, ER stress, Golgi complex disruption, and apoptosis resistance are some of the potential mechanisms for HCC development (Tanguay, Angileri, & Vogel, 2017).

3. Liver cancer staging

Liver cancer staging is the process of classifying liver cancer based on various factors. It helps determine the severity of the disease, potential prognosis and the best treatment plan after diagnosis. There are more than 18 staging systems, each with a different purpose and their own advantages and disadvantages. For example, Okuda and CLIP are score-based staging systems to predict survival whereas BCLC guides treatment decision. In addition to liver cancer staging systems, scoring systems like Child-Pugh and MELD are used to assess liver function.

3.1 Child-Pugh (CP) score

Child-Pugh or Child-Pugh-Turcotte scoring system was introduced in 1964 to classify patients undergoing liver decompression surgery for portal hypertension based on hepatic function to predict survival (Child & Turcotte, 1964; Pugh, Murray-Lyon, Dawson, Pietroni, & Williams, 1973). It was modified in 1973 to substitute the variable clinical nutritional status with prothrombin time (Pugh et al., 1973). The updated version uses blood albumin level (>35 g/L, 28–35 g/L, <28g/L), blood bilirubin level (<2, 2 − 3, >3), presence or absence of ascites (absent, slight, moderate), encephalopathy grade (absent, grade 1–2, grade 3–4) and prothrombin time (<4 s, 4–6 s, >6 s) to classify patients. Each variable can be scored 1, 2 or 3 with 1 indicating normal hepatic function and 3 indicating a compromised liver. The severity of cirrhosis is estimated based on the cumulative score—higher the score, worse the cirrhosis. A score of 5–6 is designated Child-Pugh A, 7–9 is Child-Pugh B and 10–15 is Child-Pugh C.

3.2 Model for end-stage liver disease (MELD)

MELD was introduced in 2000 to predict survival in patients undergoing transjugular intrahepatic portosystemic shunt procedure for portal hypertension (Malinchoc et al., 2000). The risk score in this method was calculated from 4 prognostic values—bilirubin level, creatinine level, International Normalized Ratio (INR), derived from prothrombin time and cause of cirrhosis. [0.957 x Log e (creatinine mg/dl) + 0.378 × Log e (bilirubin mg/dl) + 1.120 x Log e (INR) + 0.643 × 10, rounded off to the nearest integer].

Prior to 2002, the United Network for Organ Sharing which administers the national Organ Procurement and Transplantation Network, used Child-Pugh score to determine which patient qualified to receive an available donor liver. CP score is categorical and classified patients based on their liver function into status 1, 2A, 2B and 3. Within category 3 which had the sickest patients, priority was given to patients who had been on the list the longest. Hence, there was a good chance that a person that had the greatest need for a liver transplant might not receive it because they were added to the list at a later time. To overcome this problem, the CP score was replaced by the MELD score in 2002 (Freeman et al., 2002). In 2006, the MELD score was updated to include serum sodium levels and referred to as MELD-Na (Biggins et al., 2006). Another model called Albumin-Bilirubin (ALBI) score was introduced in February 2015 and uses serum albumin and bilirubin levels to assess liver function (Johnson et al., 2015).

3.3 Tumor node metastasis (TNM) system

The American Joint Committee on Cancer (AJCC) and the Union for International Cancer Control (UICC) developed the TNM staging system for multiple cancers. The first edition of the AJCC cancer staging manual was released in 1977. However, criteria for liver cancer staging was included only in the second edition in 1983. The most recent is the 8th edition of AJCC cancer staging manual, published in 2016. This is the most commonly used system in the US and includes tumor size (T), nodal involvement (N) and metastatic status (M), to stage liver cancer.

T refers to tumor size and can fall into one of the three categories (T1, T2, T3 and T4). T1 refers to a single tumor that has not grown into blood vessels. It is subcategorized based on tumor size into T1a ($\leq$2 cm) or T2a

(>2cm). T2 refers to either a single >2-cm tumor that has grown into blood vessels or more than one tumor with maximum tumor size <5-cm. More than one tumor, with at least one tumor larger than 5-cm across is considered as T3 and T4 refers to any tumor that has grown into the hepatic or portal vein. Cancer can spread through the bloodstream to distant organs or through the lymphatic system to lymph nodes. N refers to nearby nodal involvement and can fall into one of the two types—N0 when the tumor has not spread and N1 when the tumor has spread to nearby lymph nodes. M indicates the state of metastasis. M1 and M0 refer to a tumor that has spread and not spread to distant organs, respectively.

Different combinations of T, N and M are classified into the following stages—I A (T1a, N0, M0), I B (T1b, N0, M0), II (T2, N0, M0), III A (T3, N0, M0), III B (T4, N0, M0), IV A (Any T, Any N, M0) and IV B (Any T, Any N, M1). Each revision of TNM staging system is based on evaluations done on current guidelines. The current version is based on two studies that indicated there was a need to subclassify T1, which was not subclassified in the previous edition, into T1a and T1b (Shindoh et al., 2013) and that there was no difference in survival between patients with T3a and T3b tumors or between those with T3b and T4 tumors (Chan et al., 2013). Evaluation of the current system suggests that it is better in predicting patient prognosis than the previous version, but also points that a sub-stratification of T2 group would help the system perform better (Kamarajah, Frankel, Sonnenday, Cho, & Nathan, 2018). A disadvantage of the TNM classification is that it includes only tumor characteristics and does not include liver function parameters.

3.4 The Okuda system

The Okuda classification was introduced in 1984 in Japan to predict prognosis in relation to disease stage and treatment (Okuda et al., 1984). The following four criteria were used for the classification—presence of ascites, tumor $\geq$50% of the 2-dimensional size of the liver, serum albumin $\leq$30 g/L, serum bilirubin $\geq$3 mg/dL. A person that does not meet any of the above criteria is classified as stage 1, 1 or 2 criteria is stage 2 and >3 criteria is stage 3.

3.5 Cancer of the liver Italian program (CLIP) system

The CLIP system was introduced in 1998 to produce a more sensitive prognostic index for HCC patients (A new prognostic system, 1998). CLIP score is calculated by assigning a score between 0 and 2 to four clinical factors

Child-Pugh stage (A=0, B=1, C=2), AFP (<400 ng/dL=0, ≥ 400 ng/dL=1), presence of portal vein thrombosis (no=0, yes=1) and tumor morphology (uninodular and extension ≤50% of the liver=0, multinodular and extension ≤50%=1 and massive or extension >50%=2). These scores are added together to yield a CLIP score of 0–6, each score is linked to an estimated survival—0=42.5 months, 1=32.0 months, 2=16.5 months, 3=4.5 months, 4=2.5 months and 5–6=1 month.

3.6 Barcelona clinic liver cancer (BCLC) system

The BCLC system was proposed by the BCLC group from University of Barcelona in 1999 to classify HCC patients into 5-stages and link each HCC stage to a treatment option (Llovet, Bru, & Bruix, 1999). In addition to CP score and tumor size, BCLC uses the Eastern Cooperative Oncology Group (ECOG) scale to evaluate performance status (PS). PS can range from PS 0 where a person is fully active to PS4 where a person is in a chair or a bed all the time and needs complete care. There are 5-stages to the BCLC staging system. Stage 0 indicates very early stage, where the tumor is <2cm, PS 0 and CP score A. Stage A indicates early stage, with one tumor of any size or up to 3 tumors all <3cm, PS 0 and CP score A or B. Surgical ablation is the treatment option with best outcome for BCLC stages 0 and A. Stage B is intermediate stage with >3 tumors, PS 0, CP A or B and recommended Trans-Arterial Chemoembolization (TACE). Stage C is advanced with spreading of cancer into blood vessels, lymph nodes or other organs, PS 1 or 2 and CP A or B. Systemic therapy is recommended for these patients. Stage 4 is severe liver damage with CP C and PS 3 or 4, without any treatment except best supportive care.

4. Molecular classification of HCC

HCC staging is a classification based on tumor morphology and various clinical parameters to determine treatment with the most favorable outcome and patient prognosis. Despite belonging to the same HCC stage, tumors with different molecular signatures can respond differently to the same treatment. Although challenging due to molecular heterogeneity, various studies have classified HCC into a finite number of molecular categories, primarily using sequencing techniques (Table 3, Fig. 3). Molecular classification of HCC has a similar objective as HCC staging, but is based on differences at the molecular level. It can also help develop subtype-specific targeted HCC therapies.

Table 3 An overview of various established HCC molecular classification systems.

Type of classification	HCC classes	Basis for classification	Samples	Techniques	References
Gene-signature based classification	G1-G6	Loss of heterozygosity, gene mutations, promoter methylation of CDH1 and CDKN2Aand HBV DNA copy	57 HCCs and 3 hepatocellular adenomas. 63 additional HCCs (for validation)	Genome wide transcriptome microarray and unsupervised hierarchical clustering analysis	Boyault et al. (2007)
	Class 1–5	Hierarchical clustering of gene expression data	103 tumor samples From transplant patients	Single-nucleotide polymorphism arrays, statistical analysis of copy number alterations, fluorescence *in-situ* hybridization, tissue microarrays	Chiang et al. (2008)
	EpCAM + AFP +, EpCAM + AFP −, EpCAM- AFP +, EpCAM- AFP-	AFP and EpCAM expression	40 HCC samples. 349 HCC samples (for validation)	cDNA microarray, oligonucleotide microarray and immunohistochemical analyses	Yamashita et al. (2008)
	S1-S3	Expression of meta-analysis marker genes	603 patients from eight independent cohorts	Subclass mapping (SubMap), gene set enrichment analysis, microarray (for validation)	Hoshida et al. (2009)

Continued

Table 3 An overview of various established HCC molecular classification systems.—cont'd

Type of classification	HCC classes	Basis for classification	Samples	Techniques	References
	Subtype A-D	Expression of top differentially expressed genes	373 HCC samples and 50 controls	Differential gene expression analysis, enrichment analysis and gene–gene interaction	Agarwal, Narayan, Bhattacharyya, Saraswat, and Tomar (2017)
	iClust 1–3	Unsupervised clustering of data from five platforms	363 hepatocellular carcinoma (HCC) samples from various institutes	Whole exome sequencing, DNA copy number analysis, methylation profiling, mRNA-seq, miRNA-seq, and proteomic analysis	Wheeler and Roberts (2017)
Metabolic classification	iHCC 1–3	Expression status of stratifying metabolic genes identified in this study	Transcriptomics and clinical data of 369 HCC and 50 matched noncancer liver samples (TCGA)	"Functional" gene–gene networks-based characterization	Bidkhori et al. (2018)
	C1-C3	Gene expression profile of metabolic genes	Metadata set with 371 and 231 HCC samples from TCGA and the International Cancer Genome Consortium, respectively	Subsequent non-negative matrix factorization (NMF) clustering, gene set variation analysis, microenvironment cell population-counter to estimate immune filtration	Yang, Huang, Liu, Qin, and Wang (2020)

Immunological classification	Immune-high, Immune-mid and Immune-low	Immune cell infiltration pattern	158 samples from 141 patients	Multiplex IHC and cluster analysis	Kurebayashi et al. (2018)
	Immunocompetent, Immunodeficient and Immunosuppressive	Tumor microenvironment including immune status, chemokines/cytokines and cell metabolism	42 samples from eight HCC patients	Whole-exome sequencing, RNA-seq, quantitative proteomics by multiplexed tandem mass tag (TMT) MS, untargeted metabolomics by liquid chromatography (LC)-MS/MS, CyTOF analysis of immune cells	Zhang et al. (2019)
	Type I to IV	Tumor budding assessment and immune scores	423 resected HCC samples	Tumor budding evaluation, Immunohistochemistry and hematoxylin-eosin staining, Whole-exome sequencing	Wei et al. (2020)
Chromosomal classification	Poorly polyploid and highly polyploid	Percentage of polyploid hepatocytes	75 tumor and non-tumor tissue pairs from HCC surgery patients and 13 normal liver samples	ImageJ quantification of cellular and nuclear ploidy	Bou-Nader et al. (2020)

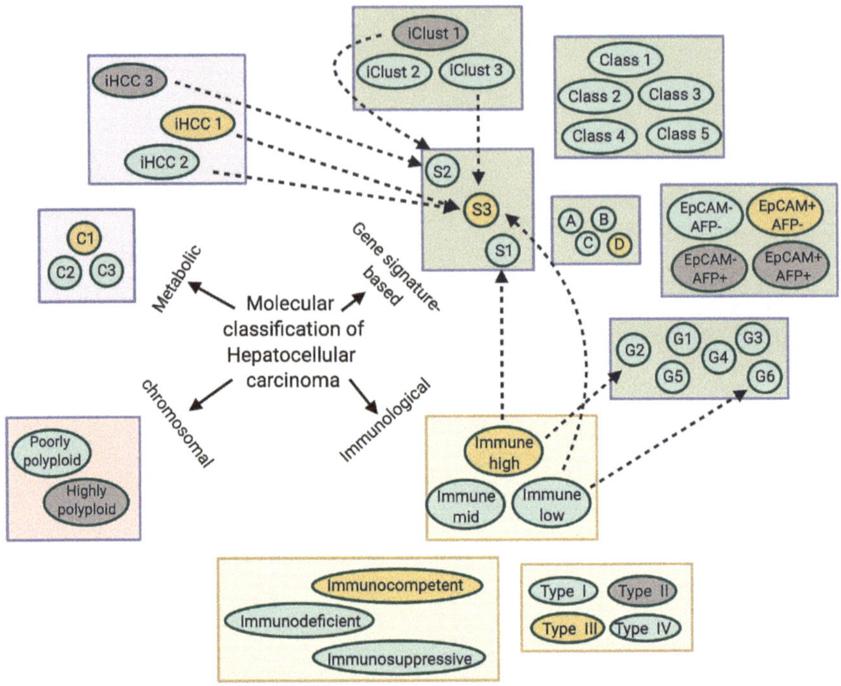

Fig. 3 An overview of various molecular classification systems for HCC. For each classification system, the subgroup with better prognosis and worse prognosis are shaded yellow and gray, respectively. Correlating subgroups between multiple classification systems are connected using dotted lines. *Created with https://biorender.com.*

4.1 Gene signature-based classification of HCC

High-resolution copy number analysis of 125 HCC tumors followed by validation using whole exome sequencing of 24 tumors and matched normal livers revealed 5 pathways that were frequently altered in HCC. The WNT/β-catenin pathway was mutated in 49.6% of tumors with three mutually exclusive mutations Catenin Beta 1 (CTNNB1) 32.8%, Axis inhibition protein 1 (AXIN1) 15.2% and adenomatous polyposis coli protein (APC) 1.6%. The second pathway most frequently mutated was p53 signaling in 28.8% of samples, specifically mutations in TP53 gene inactivation (20.8%) or cyclin dependent kinase inhibitor 2A (CDKN2A) 8%. Genes in chromatin remodeling complex (>24%) were the next frequently mutated, with AT-rich interaction domain (ARID) 1A and ARID2 mutations in 16.8% and 5.6% of samples. 9.6% of samples had a mutation in the ribosomal protein S6 kinase A3 (RPS6KA3) gene, a kinase in Ras/MAPK signaling pathway. Nuclear factor, erythroid 2 like 2 (NFE2L2), a transcription factor involved in maintaining cellular redox homeostasis was mutated in 6.4% of HCC tumors. This mutation was present with CTNNB1 mutations, potentiating alterations of the oxidative stress pathway in WNT/β-catenin signaling (Guichard et al., 2012).

4.1.1 G1–G6 classification

Boyault. et al. suggested an HCC classification system based on genome wide transcriptome microarray and unsupervised hierarchical clustering analysis of 123 tumors (Boyault et al., 2007). A combination of 16 gene signatures were used to classify HCC into 6 groups G1-G6. G1 group was associated with genes expressed during development (Myosin Heavy Chain 4, SRY-Box Transcription Factor 9), paternally imprinted genes (Insulin Like Growth Factor 2, Paternally-expressed gene 3, Paternally-expressed gene 10 and Sarcoglycan Epsilon) and high alpha-fetoprotein (AFP) expression and G2 with mutations in the phosphatidylinositol-4,5-bisphosphate 3-kinase catalytic subunit alpha (PIK3CA) gene. Both G1 and G2 have AXIN1 gene mutations and are related to HBV infection. However, G1 samples were associated with low and G2 with high copy numbers of viral DNA. G3 overexpresses genes implicated in cell cycle control including genes encoding cyclins, cell cycle checkpoint and DNA replication checkpoint proteins and has CDKN2A promoter hypermethylation. The G4 cluster comprised of the 5 samples of non-tumorous liver tissues, the 3 hepatocellular adenoma and 2 HCCs and had a T cell factor 1 (TCF1) mutation. G5 shows down-regulation of genes involved in stress and immune response. G6 tumors showed an overexpression of lymphoid enhancer binding factor 1 (LEF1), downregulation of E-cadherin, under expression of cadherin 1 (CDH1) and presence of satellite nodules around the principal tumor (Boyault et al., 2007).

In addition to the 6 classes mentioned above, the system classifies HCCs broadly into tumors with chromosomal instability (G1-G3) or chromosomal stability (G4-G6), TP53 mutated (G2-G3), CTNNB1 mutated (G5-G6) or neither mutated (G1, G4) and AKT activated (G1-G2), WNT activated (G5-G6) or neither activated (G3-G4). Unfortunately, none of the classification systems in this study were able to predict survival. However, they can act as a tool to identify patients for targeted therapy (Boyault et al., 2007). After almost a decade, the same group validated the system in another cohort and correlated the molecular alterations to HCC phenotype (Calderaro et al., 2017). CTNNB1 mutated tumors were larger, well differentiated, cholestatic, without inflammatory infiltrates and associated with intact tumor capsule and areas of micro-trabecular and pseudo-glandular histological patterns. TP53 tumors, on the other hand, were poorly differentiated, non-cholestatic, pleomorphic, multinucleated and associated with macrovascular and microvascular invasion, areas of compact histological pattern and foci of sarcomatous changes. The tumors were also associated with WHO histological classification of HCC. G1 tumors were associated with a progenitor phenotype, G3 with macro-trabecular massive subtype and G4 with steatohepatitic subtype (Calderaro et al., 2017).

This classification system was validated in other cohorts. Transcriptome analysis of 82 samples from Singapore assembled patients into similar groups using the same gene signatures. Although the distribution of patients in the 6 groups was different, overall the results were concordant. This study further examined at distinct clinical features of groups and found that subgroup G3 strongly associated with microvascular invasion (Allen. et al., 2016). Similarly, 205 Korean HCC samples were classified into the six subgroups and similar molecular characteristics were observed (Ahn et al., 2018).

4.1.2 Class 1–5 classification

Microarray analysis of gene expression profiles of 91 HCCs revealed 4 gene expression classes. Class 1 had high prevalence of CTNNB1 mutations. The tumors were >3cm in size and several liver-specific marker genes were overexpressed. Class 2 tumors had increased proliferation, high AFP, chromosomal instability (4q loss, 13q loss and 6q loss) and insulin like growth factor (IGF) 1 receptor, ribosomal protein S6 (RPS6) and Akt phosphorylation. Class 3 tumors had lower expression of IGF2, CTNNB1 target genes, over-expression of ISG marker genes and associated with tumors <3cm. Class 4 tumors contained polysomy of chromosome 7 and high expression of genes in this region like Cordon-Bleu WH2 repeat protein (COBL), claudin 15 (CLDN15), mitotic arrest deficient 1 like 1 (MAD1L1), DNA polymerase delta 2 (POLD2) and ephrin type-A receptor 1 (EPHA1). These tumors also lacked chromosome 8q gains. Class 5 consists of a group of unannotated tumors. The prognostic significance of this classification was not determined (Chiang et al., 2008).

4.1.3 EpCAM- and AFP-based classification

Epithelial cellular adhesion molecule (EpCAM) is a cell-cell adhesion molecule highly expressed in embryonic liver and hepatic stem cells (de Boer, van Krieken, Janssen-van Rhijn, & Litvinov, 1999; Schmelzer,

Wauthier, & Reid, 2006). Mature hepatocytes lack EpCAM expression (de Boer et al., 1999; Schmelzer et al., 2006) whereas premalignant livers (Kim et al., 2004) and 35% of HCC patients express EpCAM (de Boer et al., 1999; Kim et al., 2004; Ruck, Wichert, Handgretinger, & Kaiserling, 2000; Schmelzer et al., 2006). Although controversial, AFP is the most commonly used serum biomarker for HCC and expressed in 60% of HCC patients (Trevisani et al., 2001).

Gene signatures associated with EpCAM expression were identified using microarray analysis of 40 primary HCC tissues (Yamashita et al., 2008). Over-expressed genes in EpCAM-positive tumors were associated with organ development and protein synthesis and linked to the signaling networks of nine genes caldesmon 1 (CALD1), CTNNB1, epidermal growth factor (EGF), insulin (Ins)-1, N-Myelocytomatosis (MYCN), myogenic differentiation 1 (MYOD1), PPARA, transforming growth factor (TGF) beta 1, and TP53. Similarly, genes involved in mature hepatocyte functions, such as lipid metabolism and drug metabolism, were down regulated by five genes [NFE2L2, UDP glucuronosyltransferase family 1 member A6 (UGT1A6), transcription factor T cell factor 1 (TCF1), hepatocyte nuclear factor 4 Alpha (HNF4A), and plasminogen (PLG)]. EpCAM+ and EpCAM− tumors were further stratified based on the expression of AFP into EpCAM+ AFP+, EpCAM+ AFP−, EpCAM− AFP+ and EpCAM− AFP−. Although gene signatures of these four groups were not studied, morphological differences between them were. EpCAM+ AFP+, EpCAM+ AFP−, EpCAM− AFP+, EpCAM− AFP− resembled hepatic stem cells, bile duct epithelial cells, hepatocyte progenitors and mature hepatocytes, respectively. EpCAM+ AFP+ and EpCAM− AFP+ groups correlated with poor prognosis whereas EpCAM− AFP− and EpCAM+ AFP− groups corresponded to intermediate and good prognosis, respectively (Yamashita et al., 2008). This classification was validated in three independent cohorts using cDNA microarray, oligonucleotide microarray, and immunohistochemistry. This classification can be used to determine prognosis and targeted therapy.

4.1.4 S1–S3 classification

Meta-analysis of gene expression data from 603 patients from nine independent patient cohorts from around the world by Hoshida et al. revealed three HCC subclasses termed S1, S2 and S3 (Hoshida et al., 2009). S1 tumors are characterized by TGF-β activation causing β-catenin to be localized in the cytoplasm resulting in WNT activation. These tumors also showed more vascular invasion and satellite lesions and associated with risk of early recurrence. S2 tumors exhibited suppression of interferon-target genes, activation of MYC and AKT signaling and high levels of AFP expression. This group also consisted of the largest tumors in the cohort. S3 tumors were the smallest, included most of the well-differentiated tumors, had activation of p53 and p21 target genes and associated with good survival. This classification has been validated in multiple cohorts in this study, has prognostic value and can be used for targeted therapy.

4.1.5 A–D classification

Gene expression data of 373 HCC samples and 50 adjacent non-tumor samples obtained from the TCGA database were analyzed to obtain differentially expressed genes (DEGs) and enriched biological pathways (Agarwal et al., 2017). HCC patients stratified using consensus unsupervised hierarchical clustering predicted 4 subtypes A, B, C and D. HCC subtype A was the only one without any overlap with the other subtypes and showed association with 11 kinase related genes. Subtype B showed an enrichment of pathways like those involved in β-catenin degradation and repression of wnt target genes. Subtype C primarily involved fibroblast growth factor (FGF) 2, FGF Receptor 2, FGF Receptor 3 genes and drug metabolism, oxidation and PI3K pathways. HCC subtype D involved PI3K pathway and IL-34 gene and had higher survival probability. This classification can be used for subtype specific therapeutic target selection.

4.1.6 iClust1–3

Unsupervised clustering of data from multiple platform analysis of 363 HCC samples revealed three independent clusters iClust 1 ($n = 65$), 2 ($n = 55$) and 3 ($n = 63$) (Wheeler & Roberts, 2017). iClust 1 was characterized by low frequencies of CDKN2A silencing, CTNNB1 mutation and TERT promoter mutation, low TERT expression and high expression of proliferation marker genes. They were higher grade tumors with macrovascular invasion, iClust 2 consisted of lower grade tumors with less microvascular invasion and iClust 3 consisted of tumors with chromosomal instability, TP53 mutation and hypomethylation of multiple CpG sites. Although, there was no difference in overall survival between the 3 groups, iClust 1 had significantly worse prognosis than the other two groups. iClust 1 and iClust 3 corresponded to Hoshida S2 and S3 classes, respectively (Hoshida et al., 2009; Wheeler & Roberts, 2017).

4.2 Metabolic classification of HCC

Metabolic reprogramming, including enhanced glucose uptake and lactate production, overexpression of fatty acid synthase (FASN), altered lipid metabolism and oxidative metabolic imbalance gives a selective advantage to HCC cells (De Matteis et al., 2018). Metabolic classification of HCC patients can help determine patient prognosis or assist in developing specific antimetabolite-based targeted therapy (Agren et al., 2014).

4.2.1 iHCC 1–3 classification

Analysis of Genome-scale metabolic models and construction of functional gene–gene networks of 186 HCC and 50 matched normal liver samples stratified HCC patients into 3 subtypes—iHCC1 ($n=85$), iHCC2 ($n=49$), and iHCC3 ($n=52$) (Bidkhori et al., 2018). 18 stratifying metabolic genes were identified. The highest survival rate was observed in iHCC1 followed by iHCC2 and iHCC3. Differential gene expression analysis showed upregulation of inflammatory pathways in iHCC1 and WNT/β-catenin pathway in iHCC2. Gene set enrichment analysis revealed upregulation of tryptophan and indole metabolism and downregulation of noncoding RNA metabolism and ribosome biogenesis in iHCC1, up-regulation of oxidative demethylation and heme, glutamine and drug metabolism as well as down-regulation of cell development and G protein-coupled receptor signaling in iHCC2. iHCC3 was characterized by an abundance of changes like downregulation of fatty acid and lipid oxidation, small-molecule and amino-acid metabolism, drug catabolism and response to xenobiotic stimulus. In addition to metabolic changes, iHCC3 showed upregulation of multiple processes associated with cell proliferation, cell-cycle progression and mitosis, development, chromosome segregation, cytoskeleton organization, immune response, DNA replication, and recombination. Most iHCC3 tumors (96%) correspond to Hoshida S2 and iHCC1 and iHCC2 tumors (>54%) correspond to Hoshida S3 class 1 (Hoshida et al., 2009).

4.2.2 C1–C3 classification

Transcriptome analysis of a metadata set of 602 patients revealed 3 subgroups of HCC based on expression of 2752 metabolism related genes C1, C2 and C3 (Yang et al., 2020). C1 (active) was enriched in amino acid, lipid and drug metabolism-relevant signatures, expressed low AFP, pathologic stage I/II, and histologic grade G1/G2 and had good prognosis. C2 (exhausted) did not contain any distinct metabolic signatures but had high immune and stromal cell infiltration. The poor prognosis in C2 may be attributed to the combined effects of low infiltration of regulatory T cells (Treg) and Th17, high expression of immune checkpoint genes, presence of vascular invasion, advanced pathologic stage (III/IV), histologic grade (G3/G4), and high serum AFP level. C3 was enriched in hormone and proteoglycan metabolism (intermediate), high AFP level and poor prognosis.

4.3 Immunological classification of HCC

Immune evasion is a cancer hallmark promoting tumor formation and growth (Vinay et al., 2015). The first line of defense against tumors is a cell-mediated innate immune response involving the activation of receptors on immune cells including dendritic cells, macrophages, monocytes and natural killer cells by PAMPs and DAMPs (Maruyama, Selmani, Ishii, & Yamaguchi, 2011). Adaptive immunity, on the other hand, is an immune response mounted against tumor specific antigens by B and T lymphocytes. T lymphocytes can be classified into T helper (CD4+), T regulatory (CD4+) and cytotoxic T (CD8+) cells. Under normal conditions, Tregs suppress any autoimmune response. However, in a tumor environment, their expression is increased, causing an increased production of immunosuppressive cytokines, suppressing tumor immunity. B lymphocytes are activated by T helper cells to produce plasma B cells that produce antibodies. A weak immune response to tumors could be due to increased number of Tregs. Lack of CD4+ T cells can also weaken immune response by causing a depletion of CD8+ T cells (Hiraoka, 2010). Programmed death-ligand 1 (PDL1) on cancer cells and APCs can bind to Programmed cell death protein 1 (PD1) on T, B, myeloid or tumor cells to block proliferation, cytokine release and cytotoxic activities of T cells (Wu, Chen, Xu, & Gu, 2019). High levels of PD1+ CD8+ cells is seen in HCC cells (Tan, Chacko, & Chew, 2019).

4.3.1 Immune-high, mid and low classification

Cluster analysis of multiplex IHC data from 919 regions in 158 HCCs revealed three distinct HCC subtypes based on the immune microenvironment—immune-high, immune-mid and immune-low subtypes (Kurebayashi et al., 2018). Immune high group had increased B cell, plasma cell, and T cell infiltration. It was associated with poor differentiation, lower grade fibrosis, better prognosis among poorly differentiated tumors, increased PD1+CD8+T cells and high expression of PD-L1 in tumor and immune cells. The immune-mid group showed fibrosis and moderately increased infiltration with T cells and lesser infiltration with B and plasma cells. Immune-mid group is further subdivided into immune-mid-1 and immune-mid-2. Immune-mid-1 is a relatively non-homogeneous subtype including regions with increased granuloma formation, mast-cell infiltration, or neutrophil infiltration and associated with steatosis and ballooning. Immune-mid-2 is a more homogenous group. Immune-low subtype had low immune cell infiltration with well to moderately differentiated HCC. It was subdivided into immune-low-1, characterized by lower Treg/CD4 ratios and immune-low-2 with higher Treg/CD4 ratios. Immune-low-1 subtype was significantly less associated with poorly differentiated regions and fibrosis and associated with poorer prognosis whereas Immune-low-2 subtype was significantly associated with poorly differentiated regions and fibrosis. In addition to having prognostic significance, the study also identified the association of these subgroups with Hoshida's and Boyault's classification system. Immune-high group was associated with Hoshida's S1 and Boyault's G2 and Immune-low subtype with Hoshida's S3 and Boyault's G6 subclasses (Boyault et al., 2007; Hoshida et al., 2009).

4.3.2 Immunocompetent, immunodeficient and immunosuppressive classification

Evaluation of tumor heterogeneity of 42 samples from eight HCC patients using whole-exome sequencing, RNA sequencing, proteomic analysis, metabolomic analysis, cytometry, and single-cell analysis revealed three immune-based HCC subtypes—1 (immunocompetent), 2 (immunodeficient) and 3 (immunosuppressive). HCC subtype 1 patients had relatively competent antitumoral immunity with high expression of T cell recruiting molecules, normal T cell infiltration, fewer regulatory B-cell infiltration and increased urea cycle activity. Subtype 2 had reduced lymphocyte infiltration, high levels of monocytes and immune suppressive marker genes and functional genes, increased nucleotide synthesis, and low levels of chemokines and cytokines. HCC subtype 3 exhibited normal lymphocyte infiltration, increased number of immunosuppressive cells, inhibited glycolysis, increased mitochondrial respiration, over expression of genes encoding tumor suppressive chemokines/cytokines, and high expression of T cell recruiting molecules. This classification has a prognostic significance with subtype 1 having the best prognosis followed by subtype 2 and subtype 3 (Zhang et al., 2019).

4.3.3 Type I to IV classification

Tumor budding assessment and immune scores of 423 patients were used to classify HCC patients into four subgroups (Wei et al., 2020). Tumor budding refers to isolated or small groups of tumor cells present at the invasive front or in the tumor's center (Cho & Kakar, 2018). Although extensively studied as a prognostic factor for colorectal cancer, its role in HCC has only recently been studied. HCC patients were classified into three groups based on the number of tumor buds—Grade 1 (0–4), Grade 2 (5–9) and Grade 3 (≥ 10). For survival analysis, grades 1 and 2 were consolidated into one group—low grade and grade 3 was considered high grade. An immune score was assigned to each tumor based on five parameters—CD8 stromal, PD-L1 stromal, Mast-cell stromal, CD68 stromal, and FOXP3 stromal and patients were classified into immune type A and immune type B groups. According to tumor budding grade and immune type, patients were classified into four subgroups: IS_A-TB_{high} (type I), IS_B-TB_{high} (type II), IS_A-TB_{low} (type III) and IS_B-TB_{low} (type IV). Type I patients had higher frequencies of TP53 mutations, higher levels of CD8+ immune cell, and lower levels of CD68+ and PD-L1+ immune cells. Type II patients exhibited low levels of CD68+ macrophages and PD-L1+ immune cells and a higher level of CD8+ T cells. Type II and type III tumors had the worst and best overall and disease-free survivals, respectively. Type IV tumors had a high frequency of CTNNB1 mutations (Wei et al., 2020).

4.4 Chromosomal classification of HCC

Human hepatocytes exhibit nuclear ploidy, a single nuclear hepatocyte containing more than 2 sets of chromosomes, and cellular ploidy, a binuclear hepatocyte containing 2 or more than 2 sets of chromosomes (Bou-Nader et al., 2020). Analysis of the ploidy profile of 75 tumor and non-tumor tissue pairs and 13 normal liver samples revealed that although the ratio between binuclear fractions ($2 \times 2n/2 \times \geq 4n$) was similar in all three groups, the number of binuclear polyploid hepatocytes was reduced in both tumor and non-tumor tissues compared to the normal liver. Tumor tissues also had more mononuclear $4n$ and $\geq 8n$ hepatocytes compared to non-tumor tissues and normal liver tissue. Mononuclear polyploid cells were associated with TP53 mutations and had an increase in proliferation-related gene expression. Based on the ploidy profile, two groups of tumors were defined: poorly polyploid, if less than 50% of hepatocytes are polyploid and highly polyploid if >50% of hepatocytes are polyploid. Highly polyploid tumors had reduced 2-year and 5-year disease free survival.

5. Discussion

Advancements in multi-omic analyses in the past 1–2 decades have culminated in molecular classifications of HCC based on genetic, metabolic, immune and chromosomal profiles. However, the ultimate purpose of establishing a classification system is its application in a clinical setting. In order to achieve that outcome, it is important to further validate most of the reported studies in other independent HCC cohorts. Also, the etiology of the patient needs to be considered to determine if a certain subtype is more significantly correlated to one etiology than to others. This will allow a more global application of these systems, since HCC etiology varies based on geography. It will also help explain any of the observed differences between independent cohorts classified within the same system. These have been done in certain studies but clearly needs to be expanded in others.

Boyault's G1-G6 classification is probably the most significant classification system (Boyault et al., 2007). It was established in 2006 and was then validated in South east Asian and Korean populations (Ahn et al., 2018; Allen. et al., 2016). It allows classification based on various parameters such as chromosomal stability, gene mutation, gene methylation, clinical features and biological pathways. Moreover, two of the subgroups have shown a better correlation with HBV infections. As of now, clear prognostic significance of this classification system has not been observed, although, in one study, the G3 group was associated with microvascular invasion, which negatively associates with survival (Allen. et al., 2016). In the immune profile-based classification immune-high group was associated with Boyault's G2 and Immune-low subtype with Boyault's G6 subclasses (Kurebayashi et al., 2018).

Hoshida's S1–S3 classification is another well-established classification system (Hoshida et al., 2009). Several other classification systems have identified correlations between their HCC classes and Hoshida's classes. In iClust 1–3 classification, iClust 1 and iClust 3 corresponded to Hoshida S2 and S3 classes, respectively (Wheeler & Roberts, 2017). In iHCC 1–3 classification, most iHCC3 tumors (96%) corresponded to Hoshida S2 and iHCC1 and iHCC2 tumors (>54%) corresponded to Hoshida S3 class 1 (Bidkhori et al., 2018). In the immune profile-based classification immune-high group was associated with Hoshida's S1 and immune-low subtype with Hoshida's

S3 subclasses (Kurebayashi et al., 2018). The validation of a gene-based classification system in a metabolic- and immune-based classification system is promising. In addition, this system also has prognostic significance.

With respect to prognostic significance, other than Boyault's G1-G6 and the five-class system, all classification systems displayed some level of correlation with survival, although this aspect of these classification systems need further validation (Boyault et al., 2007; Chiang et al., 2008). The lack of data for correlation between etiology and HCC classes can be attributed to their small sample sizes. Increasing the sample size would provide more significant data. HCC heterogeneity makes it difficult to propose a system of molecular classification that would encompass 100% of HCC populations within a finite number of mutually exclusive groups. However, stratification of patients to any extent could have immense practical applications. Differences in disease etiology and inherent genetic differences between different populations makes it important to validate any classification system with multiple independent cohorts and diagnostic platforms that can confirm generalizability of the observations and conclusions. Such validation has been performed for a limited number of the discussed classification systems making them potentially useful in a clinical setting.

Acknowledgments

The present study was supported in part by The National Institute of Diabetes and Digestive and Kidney Diseases (NIDDK) Grant 1R01DK107451-01A1 (D.S.), National Cancer Institute (NCI) Grants 1R01CA230561-01A1 (D.S.), 1R01CA240004-01 (D.S.) and 1R01CA244993-01 (D.S. and P.B.F.), and Department of Defense (DOD) Grant CA170048 (D.S.).

References

A new prognostic system. (1998). A new prognostic system for hepatocellular carcinoma: A retrospective study of 435 patients: The Cancer of the Liver Italian Program (CLIP) investigators. *Hepatology*, *28*(3), 751–755.

Agarwal, R., Narayan, J., Bhattacharyya, A., Saraswat, M., & Tomar, A. K. (2017). Gene expression profiling, pathway analysis and subtype classification reveal molecular heterogeneity in hepatocellular carcinoma and suggest subtype specific therapeutic targets. *Cancer Genetics*, *216-217*, 37–51.

Agren, R., Mardinoglu, A., Asplund, A., Kampf, C., Uhlen, M., & Nielsen, J. (2014). Identification of anticancer drugs for hepatocellular carcinoma through personalized genome-scale metabolic modeling. *Molecular Systems Biology*, *10*, 721.

Aguilar, F., Hussain, S. P., & Cerutti, P. (1993). Aflatoxin B1 induces the transversion of G- ->T in codon 249 of the p53 tumor suppressor gene in human hepatocytes. *Proceedings of the National Academy of Sciences of the United States of America*, *90*(18), 8586–8590.

Ahn, S. M., Haq, F., Park, I., Nault, J. C., Zucman-Rossi, J., & Yu, E. (2018). The clinical implications of G1-G6 transcriptomic signature and 5-gene score in Korean patients with hepatocellular carcinoma. *BMC Cancer, 18*(1), 571.
Allen, J. C., Jr., Nault, J. C., Zhu, G., Khor, A. Y., Liu, J., Lim, T. K., et al. (2016). The transcriptomic G1-G6 signature of hepatocellular carcinoma in an Asian population: Association of G3 with microvascular invasion. *Medicine (Baltimore), 95*(47), e5263.
Amisaki, M., Tsuchiya, H., Sakabe, T., Fujiwara, Y., & Shiota, G. (2019). Identification of genes involved in the regulation of TERT in hepatocellular carcinoma. *Cancer Science, 110*(2), 550–560.
Andant, C., Puy, H., Bogard, C., Faivre, J., Soule, J. C., Nordmann, Y., et al. (2000). Hepatocellular carcinoma in patients with acute hepatic porphyria: frequency of occurrence and related factors. *Journal of Hepatology, 32*(6), 933–939.
Andersson, C., Bjersing, L., & Lithner, F. (1996). The epidemiology of hepatocellular carcinoma in patients with acute intermittent porphyria. *Journal of Internal Medicine, 240*(4), 195–201.
Antoury, C., Lopez, R., Zein, N., Stoller, J. K., & Alkhouri, N. (2015). Alpha-1 antitrypsin deficiency and the risk of hepatocellular carcinoma in end-stage liver disease. *World Journal of Hepatology*, 7(10), 1427–1432.
Ascha, M. S., Hanouneh, I. A., Lopez, R., Tamimi, T. A., Feldstein, A. F., & Zein, N. N. (2010). The incidence and risk factors of hepatocellular carcinoma in patients with nonalcoholic steatohepatitis. *Hepatology, 51*(6), 1972–1978.
Ashfaq, U. A., Javed, T., Rehman, S., Nawaz, Z., & Riazuddin, S. (2011). An overview of HCV molecular biology, replication and immune responses. *Virology Journal, 8*, 161.
Asselah, T., Rubbia-Brandt, L., Marcellin, P., & Negro, F. (2006). Steatosis in chronic hepatitis C: Why does it really matter? *Gut, 55*(1), 123–130.
Bacon, B. R., & Britton, R. S. (1989). Hepatic injury in chronic iron overload. Role of lipid peroxidation. *Chemico-Biological Interactions, 70*(3–4), 183–226.
Baravelli, C. M., Sandberg, S., Aarsand, A. K., & Tollanes, M. C. (2019). Porphyria cutanea tarda increases risk of hepatocellular carcinoma and premature death: A nationwide cohort study. *Orphanet Journal of Rare Diseases, 14*(1), 77.
Benn, J., & Schneider, R. J. (1994). Hepatitis B virus HBx protein activates Ras-GTP complex formation and establishes a Ras, Raf, MAP kinase signaling cascade. *Proceedings of the National Academy of Sciences of the United States of America, 91*(22), 10350–10354.
Bergeron, A., D'Astous, M., Timm, D. E., & Tanguay, R. M. (2001). Structural and functional analysis of missense mutations in fumarylacetoacetate hydrolase, the gene deficient in hereditary tyrosinemia type 1. *The Journal of Biological Chemistry, 276*(18), 15225–15231.
Bianchi, L. (1993). Glycogen storage disease I and hepatocellular tumours. *European Journal of Pediatrics, 152*(Suppl. 1), S63–S70.
Bidkhori, G., Benfeitas, R., Klevstig, M., Zhang, C., Nielsen, J., Uhlen, M., et al. (2018). Metabolic network-based stratification of hepatocellular carcinoma reveals three distinct tumor subtypes. *Proceedings of the National Academy of Sciences of the United States of America, 115*(50), E11874–E11883.
Biggins, S. W., Kim, W. R., Terrault, N. A., Saab, S., Balan, V., Schiano, T., et al. (2006). Evidence-based incorporation of serum sodium concentration into MELD. *Gastroenterology, 130*(6), 1652–1660.
Bou-Nader, M., Caruso, S., Donne, R., Celton-Morizur, S., Calderaro, J., Gentric, G., et al. (2020). Polyploidy spectrum: A new marker in HCC classification. *Gut, 69*(2), 355–364.
Boyault, S., Rickman, D. S., de Reynies, A., Balabaud, C., Rebouissou, S., Jeannot, E., et al. (2007). Transcriptome classification of HCC is related to gene alterations and to new therapeutic targets. *Hepatology, 45*(1), 42–52.

Bradbear, R. A., Bain, C., Siskind, V., Schofield, F. D., Webb, S., Axelsen, E. M., et al. (1985). Cohort study of internal malignancy in genetic hemochromatosis and other chronic non-alcoholic liver diseases. *The Journal of the National Cancer Institute*, *75*(1), 81–84.
Bray, F., Ferlay, J., Soerjomataram, I., Siegel, R. L., Torre, L. A., & Jemal, A. (2018). Global cancer statistics 2018: GLOBOCAN estimates of incidence and mortality worldwide for 36 cancers in 185 countries. *CA: A Cancer Journal for Clinicians*, *68*(6), 394–424.
Brooks, P. J., & Theruvathu, J. A. (2005). DNA adducts from acetaldehyde: Implications for alcohol-related carcinogenesis. *Alcohol*, *35*(3), 187–193.
Bukh, J. (2016). The history of hepatitis C virus (HCV): Basic research reveals unique features in phylogeny, evolution and the viral life cycle with new perspectives for epidemic control. *Journal of Hepatology*, *65*(1 Suppl), S2–S21.
Burda, P., & Hochuli, M. (2015). Hepatic glycogen storage disorders: What have we learned in recent years? *Current Opinion in Clinical Nutrition and Metabolic Care*, *18*(4), 415–421.
Buzzetti, E., Pinzani, M., & Tsochatzis, E. A. (2016). The multiple-hit pathogenesis of non-alcoholic fatty liver disease (NAFLD). *Metabolism*, *65*(8), 1038–1048.
Calderaro, J., Couchy, G., Imbeaud, S., Amaddeo, G., Letouzé, E., Blanc, J. F., et al. (2017). Histological subtypes of hepatocellular carcinoma are related to gene mutations and molecular tumour classification. *Journal of Hepatology*, *67*(4), 727–738.
Cao, K., Li, J., Zhao, Y., Wang, Q., Zeng, Q., He, S., et al. (2016). miR-101 inhibiting cell proliferation, migration and invasion in hepatocellular carcinoma through down-regulating Girdin. *Molecules and Cells*, *39*(2), 96–102.
Cha, M. Y., Kim, C. M., Park, Y. M., & Ryu, W. S. (2004). Hepatitis B virus X protein is essential for the activation of Wnt/beta-catenin signaling in hepatoma cells. *Hepatology*, *39*(6), 1683–1693.
Chan, A. C., Fan, S. T., Poon, R. T., Cheung, T. T., Chok, K. S., Chan, S. C., et al. (2013). Evaluation of the seventh edition of the American Joint Committee on Cancer tumour-node-metastasis (TNM) staging system for patients undergoing curative resection of hepatocellular carcinoma: Implications for the development of a refined staging system. *HPB: The Official Journal of the International Hepato Pancreato Biliary Association*, *15*(6), 439–448.
Chen, S. Y., Ma, D. N., Chen, Q. D., Zhang, J. J., Tian, Y. R., Wang, Z. C., et al. (2017). MicroRNA-200a inhibits cell growth and metastasis by targeting Foxa2 in hepatocellular carcinoma. *Journal of Cancer*, *8*(4), 617–625.
Cheng, A. S., Yu, J., Lai, P. B., Chan, H. L., & Sung, J. J. (2008). COX-2 mediates hepatitis B virus X protein abrogation of p53-induced apoptosis. *Biochemical and Biophysical Research Communications*, *374*(2), 175–180.
Chiang, D. Y., Villanueva, A., Hoshida, Y., Peix, J., Newell, P., Minguez, B., et al. (2008). Focal gains of VEGFA and molecular classification of hepatocellular carcinoma. *Cancer Research*, *68*(16), 6779–6788.
Child, C. G., & Turcotte, J. G. (1964). Surgery and portal hypertension. *Major Problems in Clinical Surgery*, *1*, 1–85.
Chirillo, P., Pagano, S., Natoli, G., Puri, P. L., Burgio, V. L., Balsano, C., et al. (1997). The hepatitis B virus X gene induces p53-mediated programmed cell death. *Proceedings of the National Academy of Sciences of the United States of America*, *94*(15), 8162–8167.
Cho, S. J., & Kakar, S. (2018). Tumor budding in colorectal carcinoma: Translating a morphologic score into clinically meaningful results. *Archives of Pathology & Laboratory Medicine*, *142*(8), 952–957.
Cho, J. H., Lee, Y. M., Bae, S. H., & Chou, J. Y. (2020). Activation of tumor-promoting pathways implicated in hepatocellular adenoma/carcinoma, a long-term complication of glycogen storage disease type Ia. *Biochemical and Biophysical Research Communications*, *522*(1), 1–7.

Cross, J. C., Wen, P., & Rutter, W. J. (1993). Transactivation by hepatitis B virus X protein is promiscuous and dependent on mitogen-activated cellular serine/threonine kinases. *Proceedings of the National Academy of Sciences of the United States of America*, *90*(17), 8078–8082.

de Boer, C. J., van Krieken, J. H., Janssen-van Rhijn, C. M., & Litvinov, S. V. (1999). Expression of Ep-CAM in normal, regenerating, metaplastic, and neoplastic liver. *The Journal of Pathology*, *188*(2), 201–206.

De Matteis, S., Ragusa, A., Marisi, G., De Domenico, S., Casadei Gardini, A., Bonafe, M., et al. (2018). Aberrant metabolism in hepatocellular carcinoma provides diagnostic and therapeutic opportunities. *Oxidative Medicine and Cellular Longevity*, *2018*, 7512159.

Demo, E., Frush, D., Gottfried, M., Koepke, J., Boney, A., Bali, D., et al. (2007). Glycogen storage disease type III-hepatocellular carcinoma a long-term complication? *Journal of Hepatology*, *46*(3), 492–498.

Ding, D., Lou, X., Hua, D., Yu, W., Li, L., Wang, J., et al. (2012). Recurrent targeted genes of hepatitis B virus in the liver cancer genomes identified by a next-generation sequencing-based approach. *PLoS Genetics*, *8*(12), e1003065.

Ding, W., Tan, H., Zhao, C., Li, X., Li, Z., Jiang, C., et al. (2016). MiR-145 suppresses cell proliferation and motility by inhibiting ROCK1 in hepatocellular carcinoma. *Tumour Biology*, *37*(5), 6255–6260.

Donato, F., Boffetta, P., & Puoti, M. (1998). A meta-analysis of epidemiological studies on the combined effect of hepatitis B and C virus infections in causing hepatocellular carcinoma. *International Journal of Cancer*, *75*(3), 347–354.

Dong, H., Zhang, L., Qian, Z., Zhu, X., Zhu, G., Chen, Y., et al. (2015). Identification of HBV-MLL4 integration and its molecular basis in Chinese hepatocellular carcinoma. *PLoS One*, *10*(4), e0123175.

Donohue, T. M., Jr. (2007). Alcohol-induced steatosis in liver cells. *World Journal of Gastroenterology*, *13*(37), 4974–4978.

Elmberg, M., Hultcrantz, R., Ekbom, A., Brandt, L., Olsson, S., Olsson, R., et al. (2003). Cancer risk in patients with hereditary hemochromatosis and in their first-degree relatives. *Gastroenterology*, *125*(6), 1733–1741.

Elmore, L. W., Hancock, A. R., Chang, S. F., Wang, X. W., Chang, S., Callahan, C. P., et al. (1997). Hepatitis B virus X protein and p53 tumor suppressor interactions in the modulation of apoptosis. *Proceedings of the National Academy of Sciences of the United States of America*, *94*(26), 14707–14712.

El-Serag, H. B., & Rudolph, K. L. (2007). Hepatocellular carcinoma: Epidemiology and molecular carcinogenesis. *Gastroenterology*, *132*(7), 2557–2576.

Elzouki, A. N., & Eriksson, S. (1996). Risk of hepatobiliary disease in adults with severe alpha 1-antitrypsin deficiency (PiZZ): Is chronic viral hepatitis B or C an additional risk factor for cirrhosis and hepatocellular carcinoma? *European Journal of Gastroenterology & Hepatology*, *8*, 989–994.

Eriksson, S., Carlson, J., & Velez, R. (1986). Risk of cirrhosis and primary liver cancer in alpha 1-antitrypsin deficiency. *The New England Journal of Medicine*, *314*(12), 736–739.

Estes, C., Razavi, H., Loomba, R., Younossi, Z., & Sanyal, A. J. (2018). Modeling the epidemic of nonalcoholic fatty liver disease demonstrates an exponential increase in burden of disease. *Hepatology*, *67*(1), 123–133.

Fan, C. G., Wang, C. M., Tian, C., Wang, Y., Li, L., Sun, W. S., et al. (2011). miR-122 inhibits viral replication and cell proliferation in hepatitis B virus-related hepatocellular carcinoma and targets NDRG3. *Oncology Reports*, *26*(5), 1281–1286.

Feng, J., Wang, J., Chen, M., Chen, G., Wu, Z., Ying, L., et al. (2015). miR-200a suppresses cell growth and migration by targeting MACC1 and predicts prognosis in hepatocellular carcinoma. *Oncology Reports*, *33*(2), 713–720.

Forgues, M., Difilippantonio, M. J., Linke, S. P., Ried, T., Nagashima, K., Feden, J., et al. (2003). Involvement of Crm1 in hepatitis B virus X protein-induced aberrant centriole replication and abnormal mitotic spindles. *Molecular and Cellular Biology, 23*(15), 5282–5292.
Fracanzani, A. L., Conte, D., Fraquelli, M., Taioli, E., Mattioli, M., Losco, A., et al. (2001). Increased cancer risk in a cohort of 230 patients with hereditary hemochromatosis in comparison to matched control patients with non-iron-related chronic liver disease. *Hepatology, 33*(3), 647–651.
Freeman, R. B., Jr., Wiesner, R. H., Harper, A., McDiarmid, S. V., Lake, J., Edwards, E., et al. (2002). The new liver allocation system: Moving toward evidence-based transplantation policy. *Liver Transplantation, 8*(9), 851–858.
Fu, Y., Liu, M., Li, F., Qian, L., Zhang, P., Lv, F., et al. (2019). MiR-221 promotes hepatocellular carcinoma cells migration via targeting PHF2. *BioMed Research International, 2019*, 4371405.
Fujii, R., Zhu, C., Wen, Y., Marusawa, H., Bailly-Maitre, B., Matsuzawa, S., et al. (2006). HBXIP, cellular target of hepatitis B virus oncoprotein, is a regulator of centrosome dynamics and cytokinesis. *Cancer Research, 66*(18), 9099–9107.
Fujimoto, A., Totoki, Y., Abe, T., Boroevich, K. A., Hosoda, F., Nguyen, H. H., et al. (2012). Whole-genome sequencing of liver cancers identifies etiological influences on mutation patterns and recurrent mutations in chromatin regulators. *Nature Genetics, 44*(7), 760–764.
Furuta, M., Tanaka, H., Shiraishi, Y., Unida, T., Imamura, M., Fujimoto, A., et al. (2018). Characterization of HBV integration patterns and timing in liver cancer and HBV-infected livers. *Oncotarget, 9*(38), 25075–25088.
Ganne-Carrie, N., & Nahon, P. (2019). Hepatocellular carcinoma in the setting of alcohol-related liver disease. *Journal of Hepatology, 70*(2), 284–293.
Gao, B., & Bataller, R. (2011). Alcoholic liver disease: Pathogenesis and new therapeutic targets. *Gastroenterology, 141*(5), 1572–1585.
Ghouse, R., Chu, A., Wang, Y., & Perlmutter, D. H. (2014). Mysteries of alpha1-antitrypsin deficiency: Emerging therapeutic strategies for a challenging disease. *Disease Models & Mechanisms*, 7(4), 411–419.
Gisbert, J. P., Garcia-Buey, L., Alonso, A., Rubio, S., Hernandez, A., Pajares, J. M., et al. (2004). Hepatocellular carcinoma risk in patients with porphyria cutanea tarda. *European Journal of Gastroenterology & Hepatology, 16*(7), 689–692.
Guichard, C., Amaddeo, G., Imbeaud, S., Ladeiro, Y., Pelletier, L., Maad, I. B., et al. (2012). Integrated analysis of somatic mutations and focal copy-number changes identifies key genes and pathways in hepatocellular carcinoma. *Nature Genetics, 44*(6), 694–698.
Hamid, A. S., Tesfamariam, I. G., Zhang, Y., & Zhang, Z. G. (2013). Aflatoxin B1-induced hepatocellular carcinoma in developing countries: Geographical distribution, mechanism of action and prevention. *Oncology Letters, 5*(4), 1087–1092.
Hiraoka, N. (2010). Tumor-infiltrating lymphocytes and hepatocellular carcinoma: Molecular biology. *International Journal of Clinical Oncology, 15*(6), 544–551.
Hoshida, Y., Nijman, S. M., Kobayashi, M., Chan, J. A., Brunet, J. P., Chiang, D. Y., et al. (2009). Integrative transcriptome analysis reveals common molecular subclasses of human hepatocellular carcinoma. *Cancer Research, 69*(18), 7385–7392.
Hsieh, A., Kim, H. S., Lim, S. O., Yu, D. Y., & Jung, G. (2011). Hepatitis B viral X protein interacts with tumor suppressor adenomatous polyposis coli to activate Wnt/beta-catenin signaling. *Cancer Letters, 300*(2), 162–172.
Hu, L., Chen, L., Yang, G., Li, L., Sun, H., Chang, Y., et al. (2011). HBx sensitizes cells to oxidative stress-induced apoptosis by accelerating the loss of Mcl-1 protein via caspase-3 cascade. *Molecular Cancer, 10*, 43.

Hua, S., Quan, Y., Zhan, M., Liao, H., Li, Y., & Lu, L. (2019). miR-125b-5p inhibits cell proliferation, migration, and invasion in hepatocellular carcinoma via targeting TXNRD1. *Cancer Cell International, 19*, 203.

Hui, A. M., Makuuchi, M., & Li, X. (1998). Cell cycle regulators and human hepatocarcinogenesis. *Hepato-Gastroenterology, 45*(23), 1635–1642.

Hyams, K. C. (1995). Risks of chronicity following acute hepatitis B virus infection: A review. *Clinical Infectious Diseases, 20*(4), 992–1000.

Innala, E., & Andersson, C. (2011). Screening for hepatocellular carcinoma in acute intermittent porphyria: A 15-year follow-up in northern Sweden. *Journal of Internal Medicine, 269*, 538–545.

Jang, H. J., Yang, H. R., Ko, J. S., Moon, J. S., Chang, J. Y., & Seo, J. K. (2020). Development of hepatocellular carcinoma in patients with glycogen storage disease: A single center retrospective study. *Journal of Korean Medical Science, 35*(1), e5.

Jin, Y., Lee, W. Y., Toh, S. T., Tennakoon, C., Toh, H. C., Chow, P. K., et al. (2019). Comprehensive analysis of transcriptome profiles in hepatocellular carcinoma. *Journal of Translational Medicine, 17*(1), 273.

Jin, Y., Wang, J., Han, J., Luo, D., & Sun, Z. (2017). MiR-122 inhibits epithelial-mesenchymal transition in hepatocellular carcinoma by targeting Snail1 and Snail2 and suppressing WNT/beta-cadherin signaling pathway. *Experimental Cell Research, 360*(2), 210–217.

Jin, Y., Wu, D., Yang, W., Weng, M., Li, Y., Wang, X., et al. (2017). Hepatitis B virus x protein induces epithelial-mesenchymal transition of hepatocellular carcinoma cells by regulating long non-coding RNA. *Virology Journal, 14*(1), 238.

Johnson, P. J., Berhane, S., Kagebayashi, C., Satomura, S., Teng, M., Reeves, H. L., et al. (2015). Assessment of liver function in patients with hepatocellular carcinoma: A new evidence-based approach-the ALBI grade. *Journal of Clinical Oncology, 33*(6), 550–558.

Kamarajah, S. K., Frankel, T. L., Sonnenday, C., Cho, C. S., & Nathan, H. (2018). Critical evaluation of the American Joint Commission on Cancer (AJCC) 8th edition staging system for patients with hepatocellular carcinoma (HCC): A Surveillance, Epidemiology, End Results (SEER) analysis. *Journal of Surgical Oncology, 117*(4), 644–650.

Kannan, M., Jayamohan, S., Moorthy, R. K., Chabattula, S. C., Ganeshan, M., & Arockiam, A. J. V. (2019). AEG-1/miR-221 Axis cooperatively regulates the progression of hepatocellular carcinoma by targeting PTEN/PI3K/AKT signaling pathway. *International Journal of Molecular Sciences, 20*(22), 5526.

Kauppinen, R., & Mustajoki, P. (1988). Acute hepatic porphyria and hepatocellular carcinoma. *British Journal of Cancer, 57*(1), 117–120.

Kew, M. C. (2003). Synergistic interaction between aflatoxin B1 and hepatitis B virus in hepatocarcinogenesis. *Liver International, 23*(6), 405–409.

Kew, M. C. (2014). Hepatic iron overload and hepatocellular carcinoma. *Liver Cancer, 3*(1), 31–40.

Kim, C. W., & Chang, K. M. (2013). Hepatitis C virus: Virology and life cycle. *Clinical and Molecular Hepatology, 19*(1), 17–25.

Kim, M., Lee, H. C., Tsedensodnom, O., Hartley, R., Lim, Y. S., Yu, E., et al. (2008). Functional interaction between Wnt3 and Frizzled-7 leads to activation of the Wnt/beta-catenin signaling pathway in hepatocellular carcinoma cells. *Journal of Hepatology, 48*(5), 780–791.

Kim, J. K., Noh, J. H., Jung, K. H., Eun, J. W., Bae, H. J., Kim, M. G., et al. (2013). Sirtuin7 oncogenic potential in human hepatocellular carcinoma and its regulation by the tumor suppressors MiR-125a-5p and MiR-125b. *Hepatology, 57*(3), 1055–1067.

Kim, S., Park, S. Y., Yong, H., Famulski, J. K., Chae, S., Lee, J. H., et al. (2008). HBV X protein targets hBubR1, which induces dysregulation of the mitotic checkpoint. *Oncogene*, *27*(24), 3457–3464.

Kim, J. W., Ye, Q., Forgues, M., Chen, Y., Budhu, A., Sime, J., et al. (2004). Cancer-associated molecular signature in the tissue samples of patients with cirrhosis. *Hepatology*, *39*(2), 518–527.

Kishnani, P. S., Austin, S. L., Arn, P., Bali, D. S., Boney, A., Case, L. E., et al. (2010). Glycogen storage disease type III diagnosis and management guidelines. *Genetics in Medicine*, *12*(7), 446–463.

Kota, J., Chivukula, R. R., O'Donnell, K. A., Wentzel, E. A., Montgomery, C. L., Hwang, H. W., et al. (2009). Therapeutic microRNA delivery suppresses tumorigenesis in a murine liver cancer model. *Cell*, *137*(6), 1005–1017.

Kowdley, K. V. (2004). Iron, hemochromatosis, and hepatocellular carcinoma. *Gastroenterology*, *127*(5 Suppl. 1), S79–S86.

Kulik, L., & El-Serag, H. B. (2019). Epidemiology and management of hepatocellular carcinoma. *Gastroenterology*, *156*(2), 477–491.e471.

Kumar, P., Mahato, D. K., Kamle, M., Mohanta, T. K., & Kang, S. G. (2016). Aflatoxins: A global concern for food safety, human health and their management. *Frontiers in Microbiology*, 7, 2170.

Kuo, T. C., & Chao, C. C. (2010). Hepatitis B virus X protein prevents apoptosis of hepatocellular carcinoma cells by upregulating SATB1 and HURP expression. *Biochemical Pharmacology*, *80*(7), 1093–1102.

Kurebayashi, Y., Ojima, H., Tsujikawa, H., Kubota, N., Maehara, J., Abe, Y., et al. (2018). Landscape of immune microenvironment in hepatocellular carcinoma and its additional impact on histological and molecular classification. *Hepatology*, *68*(3), 1025–1041.

Lackner, C., & Tiniakos, D. (2019). Fibrosis and alcohol-related liver disease. *Journal of Hepatology*, *70*(2), 294–304.

Lang, E., Schafer, M., Schwender, H., Neumann, N. J., & Frank, J. (2015). Occurrence of malignant tumours in the acute hepatic porphyrias. *JIMD Reports*, *22*, 17–22.

Larsson, C. (1978). Natural history and life expectancy in severe alpha1-antitrypsin deficiency, Pi Z. *Acta Medica Scandinavica*, *204*, 345–351.

Lee, Y. I., Kang-Park, S., Do, S. I., & Lee, Y. I. (2001). The hepatitis B virus-X protein activates a phosphatidylinositol 3-kinase-dependent survival signaling cascade. *The Journal of Biological Chemistry*, *276*(20), 16969–16977.

Lee, S. W., Lee, Y. M., Bae, S. K., Murakami, S., Yun, Y., & Kim, K. W. (2000). Human hepatitis B virus X protein is a possible mediator of hypoxia-induced angiogenesis in hepatocarcinogenesis. *Biochemical and Biophysical Research Communications*, *268*(2), 456–461.

Lee, S., Tarn, C., Wang, W. H., Chen, S., Hullinger, R. L., & Andrisani, O. M. (2002). Hepatitis B virus X protein differentially regulates cell cycle progression in X-transforming versus nontransforming hepatocyte (AML12) cell lines. *The Journal of Biological Chemistry*, *277*(10), 8730–8740.

Leong, T. Y., & Leong, A. S. (2005). Epidemiology and carcinogenesis of hepatocellular carcinoma. *HPB: The Official Journal of the International Hepato Pancreato Biliary Association*, 7(1), 5–15.

Li, C. L., Li, C. Y., Lin, Y. Y., Ho, M. C., Chen, D. S., Chen, P. J., et al. (2019). Androgen receptor enhances hepatic telomerase reverse transcriptase gene transcription after hepatitis B virus integration or point mutation in promoter region. *Hepatology*, *69*(2), 498–512.

Li, Y., Ren, M., Zhao, Y., Lu, X., Wang, M., Hu, J., et al. (2017). MicroRNA-26a inhibits proliferation and metastasis of human hepatocellular carcinoma by regulating DNMT3B-MEG3 axis. *Oncology Reports*, *37*(6), 3527–3535.

Li, Q., Wang, G., Shan, J. L., Yang, Z. X., Wang, H. Z., Feng, J., et al. (2010). MicroRNA-224 is upregulated in HepG2 cells and involved in cellular migration and invasion. *Journal of Gastroenterology and Hepatology*, *25*(1), 164–171.
Li, G. L., Yuan, J. H., Zhuang, G. D., & Wu, D. Q. (2018). miR-199b exerts tumor suppressive functions in hepatocellular carcinoma by directly targeting JAG1. *European Review for Medical and Pharmacological Sciences*, *22*(22), 7679–7687.
Liang, T. J. (2009). Hepatitis B: The virus and disease. *Hepatology*, *49*(5 Suppl), S13–S21.
Liaw, Y. F., Tai, D. I., Chu, C. M., & Chen, T. J. (1988). The development of cirrhosis in patients with chronic type B hepatitis: A prospective study. *Hepatology*, *8*(3), 493–496.
Liberal, R., Longhi, M. S., Mieli-Vergani, G., & Vergani, D. (2011). Pathogenesis of autoimmune hepatitis. *Best Practice & Research. Clinical Gastroenterology*, *25*(6), 653–664.
Lin, C. J., Gong, H. Y., Tseng, H. C., Wang, W. L., & Wu, J. L. (2008). miR-122 targets an anti-apoptotic gene, Bcl-w, in human hepatocellular carcinoma cell lines. *Biochemical and Biophysical Research Communications*, *375*(3), 315–320.
Lin, X., Li, A. M., Li, Y. H., Luo, R. C., Zou, Y. J., Liu, Y. Y., et al. (2020). Silencing MYH9 blocks HBx-induced GSK3beta ubiquitination and degradation to inhibit tumor stemness in hepatocellular carcinoma. *Signal Transduction and Targeted Therapy*, *5*(1), 13.
Liu, Z., Wang, C., Jiao, X., Zhao, S., Liu, X., Wang, Y., et al. (2016). miR-221 promotes growth and invasion of hepatocellular carcinoma cells by constitutive activation of NFkappaB. *American Journal of Translational Research*, *8*(11), 4764–4777.
Liu, Y., Wu, C., Wang, Y., Wen, S., Wang, J., Chen, Z., et al. (2014). MicroRNA-145 inhibits cell proliferation by directly targeting ADAM17 in hepatocellular carcinoma. *Oncology Reports*, *32*(5), 1923–1930.
Livezey, K. W., Negorev, D., & Simon, D. (2002). Increased chromosomal alterations and micronuclei formation in human hepatoma HepG2 cells transfected with the hepatitis B virus HBX gene. *Mutation Research*, *505*(1–2), 63–74.
Lleo, A., de Boer, Y. S., Liberal, R., & Colombo, M. (2019). The risk of liver cancer in autoimmune liver diseases. *Therapeutic Advances in Medical Oncology*, *11*. 1758835919861914.
Llovet, J. M., Bru, C., & Bruix, J. (1999). Prognosis of hepatocellular carcinoma: The BCLC staging classification. *Seminars in Liver Disease*, *19*(3), 329–338.
Lumley, S. F., McNaughton, A. L., Klenerman, P., Lythgoe, K. A., & Matthews, P. C. (2018). Hepatitis B virus adaptation to the CD8 + T cell response: Consequences for host and pathogen. *Frontiers in Immunology*, *9*, 1561.
Ma, D., Tao, X., Gao, F., Fan, C., & Wu, D. (2012). miR-224 functions as an onco-miRNA in hepatocellular carcinoma cells by activating AKT signaling. *Oncology Letters*, *4*(3), 483–488.
Malinchoc, M., Kamath, P. S., Gordon, F. D., Peine, C. J., Rank, J., & ter Borg, P. C. (2000). A model to predict poor survival in patients undergoing transjugular intrahepatic portosystemic shunts. *Hepatology*, *31*(4), 864–871.
Manns, M. P., Johnson, E. F., Griffin, K. J., Tan, E. M., & Sullivan, K. F. (1989). Major antigen of liver kidney microsomal autoantibodies in idiopathic autoimmune hepatitis is cytochrome P450db1. *The Journal of Clinical Investigation*, *83*(3), 1066–1072.
Manzia, T. M., Angelico, R., Toti, L., Cillis, A., Ciano, P., Orlando, G., et al. (2011). Glycogen storage disease type Ia and VI associated with hepatocellular carcinoma: Two case reports. *Transplantation Proceedings*, *43*(4), 1181–1183.
Marcus, N. Y., Brunt, E. M., Blomenkamp, K., Ali, F., Rudnick, D. A., Ahmad, M., et al. (2010). Characteristics of hepatocellular carcinoma in a murine model of alpha-1-antitrypsin deficiency. *Hepatology Research*, *40*(6), 641–653.
Maruyama, K., Selmani, Z., Ishii, H., & Yamaguchi, K. (2011). Innate immunity and cancer therapy. *International Immunopharmacology*, *11*(3), 350–357.

Maurice, J., & Manousou, P. (2018). Non-alcoholic fatty liver disease. *Clinical Medicine (London, England)*, *18*(3), 245–250.
Mekuria, A. N., Routledge, M. N., Gong, Y. Y., & Sisay, M. (2020). Aflatoxins as a risk factor for liver cirrhosis: A systematic review and meta-analysis. *BMC Pharmacology and Toxicology*, *21*(1), 39.
Meng, F., Henson, R., Wehbe-Janek, H., Ghoshal, K., Jacob, S. T., & Patel, T. (2007). MicroRNA-21 regulates expression of the PTEN tumor suppressor gene in human hepatocellular cancer. *Gastroenterology*, *133*(2), 647–658.
Moon, E. J., Jeong, C. H., Jeong, J. W., Kim, K. R., Yu, D. Y., Murakami, S., et al. (2004). Hepatitis B virus X protein induces angiogenesis by stabilizing hypoxia-inducible factor-1alpha. *The FASEB Journal*, *18*(2), 382–384.
Noh, J. H., Chang, Y. G., Kim, M. G., Jung, K. H., Kim, J. K., Bae, H. J., et al. (2013). MiR-145 functions as a tumor suppressor by directly targeting histone deacetylase 2 in liver cancer. *Cancer Letters*, *335*(2), 455–462.
Ogunwobi, O. O., Harricharran, T., Huaman, J., Galuza, A., Odumuwagun, O., Tan, Y., et al. (2019). Mechanisms of hepatocellular carcinoma progression. *World Journal of Gastroenterology*, *25*(19), 2279–2293.
Okuda, K., Obata, H., Nakajima, Y., Ohtsuki, T., Okazaki, N., & Ohnishi, K. (1984). Prognosis of primary hepatocellular carcinoma. *Hepatology*, *4*(1 Suppl), 3S–6S.
Ozcan, H. N., Karcaaltincaba, M., Pektas, E., Sivri, H. S., Oguz, B., Dursun, A., et al. (2019). Imaging liver nodules in tyrosinemia type-1: A retrospective review of 16 cases in a tertiary pediatric hospital. *European Journal of Radiology*, *116*, 41–46.
Ozen, H. (2007). Glycogen storage diseases: New perspectives. *World Journal of Gastroenterology*, *13*(18), 2541–2553.
Peng, X. M., Peng, W. W., & Yao, J. L. (1998). Codon 249 mutations of p53 gene in development of hepatocellular carcinoma. *World Journal of Gastroenterology*, *4*(2), 125–127.
Peoc'h, K., Manceau, H., Karim, Z., Wahlin, S., Gouya, L., Puy, H., et al. (2019). Hepatocellular carcinoma in acute hepatic porphyrias: A Damocles Sword. *Molecular Genetics and Metabolism*, *128*(3), 236–241.
Perlmutter, D. H. (2002). Liver injury in alpha1-antitrypsin deficiency: An aggregated protein induces mitochondrial injury. *The Journal of Clinical Investigation*, *110*(11), 1579–1583.
Perlmutter, D. H. (2006). Pathogenesis of chronic liver injury and hepatocellular carcinoma in alpha-1-antitrypsin deficiency. *Pediatric Research*, *60*(2), 233–238.
Petruzziello, A., Marigliano, S., Loquercio, G., Cozzolino, A., & Cacciapuoti, C. (2016). Global epidemiology of hepatitis C virus infection: An up-date of the distribution and circulation of hepatitis C virus genotypes. *World Journal of Gastroenterology*, *22*(34), 7824–7840.
Pierantonelli, I., & Svegliati-Baroni, G. (2019). Nonalcoholic fatty liver disease: Basic pathogenetic mechanisms in the progression from NAFLD to NASH. *Transplantation*, *103*(1), e1–e13.
Pietrangelo, A. (2006). Hereditary hemochromatosis. *Biochimica et Biophysica Acta*, *1763*(7), 700–710.
Plissonnier, M. L., Herzog, K., Levrero, M., & Zeisel, M. B. (2018). Non-coding RNAs and hepatitis C virus-induced hepatocellular carcinoma. *Viruses*, *10*(11), 591.
Pugh, R. N., Murray-Lyon, I. M., Dawson, J. L., Pietroni, M. C., & Williams, R. (1973). Transection of the oesophagus for bleeding oesophageal varices. *The British Journal of Surgery*, *60*(8), 646–649.
Radford-Smith, D. E., Powell, E. E., & Powell, L. W. (2018). Haemochromatosis: A clinical update for the practising physician. *Internal Medicine Journal*, *48*(5), 509–516.
Ramanujam, V. S., & Anderson, K. E. (2015). Porphyria diagnostics-part 1: A brief overview of the porphyrias. *Current Protocols in Human Genetics*, *86*, 17.20.11–17.20.26.

Rehermann, B. (2009). Hepatitis C virus versus innate and adaptive immune responses: A tale of coevolution and coexistence. *The Journal of Clinical Investigation*, *119*(7), 1745–1754.
Resaz, R., Vanni, C., Segalerba, D., Sementa, A. R., Mastracci, L., Grillo, F., et al. (2014). Development of hepatocellular adenomas and carcinomas in mice with liver-specific G6Pase-alpha deficiency. *Disease Models & Mechanisms*, 7(9), 1083–1091.
Roy, C. N., Penny, D. M., Feder, J. N., & Enns, C. A. (1999). The hereditary hemochromatosis protein, HFE, specifically regulates transferrin-mediated iron uptake in HeLa cells. *The Journal of Biological Chemistry*, *274*(13), 9022–9028.
Ruck, P., Wichert, G., Handgretinger, R., & Kaiserling, E. (2000). Ep-CAM in malignant liver tumours. *The Journal of Pathology*, *191*(1), 102–103.
Saigo, K., Yoshida, K., Ikeda, R., Sakamoto, Y., Murakami, Y., Urashima, T., et al. (2008). Integration of hepatitis B virus DNA into the myeloid/lymphoid or mixed-lineage leukemia (MLL4) gene and rearrangements of MLL4 in human hepatocellular carcinoma. *Human Mutation*, *29*(5), 703–708.
Salata, H., Cortes, J. M., de Salamanca, R. E., Oliva, H., Castro, A., Kusak, E., et al. (1985). Porphyria cutanea tarda and hepatocellular carcinoma: Frequency of occurrence and related factors. *Journal of Hepatology*, *1*(5), 477–487.
Sanz-Cameno, P., Martin-Vilchez, S., Lara-Pezzi, E., Borque, M. J., Salmeron, J., Munoz de Rueda, P., et al. (2006). Hepatitis B virus promotes angiopoietin-2 expression in liver tissue: Role of HBV x protein. *The American Journal of Pathology*, *169*(4), 1215–1222.
Sarma, U. P., Bhetaria, P. J., Devi, P., & Varma, A. (2017). Aflatoxins: Implications on health. *Indian Journal of Clinical Biochemistry*, *32*(2), 124–133.
Sartorius, K., Makarova, J., Sartorius, B., An, P., Winkler, C., Chuturgoon, A., et al. (2019). The regulatory role of MicroRNA in hepatitis-B virus-associated hepatocellular carcinoma (HBV-HCC) pathogenesis. *Cells*, *8*(12), 1504.
Schmelzer, E., Wauthier, E., & Reid, L. M. (2006). The phenotypes of pluripotent human hepatic progenitors. *Stem Cells*, *24*(8), 1852–1858.
Schneider-Yin, X., van Tuyll van Serooskerken, A. M., Went, P., Tyblewski, W., Poblete-Gutierrez, P., Minder, E. I., et al. (2010). Hepatocellular carcinoma in variegate porphyria: A serious complication. *Acta Dermato-Venereologica*, *90*(5), 512–515. https://doi.org/10.2340/00015555-0870.
Schuppan, D., Ashfaq-Khan, M., Yang, A. T., & Kim, Y. O. (2018). Liver fibrosis: Direct antifibrotic agents and targeted therapies. *Matrix Biology*, *68–69*, 435–451.
Seda Neto, J., Leite, K. M., Porta, A., Fonseca, E. A., Feier, F. H., Pugliese, R., et al. (2014). HCC prevalence and histopathological findings in liver explants of patients with hereditary tyrosinemia type 1. *Pediatric Blood & Cancer*, *61*(9), 1584–1589.
Setshedi, M., Wands, J. R., & Monte, S. M. (2010). Acetaldehyde adducts in alcoholic liver disease. *Oxidative Medicine and Cellular Longevity*, *3*(3), 178–185.
Shen, L., Zhang, X., Hu, D., Feng, T., Li, H., Lu, Y., et al. (2013). Hepatitis B virus X (HBx) play an anti-apoptosis role in hepatic progenitor cells by activating Wnt/beta-catenin pathway. *Molecular and Cellular Biochemistry*, *383*(1–2), 213–222.
Shindoh, J., Andreou, A., Aloia, T. A., Zimmitti, G., Lauwers, G. Y., Laurent, A., et al. (2013). Microvascular invasion does not predict long-term survival in hepatocellular carcinoma up to 2 cm: Reappraisal of the staging system for solitary tumors. *Annals of Surgical Oncology*, *20*(4), 1223–1229.
Shiraishi, Y., Fujimoto, A., Furuta, M., Tanaka, H., Chiba, K., Boroevich, K. A., et al. (2014). Integrated analysis of whole genome and transcriptome sequencing reveals diverse transcriptomic aberrations driven by somatic genomic changes in liver cancers. *PLoS One*, *9*(12), e114263.
Song, J., Gao, L., Yang, G., Tang, S., Xie, H., Wang, Y., et al. (2014). MiR-199a regulates cell proliferation and survival by targeting FZD7. *PLoS One*, *9*(10), e110074.

Song, S., Yang, Y., Liu, M., Liu, B., Yang, X., Yu, M., et al. (2018). MiR-125b attenuates human hepatocellular carcinoma malignancy through targeting SIRT6. *American Journal of Cancer Research*, *8*(6), 993–1007.
Stockley, R. A. (2015). The multiple facets of alpha-1-antitrypsin. *Annals of Translational Medicine*, *3*(10), 130.
Strohmeyer, G., Niederau, C., & Stremmel, W. (1988). Survival and causes of death in hemochromatosis. Observations in 163 patients. *Annals of the New York Academy of Sciences*, *526*, 245–257.
Su, F., & Schneider, R. J. (1996). Hepatitis B virus HBx protein activates transcription factor NF-kappaB by acting on multiple cytoplasmic inhibitors of rel-related proteins. *Journal of Virology*, *70*(7), 4558–4566.
Su, F., & Schneider, R. J. (1997). Hepatitis B virus HBx protein sensitizes cells to apoptotic killing by tumor necrosis factor alpha. *Proceedings of the National Academy of Sciences of the United States of America*, *94*(16), 8744–8749.
Su, H., Yang, J. R., Xu, T., Huang, J., Xu, L., Yuan, Y., et al. (2009). MicroRNA-101, down-regulated in hepatocellular carcinoma, promotes apoptosis and suppresses tumorigenicity. *Cancer Research*, *69*(3), 1135–1142.
Sucher, E., Sucher, R., Gradistanac, T., Brandacher, G., Schneeberger, S., & Berg, T. (2019). Autoimmune hepatitis-immunologically triggered liver pathogenesis-diagnostic and therapeutic strategies. *Journal of Immunology Research*, *2019*, 9437043.
Sung, W. K., Zheng, H., Li, S., Chen, R., Liu, X., Li, Y., et al. (2012). Genome-wide survey of recurrent HBV integration in hepatocellular carcinoma. *Nature Genetics*, *44*(7), 765–769.
Szabo, G., & Bala, S. (2010). Alcoholic liver disease and the gut-liver axis. *World Journal of Gastroenterology*, *16*(11), 1321–1329.
Sze, K. M., Ho, D. W., Chiu, Y. T., Tsui, Y. M., Chan, L. K., Lee, J. M., et al. (2020). HBV-TERT promoter integration harnesses host ELF4 resulting in TERT gene transcription in hepatocellular carcinoma. *Hepatology*. In press.
Takano, S., Yokosuka, O., Imazeki, F., Tagawa, M., & Omata, M. (1995). Incidence of hepatocellular carcinoma in chronic hepatitis B and C: A prospective study of 251 patients. *Hepatology*, *21*(3), 650–655.
Tan, K. W., Chacko, A. M., & Chew, V. (2019). PD-1 expression and its significance in tumour microenvironment of hepatocellular carcinoma. *Translational Gastroenterology and Hepatology*, *4*, 51.
Tan, A., Koh, S., & Bertoletti, A. (2015). Immune response in hepatitis B virus infection. *Cold Spring Harbor Perspectives in Medicine*, *5*(8), a021428.
Tanguay, R. M., Angileri, F., & Vogel, A. (2017). Molecular pathogenesis of liver injury in hereditary tyrosinemia 1. *Advances in Experimental Medicine and Biology*, *959*, 49–64.
Tarocchi, M., Polvani, S., Marroncini, G., & Galli, A. (2014). Molecular mechanism of hepatitis B virus-induced hepatocarcinogenesis. *World Journal of Gastroenterology*, *20*(33), 11630–11640.
Teckman, J. H., An, J. K., Blomenkamp, K., Schmidt, B., & Perlmutter, D. (2004). Mitochondrial autophagy and injury in the liver in alpha 1-antitrypsin deficiency. *American Journal of Physiology. Gastrointestinal and Liver Physiology*, *286*(5), G851–G862.
Topic, A., Ljujic, M., & Radojkovic, D. (2012). Alpha-1-antitrypsin in pathogenesis of hepatocellular carcinoma. *Hepatitis Monthly*, *12(10 HCC)*, e7042.
Tran, B. M., Flanagan, D. J., Ebert, G., Warner, N., Tran, H., Fifis, T., et al. (2020). The hepatitis B virus pre-core protein p22 activates Wnt signaling. *Cancers*, *12*(6), 1435.
Trevisani, F., D'Intino, P. E., Morselli-Labate, A. M., Mazzella, G., Accogli, E., Caraceni, P., et al. (2001). Serum alpha-fetoprotein for diagnosis of hepatocellular carcinoma in patients with chronic liver disease: Influence of HBsAg and anti-HCV status. *Journal of Hepatology*, *34*(4), 570–575.

Tsang, F. H., Au, S. L., Wei, L., Fan, D. N., Lee, J. M., Wong, C. C., et al. (2015). Long non-coding RNA HOTTIP is frequently up-regulated in hepatocellular carcinoma and is targeted by tumour suppressive miR-125b. *Liver International*, *35*(5), 1597–1606.

Tu, T., Budzinska, M. A., Shackel, N. A., & Urban, S. (2017). HBV DNA integration: Molecular mechanisms and clinical implications. *Viruses*, *9*(4), 75.

Turati, F., Galeone, C., Rota, M., Pelucchi, C., Negri, E., Bagnardi, V., et al. (2014). Alcohol and liver cancer: A systematic review and meta-analysis of prospective studies. *Annals of Oncology*, *25*(8), 1526–1535.

Turkez, H., & Sisman, T. (2012). The genoprotective activity of resveratrol on aflatoxin B (1)-induced DNA damage in human lymphocytes in vitro. *Toxicology and Industrial Health*, *28*(5), 474–480.

Ueda, H., Ullrich, S. J., Gangemi, J. D., Kappel, C. A., Ngo, L., Feitelson, M. A., et al. (1995). Functional inactivation but not structural mutation of p53 causes liver cancer. *Nature Genetics*, *9*(1), 41–47.

Valderrama-Trevino, A. I., Barrera-Mera, B., Ceballos-Villalva, J. C., & Montalvo-Jave, E. E. (2017). Hepatic metastasis from colorectal cancer. *Euroasian Journal of Hepatogastroenterology*, 7(2), 166–175.

Valean, S., Acalovschi, M., Dumitrascu, D. L., Ciobanu, L., Nagy, G., & Chira, R. (2019). Hepatocellular carcinoma in patients with autoimmune hepatitis—A systematic review of the literature published between 1989-2016. *Medicine and Pharmacy Reports*, *92*(2), 99–105.

van Ginkel, W. G., Pennings, J. P., & van Spronsen, F. J. (2017). Liver cancer in tyrosinemia type 1. *Advances in Experimental Medicine and Biology*, *959*, 101–109.

Villanueva, A., & Luedde, T. (2016). The transition from inflammation to cancer in the liver. *Clinical Liver Disease*, *8*(4), 89–93.

Vinay, D. S., Ryan, E. P., Pawelec, G., Talib, W. H., Stagg, J., Elkord, E., et al. (2015). Immune evasion in cancer: Mechanistic basis and therapeutic strategies. *Seminars in Cancer Biology*, *35*(Suppl), S185–S198.

von Olshausen, G., Quasdorff, M., Bester, R., Arzberger, S., Ko, C., van de Klundert, M., et al. (2018). Hepatitis B virus promotes beta-catenin-signalling and disassembly of adherens junctions in a Src kinase dependent fashion. *Oncotarget*, *9*(74), 33947–33960.

Wang, J., Chu, Y., Xu, M., Zhang, X., Zhou, Y., & Xu, M. (2019). miR-21 promotes cell migration and invasion of hepatocellular carcinoma by targeting KLF5. *Oncology Letters*, *17*(2), 2221–2227.

Wang, X. W., Forrester, K., Yeh, H., Feitelson, M. A., Gu, J. R., & Harris, C. C. (1994). Hepatitis B virus X protein inhibits p53 sequence-specific DNA binding, transcriptional activity, and association with transcription factor ERCC3. *Proceedings of the National Academy of Sciences of the United States of America*, *91*(6), 2230–2234.

Wang, X. W., Gibson, M. K., Vermeulen, W., Yeh, H., Forrester, K., Sturzbecher, H. W., et al. (1995). Abrogation of p53-induced apoptosis by the hepatitis B virus X gene. *Cancer Research*, *55*(24), 6012–6016.

Wang, Y., Hu, C., Cheng, J., Chen, B., Ke, Q., Lv, Z., et al. (2014). MicroRNA-145 suppresses hepatocellular carcinoma by targeting IRS1 and its downstream Akt signaling. *Biochemical and Biophysical Research Communications*, *446*(4), 1255–1260.

Wang, Y., Lee, A. T., Ma, J. Z., Wang, J., Ren, J., Yang, Y., et al. (2008). Profiling microRNA expression in hepatocellular carcinoma reveals microRNA-224 up-regulation and apoptosis inhibitor-5 as a microRNA-224-specific target. *The Journal of Biological Chemistry*, *283*(19), 13205–13215.

Wang, X., Lu, J., Cao, J., Ma, B., Gao, C., & Qi, F. (2018). MicroRNA-18a promotes hepatocellular carcinoma proliferation, migration, and invasion by targeting Bcl2L10. *Oncotargets and Therapy*, *11*, 7919–7934.

Wang, J., Song, W., Shen, W., Yang, X., Sun, W., Qu, S., et al. (2017). MicroRNA-200a suppresses cell invasion and migration by directly targeting GAB1 in hepatocellular carcinoma. *Oncology Research*, *25*(1), 1–10.

Wei, L., Delin, Z., Kefei, Y., Hong, W., Jiwei, H., & Yange, Z. (2020). A classification based on tumor budding and immune score for patients with hepatocellular carcinoma. *Oncoimmunology*, *9*(1), 1672495.

Wei, X., Xiang, T., Ren, G., Tan, C., Liu, R., Xu, X., et al. (2013). miR-101 is down-regulated by the hepatitis B virus x protein and induces aberrant DNA methylation by targeting DNA methyltransferase 3A. *Cellular Signalling*, *25*(2), 439–446.

Wen, Y., Golubkov, V. S., Strongin, A. Y., Jiang, W., & Reed, J. C. (2008). Interaction of hepatitis B viral oncoprotein with cellular target HBXIP dysregulates centrosome dynamics and mitotic spindle formation. *The Journal of Biological Chemistry*, *283*(5), 2793–2803.

Wheeler, D. A., & Roberts, L. R. (2017). Comprehensive and integrative genomic characterization of hepatocellular carcinoma. *Cell*, *169*(7), 1327–1341.e1323.

Wong, Q. W., Ching, A. K., Chan, A. W., Choy, K. W., To, K. F, Lai, P. B., et al. (2010). MiR-222 overexpression confers cell migratory advantages in hepatocellular carcinoma through enhancing AKT signaling. *Clinical Cancer Research*, *16*(3), 867–875.

Wong, Q. W., Lung, R. W., Law, P. T., Lai, P. B., Chan, K. Y., To, K. F., et al. (2008). MicroRNA-223 is commonly repressed in hepatocellular carcinoma and potentiates expression of Stathmin1. *Gastroenterology*, *135*(1), 257–269.

Wu, Y., Chen, W., Xu, Z. P., & Gu, W. (2019). PD-L1 distribution and perspective for cancer immunotherapy-blockade, knockdown, or inhibition. *Frontiers in Immunology*, *10*, 2022.

Wu, Y., Whitman, I., Molmenti, E., Moore, K., Hippenmeyer, P., & Perlmutter, D. H. (1994). A lag in intracellular degradation of mutant alpha 1-antitrypsin correlates with the liver disease phenotype in homozygous PiZZ alpha 1-antitrypsin deficiency. *Proceedings of the National Academy of Sciences of the United States of America*, *91*(19), 9014–9018.

Xiao, F., Zhang, W., Zhou, L., Xie, H., Xing, C., Ding, S., et al. (2013). microRNA-200a is an independent prognostic factor of hepatocellular carcinoma and induces cell cycle arrest by targeting CDK6. *Oncology Reports*, *30*(5), 2203–2210.

Xu, Y., An, Y., Wang, Y., Zhang, C., Zhang, H., Huang, C., et al. (2013). miR-101 inhibits autophagy and enhances cisplatin-induced apoptosis in hepatocellular carcinoma cells. *Oncology Reports*, *29*(5), 2019–2024.

Xu, J., An, P., Winkler, C. A., & Yu, Y. (2020). Dysregulated microRNAs in hepatitis B virus-related hepatocellular carcinoma: Potential as biomarkers and therapeutic targets. *Frontiers in Oncology*, *10*, 1271.

Xu, L., Beckebaum, S., Iacob, S., Wu, G., Kaiser, G. M., Radtke, A., et al. (2014). MicroRNA-101 inhibits human hepatocellular carcinoma progression through EZH2 downregulation and increased cytostatic drug sensitivity. *Journal of Hepatology*, *60*(3), 590–598.

Yamashita, T., Forgues, M., Wang, W., Kim, J. W., Ye, Q., Jia, H., et al. (2008). EpCAM and alpha-fetoprotein expression defines novel prognostic subtypes of hepatocellular carcinoma. *Cancer Research*, *68*(5), 1451–1461.

Yang, C., Huang, X., Liu, Z., Qin, W., & Wang, C. (2020). Metabolism-associated molecular classification of hepatocellular carcinoma. *Molecular Oncology*, *14*(4), 896–913.

Yang, X., Wang, J., Qu, S., Zhang, H., Ruan, B., Gao, Y., et al. (2015). MicroRNA-200a suppresses metastatic potential of side population cells in human hepatocellular carcinoma by decreasing ZEB2. *Oncotarget*, *6*(10), 7918–7929.

Yang, T., Zheng, Z. M., Li, X. N., Li, Z. F., Wang, Y., Geng, Y. F., et al. (2013). MiR-223 modulates multidrug resistance via downregulation of ABCB1 in hepatocellular carcinoma cells. *Experimental Biology and Medicine (Maywood, N.J.)*, *238*(9), 1024–1032.

Yongyu, Z., Lewei, Y., Jian, L., & Yuqin, S. (2018). MicroRNA-18a targets IRF2 and CBX7 to promote cell proliferation in hepatocellular carcinoma. *Oncology Research*, *26*, 1327–1334.
Younossi, Z. M. (2019). Non-alcoholic fatty liver disease—A global public health perspective. *Journal of Hepatology*, *70*(3), 531–544.
Yun, C., Cho, H., Kim, S. J., Lee, J. H., Park, S. Y., Chan, G. K., et al. (2004). Mitotic aberration coupled with centrosome amplification is induced by hepatitis B virus X oncoprotein via the Ras-mitogen-activated protein/extracellular signal-regulated kinase-mitogen-activated protein pathway. *Molecular Cancer Research*, *2*(3), 159–169.
Zanger, U. M., Hauri, H. P., Loeper, J., Homberg, J. C., & Meyer, U. A. (1988). Antibodies against human cytochrome P-450db1 in autoimmune hepatitis type II. *Proceedings of the National Academy of Sciences of the United States of America*, *85*(21), 8256–8260.
Zhang, Y., Guo, X., Xiong, L., Kong, X., Xu, Y., Liu, C., et al. (2012). MicroRNA-101 suppresses SOX9-dependent tumorigenicity and promotes favorable prognosis of human hepatocellular carcinoma. *FEBS Letters*, *586*(24), 4362–4370.
Zhang, Q., Lou, Y., Yang, J., Wang, J., Feng, J., Zhao, Y., et al. (2019). Integrated multiomic analysis reveals comprehensive tumour heterogeneity and novel immunophenotypic classification in hepatocellular carcinomas. *Gut*, *68*(11), 2019–2031.
Zhao, W. T., Lin, X. L., Liu, Y., Han, L. X., Li, J., Lin, T. Y., et al. (2019). miR-26a promotes hepatocellular carcinoma invasion and metastasis by inhibiting PTEN and inhibits cell growth by repressing EZH2. *Laboratory Investigation*, *99*(10), 1484–1500.
Zhao, L. H., Liu, X., Yan, H. X., Li, W. Y., Zeng, X., Yang, Y., et al. (2016). Genomic and oncogenic preference of HBV integration in hepatocellular carcinoma. *Nature Communications*, *7*, 12992.
Zhou, Y., Chen, E., Tang, Y., Mao, J., Shen, J., Zheng, X., et al. (2019). miR-223 overexpression inhibits doxorubicin-induced autophagy by targeting FOXO3a and reverses chemoresistance in hepatocellular carcinoma cells. *Cell Death & Disease*, *10*(11), 843.
Zoulim, F. (2005). New insight on hepatitis B virus persistence from the study of intrahepatic viral cccDNA. *Journal of Hepatology*, *42*(3), 302–308.

CHAPTER TWO

Signaling pathways in hepatocellular carcinoma

Teresa Garcia-Lezana[a], Juan Luis Lopez-Canovas[b], and Augusto Villanueva[a,c,*]

[a]Division of Liver Diseases, Liver Cancer Program, Department of Medicine, Tisch Cancer Institute, Icahn School of Medicine at Mount Sinai, New York, NY, United States
[b]Department of Cell Biology, Physiology and Immunology, Maimonides Institute of Biomedical Research of Córdoba (IMIBIC), University of Córdoba, Córdoba, Spain
[c]Division of Hematology and Medical Oncology, Department of Medicine, Icahn School of Medicine at Mount Sinai, New York, NY, United States
*Corresponding author: e-mail address: augusto.villanueva@mssm.edu

Contents

1. Introduction 64
2. Pathways involving growth factors 65
 2.1 Vascular endothelial growth factor receptor (VEGFR) 65
 2.2 Fibroblast growth factor receptors (FGFR) 66
 2.3 Transforming growth factor (TGF) 67
 2.4 Epidermal growth factor receptor (EGFR) 69
 2.5 Insulin-like growth factor (IGF) 69
 2.6 c-MET 70
 2.7 PI3K-AKT-mTOR (and PTEN) 71
 2.8 RAF/ERK/MAPK 71
3. Pathways involved in cell differentiation and development 72
 3.1 WNT-β-catenin 72
 3.2 JAK/STAT signaling pathway 75
 3.3 Hippo signaling pathway 76
 3.4 Hedgehog signaling pathway 77
 3.5 Notch signaling pathway 78
4. Nuclear signaling pathways 79
 4.1 Telomere shortening (TERT) 79
 4.2 Cell cycle (P53, retinoblastoma) 81
 4.3 Epigenetic modifications and chromatin remodeling (ARID1A, ARID2) 81
5. Emerging pathways 82
 5.1 Mechanisms involving non-coding RNA (ncRNAs) 82
 5.2 Microbiome 85
6. Conclusions 87
Acknowledgment 87
Conflict of interest 87
References 87

Advances in Cancer Research, Volume 149
ISSN 0065-230X
https://doi.org/10.1016/bs.acr.2020.10.002

Abstract

Despite the recent introduction of new effective systemic agents, the survival of patients with hepatocellular carcinoma (HCC) at advanced stages remains dismal. This underscores the need for new therapies, which has spurred extensive research on the identification of the main drivers of pathway de-regulation as a source of novel therapeutic targets. Frequently altered pathways in HCC involve growth factor receptors (e.g., VEGFR, FGFR, TGFA, EGFR, IGFR) and/or its cytoplasmic intermediates (e.g., PI3K-AKT-mTOR, RAF/ERK/MAPK) as well as key pathways in cell differentiation (e.g., Wnt/β-catenin, JAK/STAT, Hippo, Hedgehog, Notch). Somatic mutations, chromosomal aberrations and epigenetic changes are common mechanisms for pathway deregulation in HCC. Aberrant pathway activation has also been explored as a biomarker to predict response to specific therapies, but currently, these strategies are not implemented when deciding systemic therapies in HCC patients. Beyond the well-established molecular cascades, there are numerous emerging signaling pathways also deregulated in HCC (e.g., tumor microenvironment, non-coding RNA, intestinal microbiota), which have opened new avenues for therapeutic exploration.

1. Introduction

Hepatocellular carcinoma (HCC) is the most common form of liver cancer and the fourth leading cause of cancer related deaths worldwide, and it is sixth in terms of incidence. HCC usually develops in the context of an underlying chronic liver disease mostly due to hepatitis C virus (HCV), hepatitis B virus (HBV), non-alcoholic steatohepatitis or alcohol use disorder (Villanueva, 2019). The sustained presence of inflammation, fibrosis and aberrant regeneration creates a favorable environment for the accumulation of genetic and epigenetic changes that ultimately lead hepatocarcinogenesis.

Numerous signaling pathways are deregulated in HCC (Moeini, Cornellà, & Villanueva, 2012; Villanueva, Newell, Chiang, Friedman, & Llovet, 2007), and have been explored as therapeutic targets (Sarcognato, García-Lezana, & Villanueva, 2019). In fact, the most prevalent somatic mutations in HCC affect core signaling pathways in cellular physiology. The most frequent mutations in HCC affect *TERT* promoter and *TP53* that account for ~60% and ~35% of the cases, respectively (Schulze, Nault, & Villanueva, 2016). Other frequently mutated genes affect the WNT pathway, with *CTNNB1* mutations reported in 30% of cases and *AXIN1* mutations in 10%. Mutations in *ARID1A* and *ARID2*, both involved in chromatin remodeling, are present in 7% and 6% of HCC, respectively. Besides mutations, gene expression alterations also suggest significant deregulation of signaling pathways in HCC. Broadly, HCC can be classified in two major molecular subtypes (Zucman-Rossi, Villanueva, Nault, & Llovet, 2015). The "Proliferation class," more aggressive, is enriched by *TP53* mutations and activation of RAS, MAPK, AKT, mTOR and MET pathways. Tumors in the "Non-proliferation class" have genomic features resembling normal liver physiology and are enriched in activation of canonical WNT signaling. Here, we review the main signaling pathways deregulated in HCC and how they have been exploited therapeutically.

2. Pathways involving growth factors

2.1 Vascular endothelial growth factor receptor (VEGFR)

VEGFR is an important receptor family in cancer, contributing to neoangiogenesis and metastatic spread (Sia, Alsinet, Newell, & Villanueva, 2014). VEGFR has an extracellular domain for VEGF binding, a transmembrane and an intracellular domain with tyrosine kinase activity (Melincovici et al., 2018). Upon ligand binding, VEGFR activates tyrosine kinase enzyme in the intracellular receptor domain, which phosphorylates the tyrosine residues and activates several downstream intracellular signaling pathways (Neufeld, Cohen, Gengrinovitch, & Poltorak, 1999) [Fig. 1]. There are five VEGF ligands: VEGF (or VEGFA), placental growth factor (PlGF), VEGFB, VEGFC and FIGF (also known as VEGFD). There are several splicing variants in VEGF ligands, which increases the molecular heterogeneity of this pathway (Sia et al., 2014). There are three types of VEGF receptors (Neufeld et al., 1999).

High level DNA amplifications of *VEGFA* (chromosome 6p21) are found in 5–7% of human HCC (Chiang et al., 2008). Also, high VEGF levels are associated with tumoral vascular invasion and higher histological grade, as well as worse clinical outcomes (Llovet et al., 2012). Most molecular targeted agents approved by the FDA for the treatment of patients with advance stage HCC (i.e., sorafenib (Llovet et al., 2008), lenvatinib (Kudo et al., 2018), regorafenib (Bruix et al., 2017), cabozantinib (Abou-Alfa et al., 2018) and ramucirumab (Zhu et al., 2018)) target VEGFR2 signaling. Notably, other agents with a similar antiangiogenic profile were found to be toxic in HCC patients (e.g., sunitinib (Cheng et al., 2013)), suggesting that excessive interfering with VEGF signaling in cirrhosis can be detrimental (Llovet, Montal, & Villanueva, 2019).

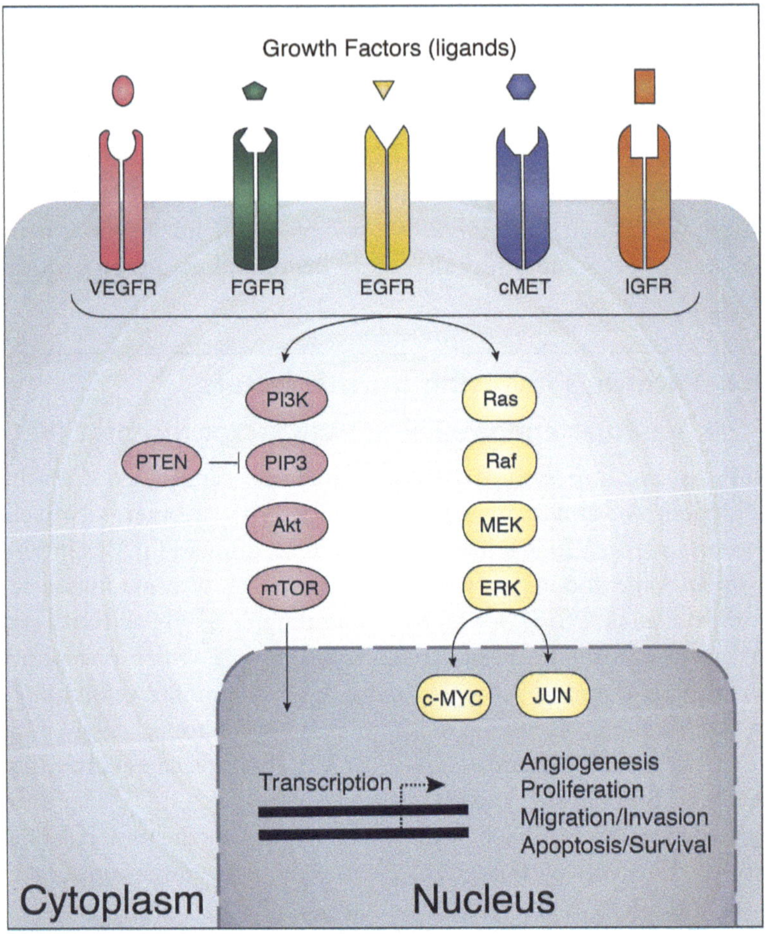

Fig. 1 Pathways involving growth factors. The binding of the growth factor (ligand) to the respective transmembrane receptor (VEGF, FGFR, EGFR, cMET and IGFR) trigger the activation of two intracellular cascades PI3K/Akt/mTOR or RAF/ERK/MAPK that ultimately provoke the regulation of gene transcription.

2.2 Fibroblast growth factor receptors (FGFR)

FGFRs are highly conserved transmembrane receptors with tyrosine kinases activity that can signal through different downstream pathways (e.g., RAS-RAF and phosphoinositol-3 kinase [PI3P]-AKT cascades) [Fig. 1]. FGFR signaling is involved in embryonic development of the liver and is key to wound healing and angiogenesis in adults. Overexpression of FGFR2 correlates with poor clinical outcomes and FGFR4 has been involved in cell proliferation and AFP secretion (Harimoto et al., 2010). Aberrant overexpression of FGF19/FGFR4 contributes to HCC progression (Zhao, Lv, et al., 2016) and *FGFR4* kinase domain mutations have been associated to acquired resistance to selective FGFR4 therapies (Hatlen et al., 2019).

In HCC, up to 8% of patients have high levels of DNA amplifications of the *FGF19/FGF4* locus (Chr. 11q13) (Chiang et al., 2008). Selective blockade of FGF19 showed antitumor activity in experimental models of HCC (Sawey et al., 2011). Selective FGFR inhibition has been further explored as a druggable oncogenic additional loop in HCC. Fisogatinib, a selective FGFR4 inhibitor, show promising clinical benefit in a phase I trial including patients with overexpression of FGF19 (Kim et al., 2019). Also, the multityrosine kinase inhibitor lenvatinib, which inhibits FGFR signaling, has shown clinical efficacy in first line after a non-inferiority phase III trial compared with sorafenib (Kudo et al., 2018). There is experimental evidence showing how lenvatinib can affect cancer stem cell biology in HCC, which aligns with data showing FGF2 and FGF19 are involved in maintaining CD44High/CD133High cancer stem cells (CSCs) pool in HCC (Shigesawa et al., 2020).

2.3 Transforming growth factor (TGF)

TGF-α is a polypeptide normally expressed in hepatocytes that promotes cellular proliferation and transformation. TGF-α is an EGFR ligand and member of the EGF family. Combined actions of different cytokines secreted in the context of chronic inflammation following hepatic injury upregulate TGF-α (Thorgeirsson & Grisham, 2002). TGF-α is involved in hepatic regeneration, hepatocyte proliferation, dysplasia, and development of HCC (Zhang, Wang, Li, & Qiao, 2004). The role of TGF-α in the pathogenesis of HCC was demonstrated by the formation of liver tumors in *MT1/TGFα* transgenic mice (Jhappan et al., 1990). Other in vivo experiments have demonstrated the cooperation between TGF-α and TGF-β pathways in HCC formation (Baek et al., 2010). Conversely, in patients, the analysis TGF-α in dysplastic nodules and HCC samples showed a decrease in TGF-α levels following disease progression (Llovet et al., 2006; Yeh et al., 2007).

TGF-β signaling plays key roles in proliferation, apoptosis, differentiation, migration and invasion. But, beyond the direct effect that TGF-β can have regulating tumor cells, it also has a relevant role controlling the surrounding stroma, immune and inflammatory processes. In many cancers, including HCC, TGF-β plays a dual role as tumor suppressor or tumor promoter (Rao & Mishra, 2019). The binding of the ligand to the TGF-β receptor activates its kinase activity resulting in the phosphorylation of SMAD proteins and adaptor proteins, such as b2-spectrin. Once phosphorylated, SMAD proteins form heterodimers that translocate into the nucleus where regulate expression of fibrosis-related genes (E-cadherin, integrins and collagen) among others (Dituri, Mancarella, Cigliano, Chieti, & Giannelli, 2019). Besides the SMAD cascade, TGF-β can activate multiple pathways such as MAPK or PI3K (Katz et al., 2013) [Fig. 2]. Genomic analysis of the TGF-β signaling pathway in a cohort of 488 human HCCs revealed genomic alterations in 40% of the cases. Subsequent evaluation of the pathways associated with patients presenting activation *TGFB* showed that the activated signature was enriched in genes regulating collagen synthesis,

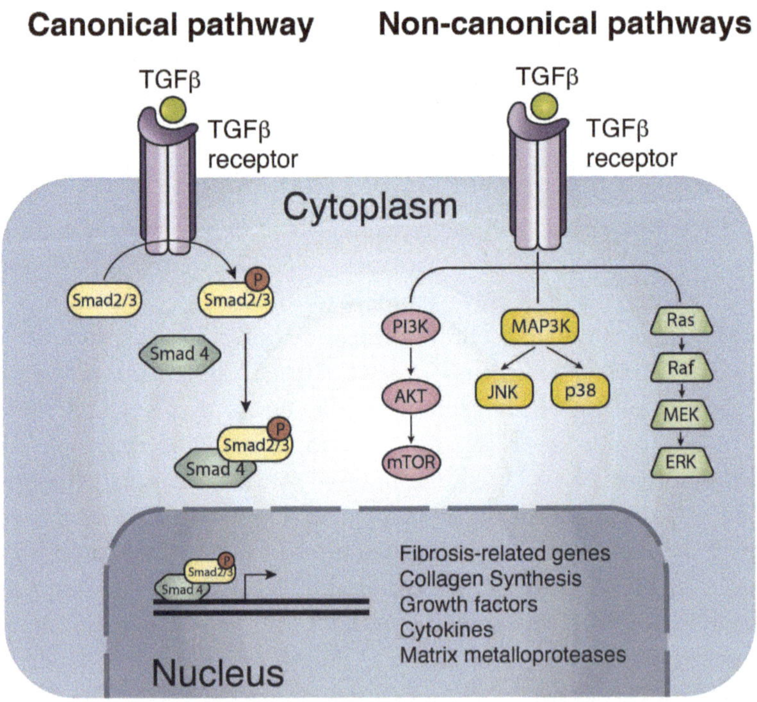

Fig. 2 Transforming growth factor-β (TGF-β) pathway. In the canonical TGF-β pathway the binding of TGF-β to its receptor provokes the activation of the tyrosine kinase activity and phosphorylation of SMAD2/3 proteins. SMAD proteins form heterodimers that translocate into the nucleus and regulate gene expression. The non-canonical TGF-β pathway include the activation of other intracellular pathways as PI3K/AKT/mTOR, MAPK/JNK-p38 or RAS/RAF/MEK/ERK.

growth factors, cytokines and matrix metalloproteases, reflecting an activated response from the microenvironment (Chen et al., 2018).

2.4 Epidermal growth factor receptor (EGFR)

EGFR (also known as ErbB1 or HER-1) is a transmembrane receptor involved in proliferation, differentiation and survival [Fig. 1]. Multiple ligands can activate EGFR including epidermal growth factor (EGF), TGF-α, amphiregulin, epiregulin, betacellulin, heparin-binding EGF and epigen (Komposch & Sibilia, 2015). EGFR is overexpressed in human cirrhotic tissue and HCC, and has been associated with tumor recurrence (Shi et al., 2019). EGFR overexpression in endothelial cells has also been linked to a higher microvessel density in human HCC samples. Further evidence suggest a HCC-specific overexpression of EGFR in sinusoidal endothelial cells (Komposch & Sibilia, 2015), which could indicate that EGFR signaling could also participate in neoangiogenesis.

EGFR inhibitors such as gefitinib, erlotinib or lapatinib, which also inhibits ErbB2, have shown antitumoral activity in experimental models of HCC (Fuchs et al., 2014; Höpfner et al., 2004; Schiffer et al., 2005). However, this has not been confirmed in clinical studies testing EGFR inhibitors, at least in combination with sorafenib (Zhu et al., 2015). This discrepancy could be partially explained by context specific actions of EGFR in different cell types. For instance, while EGFR deletion in hepatocytes favored HCC development, its deletion in Kupffer cells and liver macrophages reduced development of HCC in animal models (Lanaya et al., 2014). In fact, recent data demonstrates how PLAGL2 knockdown enhanced the response of HCC models to erlotinib, suggesting a potential mechanism to overcome resistance to EGFR inhibition (Hu et al., 2020).

2.5 Insulin-like growth factor (IGF)

IGF signaling is involved in cell proliferation and inhibition of apoptosis (Kasprzak, Kwasniewski, Adamek, & Gozdzicka-Jozefiak, 2017). IGF axis includes ligands (IGF-I and IGF-II), receptors (IGFR1 and IGFR2), binding proteins (IGFBPs) and proteases. The IGF receptors are tetrameric structures formed by two extracellular α-chains and two intracellular β-chains with tyrosine kinase activity. Alternative splicing can generate variants of IGFRs, known as IRA and IRB. The binding of IGF ligands to their receptor is modulated by IGFBPs, which ultimately triggers downstream activation of PI3K-AKT, MAPK and mTOR pathways (Pollak, 2008) [Fig. 1].

IGF-1R participates in cell transformation and its expression is regulated by steroid hormones and growth factors (e.g., FGF, PDGF, EGF). IGF pathway is activated in 20% of early HCCs and the selective inhibition of IGF-1R has antitumoral activity in experimental models of HCC (Tovar et al., 2010). In addition, IGF pathway contributes to the development of sorafenib resistance as shown in experimental models with aberrant activation of IGF1 as a potential scape mechanism (Tovar et al., 2017).

IGF-2R, which lacks tyrosine kinase function, neutralizes the effect of IGF-2 by binding and internalizing this ligand, which is then subject to lysosomal degradation. The main role of this receptor is to clear growth factors and attenuate IGF-2-activated signal transduction. Gene expression analysis of 228 human HCCs showed a >20-fold overexpression of IGF2 in 15% of the cases, which was correlated with aberrant DNA methylation of the promoter. Functional analysis of IGF2 overexpression in vivo and in vitro demonstrated how IGF2 enhances tumorigenesis by increasing proliferation and reducing survival (Martinez-Quetglas et al., 2016).

2.6 c-MET

c-Met is a receptor tyrosine kinase with one known ligand, the hepatocyte growth factor (HGF). c-Met is involved in epithelial–mesenchymal transition and plays a critical role in tissue modeling during embryogenesis. HGF activates c-MET, which leads to its phosphorylation and phosphorylation of adaptor proteins GRB2 and GAB1 (Dimri & Satyanarayana, 2020). This process activates multiple downstream signaling pathways including MAPK (Gonzalez et al., 2017), PI3K (Pascale, Feo, & Calvisi, 2016), STAT and NF-κB (Garajova, Giovannetti, Biasco, & Peters, 2015) [Fig. 1]. These pathways execute the cellular programs directed by c-Met activation including increased cell proliferation, survival, motility, invasiveness, and epithelial–mesenchymal transition (Bouattour et al., 2018). Multiple experimental models show how c-Met and/or HGF are involved in the initiation and progression of HCC (Li et al., 2009; Salvi et al., 2009; Zhang et al., 2005). Pathway deregulation is reported in up to 50% of HCC patients because of mutations (4%), gene amplification (24%) and overexpression (50%).

Inhibition of c-Met has been tested in human trials. In fact, overexpression of c-Met measured with immunohistochemistry was the first biomarker-driven phase III clinical trial reported in HCC (Rimassa et al., 2018). Based on promising results in an unselected patient population

(Santoro et al., 2013) with tivantinib, the subsequent phase III on HCC patients with high c-Met failed to meet its primary endpoint. Recent data shown that tivantinib is not a c-Met inhibitor (Rebouissou et al., 2017). c-Met has also been involved in resistance to sorafenib (Firtina Karagonlar, Koc, Iscan, Erdal, & Atabey, 2016). Selective c-Met inhibitors such as tepotinib have shown clinical potential, particularly in patients with *MET* gene amplifications (Bouattour et al., 2018).

2.7 PI3K-AKT-mTOR (and PTEN)

In normal cells, the Phosphoinositide 3-kinase (PI3K)/mammalian target of rapamycin (mTOR) pathway regulates cell survival, proliferation, and differentiation. This pathway is activated by cell-surface receptors such as receptor tyrosine kinases, G-protein-coupled receptors, and oncogenes such as *RAS*. Specifically, the activated p110 subunit of PI3K catalyzes the conversion of phosphatidylinositol 4,5-bisphosphate (PIP2) to phosphatidylinositol 3,4,5-trisphosphate (PIP3), which in turn activates Akt. Akt has multiple downstream effectors, including the mTOR complexes (mTORC). mTORC1 controls nutrient consumption and cell growth while TORC2 contributes to Akt activation via its phosphorylation at serine 473. The phosphate and tensin homolog (PTEN) antagonizes PI3K function and dephosphorylates PIP3 to PIP2 [Fig. 1]. The implication of PI3K in tumor cell survival varies across different tumor types as well as the genetic or epigenetic alterations that affect the pathway (Janku, 2017).

The mTOR pathway is crucial for HCC cell proliferation and survival and its selective inhibition has antitumoral effects in HCC experimental models (Villanueva et al., 2008). We found activation of the MTOR pathway in up to 40% of HCC patients with early stage HCC, mainly through overexpression of its ligands. PTEN loss has been reported in up to 53% of HCC patients (Subbiah et al., 2015). However, a phase III trial of the mTOR inhibitor everolimus demonstrated no survival benefit for patients with advanced HCC who had failed to sorafenib (Zhu et al., 2014); however, the participants were not selected based on PTEN or other PI3K-related abnormalities.

2.8 RAF/ERK/MAPK

The Raf/MEK/ERK pathway regulates key cellular processes, including cell proliferation, differentiation, angiogenesis and survival. *RAS* was among the first oncogenes described in humans (Gollob, Wilhelm, Carter, &

Kelley, 2006; Guo et al., 2020). MAPK activation starts with ligands binding to RTK and recruitment small G-proteins (RAS) to the cell membrane. This leads to sequential downstream activation of RAF, MEK1 and MEK2, and ultimately ERK1 and ERK2 (Guo et al., 2020). Phosphorylated ERK translocates into the nucleus where it activates, among others, two key transcription factors of the AP-1 family, c-Jun and c-Fos (Delire & Starkel, 2015) [Fig. 1].

Increased phosphorylation of MEK1/2 has been reported in HCC tissues compared to adjacent non-tumoral tissues (Huynh et al., 2003). The main mechanism that constitutively activates MAPK pathway are oncogenic mutations in RAS. Although frequent in pancreatic cancer, RAS mutations are rare in HCC (Yea et al., 2008). The main mechanisms behind MAPK activation in HCC involves aberrant epigenetic regulation and ligand overexpression (Newell et al., 2009). Sorafenib, the first systemic drug to show clinical benefit in HCC (Llovet et al., 2008), has BRAF among its targets. The MEK inhibitor refametinib was also evaluated in a phase II trial in HCC patients with RAS mutations screened using mutational analysis of cell-free DNA (Lim et al., 2014, 2018). However, RAS has been proved a very difficult target to antagonize pharmacologically.

3. Pathways involved in cell differentiation and development

3.1 WNT-β-catenin

The WNT-β-catenin pathway is highly conserved and key driver of liver development. It remains inactive in most differentiated adult cells. The WNT pathway is classified into canonical or non-canonical, depending on whether its activation involves β-catenin or not. The canonical Wnt pathway is inactivated in the absence of extracellular WNT ligands. During this state, β-catenin turnover is regulated by the β-catenin destruction complex. This complex is formed by AXIN, adenomatous polyposis coli (APC), glycogen synthase kinase 3β (GSK3β) and casein kinase I isoform α CK1α. This complex phosphorylates β-catenin at N-terminal domain, which allows it to be recognized and ubiquitinated by E3 ubiquitin ligase and degraded in the proteasome (Perugorria et al., 2019; Stamos & Weis, 2013). Thus, β-catenin localizes in the extracellular membrane, is part of the adherent junctions and acts as a connection between E-cadherin and cytoskeleton associated actin in cell-cell adhesions (Niessen & Gottardi, 2008).

In humans, the family of WNT ligands includes 19 members, which bind to the heterodimeric receptor complex formed by Frizzled (FZD) and co-receptors, low-density lipoprotein receptor related protein 5 or 6 (LPR5 or LPR6) (MacDonald, Tamai, & He, 2009). The binding of a WNT ligand to the Frizzled receptor induces its assembly with the co-receptor LPR (Cong, Schweizer, & Varmus, 2004). This induces phosphorylation of LPR, which triggers the recruitment of AXIN1 (Zeng et al., 2005). The sequestration of AXIN prevents β-catenin phosphorylation and its proteasomal degradation. Thus, β-catenin accumulates in the cytoplasm and translocates into the nucleus where it interacts with different transcription factors and/or co-activators regulating gene expression (Vlad, Röhrs, Klein-Hitpass, & Müller, 2008). The main nuclear binding factor that interacts with β-catenin is DNA-binding T cell factor (TCF), although it also interacts with hypoxia-inducible factor 1α (HIF1α), forkhead box protein O (FOXO), sex-determining region X box SOX, activating diverse target genes involved in cell proliferation and differentiation (Vlad et al., 2008) [Fig. 3A].

There are multiple ways by which Wnt-β-catenin pathway can become aberrantly activated. The most common in HCC are mutations in β-catenin gene (*CTNNB1*) in 30% of cases (Rebouissou et al., 2016) or *AXIN1/2* in 5–10% of cases (Satoh et al., 2000). WNT signaling can also be activated by aberrant methylation of *SOX1* and FZD related proteins (Tsao et al., 2012), acetylation of secreted frizzed-related proteins (Quan et al., 2014), and regulatory ncRNAs (Wang et al., 2015), or hypoxia related pathways (Xu et al., 2017). Most *CTNNB1* mutations are deletions or point mutations in exon 3 at a hotspot of threonine/serine sites. Alterations in this site prevent phosphorylation and/or degradation of β-catenin, triggering aberrant activation of WNT pathway. Other less frequent hotspot mutations have been identified in exon 7 (K335) and exon 8 (N387). A detailed report on *CTNNB1* mutations in a cohort of 220 hepatic adenomas and 373 HCC patients described three main groups of high, moderate or weak activity, according to the activation of the WNT pathway. They found that high activating *CTNNB1* mutations were enriched in HCC (56%) while moderately activating mutations were equally present in both HCC and HCA. Most weak mutations were more frequent in adenomas (60%). HCC had 20% of S45 weak mutations resulting in high activation, suggesting that the hyperactivation of WNT pathway was due to the presence of additional events. In HCCs with weak or atypical *CTNNB1* mutations (55% of the cases), patients had a second hit, a duplication of the mutated allele or an

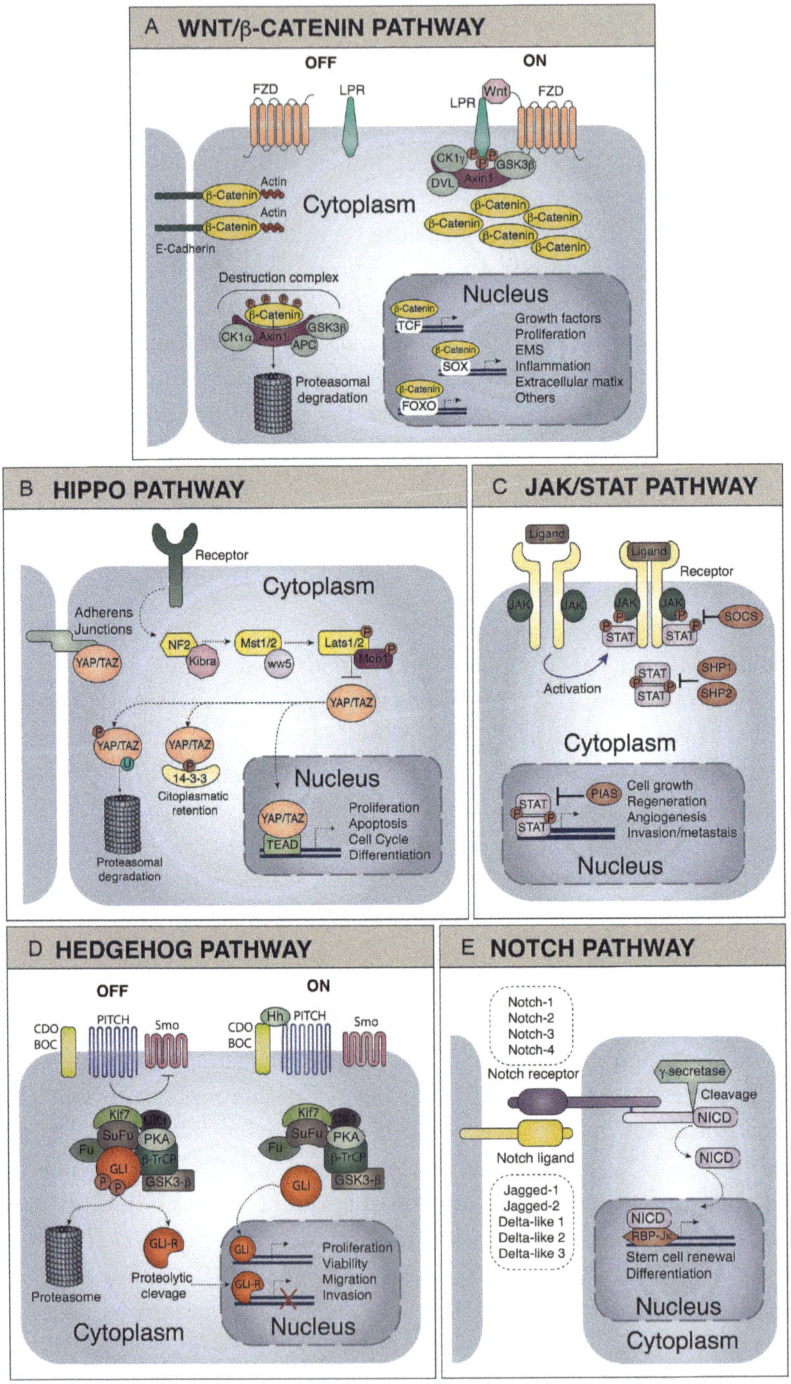

Fig. 3 Pathways involved in cell differentiation and development. (A) WNT/β-catenin pathway. (B) Hippo pathway. (C) JAK/STAT pathway. (D) Hedgehog pathway. (E) Notch pathway.

association with a more active mutation of *CTNNB1*, or inactivation of *APC* (Rebouissou et al., 2016).

Loss of function of *AXIN1* lead to activation of the WNT pathway, however, a recent study found that the 80% of *AXIN1* mutations induce weak or no WNT pathway activation. Authors divided *AXIN1*-mutated HCCs in two groups according to WNT pathway activation intensity. This suggests alternative pathways of transformation for *AXIN1* mutated tumors beyond WNT signaling (Abitbol et al., 2018). Interestingly, a recent study found that *AXIN1* mutations alone do not lead to hepatocarcinogenesis, but it cooperates with c-Met overexpression for tumor formation. This study confirmed that *AXIN1* loss is not a strong activator of the classical WNT pathway (Qiao et al., 2019). Additional data using a gene signature to categorize the different modalities of WNT signaling in HCC reinforces the notion of high heterogeneity in the role of this pathway in HCC development and progression (Lachenmayer, Alsinet, et al., 2012). To date, drugs designed to abrogate β-catenin activation have been unsuccessful in clinical trials.

3.2 JAK/STAT signaling pathway

JAK/STAT signaling plays key roles in cell proliferation, stem cell maintenance, differentiation and regulation of the immune system. Janus Kinases (JAKs) are proteins associated with the cytoplasmic domain of diverse transmembrane receptors, mainly cytokine receptors. The conformational change triggered by the ligand-receptor binding activates JAKs. Once activated, JAKs phosphorylate tyrosine residues on the receptor, creating binding sites for the recruitment of signal transducers and activators of transcription (STATs). After binding to the receptor, STATs are also phosphorylated by JAKs, this phosphorylation provokes STATs activation and subsequent dimerization. STAT dimers translocate to the nucleus where they act as transcription factors of different target genes such as CCND1, BIRC5 or Mcl-1 (Levy & Darnell, 2002). The JAK/STAT pathway is regulated at various levels: (1) tyrosine phosphatases SHP1 and SHP2 that can dephosphorylate JAK and limit downstream signaling (Frankson et al., 2017); (2) suppressors of cytokine signaling (SOCS) can compete with STATs for the binding sites at the receptors and interfere with the activity of JAKs or can target proteins for proteasomal degradation (Durham, Williams, Nasim, & Palmer, 2019); (3) inhibitor of activated STATs (PIAS) that can interfere DNA binding ability of STATs (Furqan, Mukhi, Lee, & Liu, 2013) [Fig. 3C].

Several studies confirm aberrant activation of this pathway in HCC. In a cohort of 75 HCV-related HCC, several members of the JAK/STAT pathway were found deregulated in precancerous tissues and early stages of HCC. Specifically, EPO, EPOR, CISH, STAT3, and SOCS3 were upregulated in dysplasia compared to small HCC (Wurmbach et al., 2007). JAK/STAT inhibitors are frequently downregulated or lost in human HCC resulting in persistent activation of the pathway as shown in a study including 80 human samples that reported promoter hypermethylation and loss of heterozygosity (LOH) of the JAK/STAT inhibitors CIS and SOCS1. CIS and SOC expression appeared approximately twofold decreased in surrounding non-tumoral tissue. All examined HCCs showed downregulation of at least two JAK/STAT inhibitors, suggesting that unrestrained activation of the JAK/STAT pathway is required for liver oncogenesis (Calvisi et al., 2006). The activation of IL6/JAK/STAT pathway appeared to be more frequent among steatohepatitic HCC. The mechanism was studied using a mouse model of glycine *N*-methyltransferase (GNMT) deficiency, an enzyme involved in the maintenance of DNA methylation levels that produces steatosis, fibrosis and HCC. The tumors developed showed reduced expression of JAK/STAT inhibitors caused by the hypermethylation of the promoters of SOC2 (Martínez-Chantar et al., 2008). There are no pharmacological inhibitors of JAK/STAT in late clinical development phases for HCC.

3.3 Hippo signaling pathway

Hippo signaling was first discovered as a key determinant of organ size (Dong et al., 2007). Now it is known that it has a profound impact on the physiopathology of liver diseases, being involved in different oncogenic roles in HCC such as cell proliferation, apoptosis, cell cycle and differentiation. The Hippo-YAP (Yes-associated protein) pathway is a highly conserved kinase cascade in mammals (Liu, Xu, Chen, Poon, & Luk, 2010). Transmembrane receptors transduce signals to neurofibromatosis 2 (NF2) and kidney and brain expressed protein (KIBRA) that initiate the kinase cascade (Genevet, Wehr, Brain, Thompson, & Tapon, 2010). The exact extracellular ligands and receptors regulating hippo pathway remain elusive. The key player of the pathway is MST1/2 (mammalian orthologue for hippo in Drosophila) that phosphorylates large tumor suppressor homolog 1/2 (LATS1/2) (Chan et al., 2005). The interaction between both, MST1/2 and LATS1/2, is coupled by the adaptor protein Salvador homolog 1 (WW45, the human SAV1 ortholog). This ends with the phosphorylation of YAP. YAP phosphorylation localizes it in the cytoplasm. The sequential phosphorylation of YAP results ultimately in its degradation.

When the Hippo pathway is off, the abrogation of YAP phosphorylation by MST1/2 allows YAP to accumulate into the nucleus. In the nucleus, YAP interacts with the transcription factor TEAD (Lin, Park, & Guan, 2017) to turn on the expression of target genes: *RUNXs, SMAD, P73, PAX3, PAX8, TTF-1, TBX5, PPARG*. Besides phosphorylation status, Hippo pathway is regulated by cell architecture and mechanical signals, YAP interacts with tight junction proteins, which seems to attenuate expression of target genes (Zhao et al., 2011) [Fig. 3B]. YAP activation is an early event in HCC development, but mutations within the hippo pathway are rare in human tumors (Perra et al., 2014). Genomic amplification of the genomic locus containing YAP gene (11q22) is present in 5–10% of HCC (Zender et al., 2006), and other studies show 65–85% of HCC have increased YAP activity (Li et al., 2012; Tao et al., 2014). YAP expression and activity can be regulated by multiple factors. Apart from the above mentioned regulation of YAP phosphorylation, it can also be modulated by protein phosphoatase-1 (PP1A), which dephosphorylates YAP and increases its nuclear accumulation and transcriptional activity (Wang et al., 2011). YAP together with TEAD induces interaction with FOXM1 to drive the expression of genes associated with chromosomal instability in HCC cells (Weiler et al., 2017). AXL receptor kinase has also been identified as a downstream target of YAP mediating the oncogenic functions (Xu et al., 2011). Similar to WNT, Hippo pathway is a difficult signaling pathway to drug.

3.4 Hedgehog signaling pathway

Hedgehog (Hh) signaling is initiated by a family of ligands including Desert (DHH), Indian (IHH) and Sonic Hedgehog (SHH). In the absence of a Hh ligand, phosphatidylinositol-4-phospate [PI(4)P] interacts with the PTCH receptor and inhibits the G-coupled receptor Smoothened (SMO) (Jiang et al., 2016). SMO associates with SUMO specific peptidases (SENP) triggering its ubiquitination and degradation (Ma et al., 2016). Upon an Hh ligand binding, PTCH suppressed SMO inhibition, which cleaves the glioma-associated oncogene transcription factors (GLIs) and release the active form of GLIs in the cytoplasm. The GLI family is composed of two activating members GLI1 and GLI2 and one repressor GLI3. The active form of GLIs can enter to the nucleus and regulate gene transcription (Ruiz i Altaba, Mas, & Stecca, 2007) [Fig. 3D].

There are two modes of aberrant Hh activation during carcinogenesis: (1) ligand-independent and (2) ligand-dependent. The first one is caused by inactivating mutations in PTCH and SMO, as well as alterations in GLIs. In this sense, a study analyzing hepatic adenomas identified a deletion of INHBE leading to a INHBE-GLI1 fusion and consequent Hh pathway activation (Nault et al., 2017). Also, Hh pathway may drive tumorigenesis in the context of liver injury by GLI2 overexpression as show in experimental models (Riordan et al., 2018). Interestingly, a molecular analysis of human HCCs identified an HBV insertion in the GLI2 promoter region, associated with cell proliferation (Ally et al., 2017). The ligand-dependent group includes overexpression of the ligands. For example, SHH overexpression might be involved in the transition from adenoma to HCC (Chung et al., 2016). Another regulatory mechanism is mediated by the Hedgehog interacting protein (HHIP) receptor, which can bind the three members of the Hh family and reduce pathway activation. HHIP expression is downregulated by methylation in cell lines and HCC tissue. LOH has also been described in the HHIP locus in HCC patients (Tada et al., 2008).

3.5 Notch signaling pathway

The Notch signaling pathway is critical in stem cell self-renewal and differentiation. Notch pathway includes four transmembrane receptors (Notch 1, 2, 3 and 4) that can be activated by five different ligands (Jagged-1, Jagged-2, Delta-like 1, 2 and 3). Notch receptors can also bind to non-canonical ligands such as DLK1, DLK2/EGFL9, EGFL7 and DNER, which seem to act as pathway inhibitors (D'Souza, Meloty-Kapella, & Weinmaster, 2010). The interaction between the canonical ligands and the receptor leads to the proteolytic cleavage of the intracellular domain of Notch receptors (NICD) by the γ-secretase complex. In the nucleus, NICD binds to a nuclear protein of the RBP-Jk family, which acts as a transcriptional activator and modulates the expression of *HES* and *HEY* family factors among other genes (Lu et al., 2016). Deactivation of NICD occurs by phosphorylation and proteasomal degradation and also by transcriptional repression, that is mediated by RBPJ and the recruitment of other co-repressors (Strazzabosco & Fabris, 2012). The activation of the pathway requires cell-to-cell contact [Fig. 3E].

Notch can induce hepatocytes to gain a progenitor cell like phenotype (Liu et al., 2018). Besides, in HCC, the oncogenic role of Notch is dependent on other oncogenic pathways, often related to the inflammatory microenvironment (Lim et al., 2011; Ye et al., 2019). Using gene expression and comparative genomics, Notch signaling is found aberrantly activated in 30% of human HCC across different cohorts. IGF2 and Sox9 have been identified as oncogenic partners of activated Notch (Villanueva et al., 2012). There are no Notch inhibitors being tested in late phase clinical trials for HCC.

4. Nuclear signaling pathways

4.1 Telomere shortening (TERT)

Telomeres are located at each end of chromosomes and their role is to prevent the loss of DNA material because of cell replication. Telomeres are synthetized by telomerase, which is composed of telomerase reverse transcriptase (TERT), the telomerase RNA component (TERC), and dyskerin [Fig. 4]. Telomeres are present in all mammalian cells and their

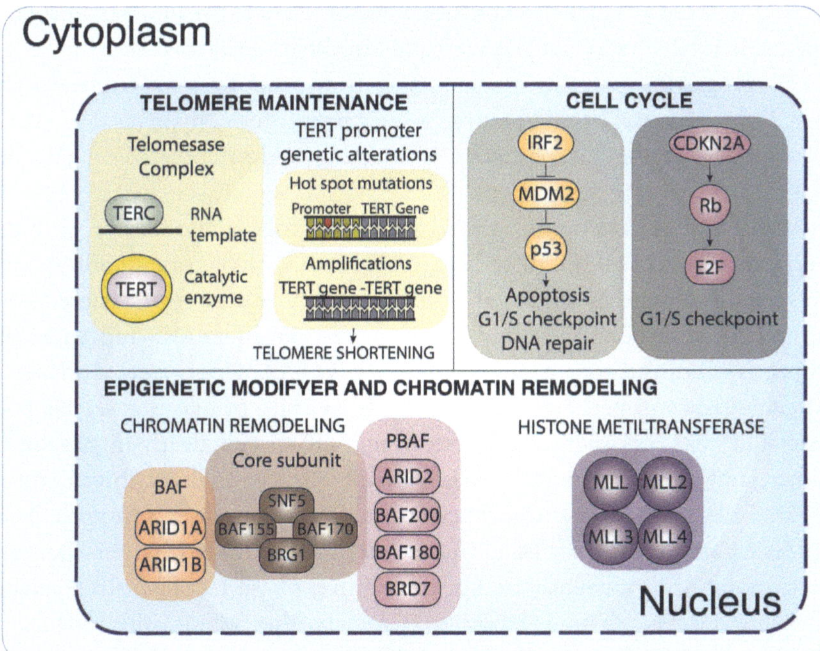

Fig. 4 Summary of nuclear signaling pathways. Telomere maintenance: telomeres are maintained by telomerase complex formed by TERC, TERT and dyskerin. Some of the most frequent alterations in this pathway involve TERT promoter mutations and TERT amplifications. Cell cycle: p53 and retinoblastoma pathways are frequently altered in HCC. Epigenetic modifications: chromatin remodeling-related genes (ex. ARID1A) and histone acetylation are frequently altered in HCC.

shortening promotes cell arrest, senescence and apoptosis. Shortening of telomeres activates the double strand break DNA damage response and lead to chromosomal instability through end-to-end chromosome fusion if tumor-suppressive mechanisms involving p53 and RB1 are defective (Maciejowski & de Lange, 2017).

In HCC, telomere length is shorter than in normal or cirrhotic hepatocytes (Ohashi et al., 1996). Moreover, TERT expression have been found higher in tumoral samples. TERT is reactivated in most HCCs through different mechanisms including mutations in the TERT promoter (30–60%), which is the most prevalent somatic mutations in human HCC (Nault et al., 2013). Other mechanisms include HBV or adeno-associated virus type 2 (AAV2) viral insertion into the promoter, which has been described in 26% of HBV-related HCCs (Sung et al., 2012; Xue et al., 2016; Zhao, Liu, et al., 2016). TERT gene amplification and chromosomal rearrangements (5–10%) (Letouzé et al., 2017; Schulze et al., 2015; Totoki et al., 2014), via fusions with highly expressed partner genes such as SLC12A7 and SLC12A2, have also been described (Barthel et al., 2017). Regardless of the mechanisms, they lead to a re-expression of TERT, increased TERT activity, which promotes cell survival and proliferation, and avoids senescence induced by telomere erosion (Barthel et al., 2017).

Many studies have demonstrated a potential role of telomerase inhibitors as new therapies in HCC. Imetelstat (GRN163L), an oligonucleotide inhibitor of telomerase, has shown antitumor activity related to telomere shortening and dysfunction, reduced tumor cell proliferation and increased apoptosis (Djojosubroto et al., 2005). BIBR1532, a non-nucleotide small molecule that competitively inhibits the active site of telomerase, has been associated with telomerase inhibition and HCC cell death in preclinical studies, but there is no data in late phase clinical trials (Tahtouh et al., 2015). TERT-based immunotherapy alone or in combination with IL-12 DNA, a potent enhancer of immune response, followed by percutaneous electroporation is currently being studied in a phase I trial in solid cancers including HCC (Phase I Study to Evaluate the Safety and Efficacy of Telomelysin (OBP-301) in Patients With Hepatocellular Carcinoma, 2014). Telomelysin (OBP-301), a TERT-driven oncolytic adenovirus that introduces the TERT promoter specifically into tumor cells, is also under investigation in a phase I/II trial in HCC (Nault, Ningarhari, Rebouissou, & Zucman-Rossi, 2019).

4.2 Cell cycle (P53, retinoblastoma)

The *TP53* gene encodes a tumor suppressor protein that functions as a stress responder and cell cycle checkpoint protein that maintains genomic integrity [Fig. 4]. The p53 protein induces cell cycle arrest, apoptosis, and senescence in response to cellular stress to maintain the integrity of the genome (Muller & Vousden, 2013). Loss of its function through genetic alteration such as mutations and deletions is a key genetic event in cancer (Parikh et al., 2014).

In HCC, *TP53* is frequently mutated, up to 45% in patients with HBV-related HCC (Schulze et al., 2015). Moreover, downregulation of p53 is linked to *TP53* gene mutations (Azer, 2018). TP53 has specific mutational hotspots, preferentially located in its DNA binding domain (Meng, Franklin, Dong, & Zhang, 2014). The presence of a characteristic *TP53* mutation in ser249, inducing C to T substitution, is tightly linked with the dietary intake of aflatoxin, a mycotoxin present in contaminated foods, first described in Africa and Asia populations (Bressac, Kew, Wands, & Ozturk, 1991; Hsu et al., 1991).

An analysis by whole-exome sequencing of advanced HCC revealed homozygous deletions and alterations of splicing in RB1 (14.6%in advance HCC vs 2.6% in early HCC). Among the 10 HCC cases with RB1 alterations 6 patients also presented TP53 mutations, suggesting that the co-occurrence of alterations promote HCC progression (Nault et al., 2020). In this sense, a study has demonstrated that NCKAP1 could act as a tumor suppressor gene that module the cell cycle of HCC cell lines by targeting RB1/p53 regulation and was correlated with a favorable outcome in HCC patients (Zhong et al., 2019). In preclinical models of HCC the use of Palbociclib, a CDK4/6 inhibitor, showed cell cycle arrest and antitumoral effects when RB1 was functionally intact (Bollard et al., 2017). On the other hand, the loss of RB1 conferred drug resistance. The analysis of RB1 status in HCC patients of the TCGA dataset identified a "RB1 loss of function" signature present in 30% of the cases (Bollard et al., 2017).

4.3 Epigenetic modifications and chromatin remodeling (ARID1A, ARID2)

The analysis of DNA methylation profiles of 304 HCC patients identified a 36 CpG methylation signature that correlates with patient survival. Among the pathways most clearly deregulated in HCC by DNA methylation were identified: IGF, PI3K, TGFβ and WNT (Villanueva et al., 2015). Some risk factors such as HBV and HCV can directly modulate DNA methylation in tumor cells and participate in carcinogenesis. For example, WDR5 is a key epigenetic determinant of HBV-induced tumorigenesis and the HBx-WDR5-H3K4me3 axis may be a potential therapeutic target in HBV-induced liver pathogenesis, since a WDR5-inhibition was able to suppress HBV transcription (Gao et al., 2020).

Generally, histone acetylation is correlated with gene activation whereas histone deacetylation is linked to gene repression and silencing. HCC progression was associated with dysregulation of histones acetylation and deacetylation [Fig. 4]. The expression levels of 11 relevant HDAC were analyzed in a cohort of 91 HCC patients. After validation in a cohort of 243 patients, 4 HDACs appeared upregulated and 2 downregulated. Upregulation of HDAC3 and HDAC5 correlated with DNA copy number gains. Functional analysis inhibiting HDACs with panobinostat in pre-clinical models showed important anti-tumoral effects (Lachenmayer, Toffanin, et al., 2012). HDACs have emerged as potential new targets for HCC. Several HDACs inhibitors are approved by FDA (e.g., vorinostat, Panobinostat) for the treatment of hematological malignancies (Dai, Zhang, Yuan, Sun, & Jiang, 2020). However, results of early phase clinical trials testing HDAC in HCC are modest.

5. Emerging pathways

5.1 Mechanisms involving non-coding RNA (ncRNAs)

In recent years, high-throughput transcriptomic analyses have revealed aberrant expression of a large number of ncRNAs across different cancer types, including HCC. ncRNAs can be classified according to size into two groups, small ncRNAs when are smaller than 200 nt (e.g., miRNA, snoRNA, piRNA) and long ncRNA when bigger than 200 nt (Derrien et al., 2012).

5.1.1 Small non-coding RNAs

miRNAs regulate gene expression at post-transcriptional level as they can bind the 3′UTR region of mRNAs to mediate deadenylation or translational blockage (Jonas & Izaurralde, 2015). Several members of the miRNA biogenesis machinery, formed by the complexes DGCR8-Drosha, exportin 5 and DICER-RISC, are commonly downregulated in HCC (Kitagawa et al., 2013). Apart from the regulation of miRNA synthesis, a large list of specific miRNA also promote oncogenesis (Wong, Tsang, & Ng, 2018) such as miR-22 (Pineau et al., 2010), miR-21 (Meng et al., 2007), and miR-222 (Wong et al., 2010) that are upregulated in HCC. miR-122 (Gramantieri et al., 2007), miR-125b (Liang et al., 2010), miR-199, miR-101 (Su et al., 2009) and let-7(Wang et al., 2010) are downregulated and act as tumor suppressors. Unsupervised clustering of miRNA expression identifies molecular classes of HCC, which correlated with clinical features and transcriptome-based subgroups (Toffanin et al., 2011).

miR-221 is associated with the activation of MET-JNK-AP1 pathway (Garofalo et al., 2009). It also targets cyclin-dependent kinase inhibitor B1 (CDKIB) and CDKN1C (Fornari et al., 2008), as well as PTEN and TIMP3 (Garofalo et al., 2009). miR-221 repress DDIT4 (DNA-damage inducible transcript 4 protein) regulating mTOR signaling and also negatively regulates Bcl2 modifying factor (Pineau et al., 2010). miR-122 is the most abundant miRNA in the liver, where it accounts for 70% of the total miRNA content (Jopling, 2012) and its downregulation promotes hepatocarcinogenesis. miR122 regulates the expression of piruvate kinase PKM2 (Wong et al., 2014), p53 (Fornari et al., 2009), or RhoA (Wang et al., 2014). miR122 is also implicated in the Wnt signaling pathway targeting WNT-1 and destabilizing β-catenin (Xu et al., 2012), besides, it targets E3 ubiquitin ligase mdm2 (MDM2), preventing p53 degradation (Simerzin et al., 2016).

5.1.2 Long non-coding RNAs

Global lncRNAs expression in human HCC is remarkably different between tumor and non-tumoral liver tissue. Diverse functions have been attributed to lncRNAs. They act as regulators of protein levels by controlling mRNA stability and can interact with transcription factors (Wong et al., 2018) [Fig. 5]. Some well-characterized and functionally relevant lncRNAs in HCC include TERC, HULC and HOTAIR (Klingenberg, Matsuda, Diederichs, & Patel, 2017). TERC is part of the catalytic center of the telomerase enzyme complex. HULC sequesters miR-200a-3p to activate zinc-finger *E*-box-binding homeobox 1 (ZEB1) and promote EMT (Li et al., 2016). HULC has also been found overexpressed in circulating tumor cells from HCC patients (D'Avola et al., 2018). HOTAIR interacts with polycomb group complex 2 (PRC2) and collaborates with H3K27me3 to mediate gene silencing of target genes (Rinn et al., 2007), and suppress p14 and p16 signaling (Fu et al., 2015).

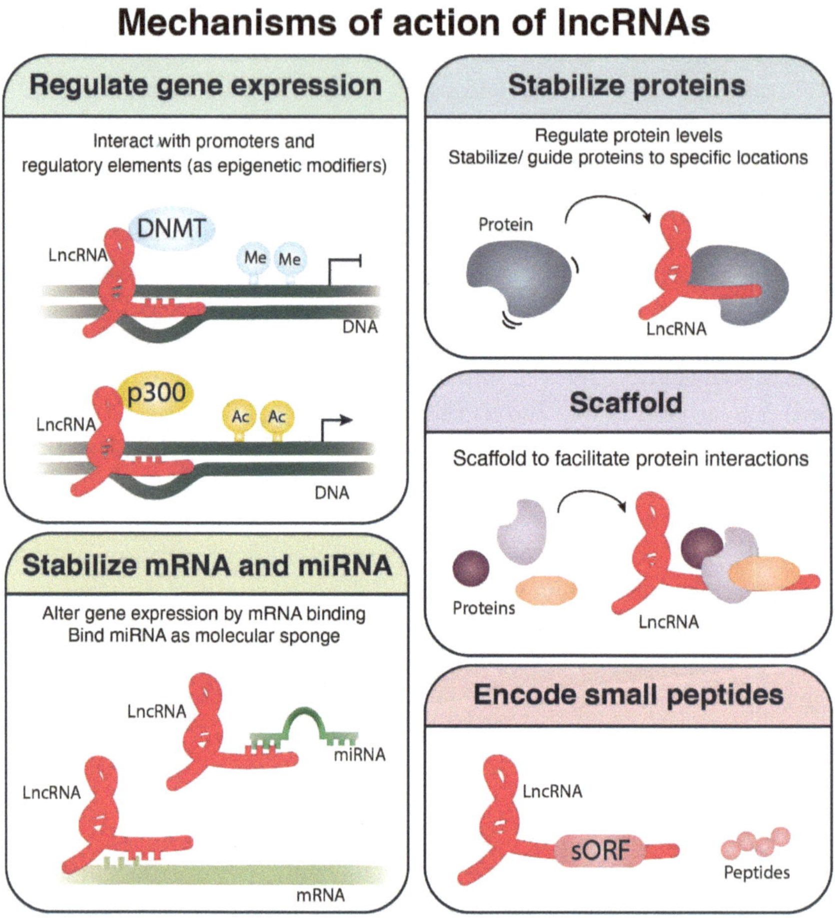

Fig. 5 Mechanism of action of long non-coding RNAs (lncRNAs). lncRNAs can regulate gene expression by interaction with promoter or regulatory elements (Ex. Recruitment of epigenetic modifiers as DNMT or p300), or stabilizing mRNAs and miRNAs. lncRNA can interact with proteins to regulate protein levels, stabilize/guide proteins to specific locations or act as scaffolds to facilitate interactions. Some lncRNAs can encode small peptides.

Other lncRNA described in HCC are MALAT1, NEAT1, H19, TOTTIP, MEG3 or GAS5, among others. MALAT1 is involved in the regulation of PI3K/Akt/mTOR axis in HCC, having effects in apoptosis and autophagy (Peng et al., 2020). It also participates in the regulation of HCC glucose metabolism by the activation of mTOR and the transcription factor TCF7L2 (Malakar et al., 2019). Interestingly, MALAT1 participates in the acquisition drug resistance (Fan et al., 2020; Yuan et al., 2016) and exhibits prognostic value after curative HCC resection (Sonohara et al., 2017). The lncRNA NEAT1 is activated by hypoxia (HIF-1α, HIF-2α) (Zhang et al., 2020; Zheng et al., 2018), and can promote HCC by the modulation of IGF-1R, TGFβ or IL6/STAT3 pathways (Liu, Shi, Yang, & Jiang, 2019; Tu et al., 2018; Wang et al., 2018). H19 has an unclear effect on HCC, it has been described as both, tumor suppressor and promotor (Tietze & Kessler, 2020). A recent study links the presence of particular haplotypes of H19 with the susceptibility to HCC (Wu et al., 2019). HOTTIP locus is close to HOXA13 and controls its expression. The levels of both, HOTTIP and HOXA13, associate with metastasis and patient survival in HCC (Quagliata et al., 2014). HOTTIP expression is modulated by two miRNAs (miR-192 and miR-204) (Ge et al., 2015). An integrated analysis of ncRNA regulatory networks in HCC identified MEG3 as core gene. Validation in vitro confirmed that the overexpression on MEG3 disturbs the expression profiles of related miRNAs and mRNAs (Tang et al., 2019). GAS5 is downregulated in HCC (Toraih et al., 2018), and has been observed that the treatment with sorafenib modulate its expression in cell lines and patients (Faranda et al., 2019).

5.2 Microbiome

The liver is the first physical target of intestinal microbiota and its derivatives, bacterial metabolites and microbe associated molecular patterns (MAMPs; e.g., lipopolysaccharide-LPS). Accumulating evidence supports an important role of intestinal microbiota in HCC development by local and distant effects (Schwabe & Jobin, 2013). Chronic liver disease facilitates the release of endotoxins, which may affect inflammation, immunity, fibrogenesis and regeneration. The communication between the intestine and the liver is also regulated by bile acids, which can influence HCC via Farnesoid X receptor (FXR) and EGFR pathways (Dossa et al., 2016; Takahashi et al., 2018) [Fig. 6]. Dysbiosis favors the release of high levels of MAMPs, such as LPS, which triggers inflammation through TLR4. In hepatocytes, TLR4 activates NF-κβ, regulating different pathways facilitating HCC progression as the upregulation of epiregulin or prevention of apoptosis (Dapito et al., 2012). Dysbiosis also modifies bile acid composition, as gut microbiota can transform intestinal bile acids into the unconjugated forms. Effects of the dysregulation of the composition of secondary bile acids (DCA) in the liver activates mTOR pathway in hepatocytes (Yamada et al., 2018). TLR2 overexpression on hepatic stellate cells results in tumor promoting senescence secretory phenotype

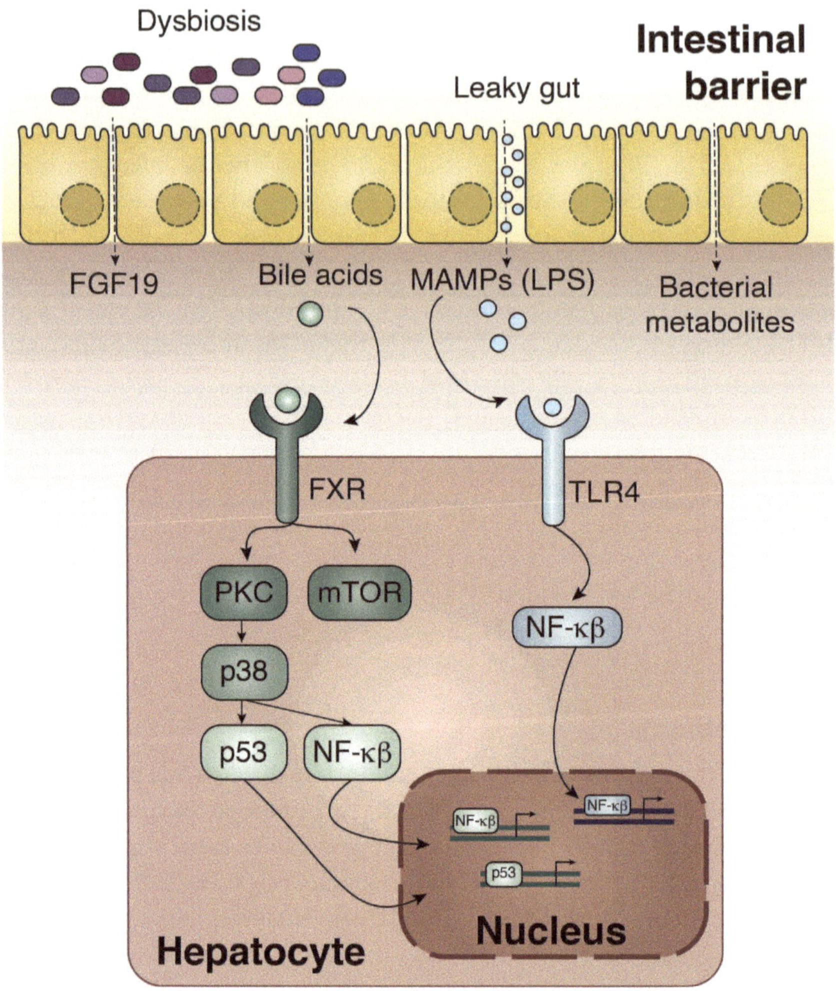

Fig. 6 Mechanism by which intestinal microbiota contributes to HCC. Dysbiosis and leaky gut facilitate liver exposure to bacterial metabolites or microbe associated molecular patterns (MAPS). Dysbiosis also modifies bile acid composition and other communication molecules as FGF19. LPS binds TLR4 in hepatocytes and activates NF-κβ that regulates different pathways affecting HCC progression. Secondary bile acids bind to FXR and activate pathways as mTOR or PKC/p38/p53-NF-κβ.

(Yoshimoto et al., 2013). Bile acids can also disrupt the plasma membrane and cause activation of PKC triggering the activation of p38 MAPK pathway resulting in p53 and NF-κB activation (Anwer, 2012). Bile acids regulate the expression of CXCL6 on sinusoidal cells with favors accumulation of

NKT cells and prevents tumor growth (Ma et al., 2018). FXR regulates bile acid synthesis and mice with FXR deficiency spontaneously develop liver tumors (Yang et al., 2007).

6. Conclusions

Advances in high-throughput sequencing technologies have improved our understanding of the key molecular alterations present in human HCC. Most of these alterations affect key signaling pathways involved in cell differentiation, proliferation and survival. Mapping the extent of pathway deregulation has been major research goal in the field in the past decade. Lessons from other tumor types suggested that key regulatory hubs of specific pathways are useful targets for new therapies. In HCC, this approach resulted in minimal success, as most deregulated pathways are difficult to antagonize pharmacologically (e.g., WNT, cell cycle, telomere stability, etc.). Clearly, targeting angiogenesis has been the cornerstone of systemic therapies in HCC, as most approved systemic therapies act on this pathway. However, recent studies demonstrate significant intra-tumoral heterogeneity in a subset of HCC patients (Losic et al., 2020). This suggest that inhibition of a single pathway may not be sufficient to warrant a durable tumor response. Based on phase III clinical trials, the combination of antiangiogenics with immune checkpoint inhibitors suggest a synergistic effect (Finn et al., 2020). This emphasizes the need to inhibit multiple pathways simultaneously to maximize antitumor efficacy. In this context, drug toxicity will be a limiting factor.

Acknowledgment

A.V. is supported by the U.S. Department of Defense (CA150272P3).

Conflict of interest

A.V. has received consulting fees from Guidepoint, Fujifilm, Boehringer Ingelheim, FirstWord, and MHLife Sciences; advisory board fees from Exact Sciences, Nucleix, Gilead and NGM Pharmaceuticals; and research support from Eisai.

References

Abitbol, S., Dahmani, R., Coulouarn, C., Ragazzon, B., Mlecnik, B., Senni, N., et al. (2018). AXIN deficiency in human and mouse hepatocytes induces hepatocellular carcinoma in the absence of β-catenin activation. *Journal of Hepatology*, *68*(6), 1203–1213. https://doi.org/10.1016/j.jhep.2017.12.018.

Abou-Alfa, G. K., Meyer, T., Cheng, A.-L., El-Khoueiry, A. B., Rimassa, L., Ryoo, B.-Y., et al. (2018). Cabozantinib in patients with advanced and progressing hepatocellular carcinoma. *The New England Journal of Medicine*, *379*(1), 54–63. https://doi.org/10.1056/NEJMoa1717002.

Ally, A., Balasundaram, M., Carlsen, R., Chuah, E., Clarke, A., Dhalla, N., et al. (2017). Comprehensive and integrative genomic characterization of hepatocellular carcinoma. *Cell*, *169*(7), 1327–1341. e23 https://doi.org/10.1016/j.cell.2017.05.046.

Anwer, M. S. (2012). Intracellular signaling by bile acids. *Journal of Bioscience*, *20*, 1–23. https://doi.org/10.3329/jbs.v20i0.17647.

Azer, S. A. (2018). MDM2-p53 interactions in human hepatocellular carcinoma: What is the role of nutlins and new therapeutic options? *Journal of Clinical Medicine Research*, 7(4), 64. https://doi.org/10.3390/jcm7040064.

Baek, J. Y., Morris, S. M., Campbell, J., Fausto, N., Yeh, M. M., & Grady, W. M. (2010). TGF-beta inactivation and TGF-alpha overexpression cooperate in an in vivo mouse model to induce hepatocellular carcinoma that recapitulates molecular features of human liver cancer. *International Journal of Cancer*, *127*(5), 1060–1071. https://doi.org/10.1002/ijc.25127.

Barthel, F. P., Wei, W., Tang, M., Martinez-Ledesma, E., Hu, X., Amin, S. B., et al. (2017). Systematic analysis of telomere length and somatic alterations in 31 cancer types. *Nature Genetics*, *49*(3), 349–357. https://doi.org/10.1038/ng.3781.

Bollard, J., Miguela, V., Ruiz de Galarreta, M., Venkatesh, A., Bian, C. B., Roberto, M. P., et al. (2017). Palbociclib (PD-0332991), a selective CDK4/6 inhibitor, restricts tumour growth in preclinical models of hepatocellular carcinoma. *Gut*, *66*(7), 1286–1296. https://doi.org/10.1136/gutjnl-2016-312268.

Bouattour, M., Raymond, E., Qin, S., Cheng, A. L., Stammberger, U., Locatelli, G., et al. (2018). Recent developments of c-Met as a therapeutic target in hepatocellular carcinoma. *Hepatology*, *67*(3), 1132–1149. https://doi.org/10.1002/hep.29496.

Bressac, B., Kew, M., Wands, J., & Ozturk, M. (1991). Selective G to T mutations of p53 gene in hepatocellular carcinoma from southern Africa. *Nature*, *350*(6317), 429–431. https://doi.org/10.1038/350429a0.

Bruix, J., Qin, S., Merle, P., Granito, A., Huang, Y.-H., Bodoky, G., et al. (2017). Regorafenib for patients with hepatocellular carcinoma who progressed on sorafenib treatment (RESORCE): A randomised, double-blind, placebo-controlled, phase 3 trial. *The Lancet*, *389*(10064), 56–66. https://doi.org/10.1016/S0140-6736(16)32453-9.

Calvisi, D. F., Ladu, S., Gorden, A., Farina, M., Conner, E. A., Lee, J. S., et al. (2006). Ubiquitous activation of Ras and Jak/Stat pathways in human HCC. *Gastroenterology*, *130*(4), 1117–1128. https://doi.org/10.1053/j.gastro.2006.01.006.

Chan, E. H. Y., Nousiainen, M., Chalamalasetty, R. B., Schäfer, A., Nigg, E. A., & Sillje, H. H. W. (2005). The Ste20-like kinase Mst2 activates the human large tumor suppressor kinase Lats1. *Oncogene*, *24*(12), 2076–2086. https://doi.org/10.1038/sj.onc.1208445.

Chen, J., Zaidi, S., Rao, S., Chen, J.-S., Phan, L., Farci, P., et al. (2018). Analysis of genomes and transcriptomes of hepatocellular carcinomas identifies mutations and gene expression changes in the transforming growth factor-β pathway. *Gastroenterology*, *154*(1), 195–210. https://doi.org/10.1053/j.gastro.2017.09.007.

Cheng, A.-L., Kang, Y.-K., Lin, D.-Y., Park, J.-W., Kudo, M., Qin, S., et al. (2013). Sunitinib versus sorafenib in advanced hepatocellular cancer: Results of a randomized phase III trial. *Journal of Clinical Oncology: Official Journal of the American Society of Clinical Oncology*, *31*(32), 4067–4075. https://doi.org/10.1200/JCO.2012.45.8372.

Chiang, D. Y., Villanueva, A., Hoshida, Y., Peix, J., Newell, P., Minguez, B., et al. (2008). Focal gains of VEGFA and molecular classification of hepatocellular carcinoma. *Cancer Research*, *68*(16), 6779–6788. https://doi.org/10.1158/0008-5472.CAN-08-0742.

Chung, S. I., Moon, H., Ju, H. L., Cho, K. J., Kim, D. Y., Han, K. H., et al. (2016). Hepatic expression of sonic hedgehog induces liver fibrosis and promotes hepatocarcinogenesis in a transgenic mouse model. *Journal of Hepatology, 64*(3), 618–627. https://doi.org/10.1016/j.jhep.2015.10.007.

Cong, F., Schweizer, L., & Varmus, H. (2004). Wnt signals across the plasma membrane to activate the β-catenin pathway by forming oligomers containing its receptors, frizzled and LRP. *Development, 131*(20), 5103–5115. https://doi.org/10.1242/dev.01318.

Dai, Q., Zhang, C., Yuan, Z., Sun, Q., & Jiang, Y. (2020). Current discovery strategies for hepatocellular carcinoma therapeutics. *Expert Opinion on Drug Discovery, 15*(2), 243–258. https://doi.org/10.1080/17460441.2020.1696769.

Dapito, D. H., Mencin, A., Gwak, G. Y., Pradere, J. P., Jang, M. K., Mederacke, I., et al. (2012). Promotion of hepatocellular carcinoma by the intestinal microbiota and TLR4. *Cancer Cell, 21*(4), 504–516. https://doi.org/10.1016/j.ccr.2012.02.007.

D'Avola, D., Villacorta-Martin, C., Martins-Filho, S. N., Craig, A., Labgaa, I., von Felden, J., et al. (2018). High-density single cell mRNA sequencing to characterize circulating tumor cells in hepatocellular carcinoma. *Scientific Reports, 8*(1), 11570. https://doi.org/10.1038/s41598-018-30047-y.

Delire, B., & Starkel, P. (2015). The Ras/MAPK pathway and hepatocarcinoma: Pathogenesis and therapeutic implications. *European Journal of Clinical Investigation, 45*(6), 609–623. https://doi.org/10.1111/eci.12441.

Derrien, T., Johnson, R., Bussotti, G., Tanzer, A., Djebali, S., Tilgner, H., et al. (2012). The GENCODE v7 catalog of human long noncoding RNAs: Analysis of their gene structure, evolution, and expression. *Genome Research, 22*(9), 1775–1789. https://doi.org/10.1101/gr.132159.111.

Dimri, M., & Satyanarayana, A. (2020). Molecular signaling pathways and therapeutic targets in hepatocellular carcinoma. *Cancers, 12*(2). https://doi.org/10.3390/cancers12020491.

Dituri, F., Mancarella, S., Cigliano, A., Chieti, A., & Giannelli, G. (2019). TGF-β as multifaceted orchestrator in HCC progression: Signaling, EMT, immune microenvironment, and novel therapeutic perspectives. *Seminars in Liver Disease, 39*(01), 053–069. https://doi.org/10.1055/s-0038-1676121.

Djojosubroto, M. W., Chin, A. C., Go, N., Schaetzlein, S., Manns, M. P., Gryaznov, S., et al. (2005). Telomerase antagonists GRN163 and GRN163L inhibit tumor growth and increase chemosensitivity of human hepatoma. *Hepatology, 42*(5), 1127–1136. https://doi.org/10.1002/hep.20822.

Dong, J., Feldmann, G., Huang, J., Wu, S., Zhang, N., Comerford, S. A., et al. (2007). Elucidation of a universal size-control mechanism in Drosophila and mammals. *Cell, 130*(6), 1120–1133. https://doi.org/10.1016/j.cell.2007.07.019.

Dossa, A. Y., Escobar, O., Golden, J., Frey, M. R., Ford, H. R., & Gayer, C. P. (2016). Bile acids regulate intestinal cell proliferation by modulating EGFR and FXR signaling. *American Journal of Physiology-Gastrointestinal and Liver Physiology, 310*(2), G81–G92. https://doi.org/10.1152/ajpgi.00065.2015.

D'Souza, B., Meloty-Kapella, L., & Weinmaster, G. (2010). *Canonical and non-canonical notch ligands. Vol. 92.* Retrieved from https://pubmed.ncbi.nlm.nih.gov/20816393/.

Durham, G. A., Williams, J. J. L., Nasim, M. T., & Palmer, T. M. (2019). Targeting SOCS proteins to control JAK-STAT signalling in disease. *Trends in Pharmacological Sciences, 40*(5), 298–308. https://doi.org/10.1016/j.tips.2019.03.001.

Fan, L., Huang, X., Chen, J., Zhang, K., Gu, Y.-H., Sun, J., et al. (2020). Long noncoding RNA MALAT1 contributes to sorafenib resistance by targeting miR-140-5p/Aurora-A signaling in hepatocellular carcinoma. *Molecular Cancer Therapeutics, 19*(5), 1197–1209. https://doi.org/10.1158/1535-7163.MCT-19-0203.

Faranda, T., Grossi, I., Manganelli, M., Marchina, E., Baiocchi, G., Portolani, N., et al. (2019). Differential expression profiling of long non-coding RNA GAS5 and miR-126-3p in human cancer cells in response to sorafenib. *Scientific Reports, 9*(1), 9118. https://doi.org/10.1038/s41598-019-45604-2.

Finn, R. S., Qin, S., Ikeda, M., Galle, P. R., Ducreux, M., Kim, T.-Y., et al. (2020). Atezolizumab plus bevacizumab in unresectable hepatocellular carcinoma. *The New England Journal of Medicine*, *382*(20), 1894–1905. https://doi.org/10.1056/NEJMoa1915745.

Firtina Karagonlar, Z., Koc, D., Iscan, E., Erdal, E., & Atabey, N. (2016). Elevated hepatocyte growth factor expression as an autocrine c-Met activation mechanism in acquired resistance to sorafenib in hepatocellular carcinoma cells. *Cancer Science*, *107*(4), 407–416. https://doi.org/10.1111/cas.12891.

Fornari, F., Gramantieri, L., Ferracin, M., Veronese, A., Sabbioni, S., Calin, G. A., et al. (2008). MiR-221 controls CDKN1C/p57 and CDKN1B/p27 expression in human hepatocellular carcinoma. *Oncogene*, *27*(43), 5651–5661. https://doi.org/10.1038/onc.2008.178.

Fornari, F., Gramantieri, L., Giovannini, C., Veronese, A., Ferracin, M., Sabbioni, S., et al. (2009). MiR-122/cyclin G1 interaction modulates p53 activity and affects doxorubicin sensitivity of human hepatocarcinoma cells. *Cancer Research*, *69*(14), 5761–5767. https://doi.org/10.1158/0008-5472.CAN-08-4797.

Frankson, R., Yu, Z. H., Bai, Y., Li, Q., Zhang, R. Y., & Zhang, Z. Y. (2017). Therapeutic targeting of oncogenic tyrosine phosphatases. *Cancer Research*, 77(21), 5701–5705. https://doi.org/10.1158/0008-5472.CAN-17-1510.

Fu, W.-M., Zhu, X., Wang, W.-M., Lu, Y.-F., Hu, B.-G., Wang, H., et al. (2015). Hotair mediates hepatocarcinogenesis through suppressing miRNA-218 expression and activating P14 and P16 signaling. *Journal of Hepatology*, *63*(4), 886–895. https://doi.org/10.1016/j.jhep.2015.05.016.

Fuchs, B. C., Hoshida, Y., Fujii, T., Wei, L., Yamada, S., Lauwers, G. Y., et al. (2014). Epidermal growth factor receptor inhibition attenuates liver fibrosis and development of hepatocellular carcinoma. *Hepatology*, *59*(4), 1577–1590. https://doi.org/10.1002/hep.26898.

Furqan, M., Mukhi, N., Lee, B., & Liu, D. (2013). Dysregulation of JAK-STAT pathway in hematological malignancies and JAK inhibitors for clinical application. *Biomarker Research*, *1*(1), 1–10. https://doi.org/10.1186/2050-7771-1-5.

Gao, W., Jia, Z., Tian, Y., Yang, P., Sun, H., Wang, C., et al. (2020). HBx protein contributes to liver carcinogenesis by H3K4me3 modification through stabilizing WD repeat domain 5 protein. *Hepatology*, *71*(5), 1678–1695. https://doi.org/10.1002/hep.30947.

Garajova, I., Giovannetti, E., Biasco, G., & Peters, G. J. (2015). c-Met as a target for personalized therapy. *Translational Oncogenomics*, 7(Suppl. 1), 13–31. https://doi.org/10.4137/TOG.S30534.

Garofalo, M., Di Leva, G., Romano, G., Nuovo, G., Suh, S. S., Ngankeu, A., et al. (2009). miR-221&222 regulate TRAIL resistance and enhance tumorigenicity through PTEN and TIMP3 downregulation. *Cancer Cell*, *16*(6), 498–509. https://doi.org/10.1016/j.ccr.2009.10.014.

Ge, Y., Yan, X., Jin, Y., Yang, X., Yu, X., Zhou, L., et al. (2015). MiRNA-192 and miRNA-204 directly suppress lncRNA HOTTIP and interrupt GLS1-mediated glutaminolysis in hepatocellular carcinoma. *PLoS Genetics*, *11*(12), e1005726. https://doi.org/10.1371/journal.pgen.1005726.

Genevet, A., Wehr, M. C., Brain, R., Thompson, B. J., & Tapon, N. (2010). Kibra is a regulator of the salvador/warts/hippo signaling network. *Developmental Cell*, *18*(2), 300–308. https://doi.org/10.1016/j.devcel.2009.12.011.

Gollob, J. A., Wilhelm, S., Carter, C., & Kelley, S. L. (2006). Role of Raf kinase in cancer: Therapeutic potential of targeting the Raf/MEK/ERK signal transduction pathway. *Seminars in Oncology*, *33*(4), 392–406. https://doi.org/10.1053/j.seminoncol.2006.04.002.

Gonzalez, M. N., de Mello, W., Butler-Browne, G. S., Silva-Barbosa, S. D., Mouly, V., Savino, W., et al. (2017). HGF potentiates extracellular matrix-driven migration of human myoblasts: Involvement of matrix metalloproteinases and MAPK/ERK pathway. *Skeletal Muscle*, 7(1), 20. https://doi.org/10.1186/s13395-017-0138-6.

Gramantieri, L., Ferracin, M., Fornari, F., Veronese, A., Sabbioni, S., Liu, C. G., et al. (2007). Cyclin G1 is a target of miR-122a, a MicroRNA frequently down-regulated in human hepatocellular carcinoma. *Cancer Research*, *67*(13), 6092–6099. https://doi.org/10.1158/0008-5472.CAN-06-4607.

Guo, Y. J., Pan, W. W., Liu, S. B., Shen, Z. F., Xu, Y., & Hu, L. L. (2020). ERK/MAPK signalling pathway and tumorigenesis. *Experimental and Therapeutic Medicine*, *19*(3), 1997–2007. https://doi.org/10.3892/etm.2020.8454.

Harimoto, N., Taguchi, K., Shirabe, K., Adachi, E., Sakaguchi, Y., Toh, Y., et al. (2010). The significance of fibroblast growth factor receptor 2 expression in differentiation of hepatocellular carcinoma. *Oncology*, *78*(5–6), 361–368. https://doi.org/10.1159/000320463.

Hatlen, M. A., Schmidt-Kittler, O., Sherwin, C. A., Rozsahegyi, E., Rubin, N., Sheets, M. P., et al. (2019). Acquired on-target clinical resistance validates FGFR4 as a driver of hepatocellular carcinoma. *Cancer Discovery*, *9*(12), 1686–1695. https://doi.org/10.1158/2159-8290.CD-19-0367.

Höpfner, M., Sutter, A. P., Huether, A., Schuppan, D., Zeitz, M., & Scherübl, H. (2004). Targeting the epidermal growth factor receptor by gefitinib for treatment of hepatocellular carcinoma. *Journal of Hepatology*, *41*(6), 1008–1016. https://doi.org/10.1016/j.jhep.2004.08.024.

Hsu, I. C., Metcalf, R. A., Sun, T., Welsh, J. A., Wang, N. J., & Harris, C. C. (1991). Mutational hotspot in the p53 gene in human hepatocellular carcinomas. *Nature*, *350*(6317), 427–428. https://doi.org/10.1038/350427a0.

Hu, W., Zheng, S., Guo, H., Dai, B., Ni, J., Shi, Y., et al. (2020). PLAGL2-EGFR-HIF-1/2alpha signaling loop promotes HCC progression and Erlotinib insensitivity. *Hepatology*. https://doi.org/10.1002/hep.31293. 2020/04/27.

Huynh, H., Nguyen, T. T. T., Chow, K.-H. K.-P., Tan, P. H., Soo, K. C., & Tran, E. (2003). Over-expression of the mitogen-activated protein kinase (MAPK) kinase (MEK)-MAPK in hepatocellular carcinoma: Its role in tumor progression and apoptosis. *BMC Gastroenterology*, *3*(1), 19. https://doi.org/10.1186/1471-230X-3-19.

Janku, F. (2017). Phosphoinositide 3-kinase (PI3K) pathway inhibitors in solid tumors: From laboratory to patients. *Cancer Treatment Reviews*, *59*, 93–101. https://doi.org/10.1016/j.ctrv.2017.07.005.

Jhappan, C., Stahle, C., Harkins, R. N., Fausto, N., Smith, G. H., & Merlino, G. T. (1990). TGF alpha overexpression in transgenic mice induces liver neoplasia and abnormal development of the mammary gland and pancreas. *Cell*, *61*(6), 1137–1146. https://doi.org/10.1016/0092-8674(90)90076-q.

Jiang, K., Liu, Y., Fan, J., Zhang, J., Li, X.-A., Evers, B. M., et al. (2016). PI(4)P promotes phosphorylation and conformational change of smoothened through interaction with its C-terminal tail. *PLoS Biology*, *14*(2), e1002375. https://doi.org/10.1371/journal.pbio.1002375.

Jonas, S., & Izaurralde, E. (2015). Towards a molecular understanding of microRNA-mediated gene silencing. *Nature Reviews. Genetics*, *16*(7), 421–433. https://doi.org/10.1038/nrg3965.

Jopling, C. L. (2012). Liver-specific microRNA-122: Biogenesis and function. *RNA Biology*, *9*(2), 137–142. https://doi.org/10.4161/rna.18827.

Kasprzak, A., Kwasniewski, W., Adamek, A., & Gozdzicka-Jozefiak, A. (2017). Insulin-like growth factor (IGF) axis in cancerogenesis. *Mutation Research-Reviews in Mutation Research*, *772*, 78–104. https://doi.org/10.1016/j.mrrev.2016.08.007.

Katz, L. H., Li, Y., Chen, J.-S., Muñoz, N. M., Majumdar, A., Chen, J., et al. (2013). Targeting TGF-β signaling in cancer. *Expert Opinion on Therapeutic Targets*, *17*(7), 743–760. https://doi.org/10.1517/14728222.2013.782287.

Kim, R. D., Sarker, D., Meyer, T., Yau, T., Macarulla, T., Park, J.-W., et al. (2019). First-in-human phase I study of fisogatinib (BLU-554) validates aberrant FGF19 signaling as a driver event in hepatocellular carcinoma. *Cancer Discovery*, *9*(12), 1696–1707. https://doi.org/10.1158/2159-8290.CD-19-0555.

Kitagawa, N., Ojima, H., Shirakihara, T., Shimizu, H., Kokubu, A., Urushidate, T., et al. (2013). Downregulation of the microRNA biogenesis components and its association with poor prognosis in hepatocellular carcinoma. *Cancer Science*, *104*(5), 543–551. https://doi.org/10.1111/cas.12126.

Klingenberg, M., Matsuda, A., Diederichs, S., & Patel, T. (2017). Non-coding RNA in hepatocellular carcinoma: Mechanisms, biomarkers and therapeutic targets. *Journal of Hepatology*, *67*(3), 603–618. https://doi.org/10.1016/j.jhep.2017.04.009.

Komposch, K., & Sibilia, M. (2015). EGFR signaling in liver diseases. *International Journal of Molecular Sciences*, *17*(1), 30. https://doi.org/10.3390/ijms17010030.

Kudo, M., Finn, R. S., Qin, S., Han, K.-H., Ikeda, K., Piscaglia, F., et al. (2018). Lenvatinib versus sorafenib in first-line treatment of patients with unresectable hepatocellular carcinoma: A randomised phase 3 non-inferiority trial. *Lancet (London, England)*, *391*(10126), 1163–1173. https://doi.org/10.1016/S0140-6736(18)30207-1.

Lachenmayer, A., Alsinet, C., Savic, R., Cabellos, L., Toffanin, S., Hoshida, Y., et al. (2012). Wnt-pathway activation in two molecular classes of hepatocellular carcinoma and experimental modulation by Sorafenib. *Clinical Cancer Research*, *18*(18), 4997–5007. https://doi.org/10.1158/1078-0432.CCR-11-2322.

Lachenmayer, A., Toffanin, S., Cabellos, L., Alsinet, C., Hoshida, Y., Villanueva, A., et al. (2012). Combination therapy for hepatocellular carcinoma: Additive preclinical efficacy of the HDAC inhibitor panobinostat with sorafenib. *Journal of Hepatology*, *56*(6), 1343–1350. https://doi.org/10.1016/j.jhep.2012.01.009.

Lanaya, H., Natarajan, A., Komposch, K., Li, L., Amberg, N., Chen, L., et al. (2014). EGFR has a tumour-promoting role in liver macrophages during hepatocellular carcinoma formation. *Nature Cell Biology*, *16*(10), 972–977. https://doi.org/10.1038/ncb3031.

Letouzé, E., Shinde, J., Renault, V., Couchy, G., Blanc, J.-F., Tubacher, E., et al. (2017). Mutational signatures reveal the dynamic interplay of risk factors and cellular processes during liver tumorigenesis. *Nature Communications*, *8*(1), 1315. https://doi.org/10.1038/s41467-017-01358-x.

Levy, D. E., & Darnell, J. E. (2002). STATs: Transcriptional control and biological impact. *Nature Reviews. Molecular Cell Biology*, *3*(9), 651–662. https://doi.org/10.1038/nrm909.

Li, N., Fu, H., Tie, Y., Hu, Z., Kong, W., Wu, Y., et al. (2009). miR-34a inhibits migration and invasion by down-regulation of c-met expression in human hepatocellular carcinoma cells. *Cancer Letters*, *275*(1), 44–53. https://doi.org/10.1016/j.canlet.2008.09.035.

Li, H., Wolfe, A., Septer, S., Edwards, G., Zhong, X., Bashar Abdulkarim, A., et al. (2012). Deregulation of hippo kinase signalling in human hepatic malignancies. *Liver International: Official Journal of the International Association for the Study of the Liver*, *32*(1), 38–47. https://doi.org/10.1111/j.1478-3231.2011.02646.x.

Li, S. P., Xu, H. X., Yu, Y., He, J. D., Wang, Z., Xu, Y. J., et al. (2016). LncRNA HULC enhances epithelial-mesenchymal transition to promote tumorigenesis and metastasis of hepatocellular carcinoma via the miR-200a-3p/ZEB1 signaling pathway. *Oncotarget*, 7(27), 42431–42446. https://doi.org/10.18632/oncotarget.9883.

Liang, L., Wong, C. M., Ying, Q., Fan, D. N. Y., Huang, S., Ding, J., et al. (2010). MicroRNA-125b suppressesed human liver cancer cell proliferation and metastasis by directly targeting oncogene LIN28B2. *Hepatology*, *52*(5), 1731–1740. https://doi.org/10.1002/hep.23904.

Lim, H. Y., Heo, J., Choi, H. J., Lin, C.-Y., Yoon, J.-H., Hsu, C., et al. (2014). A phase II study of the efficacy and safety of the combination therapy of the MEK inhibitor refametinib (BAY 86-9766) plus sorafenib for Asian patients with unresectable hepatocellular carcinoma. *Clinical Cancer Research: An Official Journal of the American Association for Cancer Research, 20*(23), 5976–5985. https://doi.org/10.1158/1078-0432.CCR-13-3445.

Lim, H. Y., Merle, P., Weiss, K. H., Yau, T., Ross, P., Mazzaferro, V., et al. (2018). Phase II studies with refametinib or refametinib plus sorafenib in patients with RAS-mutated hepatocellular carcinoma. *Clinical Cancer Research: An Official Journal of the American Association for Cancer Research, 24*(19), 4650–4661. https://doi.org/10.1158/1078-0432.CCR-17-3588.

Lim, S. O., Park, Y. M., Kim, H. S., Quan, X., Yoo, J. E., Park, Y. N., et al. (2011). Notch1 differentially regulates oncogenesis by wildtype p53 overexpression and p53 mutation in grade III hepatocellular carcinoma. *Hepatology, 53*(4), 1352–1362. https://doi.org/10.1002/hep.24208.

Lin, K. C., Park, H. W., & Guan, K.-L. (2017). Regulation of the hippo pathway transcription factor TEAD. *Trends in Biochemical Sciences, 42*(11), 862–872. https://doi.org/10.1016/j.tibs.2017.09.003.

Liu, C., Liu, L., Chen, X., Cheng, J., Zhang, H., Zhang, C., et al. (2018). LSD1 stimulates cancer-associated fibroblasts to drive Notch3-dependent self-renewal of liver cancer stem–like cells. *Cancer Research, 78*(4), 938–949. https://doi.org/10.1158/0008-5472.CAN-17-1236.

Liu, Q., Shi, H., Yang, J., & Jiang, N. (2019). Long non-coding RNA NEAT1 promoted hepatocellular carcinoma cell proliferation and reduced apoptosis through the regulation of let-7b-IGF-1R Axis. *Oncotargets and Therapy, 12*, 10401–10413. https://doi.org/10.2147/OTT.S217763.

Liu, A. M., Xu, M. Z., Chen, J., Poon, R. T., & Luk, J. M. (2010). Targeting YAP and hippo signaling pathway in liver cancer. *Expert Opinion on Therapeutic Targets, 14*(8), 855–868. https://doi.org/10.1517/14728222.2010.499361.

Llovet, J. M., Chen, Y., Wurmbach, E., Roayaie, S., Fiel, M. I., Schwartz, M., et al. (2006). A molecular signature to discriminate dysplastic nodules from early hepatocellular carcinoma in HCV cirrhosis. *Gastroenterology, 131*(6), 1758–1767. https://doi.org/10.1053/j.gastro.2006.09.014.

Llovet, J. M., Montal, R., & Villanueva, A. (2019). Randomized trials and endpoints in advanced HCC: Role of PFS as a surrogate of survival. *Journal of Hepatology, 70*(6), 1262–1277. https://doi.org/10.1016/j.jhep.2019.01.028.

Llovet, J. M., Peña, C. E. A., Lathia, C. D., Shan, M., Meinhardt, G., Bruix, J., et al. (2012). Plasma biomarkers as predictors of outcome in patients with advanced hepatocellular carcinoma. *Clinical Cancer Research: An Official Journal of the American Association for Cancer Research, 18*(8), 2290–2300. https://doi.org/10.1158/1078-0432.CCR-11-2175.

Llovet, J. M., Ricci, S., Mazzaferro, V., Hilgard, P., Gane, E., Blanc, J.-F., et al. (2008). Sorafenib in advanced hepatocellular carcinoma. *New England Journal of Medicine, 359*(4), 378–390. https://doi.org/10.1056/NEJMoa0708857.

Losic, B., Craig, A. J., Villacorta-Martin, C., Martins-Filho, S. N., Akers, N., Chen, X., et al. (2020). Intratumoral heterogeneity and clonal evolution in liver cancer. *Nature Communications, 11*(1), 291. https://doi.org/10.1038/s41467-019-14050-z.

Lu, J., Xia, Y., Chen, K., Zheng, Y., Wang, J., Lu, W., et al. (2016). Oncogenic role of the notch pathway in primary liver cancer. *Oncology Letters, 12*(1), 3–10. https://doi.org/10.3892/ol.2016.4609.

Ma, C., Han, M., Heinrich, B., Fu, Q., Zhang, Q., Sandhu, M., et al. (2018). Gut microbiome-mediated bile acid metabolism regulates liver cancer via NKT cells. *Science, 360*(6391), 858–859. https://doi.org/10.1126/science.aan5931.

Ma, G., Li, S., Han, Y., Li, S., Yue, T., Wang, B., et al. (2016). Regulation of smoothened trafficking and hedgehog signaling by the SUMO pathway. *Developmental Cell, 39*(4), 438–451. https://doi.org/10.1016/j.devcel.2016.09.014.

MacDonald, B. T., Tamai, K., & He, X. (2009). Wnt/β-catenin signaling: Components, mechanisms, and diseases. *Developmental Cell*, *17*(1), 9–26. https://doi.org/10.1016/j.devcel.2009.06.016.

Maciejowski, J., & de Lange, T. (2017). Telomeres in cancer: Tumour suppression and genome instability. *Nature Reviews. Molecular Cell Biology*, *18*(3), 175–186. https://doi.org/10.1038/nrm.2016.171.

Malakar, P., Stein, I., Saragovi, A., Winkler, R., Stern-Ginossar, N., Berger, M., et al. (2019). Long noncoding RNA MALAT1 regulates cancer glucose metabolism by enhancing mTOR-mediated translation of TCF7L2. *Cancer Research*, *79*(10), 2480–2493. https://doi.org/10.1158/0008-5472.CAN-18-1432.

Martínez-Chantar, M. L., Vázquez-Chantada, M., Ariz, U., Martínez, N., Varela, M., Luka, Z., et al. (2008). Loss of the glycine N-methyltransferase gene leads to steatosis and hepatocellular carcinoma in mice. *Hepatology*, *47*(4), 1191–1199. https://doi.org/10.1002/hep.22159.

Martinez-Quetglas, I., Pinyol, R., Dauch, D., Torrecilla, S., Tovar, V., Moeini, A., et al. (2016). IGF2 is up-regulated by epigenetic mechanisms in hepatocellular carcinomas and is an actionable oncogene product in experimental models. *Gastroenterology*, *151*(6), 1192–1205. https://doi.org/10.1053/j.gastro.2016.09.001.

Melincovici, C. S., Boşca, A. B., Şuşman, S., Mărginean, M., Mihu, C., Istrate, M., et al. (2018). Vascular endothelial growth factor (VEGF)—Key factor in normal and pathological angiogenesis. *Romanian Journal of Morphology and Embryology*, *59*(2), 455–467. Retrieved from https://www.ncbi.nlm.nih.gov/pubmed/30173249.

Meng, X., Franklin, D. A., Dong, J., & Zhang, Y. (2014). MDM2-p53 pathway in hepatocellular carcinoma. *Cancer Research*, *74*(24), 7161–7167. https://doi.org/10.1158/0008-5472.CAN-14-1446.

Meng, F., Henson, R., Wehbe-Janek, H., Ghoshal, K., Jacob, S. T., & Patel, T. (2007). MicroRNA-21 regulates expression of the PTEN tumor suppressor gene in human hepatocellular cancer. *Gastroenterology*, *133*(2), 647–658. https://doi.org/10.1053/j.gastro.2007.05.022.

Moeini, A., Cornellà, H., & Villanueva, A. (2012). Emerging signaling pathways in hepatocellular carcinoma. *Liver Cancer*, *1*(2), 83–93. https://doi.org/10.1159/000342405.

Muller, P. A., & Vousden, K. H. (2013). p53 mutations in cancer. *Nature Cell Biology*, *15*(1), 2–8. https://doi.org/10.1038/ncb2641.

Nault, J.-C., Couchy, G., Balabaud, C., Morcrette, G., Caruso, S., Blanc, J. F., et al. (2017). Molecular classification of hepatocellular adenoma associates with risk factors, bleeding, and malignant transformation. *Gastroenterology*, *152*(4), 880–894.e6. https://doi.org/10.1053/j.gastro.2016.11.042.

Nault, J.-C., Mallet, M., Pilati, C., Calderaro, J., Bioulac-Sage, P., Laurent, C., et al. (2013). High frequency of telomerase reverse-transcriptase promoter somatic mutations in hepatocellular carcinoma and preneoplastic lesions. *Nature Communications*, *4*, 2218. https://doi.org/10.1038/ncomms3218.

Nault, J.-C., Martin, Y., Caruso, S., Hirsch, T. Z., Bayard, Q., Calderaro, J., et al. (2020). Clinical impact of genomic diversity from early to advanced hepatocellular carcinoma. *Hepatology*, *71*(1), 164–182. https://doi.org/10.1002/hep.30811.

Nault, J.-C., Ningarhari, M., Rebouissou, S., & Zucman-Rossi, J. (2019). The role of telomeres and telomerase in cirrhosis and liver cancer. *Nature Reviews. Gastroenterology & Hepatology*, *16*(9), 544–558. https://doi.org/10.1038/s41575-019-0165-3.

Neufeld, G., Cohen, T., Gengrinovitch, S., & Poltorak, Z. (1999). Vascular endothelial growth factor (VEGF) and its receptors. *FASEB Journal: Official Publication of the Federation of American Societies for Experimental Biology*, *13*(1), 9–22. Retrieved from https://www.ncbi.nlm.nih.gov/pubmed/9872925.

Newell, P., Toffanin, S., Villanueva, A., Chiang, D. Y., Minguez, B., Cabellos, L., et al. (2009). Ras pathway activation in hepatocellular carcinoma and anti-tumoral effect of combined sorafenib and rapamycin in vivo. *Journal of Hepatology, 51*(4), 725–733. https://doi.org/10.1016/j.jhep.2009.03.028.

Niessen, C. M., & Gottardi, C. J. (2008). Molecular components of the adherens junction. *Biochimica et Biophysica Acta—Biomembranes, 1778*(3), 562–571. https://doi.org/10.1016/j.bbamem.2007.12.015.

Ohashi, K., Tsutsumi, M., Nakajima, Y., Kobitsu, K., Nakano, H., & Konishi, Y. (1996). Telomere changes in human hepatocellular carcinomas and hepatitis virus infected noncancerous livers. *Cancer,* 77(8 Suppl), 1747–1751. https://doi.org/10.1002/(SICI)1097-0142(19960415)77:8<1747::AID-CNCR50>3.0.CO;2-W.

Parikh, N., Hilsenbeck, S., Creighton, C. J., Dayaram, T., Shuck, R., Shinbrot, E., et al. (2014). Effects of TP53 mutational status on gene expression patterns across 10 human cancer types. *The Journal of Pathology, 232*(5), 522–533. https://doi.org/10.1002/path.4321.

Pascale, R. M., Feo, F., & Calvisi, D. F. (2016). An infernal cross-talk between oncogenic beta-catenin and c-Met in hepatocellular carcinoma: Evidence from mouse modeling. *Hepatology, 64*(5), 1421–1423. https://doi.org/10.1002/hep.28790.

Peng, N., He, J., Li, J., Huang, H., Huang, W., Liao, Y., et al. (2020). Long noncoding RNA MALAT1 inhibits the apoptosis and autophagy of hepatocellular carcinoma cell by targeting the microRNA-146a/PI3K/Akt/mTOR axis. *Cancer Cell International, 20,* 165. https://doi.org/10.1186/s12935-020-01231-w.

Perra, A., Kowalik, M. A., Ghiso, E., Ledda-Columbano, G. M., Di Tommaso, L., Angioni, M. M., et al. (2014). YAP activation is an early event and a potential therapeutic target in liver cancer development. *Journal of Hepatology, 61*(5), 1088–1096. https://doi.org/10.1016/j.jhep.2014.06.033.

Perugorria, M. J., Olaizola, P., Labiano, I., Esparza-Baquer, A., Marzioni, M., Marin, J. J. G., et al. (2019). Wnt–β-catenin signalling in liver development, health and disease. *Nature Reviews. Gastroenterology & Hepatology, 16*(2), 121–136. https://doi.org/10.1038/s41575-018-0075-9.

Phase I Study to Evaluate the Safety and Efficacy of Telomelysin (OBP-301) in Patients With Hepatocellular Carcinoma. (2014). Retrieved September 17, 2020, from https://clinicaltrials.gov/website:https://clinicaltrials.gov/ct2/show/NCT02293850.

Pineau, P., Volinia, S., McJunkin, K., Marchio, A., Battiston, C., Terris, B., et al. (2010). miR-221 overexpression contributes to liver tumorigenesis. *Proceedings of the National Academy of Sciences of the United States of America, 107*(1), 264–269. https://doi.org/10.1073/pnas.0907904107.

Pollak, M. (2008). Insulin and insulin-like growth factor signalling in neoplasia. *Nature Reviews Cancer, 8*(12), 915–928. https://doi.org/10.1038/nrc2536.

Qiao, Y., Wang, J., Karagoz, E., Liang, B., Song, X., Shang, R., et al. (2019). Axis inhibition protein 1 (Axin1) deletion–induced hepatocarcinogenesis requires intact β-catenin but not notch cascade in mice. *Hepatology, 70*(6), 2003–2017. https://doi.org/10.1002/hep.30556.

Quagliata, L., Matter, M. S., Piscuoglio, S., Arabi, L., Ruiz, C., Procino, A., et al. (2014). Long noncoding RNA HOTTIP/HOXA13 expression is associated with disease progression and predicts outcome in hepatocellular carcinoma patients. *Hepatology, 59*(3), 911–923. https://doi.org/10.1002/hep.26740.

Quan, H., Zhou, F., Nie, D., Chen, Q., Cai, X., Shan, X., et al. (2014). Hepatitis C virus core protein epigenetically silences SFRP1 and enhances HCC aggressiveness by inducing epithelial-mesenchymal transition. *Oncogene, 33*(22), 2826–2835. https://doi.org/10.1038/onc.2013.225.

Rao, S., & Mishra, L. (2019). Targeting transforming growth factor beta signaling in liver cancer. *Hepatology*, *69*(4), 1375–1378. https://doi.org/10.1002/hep.30426.

Rebouissou, S., Franconi, A., Calderaro, J., Letouzé, E., Imbeaud, S., Pilati, C., et al. (2016). Genotype-phenotype correlation of CTNNB1 mutations reveals different ß-catenin activity associated with liver tumor progression. *Hepatology*, *64*(6), 2047–2061. https://doi.org/10.1002/hep.28638.

Rebouissou, S., La Bella, T., Rekik, S., Imbeaud, S., Calatayud, A.-L., Rohr-Udilova, N., et al. (2017). Proliferation markers are associated with MET expression in hepatocellular carcinoma and predict tivantinib sensitivity *In Vitro*. *Clinical Cancer Research*, *23*(15), 4364–4375. https://doi.org/10.1158/1078-0432.CCR-16-3118.

Rimassa, L., Assenat, E., Peck-Radosavljevic, M., Pracht, M., Zagonel, V., Mathurin, P., et al. (2018). Tivantinib for second-line treatment of MET-high, advanced hepatocellular carcinoma (METIV-HCC): A final analysis of a phase 3, randomised, placebo-controlled study. *The Lancet. Oncology*, *19*(5), 682–693. https://doi.org/10.1016/S1470-2045(18)30146-3.

Rinn, J. L., Kertesz, M., Wang, J. K., Squazzo, S. L., Xu, X., Brugmann, S. A., et al. (2007). Functional demarcation of active and silent chromatin domains in human HOX loci by noncoding RNAs. *Cell*, *129*(7), 1311–1323. https://doi.org/10.1016/j.cell.2007.05.022.

Riordan, J. D., Feddersen, C. R., Tschida, B. R., Beckmann, P. J., Keng, V. W., Linden, M. A., et al. (2018). Chronic liver injury alters driver mutation profiles in hepatocellular carcinoma in mice. *Hepatology*, *67*(3), 924–939. https://doi.org/10.1002/hep.29565.

Ruiz i Altaba, A., Mas, C., & Stecca, B. (2007). The Gli code: An information nexus regulating cell fate, stemness and cancer. *Trends in Cell Biology*, *17*(9), 438–447. https://doi.org/10.1016/j.tcb.2007.06.007.

Salvi, A., Sabelli, C., Moncini, S., Venturin, M., Arici, B., Riva, P., et al. (2009). MicroRNA-23b mediates urokinase and c-met downmodulation and a decreased migration of human hepatocellular carcinoma cells. *The FEBS Journal*, *276*(11), 2966–2982. https://doi.org/10.1111/j.1742-4658.2009.07014.x.

Santoro, A., Rimassa, L., Borbath, I., Daniele, B., Salvagni, S., Van Laethem, J. L., et al. (2013). Tivantinib for second-line treatment of advanced hepatocellular carcinoma: A randomised, placebo-controlled phase 2 study. *The Lancet Oncology*, *14*(1), 55–63. https://doi.org/10.1016/S1470-2045(12)70490-4.

Sarcognato, S., García-Lezana, T., & Villanueva, A. (2019). Mechanisms of action of drugs effective in hepatocellular carcinoma. *Clinics in Liver Disease*, *14*(2), 62–65. https://doi.org/10.1002/cld.810.

Satoh, S., Daigo, Y., Furukawa, Y., Kato, T., Miwa, N., Nishiwaki, T., et al. (2000). AXIN1 mutations in hepatocellular carcinomas, and growth suppression in cancer cells by virus-mediated transfer of AXIN1. *Nature Genetics*, *24*(3), 245–250. https://doi.org/10.1038/73448.

Sawey, E. T., Chanrion, M., Cai, C., Wu, G., Zhang, J., Zender, L., et al. (2011). Identification of a therapeutic strategy targeting amplified FGF19 in liver cancer by oncogenomic screening. *Cancer Cell*, *19*(3), 347–358. https://doi.org/10.1016/j.ccr.2011.01.040.

Schiffer, E., Housset, C., Cacheux, W., Wendum, D., Desbois-Mouthon, C., Rey, C., et al. (2005). Gefitinib, an EGFR inhibitor, prevents hepatocellular carcinoma development in the rat liver with cirrhosis. *Hepatology*, *41*(2), 307–314. https://doi.org/10.1002/hep.20538.

Schulze, K., Imbeaud, S., Letouze, E., Alexandrov, L. B., Calderaro, J., Rebouissou, S., et al. (2015). Exome sequencing of hepatocellular carcinomas identifies new mutational signatures and potential therapeutic targets. *Nature Genetics*, *47*(5), 505–511. https://doi.org/10.1038/ng.3252.

Schulze, K., Nault, J.-C., & Villanueva, A. (2016). Genetic profiling of hepatocellular carcinoma using next-generation sequencing. *Journal of Hepatology, 65*(5), 1031–1042. https://doi.org/10.1016/j.jhep.2016.05.035.

Schwabe, R. F., & Jobin, C. (2013). The microbiome and cancer. *Nature Reviews. Cancer, 13*(11), 800–812. https://doi.org/10.1038/nrc3610.

Shi, J. H., Guo, W. Z., Jin, Y., Zhang, H. P., Pang, C., Li, J., et al. (2019). Recognition of HER2 expression in hepatocellular carcinoma and its significance in postoperative tumor recurrence. *Cancer Medicine, 8*(3), 1269–1278. https://doi.org/10.1002/cam4.2006.

Shigesawa, T., Maehara, O., Suda, G., Natsuizaka, M., Kimura, M., Shimazaki, T., et al. (2020). Lenvatinib suppresses cancer stem-like cells in HCC by inhibiting FGFR 1-3 signaling, but not FGFR4 signaling. *Carcinogenesis*. https://doi.org/10.1093/carcin/bgaa049. 2020/05/26.

Sia, D., Alsinet, C., Newell, P., & Villanueva, A. (2014). VEGF signaling in cancer treatment. *Current Pharmaceutical Design, 20*(17), 2834–2842. https://doi.org/10.2174/13816128113199990590.

Simerzin, A., Zorde-Khvalevsky, E., Rivkin, M., Adar, R., Zucman-Rossi, J., Couchy, G., et al. (2016). The liver-specific microRNA-122*, the complementary strand of microRNA-122, acts as a tumor suppressor by modulating the p53/mouse double minute 2 homolog circuitry. *Hepatology, 64*(5), 1623–1636. https://doi.org/10.1002/hep.28679.

Sonohara, F., Inokawa, Y., Hayashi, M., Yamada, S., Sugimoto, H., Fujii, T., et al. (2017). Prognostic value of long non-coding RNA HULC and MALAT1 following the curative resection of hepatocellular carcinoma. *Scientific Reports*, 7(1), 16142. https://doi.org/10.1038/s41598-017-16260-1.

Stamos, J. L., & Weis, W. I. (2013). The β-catenin destruction complex. *Cold Spring Harbor Perspectives in Biology, 5*(1), 1. https://doi.org/10.1101/cshperspect.a007898.

Strazzabosco, M., & Fabris, L. (2012). Notch signaling in hepatocellular carcinoma: Guilty in association! *Gastroenterology, 143*(6), 1430–1434. https://doi.org/10.1053/j.gastro.2012.10.025.

Su, H., Yang, J. R., Xu, T., Huang, J., Xu, L., Yuan, Y., et al. (2009). MicroRNA-101, down-regulated in hepatocellular carcinoma, promotes apoptosis and suppresses tumorigenicity. *Cancer Research, 69*(3), 1135–1142. https://doi.org/10.1158/0008-5472.CAN-08-2886.

Subbiah, I. M., Falchook, G. S., Kaseb, A. O., Hess, K. R., Tsimberidou, A. M., Fu, S., et al. (2015). Exploring response signals and targets in aggressive unresectable hepatocellular carcinoma: An analysis of targeted therapy phase 1 trials. *Oncotarget, 6*(29), 28453–28462. https://doi.org/10.18632/oncotarget.4601.

Sung, W.-K., Zheng, H., Li, S., Chen, R., Liu, X., Li, Y., et al. (2012). Genome-wide survey of recurrent HBV integration in hepatocellular carcinoma. *Nature Genetics, 44*(7), 765–769. https://doi.org/10.1038/ng.2295.

Tada, M., Kanai, F., Tanaka, Y., Tateishi, K., Ohta, M., Asaoka, Y., et al. (2008). Down-regulation of hedgehog-interacting protein through genetic and epigenetic alterations in human hepatocellular carcinoma. *Clinical Cancer Research: An Official Journal of the American Association for Cancer Research, 14*(12), 3768–3776. https://doi.org/10.1158/1078-0432.CCR-07-1181.

Tahtouh, R., Azzi, A.-S., Alaaeddine, N., Chamat, S., Bouharoun-Tayoun, H., Wardi, L., et al. (2015). Telomerase inhibition decreases alpha-fetoprotein expression and secretion by hepatocellular carcinoma cell lines: In vitro and in vivo study. *PLoS One, 10*(3), e0119512. https://doi.org/10.1371/journal.pone.0119512.

Takahashi, S., Tanaka, N., Fukami, T., Xie, C., Yagai, T., Kim, D., et al. (2018). Role of Farnesoid X receptor and bile acids in hepatic tumor development. *Hepatology Communications, 2*(12), 1567–1582. https://doi.org/10.1002/hep4.1263.

Tang, X., Feng, D., Li, M., Zhou, J., Li, X., Zhao, D., et al. (2019). Transcriptomic analysis of mRNA-lncRNA-miRNA interactions in hepatocellular carcinoma. *Scientific Reports*, *9*(1), 16096. https://doi.org/10.1038/s41598-019-52559-x.

Tao, J., Calvisi, D. F., Ranganathan, S., Cigliano, A., Zhou, L., Singh, S., et al. (2014). Activation of β-catenin and Yap1 in human hepatoblastoma and induction of hepatocarcinogenesis in mice. *Gastroenterology*, *147*(3), 690–701. https://doi.org/10.1053/j.gastro.2014.05.004.

Thorgeirsson, S. S., & Grisham, J. W. (2002). Molecular pathogenesis of human hepatocellular carcinoma. *Nature Genetics*, *31*(4), 339–346. https://doi.org/10.1038/ng0802-339.

Tietze, L., & Kessler, S. M. (2020). The good, the bad, the question-H19 in hepatocellular carcinoma. *Cancers*, *12*(5), 1261. https://doi.org/10.3390/cancers12051261.

Toffanin, S., Hoshida, Y., Lachenmayer, A., Villanueva, A., Cabellos, L., Minguez, B., et al. (2011). MicroRNA-based classification of hepatocellular carcinoma and oncogenic role of miR-517a. *Gastroenterology*, *140*(5), 1618–1628. e16 https://doi.org/10.1053/j.gastro.2011.02.009.

Toraih, E. A., Alghamdi, S. A., El-Wazir, A., Hosny, M. M., Hussein, M. H., Khashana, M. S., et al. (2018). Dual biomarkers long non-coding RNA GAS5 and microRNA-34a co-expression signature in common solid tumors. *PLoS One*, *13*(10), e0198231. https://doi.org/10.1371/journal.pone.0198231.

Totoki, Y., Tatsuno, K., Covington, K. R., Ueda, H., Creighton, C. J., Kato, M., et al. (2014). Trans-ancestry mutational landscape of hepatocellular carcinoma genomes. *Nature Genetics*, *46*(12), 1267–1273. https://doi.org/10.1038/ng.3126.

Tovar, V., Alsinet, C., Villanueva, A., Hoshida, Y., Chiang, D. Y., Solé, M., et al. (2010). IGF activation in a molecular subclass of hepatocellular carcinoma and pre-clinical efficacy of IGF-1R blockage. *Journal of Hepatology*, *52*(4), 550–559. https://doi.org/10.1016/j.jhep.2010.01.015.

Tovar, V., Cornella, H., Moeini, A., Vidal, S., Hoshida, Y., Sia, D., et al. (2017). Tumour initiating cells and IGF/FGF signalling contribute to sorafenib resistance in hepatocellular carcinoma. *Gut*, *66*(3), 530–540. https://doi.org/10.1136/gutjnl-2015-309501.

Tsao, C.-M., Yan, M.-D., Shih, Y.-L., Yu, P.-N., Kuo, C.-C., Lin, W.-C., et al. (2012). SOX1 functions as a tumor suppressor by antagonizing the WNT/β-catenin signaling pathway in hepatocellular carcinoma. *Hepatology*, *56*(6), 2277–2287. https://doi.org/10.1002/hep.25933.

Tu, J., Zhao, Z., Xu, M., Lu, X., Chang, L., & Ji, J. (2018). NEAT1 upregulates TGF-β1 to induce hepatocellular carcinoma progression by sponging hsa-mir-139-5p. *Journal of Cellular Physiology*, *233*(11), 8578–8587. https://doi.org/10.1002/jcp.26524.

Villanueva, A. (2019). Hepatocellular carcinoma. *The New England Journal of Medicine*, *380*(15), 1450–1462. https://doi.org/10.1056/NEJMra1713263.

Villanueva, A., Alsinet, C., Yanger, K., Hoshida, Y., Zong, Y., Toffanin, S., et al. (2012). Notch signaling is activated in human hepatocellular carcinoma and induces tumor formation in mice. *Gastroenterology*, *143*, 1660–1669.e7. https://doi.org/10.1053/j.gastro.2012.09.002.

Villanueva, A., Chiang, D. Y., Newell, P., Peix, J., Thung, S., Alsinet, C., et al. (2008). Pivotal role of mTOR signaling in hepatocellular carcinoma. *Gastroenterology*, *135*(6), 1972–1983. 1983.e1-11 https://doi.org/10.1053/j.gastro.2008.08.008.

Villanueva, A., Newell, P., Chiang, D. Y., Friedman, S. L., & Llovet, J. M. (2007). Genomics and signaling pathways in hepatocellular carcinoma. *Seminars in Liver Disease*, *27*(1), 55–76. https://doi.org/10.1055/s-2006-960171.

Villanueva, A., Portela, A., Sayols, S., Battiston, C., Hoshida, Y., Méndez-González, J., et al. (2015). DNA methylation-based prognosis and epidrivers in hepatocellular carcinoma. *Hepatology*, *61*(6), 1945–1956. https://doi.org/10.1002/hep.27732.

Vlad, A., Röhrs, S., Klein-Hitpass, L., & Müller, O. (2008). The first five years of the Wnt targetome. *Cellular Signalling*, *20*(5), 795–802. https://doi.org/10.1016/j.cellsig.2007.10.031.

Wang, P., Bai, Y., Song, B., Wang, Y., Liu, D., Lai, Y., et al. (2011). PP1A-mediated dephosphorylation positively regulates YAP2 activity. *PLoS One*, *6*(9), e24288. https://doi.org/10.1371/journal.pone.0024288.

Wang, Y., He, L., Du, Y., Zhu, P., Huang, G., Luo, J., et al. (2015). The long noncoding RNA lncTCF7 promotes self-renewal of human liver cancer stem cells through activation of Wnt signaling. *Cell Stem Cell*, *16*(4), 413–425. https://doi.org/10.1016/j.stem.2015.03.003.

Wang, S.-C., Lin, X. L., Li, J., Zhang, T. T., Wang, H. Y., Shi, J. W., et al. (2014). MicroRNA-122 triggers mesenchymal-epithelial transition and suppresses hepatocellular carcinoma cell motility and invasion by targeting RhoA. *PLoS One*, *9*(7), e101330. https://doi.org/10.1371/journal.pone.0101330.

Wang, Y., Lu, Y., Toh, S. T., Sung, W. K., Tan, P., Chow, P., et al. (2010). Lethal-7 is down-regulated by the hepatitis B virus x protein and targets signal transducer and activator of transcription 3. *Journal of Hepatology*, *53*(1), 57–66. https://doi.org/10.1016/j.jhep.2009.12.043.

Wang, S., Zhang, Q., Wang, Q., Shen, Q., Chen, X., Li, Z., et al. (2018). NEAT1 paraspeckle promotes human hepatocellular carcinoma progression by strengthening IL-6/STAT3 signaling. *Oncoimmunology*, 7(11), e1503913. https://doi.org/10.1080/2162402X.2018.1503913.

Weiler, S. M. E., Pinna, F., Wolf, T., Lutz, T., Geldiyev, A., Sticht, C., et al. (2017). Induction of chromosome instability by activation of yes-associated protein and forkhead box M1 in liver cancer. *Gastroenterology*, *152*(8), 2037–2051.e22. https://doi.org/10.1053/j.gastro.2017.02.018.

Wong, C. C.-L., Au, S. L. K., Tse, A. P. W., Xu, I. M. J., Lai, R. K. H., Chiu, D. K. C., et al. (2014). Switching of pyruvate kinase isoform 1 to M2 promotes metabolic reprogramming in hepatocarcinogenesis. *PLoS One*, *9*(12), e115036. https://doi.org/10.1371/journal.pone.0115036.

Wong, Q. W. L., Ching, A. K. K., Chan, A. W. H., Choy, K. W., To, K. F., Lai, P. B. S., et al. (2010). MiR-222 overexpression confers cell migratory advantages in hepatocellular carcinoma through enhancing AKT signaling. *Clinical Cancer Research: An Official Journal of the American Association for Cancer Research*, *16*(3), 867–875. https://doi.org/10.1158/1078-0432.CCR-09-1840.

Wong, C. M., Tsang, F. H., & Ng, I. O. L. (2018). Non-coding RNAs in hepatocellular carcinoma: Molecular functions and pathological implications. *Nature Reviews. Gastroenterology & Hepatology*, *15*(3), 137–151. https://doi.org/10.1038/nrgastro.2017.169.

Wu, E.-R., Chou, Y.-E., Liu, Y.-F., Hsueh, K.-C., Lee, H.-L., Yang, S.-F., et al. (2019). Association of lncRNA H19 gene polymorphisms with the occurrence of hepatocellular carcinoma. *Genes*, *10*(7), 506. https://doi.org/10.3390/genes10070506.

Wurmbach, E., Chen, Y. B., Khitrov, G., Zhang, W., Roayaie, S., Schwartz, M., et al. (2007). Genome-wide molecular profiles of HCV-induced dysplasia and hepatocellular carcinoma. *Hepatology*, *45*(4), 938–947. https://doi.org/10.1002/hep.21622.

Xu, M. Z., Chan, S. W., Liu, A. M., Wong, K. F., Fan, S. T., Chen, J., et al. (2011). AXL receptor kinase is a mediator of YAP-dependent oncogenic functions in hepatocellular carcinoma. *Oncogene*, *30*(10), 1229–1240. https://doi.org/10.1038/onc.2010.504.

Xu, W., Zhou, W., Cheng, M., Wang, J., Liu, Z., He, S., et al. (2017). Hypoxia activates Wnt/β-catenin signaling by regulating the expression of BCL9 in human hepatocellular carcinoma. *Scientific Reports*, 7, 40446. https://doi.org/10.1038/srep40446.

Xu, J., Zhu, X., Wu, L., Yang, R., Yang, Z., Wang, Q., et al. (2012). MicroRNA-122 suppresses cell proliferation and induces cell apoptosis in hepatocellular carcinoma by directly targeting Wnt/β-catenin pathway. *Liver International: Official Journal of the International Association for the Study of the Liver*, *32*(5), 752–760. https://doi.org/10.1111/j.1478-3231.2011.02750.x.

Xue, R., Li, R., Guo, H., Guo, L., Su, Z., Ni, X., et al. (2016). Variable intra-tumor genomic heterogeneity of multiple lesions in patients with hepatocellular carcinoma. *Gastroenterology*, *150*(4), 998–1008. https://doi.org/10.1053/j.gastro.2015.12.033.

Yamada, S., Takashina, Y., Watanabe, M., Nagamine, R., Saito, Y., Kamada, N., et al. (2018). Bile acid metabolism regulated by the gut microbiota promotes non-alcoholic steatohepatitis-associated hepatocellular carcinoma in mice. *Oncotarget*, *9*(11), 9925–9939. https://doi.org/10.18632/oncotarget.24066.

Yang, F., Huang, X., Yi, T., Yen, Y., Moore, D. D., & Huang, W. (2007). Spontaneous development of liver tumors in the absence of the bile acid receptor farnesoid X receptor. *Cancer Research*, *67*(3), 863–867. https://doi.org/10.1158/0008-5472.CAN-06-1078.

Ye, Y. C., Zhao, J. L., Lu, Y. T., Gao, C. C., Yang, Y., Liang, S. Q., et al. (2019). Notch signaling via Wnt regulates the proliferation of alternative, CCR2-independent tumor-associated macrophages in hepatocellular carcinoma. *Cancer Research*, *79*(16), 4160–4172. https://doi.org/10.1158/0008-5472.CAN-18-1691.

Yea, S., Narla, G., Zhao, X., Garg, R., Tal-Kremer, S., Hod, E., et al. (2008). Ras promotes growth by alternative splicing-mediated inactivation of the KLF6 tumor suppressor in hepatocellular carcinoma. *Gastroenterology*, *134*(5), 1521–1531. https://doi.org/10.1053/j.gastro.2008.02.015.

Yeh, M. M., Larson, A. M., Campbell, J. S., Fausto, N., Rulyak, S. J., & Swanson, P. E. (2007). The expression of transforming growth factor-alpha in cirrhosis, dysplastic nodules, and hepatocellular carcinoma: An immunohistochemical study of 70 cases. *The American Journal of Surgical Pathology*, *31*(5), 681–689. https://doi.org/10.1097/PAS.0b013e31802ff7aa.

Yoshimoto, S., Loo, T. M., Atarashi, K., Kanda, H., Sato, S., Oyadomari, S., et al. (2013). Obesity-induced gut microbial metabolite promotes liver cancer through senescence secretome. *Nature*, *499*(7456), 97–101. https://doi.org/10.1038/nature12347.

Yuan, P., Cao, W., Zang, Q., Li, G., Guo, X., & Fan, J. (2016). The HIF-2α-MALAT1-miR-216b axis regulates multi-drug resistance of hepatocellular carcinoma cells via modulating autophagy. *Biochemical and Biophysical Research Communications*, *478*(3), 1067–1073. https://doi.org/10.1016/j.bbrc.2016.08.065.

Zender, L., Spector, M. S., Xue, W., Flemming, P., Cordon-Cardo, C., Silke, J., et al. (2006). Identification and validation of oncogenes in liver cancer using an integrative oncogenomic approach. *Cell*, *125*(7), 1253–1267. https://doi.org/10.1016/j.cell.2006.05.030.

Zeng, X., Tamai, K., Doble, B., Li, S., Huang, H., Habas, R., et al. (2005). A dual-kinase mechanism for Wnt co-receptor phosphorylation and activation. *Nature*, *438*(7069), 873–877. https://doi.org/10.1038/nature04185.

Zhang, X., Kang, Z., Xie, X., Qiao, W., Zhang, L., Gong, Z., et al. (2020). Silencing of HIF-1α inhibited the expression of lncRNA NEAT1 to suppress development of hepatocellular carcinoma under hypoxia. *American Journal of Translational Research*, *12*(7), 3871–3883. Retrieved from https://www.ncbi.nlm.nih.gov/pubmed/32774741.

Zhang, S.-Z., Pan, F.-Y., Xu, J.-F., Yuan, J., Guo, S.-Y., Dai, G., et al. (2005). Knockdown of c-Met by adenovirus-delivered small interfering RNA inhibits hepatocellular carcinoma growth in vitro and in vivo. *Molecular Cancer Therapeutics*, *4*(10), 1577–1584. https://doi.org/10.1158/1535-7163.MCT-05-0106.

Zhang, J., Wang, W. L., Li, Q., & Qiao, Q. (2004). Expression of transforming growth factor-alpha and hepatitis B surface antigen in human hepatocellular carcinoma tissues and its significance. *World journal of gastroenterology: WJG, 10*(6), 830–833. https://doi.org/10.3748/wjg.v10.i6.830.

Zhao, B., Li, L., Lu, Q., Wang, L. H., Liu, C. Y., Lei, Q., et al. (2011). Angiomotin is a novel hippo pathway component that inhibits YAP oncoprotein. *Genes and Development, 25*(1), 51–63. https://doi.org/10.1101/gad.2000111.

Zhao, L.-H., Liu, X., Yan, H.-X., Li, W.-Y., Zeng, X., Yang, Y., et al. (2016). Genomic and oncogenic preference of HBV integration in hepatocellular carcinoma. *Nature Communications*, 7(1), 12992. https://doi.org/10.1038/ncomms12992.

Zhao, H., Lv, F., Liang, G., Huang, X., Wu, G., Zhang, W., et al. (2016). FGF19 promotes epithelial-mesenchymal transition in hepatocellular carcinoma cells by modulating the GSK3β/β-catenin signaling cascade via FGFR4 activation. *Oncotarget*, 7(12), 13575. Retrieved from https://www.ncbi.nlm.nih.gov/pmc/articles/pmc4924662.

Zheng, X., Zhang, Y., Liu, Y., Fang, L., Li, L., Sun, J., et al. (2018). HIF-2α activated lncRNA NEAT1 promotes hepatocellular carcinoma cell invasion and metastasis by affecting the epithelial-mesenchymal transition. *Journal of Cellular Biochemistry, 119*(4), 3247–3256. https://doi.org/10.1002/jcb.26481.

Zhong, X. P., Kan, A., Ling, Y. H., Lu, L. H., Mei, J., Wei, W., et al. (2019). NCKAP1 improves patient outcome and inhibits cell growth by enhancing Rb1/p53 activation in hepatocellular carcinoma. *Cell Death & Disease, 10*(5), 369. https://doi.org/10.1038/s41419-019-1603-4.

Zhu, A. X., Kang, Y.-K., Yen, C.-J., Finn, R. S., Galle, P. R., Llovet, J. M., et al. (2018). REACH-2: A randomized, double-blind, placebo-controlled phase 3 study of ramucirumab versus placebo as second-line treatment in patients with advanced hepatocellular carcinoma (HCC) and elevated baseline alpha-fetoprotein (AFP) following first-line sorafenib. *Journal of Clinical Oncology: Official Journal of the American Society of Clinical Oncology, 36*(15_Suppl), 4003. https://doi.org/10.1200/JCO.2018.36.15_suppl.4003.

Zhu, A. X., Kudo, M., Assenat, E., Cattan, S., Kang, Y.-K., Lim, H. Y., et al. (2014). Effect of everolimus on survival in advanced hepatocellular carcinoma after failure of sorafenib: The EVOLVE-1 randomized clinical trial. *JAMA, 312*(1), 57–67. https://doi.org/10.1001/jama.2014.7189.

Zhu, A. X., Rosmorduc, O., Jeffry Evans, T. R., Ross, P. J., Santoro, A., Carrilho, F. J., et al. (2015). SEARCH: A phase III, randomized, double-blind, placebo-controlled trial of Sorafenib plus erlotinib in patients with advanced hepatocellular carcinoma. *Journal of Clinical Oncology, 33*(6), 559–566. https://doi.org/10.1200/jco.2013.53.7746.

Zucman-Rossi, J., Villanueva, A., Nault, J. C., & Llovet, J. M. (2015). Genetic landscape and biomarkers of hepatocellular carcinoma. *Gastroenterology, 149*(5), 1226–1239.e4. https://doi.org/10.1053/j.gastro.2015.05.061.

CHAPTER THREE

Hepatitis C virus associated hepatocellular carcinoma

Mousumi Khatun[a], Ranjit Ray[b], and Ratna B. Ray[a,*]
[a]Department of Pathology, Saint Louis University, St. Louis, MO, United States
[b]Department of Internal Medicine, Saint Louis University, St. Louis, MO, United States
*Corresponding author: e-mail address: ratna.ray@health.slu.edu

Contents

1. Introduction 104
   1.1 HCV genome 105
2. HCV infection and immune responses 106
3. Risk factors for development of HCC from chronic infection 108
4. HCV regulation on metabolic pathways 109
5. Mechanisms of HCV associated HCC 111
   5.1 Viral factors involved in cellular proliferation and survival pathways 111
   5.2 HCV associated inflammation and fibrosis 112
   5.3 Oxidative stress 114
   5.4 Genomic instability and epigenetic alteration in HCV associated HCC 115
   5.5 Angiogenesis in HCV associated HCC 116
   5.6 Impact of non-coding RNAs in HCV associated HCC 117
   5.7 MicroRNAs (miRNAs) 117
   5.8 Long non-coding RNAs (LncRNAs) 120
6. Biomarkers for detection of HCV associated HCC 122
7. Therapeutic strategies for HCV associated HCC 123
8. Perspectives 125
Acknowledgments 127
References 127

Abstract

Liver cancer is a global problem and hepatocellular carcinoma (HCC) accounts for about 85% of this cancer. In the USA, etiologies and risk factors for HCC include chronic hepatitis C virus (HCV) infection, diabetes, non-alcoholic steatohepatitis (NASH), obesity, excessive alcohol drinking, exposure to tobacco smoke, and genetic factors. Chronic HCV infection appears to be associated with about 30% of HCC. Chronic HCV infection induces multistep changes in liver, involving metabolic disorders, steatosis, cirrhosis and HCC. Liver carcinogenesis requires initiation of neoplastic clones, and progression to clinically diagnose malignancy. Tumor progression associates with profound exhaustion of tumor-antigen-specific $CD8^{+}$T cells, and accumulation of PD-1^{hi} $CD8^{+}$T cells and Tregs. In this chapter, we provide a brief description of HCV and environmental/genetic

Advances in Cancer Research, Volume 149
ISSN 0065-230X
https://doi.org/10.1016/bs.acr.2020.10.003

factors, immune regulation, and highlight mechanisms of HCV associated HCC. We also underscore HCV treatment and recent paradigm of HCC progression, highlighted the current treatment and potential future therapeutic opportunities.

1. Introduction

Two major forms of primary liver cancer occur—cholangiocarcinoma (CCC) and hepatocellular carcinoma (HCC). Rates of HCC are 3–4 times higher than those of CCC. Here, we have focused on HCC as it is the second most frequent cause of cancer mortality worldwide. HCC is the second most common cause of cancer-related deaths in males, for both sexes it is the fourth (Bray et al., 2018). Disparities are observed in cancer incidence and survival among racial/ethnic minority populations (Saab, 2014). The incidence of liver cancer in the United States has tripled over the last 3 decades. The prognosis of HCC is poor, with a 5-year survival rate between 10% and 20%, with only modest improvement in the last 2 decades (Allemani et al., 2015). HCC is one of the very few major cancers in which the actual etiology can be identified in individual patients. Thus, in contrast to some cancers where risk factors (diet, smoking, genetic factors) increase the risk of a cancer, the etiology of underlying liver disease in most patients with HCC is readily identifiable. Although potentially excellent curative therapies are available for patients diagnosed at an early stage (typically <5 cm in tumor diameter), HCC is initially a silent disease, and often diagnosed at a late stage.

In the USA, etiologies for HCC include chronic infection with HCV, diabetes, NASH, obesity, excessive alcohol drinking, exposure to tobacco smoke, and genetic factors (Di Bisceglie et al., 2003). In contrast, HCC cases from Southeast Asia and sub-Saharan African countries arise from hepatitis B virus (HBV) infection and dietary exposure to aflatoxin (Boffetta & Islami, 2014). In a global picture, it is estimated that 71 million people are chronically infected with HCV and an estimated 580,000 people died from HCV in 2017 (Lazarus, Roel, & Elsharkawy, 2020). In addition, men are about three times more likely to develop liver cancer and more than twice as likely as women to die, so there are marked gender disparities.

Chronic HCV infection appears to be associated with about 30% of HCCs worldwide, but in some regions including Europe and the United States it is the leading underlying cause of HCC (de Martel, Maucort-Boulch, Plummer, & Franceschi, 2015). Measures to prevent HCV-related HCC include prevention of virus infection in the first place (primary prevention), control or elimination of HCV infection (secondary prevention) and treatment of HCV after HCC has already been developed and treated with the hope of preventing new tumors from developing (tertiary prevention). Public health efforts to reduce the incidence of HCV infection have been remarkably successful. Post-transfusion HCV is now an extremely rare event in the developed Western world, whereas transfusions of blood and blood products had been a major source of infection as recently as 3 decades ago (Hossain et al., 2017). Unfortunately, there may be a recent spike in incidence of acute hepatitis C related to the ongoing epidemic of heroin use in the United States, but incidence rates are still way below those seen in previous decades (Midgard et al., 2016). The evidence is now substantial that effective treatment of hepatitis C can at least partially reduce the risk of HCC, although treated patients with cirrhosis remain at risk of HCC (Conti, Heckman, & Pinto, 2016; Mehta & Yao, 2017; Reig et al., 2016).

1.1 HCV genome

HCV is a positive strand RNA virus and encodes a polyprotein of ~3000 amino acids which is cleaved by cellular and viral proteases (Fig. 1).

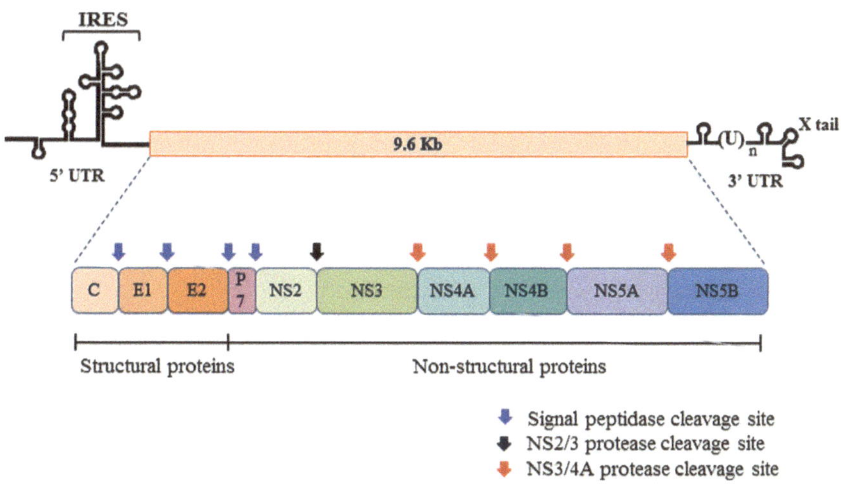

Fig. 1 Schematic structure of HCV genome and proteins. The positive-strand HCV RNA of ~9.6 kb genome is shown with the 5′ untranslated region (UTR) contains an internal ribosome entry site (IRES). The 3′ UTR contains a variable poly(U/UC) tract, followed by a conserved ~100 nucleotides that constitute the 3′ X-tail sequence. The HCV genome encodes a single polyprotein, which is processed to 10 viral proteins by cellular proteases (shown by blue, black and red arrows) and viral protease (shown by blue or green arrows).

There are three structural proteins (Core, E1 and E2) which are cleaved by cellular proteases, and the seven nonstructural proteins (p7, NS2, NS3, NS4A, NS4B, NS5A and NS5B, cleaved by virus encoded proteases (Lindenbach & Rice, 2013; Paul, Madan, & Bartenschlager, 2014). HCV Core protein encapsulates the positive strand viral genome. HCV envelope glycoproteins, E1 and E2, bind to the host proteins, designated as binding factors, co-receptors or receptors, and promote virus entry into cells. P7 is a viroporin and involved in HCV assembly. The nonstructural proteins, NS3 (protease and helicase), NS4A (protease cofactor of NS3), NS4B (integral membrane protein), NS5A (phosphoprotein) and NS5B (RNA-dependent RNA polymerase) are required for RNA replication and assembly.

2. HCV infection and immune responses

Several studies of acute HCV infection demonstrate virus clearance is associated with broad and potent T-cell activation (Neumann-Haefelin & Thimme, 2013; Rehermann, 2013) and a rapid induction of cross-reactive neutralizing antibody responses (Keck et al., 2019; Osburn et al., 2010, 2014; Pestka et al., 2007). Individuals who spontaneously clear HCV display broad $CD4^{+}$ T-cell responses, stronger T-cell proliferation, and IL-2, IFN-γ, and TNF-α production than individuals who develop chronic infection dominated by Tregs and IL-10 production (Diepolder et al., 1995; Kared, Fabre, Bedard, Bruneau, & Shoukry, 2013; Keoshkerian et al., 2016; Missale et al., 1996; Smyk-Pearson et al., 2008; Urbani et al., 2006). Strong HCV-specific $CD8^{+}$ T-cell responses are generated and HCV-specific memory T cells persist after recovery from infection (Lauer, 2013; Lechner et al., 2000; Shata et al., 2002; Takaki et al., 2000).

HCC development closely associates with inflammation and the complement system is an important component of the inflammatory response, and involved in various stages of cancer progression. Complement activation also regulates the adaptive immune response, and has a role in regulating T-cell response to cancer growth. A prominent role for complement has been identified in the linkage between innate and adaptive immunity (Patra, Ray, & Ray, 2019; Stoermer & Morrison, 2011). Studies using HCV infected hepatocytes as a surrogate model and specimens from chronically HCV infected patients displayed modulation of complement protein synthesis, and a significant up-regulation of complement regulatory protein on virus infected cell surface (Fig. 2). Complement effectors, such as C5a and

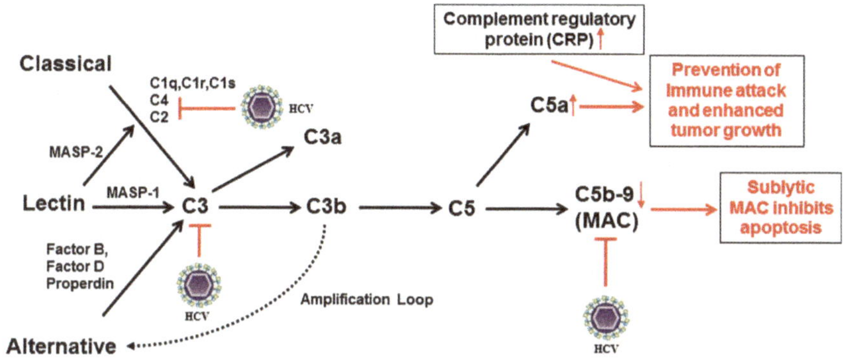

Fig. 2 Multiple complement component inhibition by HCV for prevention of immune attack and enhanced cell growth.

its receptor C5a-R1, associates with anti-tumor T-cell responses. Hepatocytes are the primary sites for complement components synthesis and we have shown that HCV impairs C3 and C4 synthesis (Banerjee et al., 2011; Kim, Meyer, Di Bisceglie, & Ray, 2013; Mazumdar et al., 2012), and enhances C5a and Factor H. C9 is significantly reduced ($\geq$20-fold) at the mRNA level in chronically HCV infected liver biopsy specimens. Clearance of HCV by treatment may help in recovering some of the complement components, but the underlying disease associated with HCV infection often continues (El-Fatah Fahmy Hanno, Mohiedeen, Deghedy, & Sayed, 2014). HCV infection may generate an overall state of immune suppression (Moorman et al., 2011; Shi et al., 2014). Complement elements (C1q, C3 derived fragments, and C5a) are recognized as major modulators of tumor progression, and gaining these complement functions may dampen cancer immunity (El-Shamy, Branch, Schiano, & Gorevic, 2018; Pio, Ajona, Ortiz-Espinosa, Mantovani, & Lambris, 2019). We have identified that DAF/CD55, a complement regulatory protein, and its isoform sDAF are generated by alternate splicing, and secreted into HCV infected cell culture medium, which inhibits C3 convertase activity (Kwon, Kim, Meyer, Di Bisceglie, & Ray, 2016; Mazumdar et al., 2013).

HCC requires initiation of neoplastic clones, and progression to clinically diagnosed malignancy (Goossens & Hoshida, 2015). Tumor progression associates with profound exhaustion of tumor-antigen-specific $CD8^{+}$T cells, and accumulation of PD-1^{hi} $CD8^{+}$T cells and Tregs (Greten, Wang, & Korangy, 2015). Indoleamine 2,3-dioxygenase 1 (IDO1) protein expression is observed in a wide variety of human cancers.

IDO1 expression may provide tumors with a mechanism of immune escape (Munn & Mellor, 2016; Prendergast et al., 2014). IDO1 expression can result in a shift in local tryptophan and kynurenine levels, and may directly suppress cytotoxic lymphocytes (CTL), natural killer (NK) cells, and activate effector pathways that may further suppress these cells through generation, recruitment, and activation of immune-suppressing regulatory T cells (Treg), and recruitment/activation of immune-suppressing MDSC (Holmgaard et al., 2015; Munn & Mellor, 2016; Platten, von Knebel Doeberitz, Oezen, Wick, & Ochs, 2014; Prendergast et al., 2014). HCV modulates complement components and regulatory proteins, and limits immune attack on neoplastic hepatocytes (Kwon et al., 2016). C5a and IDO1 play important roles in neoplastic cell growth and immune evasion (El-Shamy et al., 2018; Pio et al., 2019), suggesting that the strategies adopted by HCV in modulating complement related proteins/pathways, and IDO1 may have a significant role on neoplastic hepatocyte survival and growth advantage. Immunotherapies for HCC are in early phase and this research area is waiting for further development (Hou, Zhang, Sun, & Karin, 2020).

3. Risk factors for development of HCC from chronic infection

In chronic HCV-infection, factors related to the host, environment, or both are important than viral factors in determining the progression to HCC. HCV has seven genotypes and several sub-types. Certain HCV genotypes are associated with higher risk of HCC. An 80% higher risk of developing HCC was observed in patients infected with HCV genotype 1 and genotype 3 (Kanwal, Kramer, Ilyas, Duan, & El-Serag, 2014). HCV patients infected with genotype 3 are at higher risk for end-stage liver disease (Kanwal et al., 2014; McMahon et al., 2017). HCV genotype 1 when compared with other genotypes, may have a particular oncogenic mechanism, leading to HCC development even in non-cirrhotic patients (Mattos, Marcon Pdos, Araujo, Coral, & Tovo, 2015). The risk of HCC remains high even after eradication of HCV genotype 3 (Baumert, Juhling, Ono, & Hoshida, 2017; El-Serag, Kanwal, Richardson, & Kramer, 2016; Janjua et al., 2017). HCV genotype 6 is prevalent in Southeast Asia, and has also been shown to confer increased risk for development of HCC (Lee et al., 2017).

Diabetes is associated with increased levels of insulin-like factors that are potential carcinogenic (Moore, Park, & Tsuda, 1998). Obesity, which frequently accompanies diabetes, increase the risk for hepatic steatosis and fibrosis in HCV-infected patients. HCV infection has high comorbidity with diabetes mellitus and confers a two- to threefold increase in HCC risk Huang et al., 2017; Raff et al., 2015).

Alcohol abuse, smoking and coffee ingestion are associated with accelerated progression to HCC in chronic HCV infection. HCV infection and alcohol operate together to increase the risk for HCC by more actively promoting cirrhosis through the process of fibrosis (Chiba et al., 1996; Donato et al., 1997, 2002; Miyakawa, Izumi, Marumo, & Sato, 1996; Simonetti et al., 1992). A significant increase in relative risk (about three times) of HCC among the smokers with HCV infections compared to the non-smokers with HCV was reported (Chuang et al., 2010).

Co-infection with HBV and human immunodeficiency virus (HIV) significantly upsurges the danger of HCC in chronic HCV infected individuals. Chronic HCV infected individuals with active HBV infection have twice the risk of HCC and a 21% increase in mortality compared to those with latent HBV and HCV (Kruse et al., 2014). On the other hand, there is a substantial increase in HCC prevalence in HIV-HCV co-infected patients compared with only HCV infected patients, with HCC occurring in the younger population (Kramer, Kowalkowski, Duan, & Chiao, 2015). This increase is attributed to decreased immune response in HIV infection leading to accelerated HCV replication and a quick development towards HCC (Ioannou et al., 2013).

4. HCV regulation on metabolic pathways

Chronic HCV infection contributes to perturbation of host metabolic pathways resulting in HCC progression. HCV influences mostly glucose and lipid metabolisms among the different host metabolic pathways. The deregulated glucose metabolism leads to both insulin resistance and diabetes; whereas HCV induced steatosis is ultimate effects of enhanced lipogenesis, impaired degradation and export.

HCV replication suppressed by IFN treatment alters the gluconeogenesis pathways (Deng et al., 2011). The association between HCV infection and glucose abnormalities holds true with the occurrence of prediabetic conditions, such as insulin resistance (Hui et al., 2003; Moucari et al., 2008). Further, HCV promotes hepatic gluconeogenesis *via* an NS5A-mediated, forkhead box O1 (FoxO1)-dependent pathway, resulting in an increased cellular glucose production in hepatocytes (Banerjee, Meyer, Mazumdar, Ray, & Ray, 2010; Deng et al., 2011). HCV NS5A is directly linked to cellular hexokinase 2, the first rate-limiting enzyme of glycolysis (Ramiere et al., 2014). Insulin-mediated PI3K pathway and mammalian target of rapamycin (mTOR) signaling activate p70 ribosomal S6 kinase, which synthesizes proteins and modulates the mammalian translation machinery (Raught, Gingras, & Sonenberg, 2001).We and others have demonstrated that HCV modulates the protein kinase B/mammalian target of rapamycin/S6 kinase 1 (Akt/mTOR/S6K1) signaling cascade by increasing tumor necrosis factor production, while enhancing the activity of suppressor of cytokine signaling 3 (SOCS-3) in inhibiting insulin receptor substrate (IRS) function to perturb glucose metabolism *via* the down-regulation of GLUT4 and the upregulation of PEPCK for insulin resistance (Bose, Shrivastava, Meyer, Ray, & Ray, 2012; Pazienza, Vinciguerra, Andriulli, & Mangia, 2010; Persico et al., 2009; Sheikh, Choi, Qadri, Friedman, & Sanyal, 2008). Glucose is transported into hepatocytes by glucose transporter 2 (GLUT2). HCV replication suppresses cellular glucose uptake through down-regulation of cell surface expression of GLUT2 (Kasai et al., 2009). HCV core protein induces proteasomal degradation of IRS-1 and IRS-2, blocking intracellular insulin signaling (Bose et al., 2012; Sheikh et al., 2008). HCV genotype-specific impairment of insulin signaling was observed. Expression of HCV genotype 3 core protein led to PPARγ down-regulation and SOCS-7 up-regulation whereas HCV genotype 1 core protein activates mTOR signaling (Bugianesi, Salamone, & Negro, 2012). Increased serine phosphorylation of IRS-1 was also observed in the presence of HCV NS5A for contributing to insulin resistance (Parvaiz et al., 2014).

HCV infection influences lipid accumulation in hepatocytes by three ways, causing increased lipid synthesis, reduced lipogenolysis and impaired lipid transport (Syed, Amako, & Siddiqui, 2010). Fatty acid synthase (FASN), an enzyme primarily involved in the *de novo* synthesis of fatty acids, is up-regulated during HCV infection, and the inhibition of FASN activity inhibits virus replication and release (Su et al., 2002; Yang et al., 2008). HCV infection decreases the expression and activity of PPAR-α/γ in hepatocytes (Dharancy, Lemoine, Mathurin, Serfaty, & Dubuquoy, 2009). HCV core protein directly interacts with apolipoprotein AII to facilitate steatosis mechanism (Barba et al., 1997). PPARα, one of the important transcription regulator for lipid metabolism, is significantly reduced in the liver of patients infected with HCV genotype 3 as compared to genotype 1 (de Gottardi et al., 2006), and down-regulates PPARα mRNA (Dharancy et al., 2005). Our study also demonstrated that HCV genotype 2a- or 3a-induces lipogenic pathway, exhibiting higher triglyceride synthesis and a decreased lipolytic mechanism by PPARα inductive mechanism (Patra, Sasaki, Meyer, Ray, & Ray, 2019). TGF-β acts as a regulatory molecule for lipogenic pathways among HCV genotype-specific infections (Patra, Sasaki, et al., 2019). HCV core protein suppresses microsomal triglyceride transfer protein (MTTP) activity and interferes with hepatic assembly and secretion of triglyceride-rich very low density lipoproteins (VLDL), further contributing to steatosis (Perlemuter et al., 2002). HCV core protein co-localizes with apolipoprotein A2 on the surface of triglyceride containing lipid droplets *in vitro* and *in vivo*, suggesting its association with lipid metabolism (Abid et al., 2005). Transgenic mice expressing HCV core protein develop progressive steatosis and HCC in the liver (Barba et al., 1997; Moriya et al., 1997; Serfaty et al., 2001). Insulin resistance is another feature of the HCV core transgenic mice, which results in lipid accumulation in the liver (Perlemuter et al., 2002). Hepatic steatosis is most common in patients infected with HCV genotype 3 possibly due to the direct effects of HCV core protein (Abid et al., 2005). Involvement of other viral proteins also influences HCV-induced steatosis. An interaction between HCV NS5A and apolipoprotein occurs *in vitro*, and core protein/NS5A co-localization is observed in the cytoplasmic lipid droplets after transfection. The expression of HCV NS5A in human hepatoma cells increased lipid droplet formation through enhanced lipogenesis and the transcriptional expression of PPARγ (Castera et al., 2005; Parvaiz et al., 2014; Shi et al., 2002). Transgenic mice expressing HCV polyprotein develop hepatic steatosis with an excess iron diet, and exhibit ultrastructural alterations of mitochondria and HCC, accompanied with elevated hepatic 8-hydroxy-20-deoxyguanosine levels (Furutani et al., 2006). Our recent unpublished observation suggested that long non-coding RNA Linc-Pint inhibition by HCV enhances lipogenesis. Thus, HCV infection modulates the metabolic activities in hepatocytes.

5. Mechanisms of HCV associated HCC

5.1 Viral factors involved in cellular proliferation and survival pathways

HCV genome in concert of host factors help in replication to generate structural and non-structural proteins for assembly and egress. Interestingly, some of these viral proteins modulate host cellular signaling and metabolic pathways (Kwon, Ray, & Ray, 2014). Most studied HCV proteins are Core and NS5A. HCV Core protein plays a crucial role in the modulation of several major cell signaling pathways, including transforming growth factor β (TGF-β), nuclear factor κB (NF-κB), tumor necrosis factor α (TNF-α), cyclooxygenase-2 (COX-2), Wnt/β-catenin (WNT), vascular endothelial growth factor (VEGF), peroxisome proliferator activated receptor α (PPARα), involved in promotion of hepatocyte growth (Banerjee, Ray, & Ray, 2010; Mahmoudvand, Shokri, Taherkhani, & Farshadpour, 2019; Meyer, Basu, Saito, Ray, & Ray, 2005; Ray & Ray, 2019). Core protein acts as a positive regulator in androgen receptor signaling and enhances VEGF expression in hepatocytes providing further insight into oncogenic potential in the development of HCC in HCV infected individuals (Kanda, Steele, Ray, & Ray, 2008; Kwon et al., 2014).

Similarly, HCV NS5A protein acts as transcription factor activator and interacts with various signaling pathways including cell cycle/apoptosis, lipid metabolism. NS5A inhibits TGF-β signaling by preventing nuclear translocation of SMAD proteins resulting in downregulation of tumor-suppressor p21 expression in hepatoma cell lines (Choi & Hwang, 2006; Han, Niu, Wang, & Li, 2016; Jiang et al., 2011; Lin, King, & Chung, 2015; Majumder, Ghosh, Steele, Ray, & Ray, 2001). HCV Core, NS2, NS5A and NS5B interact with tumor suppressor p53 and Rb to dysregulate their function (Bittar, Shrivastava, Bhanja Chowdhury, Rahal, & Ray, 2013; Han et al., 2016; Irshad, Gupta, & Irshad, 2017; Jiang et al., 2011; Kwon et al., 2014; Lin et al., 2015; Majumder et al., 2001; McGivern & Lemon, 2009).

5.2 HCV associated inflammation and fibrosis

Chronic HCV infection is associated with increased expression of different pro-inflammatory cytokines and chemokines from hepatocytes, liver residential macrophages (Kupffer cells) and different immune cells (macrophages, natural killer cells and dendritic cells) recruited to the liver (Capone et al., 2014; Shrivastava, Mukherjee, Ray, & Ray, 2013). HCV exposed macrophages, including Kupffer cells, induce inflammasome formation. Inflammasome is a multiprotein cytoplasmic complex that sense viral pathogen associated molecular patterns (PAMPs) through Nod-like receptors (NLRs) especially with the NLR Family Pyrin Domain Containing 3 (NLRP3) (Negash et al., 2013; Shrivastava, Mukherjee, et al., 2013) and up-regulates IL-1β and IL-18, two major cytokines produced by inflammasome pathways (van de Veerdonk et al., 2011). Inflammatory response in general is beneficial for the host as it is a protective mechanism against viral infection. However, inefficient immune response against HCV contributes to the establishment and maintenance of persistent infection. Immune mediated inflammation caused during chronic HCV infection indirectly triggers hepatocarcinogenesis, possibly by the activation of several pathways involved in hepatocyte metabolism, reactive oxygen species (ROS) production, DNA repair and apoptosis (Kanda, Goto, Hirotsu, Moriyama, & Omata, 2019). On the other hand, cytokine production reflects the degree of inflammation in the liver of patients with chronic hepatitis C, which may possibly be related to HCC development (Kanda et al., 2019). HCV proteins also disrupt innate immune pathways. The NS3 protein suppresses innate immunity by cleavage of the mitochondrial antiviral signaling protein (MAVS) responsible for induction of type-I interferon (Li et al., 2005). The E2 protein may contribute to immune evasion and establishment of chronic infection by inhibition of natural killer cells (Tseng & Klimpel, 2002).

A complex interplay between various cell types in the liver, including hepatic stellate cells (HSCs), Kupffer cells, hepatocytes, cholangiocytes, sinusoidal endothelial cells, and infiltrating immune cells in response to chronic liver injury results in increased production and deposition of extracellular matrix (ECM). HCV infection causes a chronic inflammatory state, induces collagen production and accumulation, which eventually lead to an alteration of the structure of the hepatic parenchyma resulting in liver fibrosis (Khatun & Ray, 2019). Severity of liver fibrosis is tightly correlated with an increased risk of HCC in patients with chronic HCV infection, suggesting that fibrosis/cirrhosis associate with carcinogenesis. The increase in hepatocyte turnover due to the continuous processes of cell death and regeneration, as well as initiation of pre-neoplastic clones accompanied by irreversible somatic genetic/epigenetic changes, lead to a high probability of genetic alterations, whose accumulation leads to the formation and proliferation of cells favoring HCC development (Koike, 2007). Furthermore, hepatocytes death can amplify the fibrogenic signal, thus stimulating the activation of stellate cells and causing the progression of hepatic fibrosis (Gieseler et al., 2011). Transforming growth factor beta (TGF-β), a potent cytokine involved in HSCs activation and progression of liver fibrosis, also known to possess oncogenic characteristics. HCV infection is also able to modify the intracellular signaling pathways of TGF-β signaling, thus accelerating the progression of liver injury and increasing the risk of HCC (Matsuzaki et al., 2007; Vescovo, Refolo, Vitagliano, Fimia, & Piacentini, 2016). At the same time, HSCs activation may increase by cumulative effect of inflammation, ROS, steatosis and insulin resistance caused during chronic HCV infection. The activated HSCs, under the effect of fibrogenic cytokines undergo epithelial to mesenchymal trans-differentiation (EMT) into myofibroblast like cells (Bartosch, Thimme, Blum, & Zoulim, 2009; Irshad et al., 2017). The process of fibrogenesis is also regulated by several signaling pathways, altered in chronic HCV infection, including SMADs, phosphatidylinositide 3-kinases (PI3K), protein kinase (Akt), mitogen-activated protein kinase (MAPK) and c-Jun N-terminal kinases (JNKs) (Goossens & Hoshida, 2015). JNK activation by IL1-β cytokine increases fibrogenesis, oncogenesis and cell motility (Matsuzaki et al., 2007). The risk of HCC gradually increases as liver fibrosis progresses in patients with chronic HCV infection. Once cirrhosis is established, the annual incidence of HCC is detected higher (1–7% per year), although HCC sometimes develops in less fibrotic livers. HCV eradication does not eliminate the risk of HCC, especially when the patients already have advanced liver fibrosis, suggesting the necessity of anti-fibrotic therapies to prevent HCC (Hoshida, Fuchs, Bardeesy, Baumert, & Chung, 2014; Meringer, Shibolet, & Deutsch, 2019). Thus, cumulative effects of all these liver alterations produce a suitable environment for HCC development in chronic HCV infection.

5.3 Oxidative stress

Oxidative damage is considered a key pathway in HCC progression (Fitian & Cabrera, 2017). Oxidative stress closely correlates with HCV- and NASH-related HCC, but relatively weak with HBV-related HCC (Takaki & Yamamoto, 2015). During HCV replication, HCV core protein localizes to the endoplasmic reticulum (ER) (Duvet et al., 1998), facilitates release of calcium from the ER and uptake of calcium into the outer mitochondrial membrane and modulates mitochondrial membrane permeability. Transition of mitochondrial membrane potential stimulates the electron transport, which elevates ROS production in hepatocytes (Korenaga et al., 2005; Lim, Myers, Hamilton, Davis, & Lau, 1995). The production of oxidative stress is increased in HCV core protein transgenic mouse model causing inflammation in mouse liver (Lim et al., 1995; Machida et al., 2006). HCV NS3 and NS5A act as key mediators in the induction of oxidative stress. The association of NS5A with the ER may stimulate ROS production by generating ER stress (Gong, Waris, Tanveer, & Siddiqui, 2001).

In addition, NS3 activates nitric oxide (NO) synthase, generates NO and dislodges cytochrome C stimulating production of oxidative stress in the liver (Adachi et al., 2005; Bataller et al., 2003). Glutathione is a non-enzymatic tripeptide that plays a central role in the cellular antioxidant defense system, and represents the most abundant antioxidant in hepatocytes. The reduced form of glutathione (GSH) is considered as an important molecule for balancing redox state of the cells (Shimomura et al., 2017). HCV replication on the cytosolic side of the ER membrane causes accumulation of the chaperones, reducing GSH level (Tardif, Mori, & Siddiqui, 2002). Depletion of GSH increases ROS level leading to liver damage (Li et al., 2015). The overproduction of oxidative stress results in DNA damage contributing to hepatocarcinogenesis in HCV infection.

5.4 Genomic instability and epigenetic alteration in HCV associated HCC

Genomic instability is a common feature associated with HCC. Different mechanisms are believed to contribute to this instability, including telomere erosion, chromosome segregation defects, and alterations in the DNA damage-response pathways. HCV proteins interact with many host cell factors and affect a wide range of cellular and immune activities, including cell signaling, transcriptional modulation, transformation, apoptosis, membrane rearrangements, vesicular trafficking and translational regulation, cytokine and chemokine production. HCV can induce genomic instability in at least three ways: (i) by directly interfering with mitotic regulator proteins, (ii) by integrating viral genome into various sites in the host genome, and (iii) by causing inflammation in the liver. HCV NS5A protein induces mitotic dysregulation by downregulating abnormal spindle-like, microcephaly-associated protein, a regulator of mitotic spindle, and thus causes chromosomal instability (Rao, Asch, & Yamada, 2017; Wu, Chang, Wu, Liao, & Chang, 2008). Structural alterations of host genomic DNA, including somatic oncogenic mutations and deletions of tumor suppressor genes, are major drivers of carcinogenesis. Reduction in the size of telomere triggers cellular senescence. Activation mutation in the telomerase reverse transcriptase (TERT) promoter gene has been detected in HCC induced by HCV infection. HCV core protein downregulates CDKN2A expression to overcome hepatocyte senescence, increases telomerase activity, a characteristic of transformation prone cells in transfected primary human hepatocytes that acquires an immortalized phenotype (Basu, Meyer, Ray, & Ray, 2002).

HCV genome does not integrate into the human genome, although can cause epigenetic changes that favor its own replication through oncogenic events associated with the development of liver cancer. Several histone post-translational modifications also play an important role in altering the transcription of cellular genes, including classic tumor suppressors that are commonly known to epigenetically promote malignant transformation. Epigenetic-related modification observed during HCV infection is the inhibition of histone H4 arginine methyltransferase 1, which is associated with the virus-mediated up-regulation of protein phosphatase 2A. HCV core protein inhibits the expression of the CDKN2A gene, which encodes p16INK and p14Arf, two inhibitors of cell proliferation, by up-regulating the methyltransferases DNMT1 and DNMT3b (Vescovo et al., 2016). Various tumor suppressor genes, including CDKN2A, GSTP1, RUNX3, APC, SOCS-1 and RASSF1A, are highly methylated in HCC caused by HCV infection. Once epigenetic changes have occurred following infection, the persistent HCV-specific gene expression pattern is maintained in cured cells and the presence of the virus is no longer required for its oncogenic effect on the host cell (Perez et al., 2019). LINE-1 and Alu elements associated with HCV-HCC located mainly in intronic and intergenic regions, preferentially enriched in H3K27me3 marks, and positively correlated with proximal gene expression (Zheng et al., 2019). Thus, genetic instability and epigenetic modifications caused by chronic HCV infection are associated with HCC risk and can persist after HCV cure.

5.5 Angiogenesis in HCV associated HCC

Solid tumors require an extensive vascular network to ensure enough oxygen and nutrient delivery to the cells for further growth (Ronca, Benkheil, Mitola, Struyf, & Liekens, 2017). Angiogenesis serves as a major determinant of tumor malignancy. HCV associated HCC is more vascularized compared to HCC with other underlying etiologies, indicating a direct relationship between HCV infection and angiogenesis (Mazzanti et al., 1997; Messerini, Novelli, & Comin, 2004). HCV core protein upregulates the expression of hypoxia inducible factor Iα (HIF1α) (Abe et al., 2012). HIF1α, which transcriptionally upregulate the expression of VEGF and cyclooxygenase 2, MMP-2 and MMP-9 (Abe et al., 2012; Vrancken, Paeshuyse, & Liekens, 2012). Angiopoietin-2 (Ang2) is another important factor for angiogenesis which binds with Tie-2 receptor to destabilize vascular structures. It could be released from the endothelial cells under the stimulation of VEGF and allows neovascularization or vessel regression depending on local VEGF concentration (Tsigkos et al., 2006). Ang2 is up-regulated by HCV infection (Hernandez-Bartolome et al., 2017). High serum levels of CCL20, also known as liver activation regulated chemokine (LARC) in patients with HCC is induced by the HCV (Benkheil et al., 2018). HCV-induced CCL20 interacts with the endothelial cells specifically expressing CCR6 receptors and induces angiogenesis and promote the hyper-vascular state of HCV associated HCC (Benkheil et al., 2018). Endoglin is highly expressed by activated endothelial cells and plays a vital role in angiogenesis, and is an important protein for tumor growth, survival, and cancer cell metastasis (Arthur et al., 2000). Endoglin enhances tumor growth by modulating TGF-signaling, and could be linked to HCC development (Sanchez-Elsner, Botella, Velasco, Langa, & Bernabeu, 2002). Endoglin expression is increased on HCV core-expressing cell surface of human hepatocytes along with the activation of phosphoSMAD1/5 and ID1 downstream signaling molecules (Kwon, Sasaki, Meyer, & Ray, 2017). Endoglin mRNA expression is also upregulated in chronically HCV infected patient's liver biopsy samples.

5.6 Impact of non-coding RNAs in HCV associated HCC

Transcriptome studies and genome tiling arrays suggested a large number of genome (more than 90%) transcribes non-coding RNAs (ncRNA) (Iyer et al., 2015; Rinn & Chang, 2012; Spizzo, Almeida, Colombatti, & Calin, 2012). Next-generation sequencing technology has revealed that many non-coding RNAs (ncRNAs) play important roles in biological processes, such as proliferation, differentiation and cell death. Aberrant expression of ncRNAs is associated with metastasis, invasion and recurrence of many cancers, including HCC. Based on their length, ncRNAs are classified into small ncRNAs (sncRNAs) of less than 200 nucleotides, and long ncRNAs (lncRNAs) of more than 200 nucleotides. Here we are focusing on microRNA and long non-coding RNAs in relation with HCV infection (Table 1).

5.7 MicroRNAs (miRNAs)

miRNAs, among the ncRNA, have been studied extensively with respect to pathological and physiological processes of HCV associated HCC. miRNAs are an important class of ncRNAs of 18–22 nucleotides that can bind to the 3′ untranslated region (3′UTR) of complementary sequences of mRNAs

Table 1 Role of microRNAs and long non-coding RNAs in HCV associated HCC.

MicroRNAs	Expression	Oncogenic/ tumor suppressor	Potential function in HCV associated HCC
miR-122	Upregulated	Tumor suppressor	Plays major role in HCV infection and replication
miR-29	Downregulated	Tumor suppressor	Reduction in HCV replication in cell culture system
Mir-155	Upregulated	Oncogenic	Induce Wnt signaling pathway to promote hepatocyte proliferation and tumorigenesis
miR-135a-5p	Upregulated	Oncogenic	Promotes HCC by inducing HCV induced STAT3 transcription factor
miR-146a-5p	Upregulated	Oncogenic	Induce HCC development by modulating fatty acid metabolism and proteasome pathway
miR-181c	Downregulation	Tumor suppressor	Reduce HCC progression by targeting homeobox A1
miR-19a	Upregulated	Oncogenic	Found in exosomes in HCV infected patients and activates hepatic stellate cells by STAT3–TGF-β pathway
miR-30e	Downregulation	Tumor suppressor	Inhibits HCC progression by downregulating *de-novo* lipogenesis pathway
miR-20a	Upregulated	Oncogenic	Upregulated in serum of HCV associated patients with liver fibrosis
miR223	Downregulated	Tumor suppressor	Downregulated in HCV infected sera and liver biopsy samples

LncRNAs	Expression	Oncogenic/ tumor suppressor	Potential function in HCV associated HCC
MALAT1	Upregulated	Oncogenic	Induce HCC development
UCA1	Upregulated	Oncogenic	Induce EMT by activating Snail2 expression, induce cell proliferation, migration and invasion
WRAP53	Upregulated	Oncogenic	High expression in HCV associated HCC patient's serum and liver tissue samples

Table 1 Role of microRNAs and long non-coding RNAs in HCV associated HCC.—cont'd

LncRNAs	Expression	Oncogenic/ tumor suppressor	Potential function in HCV associated HCC
PVT1	Upregulated	Oncogenic	Induce MYC expression, inhibit apoptosis by recruiting EZH2
NORAD	Upregulated	Oncogenic	MiR-373 binds with NORAD and induce Wee1 release resulting in uncontrolled cell proliferation
EGOT	Upregulated	Oncogenic	Promotes HCC by inducing HMGA2 expression by inhibiting miR-33a-5p
HEIH	Upregulated	Oncogenic	High expression in serum and exosomes in HCV associated HCC patients
LINC01419	Upregulated	Oncogenic	Elevated in patients with early stage of HCC
AK021443	Upregulated	Oncogenic	Elevated in patients with advanced stage of HCC
Linc-Pint	Downregulated	Tumor suppressor	Suppress HCV replication by inhibiting *de-novo* lipogenesis pathway

through RNA-induced silencing complex (RISC) to regulate gene expression at the post-transcriptional level (Bartel, 2004; Ha & Kim, 2014). miRNAs play crucial roles in a variety of disease, including liver disease (Pea, Jamieson, & Braconi, 2020; Wei et al., 2019). miRNAs are involved in virus replication and life-cycle and in cell cycle. One of the hallmarks of HCV replication is its dependency on miR-122, the most abundant miRNA in the liver, which plays a major role in liver physiology by regulating metabolic pathways and acts as a tumor suppressor (Bandiera, Pfeffer, Baumert, & Zeisel, 2015; Jopling, Yi, Lancaster, Lemon, & Sarnow, 2005). miR-29 is downregulated in most of the patients with HCV infection and upregulation of miR-29 leads to the reduction in HCV replication in cell culture system (Bandyopadhyay et al., 2011). The expression of oncogenic miR-155 is increased in patients infected with HCV and activates the Wnt signaling pathway to promote hepatocyte proliferation and tumorigenesis (Zhang et al., 2012). Similarly, miR-135a-5p is upregulated in

HCV-infected patients and promotes HCV-induced STAT3 transcriptional factor by suppressing its regulator protein tyrosine phosphatase receptor delta, resulting in the induction of malignant transformation of liver (Van Renne et al., 2018). miR-146a-5p associates with a modulation of the proteasome pathway, fatty acid metabolism, and is upregulated in HCV infection favoring both HCV infection and liver disease progression (Bandiera et al., 2016). HCV infection can transcriptionally downregulate miR-181c expression by modulating CCAAT/enhancer binding protein β (C/EBP-β) leading to upregulation of homeobox A1 (HOXA1) resulting in hepatocyte growth promotion (Mukherjee, Shrivastava, Bhanja Chowdhury, Ray, & Ray, 2014). Further study revealed that HCV mediated suppression of miR-181C results in activation of ATM signaling for cell cycle promotion (Patra, Meyer, Ray, & Ray, 2020b). miR-19a secreted from HCV infected hepatocytes activate hepatic stellate cells (Devhare et al., 2017). HCV also transcriptionally downregulates miR-30e expression and leads to advanced liver disease progression by inducing *de-novo* lipogenesis pathway (Sasaki, Sur, Cheng, Steele, & Ray, 2019). miR-20a is specifically upregulated in HCV associated liver disease and positively correlates with disease severity (Shrivastava et al., 2013). On the other hand, miR-30e and miR-223 are significantly downregulated in HCV infected patient sera compared to chronic liver disease patient sera (Bhattacharya et al., 2016).

5.8 Long non-coding RNAs (LncRNAs)

Long non-coding RNAs (lncRNAs) are greater than 200 nucleotides in length with no significant protein-coding potential. Emerging evidence suggests that lncRNAs play important roles during virus infection, antiviral immune responses and pathogenesis (Agliano, Rathinam, Medvedev, Vanaja, & Vella, 2019; Damas, Fossat, & Scheel, 2019; Landeras-Bueno & Ortin, 2016). LncRNAs can bind to DNA, RNA, or proteins to elicit functions in regulating transcription, mRNA and protein stabilization. Aberrant lncRNA expression has been observed in various diseases (Balas & Johnson, 2018; Cipolla et al., 2018; Lin & Yang, 2018). In recent years, our knowledge is expanding on gene expression regulation mediated by ncRNAs, and such mechanisms play central regulatory roles of cellular response in viral infection. Interaction with cellular ncRNAs is further exploited by RNA viruses to their own advantage, and this could provide new modalities for therapy.

In HCV associated HCC patients, high serum level of oncogenic lncRNA, metastasis-associated lung adenocarcinoma transcript 1 (MALAT1), is related to disease outcome (Toraih et al., 2018). MALAT1 can induce HCC cell proliferation and inhibits apoptosis and autophagy by sponging miR-146a which directly target the 3′-untranslated region of PI3K resulting in HCC development (Peng et al., 2020). Another study revealed that MALAT1 promotes HCC progression by modulating the expression of Forkhead Box M1 (FOXM1) *via* sponging miR-125a-3p (Liu et al., 2019). However, the role of MALAT1 in HCV associated HCC remains to be elucidated. LncRNAs, urothelial carcinoma associated-1 (UCA1) together with WD repeat containing antisense to TP53 (WRAP53) are increased in the liver and serum of HCV-derived HCC compared to HCV-derived cirrhotic tissues and healthy livers (Kamel et al., 2016). LncRNA UCA1 activates transcription factor Snail2 by sponging with miR-203 to increase epithelial-to-mesenchymal transition in HCC (Xiao et al., 2017). UCA1 also downregulates miR-124 resulting in increased cancer cell proliferation, migration and invasion (Zhao et al., 2019). HCV infection can also increase the level of oncogenic lncRNA, plasmocytoma variant translocation 1 gene (PVT1) by NS5A-mediated induction of MYC (Carnero et al., 2016; Higgs, Lerat, & Pawlotsky, 2013). Although PVT1's oncogenic role is closely associated with MYC, it could promote tumorigenesis by participating in the regulation of various oncogenes and tumor suppressors through DNA rearrangements (Cui et al., 2016). Furthermore, Guo and colleagues demonstrated that the lncRNA PVT1 is upregulated in HCC, promoting propagation of HCC cells and inhibiting apoptosis by recruiting EZH2 (Guo, Hao, Wang, & Li, 2018). HCV infection upregulates the expression of miR-373 and its target protein Wee1, a pivotal player in the G2/M transition in the cell cycle (Mukherjee, Di Bisceglie, & Ray, 2015). miR-373 forms a complex with the lncRNA NORAD, resulting in release of Wee1, in HCV-infected cells, and the promotion of uncontrolled cell growth (Sur et al., 2018). LncRNA EGOT expression was found to be significantly upregulated in HCV infected HCC patients. EGOT promotes HCC cell proliferation, migration and invasion by inhibition of miR-33a-5p to induce high mobility group protein A2 (HMGA2) gene expression (Wu, Ai, Zhang, Yun, & Xie, 2019). LINC01419 and AK021443 transcripts were expressed at higher levels in early and in advanced stage HCC respectively, and regulate the expression of genes associated with cell cycle progression for progression of HCV associated HCC (Zhang et al., 2015). We recently observed that HCV mediated downregulation of Linc-Pint enhances lipogenesis favoring virus replication and liver disease progression (unpublished publication). The higher expression of EZH2-associated long non-coding RNA (HEIH) was detected in serum and exosomes from patients with HCV-related HCC (Zhang et al., 2018). All these studies suggested significant involvement of ncRNAs in the progression of HCV associated HCC.

6. Biomarkers for detection of HCV associated HCC

Chronically HCV infected patients are asymptomatic in most of the cases for many years and may often lead to development of HCC. Early detection of HCC is urgently needed since it may increase the chance of effective therapy and patient survival. Due to the challenge of detecting small liver tumors using ultrasound and the absence of a robust HCC serum marker, molecular signatures using non-coding RNA expression profiles and circulating non-coding RNAs hold promise as non-invasive biomarkers for the early detection of HCC.

AFP is the widely used diagnostic serum marker for HCC. Several studies have also suggested the potential role of miRNAs as important non-invasive biomarkers in HCV associated HCC (Plissonnier, Herzog, Levrero, & Zeisel, 2018; Sadri Nahand et al., 2019; Shrivastava, Steele, Ray, & Ray, 2015). A miRNA panel composed of miR-122, miR-885-5p, and miR-29b in association with AFP may work as a new biomarker for the early detection of HCC in normal population, while a different miRNA panel composed of miR-122, miR-885-5p, miR-221, and miR-22 in association with AFP also provided a high accuracy for the early detection of HCC. On the other hand, miRNA panel of miR-22 and miR-199a-3p with AFP provided high diagnostic accuracy for early detection of HCC in chronic HCV infected patients (Zekri et al., 2016).

A panel of nine liver-associated miRNAs (miR-21, miR-30c, miR-93, miR-122, miR-125b, miR-126, miR-130a, miR-193b, and miR-222) was reported to discriminate between healthy individuals and HCV associated HCC (Ali et al., 2017; Huang et al., 2017). As discussed above, lncRNAs UCA1, HEIH, MALAT1and WRAP53 have been suggested as diagnostic and prognostic markers of HCV associated HCC in association with AFP (Kamel et al., 2016). Further investigations are indeed necessary with larger patient populations to fully understand the impact of non-coding RNAs as potential non-invasive biomarkers for early detection of HCV associated HCC.

Growing interest is directed for the identification of genetic markers, such as intronic single nucleotide polymorphism (SNPs) influencing the risk of developing HCV-associated HCC using high-throughput genomic technology, in a genome-wide association study (GWAS). For examples, SNPs rs1012068 and rs5998152 in DEPDC5 gene are described as markers for HCC progression among chronic HCV-infected Japanese patients and Saudi Arabian population (Al-Anazi et al., 2014; Miki et al., 2011). Another SNP rs2596542, located 4.7 kb upstream of MHC class I polypeptide-related sequence A (MICA) gene is strongly associated with HCV associated HCC (Kumar et al., 2011). Thus, future studies are needed to focus on efforts to combine the conventional biomarkers for HCC together with different novel biomarkers for the achievement of maximum diagnostic and predictive ability at an early stage of HCV associated HCC.

7. Therapeutic strategies for HCV associated HCC

The treatment with direct acting antivirals (DAA) in HCV infected patients results in a sustained virological response (SVR) rate in majority of the patients, however, asymptomatic HCV infection causes delayed diagnosis and treatment in communities. The impact of DAA-based regimens on the occurrence of HCC in patients with cirrhosis, and the recurrence of HCC following successful curative treatment is controversial (Roche, Coilly, Duclos-Vallee, & Samuel, 2018). Several studies have observed an unexpectedly high incidence of HCC recurrence among patients during and after the DAA therapy (Cardoso et al., 2016; Cheung et al., 2016; Kozbial et al., 2016; Zavaglia et al., 2017). Modulation of immune cell response, cytokine, and chemokine signaling by DAA may aggravate occurrence of HCC during and after therapy (Sasaki, Kanda, Kato, Yokosuka, & Moriyama, 2018; Villani, Vendemiale, & Serviddio, 2018). Furthermore, HCV induces epigenetic alterations which persists after cure by DAA and can be a unique reason for HCC risk. On the other hand, genome wide ChIPmentation-based ChIP-Seq and RNA-seq analyses of liver tissue from patients without HCV infection (controls), with chronic HCV infection, with chronic HCV infection cured by DAA treatment and with chronic HCV infection cured by interferon therapy suggested a specific genome-wide change in H3K27ac contributes to an increased risk of HCC (Hamdane et al., 2019). In fact, DAA treatment may have several benefits in reducing HCC, unfortunately, individuals infected with HCV often do not show symptoms. Therefore, late diagnosis of HCV infection and elimination by DAA may not be sufficient to protect end stage liver disease including HCC.

Treatment of HCC is limited for last several years. Sorafenib, a multi-kinase inhibitor, was introduced in 2007 which extends survival by a maximum of 3-months and most treatment drops because of disease progression or developing resistance. Newer FDA approved drugs, regorafenib—the congener of sorafenib, responds only for a subset of sorafenib resistant patients (~11%). FDA also approved nivolumab—a programmed cell death 1 (PD-1) inhibitor antibody, for second line treatment in patients who became resistance or intolerant to sorafenib. The potentially curative options for treating HCC are surgical resection and orthotropic liver transplantation. However, liver transplantation approach is limited to one-third of patients due to the low donor organ availability. Moreover, the 5-year recurrence of HCC (~70%) occurs after surgical resection approaches (Villanueva, 2019). Therefore, targeting for elimination of tumor cells by most effective approach will be highly significant for inhibition of neoplastic liver disease progression.

We reported that cancer stem cells from HCV associated transformed human hepatocytes display higher c-Kit expression (Kwon et al., 2015). We also observed recently that chronically HCV infected HCC patients had a higher c-Kit expression status (Nazzal et al., 2020). c-Kit mRNA and protein expression were also detected in ~27% of HCV associated HCC from a Taiwanese cohort (Chung et al., 2005). We further observed treatment of HCC-PDX xenograft tumor bearing mice with c-Kit inhibitor, imatinib, significantly reduced tumor growth (Nazzal et al., 2020), further suggesting the importance of targeted therapy.

Markedly different pathological processes are involved in chronic hepatitis C infection and the development of HCC. There are some similarities between the two when considering the role of the immune system, with T cell exhaustion implicated in both (Moreno-Cubero & Larrubia, 2016). $CD8^+$ T cells are integral in targeting and destroying both tumor cells and HCV infected cells. T cell exhaustion exists to protect from tissue damage due to persistent and overzealous immunological response to antigens and is driven by upregulation of negative co-stimulatory pathways. With these pathways in action, key T cell effector processes are disrupted, they become tolerant of antigenic stimuli and ultimately apoptosis (Moreno-Cubero & Larrubia, 2016). T cell exhaustion occurs in the liver, causing an immune-tolerant state required in an organ that encounters many antigenic threats. Chronic inflammation and development of cancer are associated with T cell exhaustion. These negative co-stimulatory pathways include PD1, CTLA4, Tim3 and LAG3, and are targets for exploring new pharmacological options for HCC treatment. More studies assessing the impact of anti-PD1 immunotherapy on active HCV infection with HCC are required to fully understand its role in viral response. Multiple ongoing clinical trials using anti-PD1 antibodies are currently allowing patients with untreated HCV to enroll, which may go some way to answering this question. Nivolumab, a human monoclonal IgG4 antibody against PD1, shows promise as a treatment in advanced HCC (El-Khoueiry et al., 2017), and showed persistent suppression of HCV RNA in a subgroup of chronically infected patients (Gardiner et al., 2013). Pembrolizumab, another anti-PD1 monoclonal antibody, has also been granted accelerated FDA approval for treatment of advanced HCC, with promising results in the Keynote-224 study (Zhu et al., 2018). Clearly, we must expand our basic and clinical knowledge about how modulating and fine tuning these therapeutic aspects will help to avoid adverse events, and how appropriate patient selection will help to respond to these kinds of treatment. We must also learn how to best use these approaches in combination with other therapies to obtain the maximum clinical benefit.

8. Perspectives

Chronic HCV infection triggers the oncogenic signaling pathways, since several HCV proteins display cell growth regulatory events. Immunological deficiency in HCV clearance from liver, and an interaction of viral proteins with host cellular proteins involved in signaling cascade promotes liver disease progression. In the process of activating cell growth regulation, immune system dysregulation occurs, which initiates fibrosis process. During HCV infection, Kupffer cells get activated and release several cytokines/chemokines, which in turn activate hepatic stellate cells (Sasaki, Devhare, Steele, Ray, & Ray, 2017). On the other hand, exosomes secreted from HCV infected hepatocytes carry microRNAs and activate hepatic stellate cells (Devhare et al., 2017; Kim, Lee, & Lee, 2019). Activate hepatic stellate cells secretes several soluble factors including TGF-β, which may influence preneoplastic hepatocytes toward neoplasia. In general HCC, including HCV associated liver disease, occur in presence of late stage of fibrosis (Fig. 3(. We postulate that transformed hepatocytes in their early stage is difficult to diagnose. Fibrosis occurs much faster since

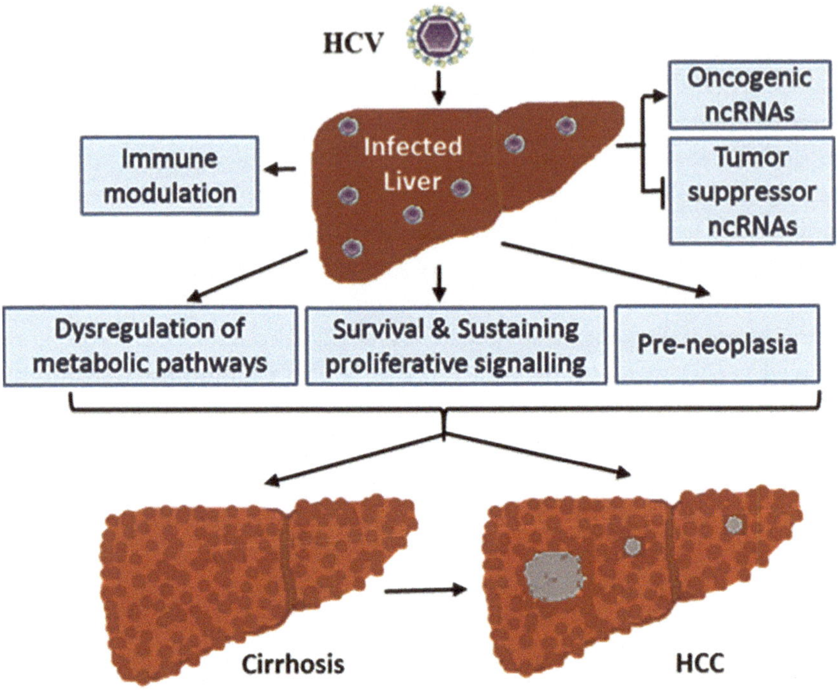

Fig. 3 A simple schematic view of the steps for HCV associated cirrhosis and HCC.

there are several stimulators present in HCV infection and relatively easy to diagnose. Further, current diagnostic method cannot detect HCC until they are more than 1 cm in size by MRI, and very difficult to get liver biopsy specimens for detection of HCC at early stage. Thus, there are urgent need to develop HCC biomarkers for early diagnosis and surveillance for this devastating disease. Non-coding RNAs are gaining interest in cancer therapy. Several microRNAs and lncRNAs are in clinical trials for several cancers including HCC for either as biomarker or therapeutic agent (https://www.clinicaltrials.gov).

Curative approaches are limited for patients diagnosed with HCC at an intermediate or advanced disease stage (Patra, Meyer, Ray, & Ray, 2020a). Several signaling pathways are dysregulated in cancer cells and cause interruption in the homeostatic environments that usually contribute to the aberrant growth of the cells. A combination of anti-cancer drugs targeting diverse signaling pathways should enhance efficacy compared to the monotherapy approach, because they normally impact in a characteristically additive or synergistic manner. Combination therapy for cancer treatment is a cornerstone modality that includes multiple therapeutic agents, and will provide anti-cancer benefits like inhibiting tumor growth/metastatic potential, reducing cancer stem cell population and drug resistance.

Since treatment for HCC is very limited and none of the current regimen has given >50% efficacy, molecular testing and targeted treatment will likely be most helpful. Our understanding of tumors' biology to create highly effective therapies for precision medicine has leveraged novel therapeutic modalities for other tumors. However, these breakthroughs have been elusive in liver cancer. There is a critical need to understand the disease process, and to identify better preventive and/or therapeutic strategies for successful management of these patients.

Acknowledgments

We thank Tapas Patra and Mustafa Nazzal for helpful comments while preparing the article. The study was supported by grants DK080812, DK081817 from the National Institutes of Health, and a grant from the Lottie Caroline Hardy Charitable Trust.

References

Abe, M., Koga, H., Yoshida, T., Masuda, H., Iwamoto, H., Sakata, M., et al. (2012). Hepatitis C virus core protein upregulates the expression of vascular endothelial growth factor via the nuclear factor-kappaB/hypoxia-inducible factor-1alpha axis under hypoxic conditions. *Hepatology Research*, *42*(6), 591–600. https://doi.org/10.1111/j.1872-034X.2011.00953.x.

Abid, K., Pazienza, V., de Gottardi, A., Rubbia-Brandt, L., Conne, B., Pugnale, P., et al. (2005). An in vitro model of hepatitis C virus genotype 3a-associated triglycerides accumulation. *Journal of Hepatology*, *42*(5), 744–751. https://doi.org/10.1016/j.jhep.2004.12.034.

Adachi, T., Togashi, H., Suzuki, A., Kasai, S., Ito, J., Sugahara, K., et al. (2005). NAD(P)H oxidase plays a crucial role in PDGF-induced proliferation of hepatic stellate cells. *Hepatology*, *41*(6), 1272–1281. https://doi.org/10.1002/hep.20719.

Agliano, F., Rathinam, V. A., Medvedev, A. E., Vanaja, S. K., & Vella, A. T. (2019). Long noncoding RNAs in host-pathogen interactions. *Trends in Immunology*, *40*(6), 492–510. https://doi.org/10.1016/j.it.2019.04.001.

Al-Anazi, M. R., Matou-Nasri, S., Abdo, A. A., Sanai, F. M., Khan, M. Q., Albenmousa, A., et al. (2014). Variations in DEPDC5 gene and its association with chronic hepatitis C virus infection in Saudi Arabia. *BMC Infectious Diseases*, *14*, 632. https://doi.org/10.1186/s12879-014-0632-y.

Ali, H. E. A., Abdel Hameed, R., Effat, H., Ahmed, E. K., Atef, A. A., Sharawi, S. K., et al. (2017). Circulating microRNAs panel as a diagnostic tool for discrimination of HCV-associated hepatocellular carcinoma. *Clinics and Research in Hepatology and Gastroenterology*, *41*(4), e51–e62. https://doi.org/10.1016/j.clinre.2017.06.004.

Allemani, C., Weir, H. K., Carreira, H., Harewood, R., Spika, D., Wang, X. S., et al. (2015). Global surveillance of cancer survival 1995-2009: Analysis of individual data for 25,676,887 patients from 279 population-based registries in 67 countries (CONCORD-2). *Lancet*, *385*(9972), 977–1010. https://doi.org/10.1016/S0140-6736(14)62038-9.

Arthur, H. M., Ure, J., Smith, A. J., Renforth, G., Wilson, D. I., Torsney, E., et al. (2000). Endoglin, an ancillary TGFbeta receptor, is required for extraembryonic angiogenesis and plays a key role in heart development. *Developmental Biology*, *217*(1), 42–53. https://doi.org/10.1006/dbio.1999.9534.

Balas, M. M., & Johnson, A. M. (2018). Exploring the mechanisms behind long noncoding RNAs and cancer. *Noncoding RNA Research*, *3*(3), 108–117. https://doi.org/10.1016/j.ncrna.2018.03.001.

Bandiera, S., Pernot, S., El Saghire, H., Durand, S. C., Thumann, C., Crouchet, E., et al. (2016). Hepatitis C virus-induced upregulation of microRNA miR-146a-5p in hepatocytes promotes viral infection and deregulates metabolic pathways associated with liver disease pathogenesis. *Journal of Virology*, *90*(14), 6387–6400. https://doi.org/10.1128/JVI.00619-16.

Bandiera, S., Pfeffer, S., Baumert, T. F., & Zeisel, M. B. (2015). miR-122- -a key factor and therapeutic target in liver disease. *Journal of Hepatology*, *62*(2), 448–457. https://doi.org/10.1016/j.jhep.2014.10.004.

Bandyopadhyay, S., Friedman, R. C., Marquez, R. T., Keck, K., Kong, B., Icardi, M. S., et al. (2011). Hepatitis C virus infection and hepatic stellate cell activation downregulate miR-29: miR-29 overexpression reduces hepatitis C viral abundance in culture. *The Journal of Infectious Diseases*, *203*(12), 1753–1762. https://doi.org/10.1093/infdis/jir186.

Banerjee, A., Mazumdar, B., Meyer, K., Di Bisceglie, A. M., Ray, R. B., & Ray, R. (2011). Transcriptional repression of C4 complement by hepatitis C virus proteins. *Journal of Virology*, *85*(9), 4157–4166. https://doi.org/10.1128/JVI.02449-10.

Banerjee, A., Meyer, K., Mazumdar, B., Ray, R. B., & Ray, R. (2010). Hepatitis C virus differentially modulates activation of forkhead transcription factors and insulin-induced metabolic gene expression. *Journal of Virology*, *84*(12), 5936–5946. https://doi.org/10.1128/JVI.02344-09.

Banerjee, A., Ray, R. B., & Ray, R. (2010). Oncogenic potential of hepatitis C virus proteins. *Viruses*, *2*(9), 2108–2133. https://doi.org/10.3390/v2092108.

Barba, G., Harper, F., Harada, T., Kohara, M., Goulinet, S., Matsuura, Y., et al. (1997). Hepatitis C virus core protein shows a cytoplasmic localization and associates to cellular lipid storage droplets. *Proceedings of the National Academy of Sciences of the United States of America*, *94*(4), 1200–1205. https://doi.org/10.1073/pnas.94.4.1200.

Bartel, D. P. (2004). MicroRNAs: Genomics, biogenesis, mechanism, and function. *Cell*, *116*(2), 281–297. https://doi.org/10.1016/s0092-8674(04)00045-5.

Bartosch, B., Thimme, R., Blum, H. E., & Zoulim, F. (2009). Hepatitis C virus-induced hepatocarcinogenesis. *Journal of Hepatology*, *51*(4), 810–820. https://doi.org/10.1016/j.jhep.2009.05.008.

Basu, A., Meyer, K., Ray, R. B., & Ray, R. (2002). Hepatitis C virus core protein is necessary for the maintenance of immortalized human hepatocytes. *Virology*, *298*(1), 53–62. https://doi.org/10.1006/viro.2002.1460.

Bataller, R., Schwabe, R. F., Choi, Y. H., Yang, L., Paik, Y. H., Lindquist, J., et al. (2003). NADPH oxidase signal transduces angiotensin II in hepatic stellate cells and is critical in hepatic fibrosis. *The Journal of Clinical Investigation*, *112*(9), 1383–1394. https://doi.org/10.1172/JCI18212.

Baumert, T. F., Juhling, F., Ono, A., & Hoshida, Y. (2017). Hepatitis C-related hepatocellular carcinoma in the era of new generation antivirals. *BMC Medicine*, *15*(1), 52. https://doi.org/10.1186/s12916-017-0815-7.

Benkheil, M., Van Haele, M., Roskams, T., Laporte, M., Noppen, S., Abbasi, K., et al. (2018). CCL20, a direct-acting pro-angiogenic chemokine induced by hepatitis C virus (HCV): Potential role in HCV-related liver cancer. *Experimental Cell Research*, *372*(2), 168–177. https://doi.org/10.1016/j.yexcr.2018.09.023.

Bhattacharya, S., Steele, R., Shrivastava, S., Chakraborty, S., Di Bisceglie, A. M., & Ray, R. B. (2016). Serum miR-30e and miR-223 as novel noninvasive biomarkers for hepatocellular carcinoma. *The American Journal of Pathology, 186*(2), 242–247. https://doi.org/10.1016/j.ajpath.2015.10.003.

Bittar, C., Shrivastava, S., Bhanja Chowdhury, J., Rahal, P., & Ray, R. B. (2013). Hepatitis C virus NS2 protein inhibits DNA damage pathway by sequestering p53 to the cytoplasm. *PLoS One, 8*(4), e62581. https://doi.org/10.1371/journal.pone.0062581.

Boffetta, P., & Islami, F. (2014). Cancer in low- and medium-income countries. *Annals of Global Health, 80*(5), 345. https://doi.org/10.1016/j.aogh.2014.09.008.

Bose, S. K., Shrivastava, S., Meyer, K., Ray, R. B., & Ray, R. (2012). Hepatitis C virus activates the mTOR/S6K1 signaling pathway in inhibiting IRS-1 function for insulin resistance. *Journal of Virology, 86*(11), 6315–6322. https://doi.org/10.1128/JVI.00050-12.

Bray, F., Ferlay, J., Soerjomataram, I., Siegel, R. L., Torre, L. A., & Jemal, A. (2018). Global cancer statistics 2018: GLOBOCAN estimates of incidence and mortality worldwide for 36 cancers in 185 countries. *CA: A Cancer Journal for Clinicians, 68*(6), 394–424. https://doi.org/10.3322/caac.21492.

Bugianesi, E., Salamone, F., & Negro, F. (2012). The interaction of metabolic factors with HCV infection: Does it matter? *Journal of Hepatology, 56*(Suppl 1), S56–S65. https://doi.org/10.1016/S0168-8278(12)60007-5.

Capone, F., Guerriero, E., Colonna, G., Maio, P., Mangia, A., Castello, G., et al. (2014). Cytokinome profile evaluation in patients with hepatitis C virus infection. *World Journal of Gastroenterology, 20*(28), 9261–9269. https://doi.org/10.3748/wjg.v20.i28.9261.

Cardoso, H., Vale, A. M., Rodrigues, S., Goncalves, R., Albuquerque, A., Pereira, P., et al. (2016). High incidence of hepatocellular carcinoma following successful interferon-free antiviral therapy for hepatitis C associated cirrhosis. *Journal of Hepatology, 65*(5), 1070–1071. https://doi.org/10.1016/j.jhep.2016.07.027.

Carnero, E., Barriocanal, M., Prior, C., Pablo Unfried, J., Segura, V., Guruceaga, E., et al. (2016). Long noncoding RNA EGOT negatively affects the antiviral response and favors HCV replication. *EMBO Reports, 17*(7), 1013–1028. https://doi.org/10.15252/embr.201541763.

Castera, L., Chouteau, P., Hezode, C., Zafrani, E. S., Dhumeaux, D., & Pawlotsky, J. M. (2005). Hepatitis C virus-induced hepatocellular steatosis. *The American Journal of Gastroenterology, 100*(3), 711–715. https://doi.org/10.1111/j.1572-0241.2005.40898.x.

Cheung, M. C. M., Walker, A. J., Hudson, B. E., Verma, S., McLauchlan, J., Mutimer, D. J., et al. (2016). Outcomes after successful direct-acting antiviral therapy for patients with chronic hepatitis C and decompensated cirrhosis. *Journal of Hepatology, 65*(4), 741–747. https://doi.org/10.1016/j.jhep.2016.06.019.

Chiba, T., Matsuzaki, Y., Abei, M., Shoda, J., Aikawa, T., Tanaka, N., et al. (1996). Multivariate analysis of risk factors for hepatocellular carcinoma in patients with hepatitis C virus-related liver cirrhosis. *Journal of Gastroenterology, 31*(4), 552–558. https://doi.org/10.1007/BF02355056.

Choi, S. H., & Hwang, S. B. (2006). Modulation of the transforming growth factor-beta signal transduction pathway by hepatitis C virus nonstructural 5A protein. *The Journal of Biological Chemistry, 281*(11), 7468–7478. https://doi.org/10.1074/jbc.M512438200.

Chuang, S. C., Lee, Y. C., Hashibe, M., Dai, M., Zheng, T., & Boffetta, P. (2010). Interaction between cigarette smoking and hepatitis B and C virus infection on the risk of liver cancer: A meta-analysis. *Cancer Epidemiology, Biomarkers & Prevention, 19*(5), 1261–1268. https://doi.org/10.1158/1055-9965.EPI-09-1297.

Chung, C. Y., Yeh, K. T., Hsu, N. C., Chang, J. H., Lin, J. T., Horng, H. C., et al. (2005). Expression of c-kit protooncogene in human hepatocellular carcinoma. *Cancer Letters*, *217*(2), 231–236. https://doi.org/10.1016/j.canlet.2004.06.045.

Cipolla, G. A., de Oliveira, J. C., Salviano-Silva, A., Lobo-Alves, S. C., Lemos, D. S., Oliveira, L. C., et al. (2018). Long non-coding RNAs in multifactorial diseases: Another layer of complexity. *Noncoding RNA*, *4*(2), 13. https://doi.org/10.3390/ncrna4020013.

Conti, G., Heckman, J., & Pinto, R. (2016). The effects of two influential early childhood interventions on health and healthy behaviour. *Economic Journal*, *126*(596), F28–F65. https://doi.org/10.1111/ecoj.12420.

Cui, M., You, L., Ren, X., Zhao, W., Liao, Q., & Zhao, Y. (2016). Long non-coding RNA PVT1 and cancer. *Biochemical and Biophysical Research Communications*, *471*(1), 10–14. https://doi.org/10.1016/j.bbrc.2015.12.101.

Damas, N. D., Fossat, N., & Scheel, T. K. H. (2019). Functional interplay between RNA viruses and non-coding RNA in mammals. *Noncoding RNA*, *5*(1). https://doi.org/10.3390/ncrna5010007.

de Gottardi, A., Pazienza, V., Pugnale, P., Bruttin, F., Rubbia-Brandt, L., Juge-Aubry, C. E., et al. (2006). Peroxisome proliferator-activated receptor-alpha and -gamma mRNA levels are reduced in chronic hepatitis C with steatosis and genotype 3 infection. *Alimentary Pharmacology & Therapeutics*, *23*(1), 107–114. https://doi.org/10.1111/j.1365-2036.2006.02729.x.

de Martel, C., Maucort-Boulch, D., Plummer, M., & Franceschi, S. (2015). World-wide relative contribution of hepatitis B and C viruses in hepatocellular carcinoma. *Hepatology*, *62*(4), 1190–1200. https://doi.org/10.1002/hep.27969.

Deng, L., Shoji, I., Ogawa, W., Kaneda, S., Soga, T., Jiang, D. P., et al. (2011). Hepatitis C virus infection promotes hepatic gluconeogenesis through an NS5A-mediated, FoxO1-dependent pathway. *Journal of Virology*, *85*(17), 8556–8568. https://doi.org/10.1128/JVI.00146-11.

Devhare, P. B., Sasaki, R., Shrivastava, S., Di Bisceglie, A. M., Ray, R., & Ray, R. B. (2017). Exosome-mediated intercellular communication between hepatitis C virus-infected hepatocytes and hepatic stellate cells. *Journal of Virology*, *91*(6), e02225-16. https://doi.org/10.1128/JVI.02225-16.

Dharancy, S., Lemoine, M., Mathurin, P., Serfaty, L., & Dubuquoy, L. (2009). Peroxisome proliferator-activated receptors in HCV-related infection. *PPAR Research*, *2009*, 357204. https://doi.org/10.1155/2009/357204.

Dharancy, S., Malapel, M., Perlemuter, G., Roskams, T., Cheng, Y., Dubuquoy, L., et al. (2005). Impaired expression of the peroxisome proliferator-activated receptor alpha during hepatitis C virus infection. *Gastroenterology*, *128*(2), 334–342. https://doi.org/10.1053/j.gastro.2004.11.016.

Di Bisceglie, A. M., Lyra, A. C., Schwartz, M., Reddy, R. K., Martin, P., Gores, G., et al. (2003). Hepatitis C-related hepatocellular carcinoma in the United States: Influence of ethnic status. *The American Journal of Gastroenterology*, *98*(9), 2060–2063. https://doi.org/10.1111/j.1572-0241.2003.t01-1-07641.x.

Diepolder, H. M., Zachoval, R., Hoffmann, R. M., Wierenga, E. A., Santantonio, T., Jung, M. C., et al. (1995). Possible mechanism involving T-lymphocyte response to non-structural protein 3 in viral clearance in acute hepatitis C virus infection. *Lancet*, *346*(8981), 1006–1007. https://doi.org/10.1016/s0140-6736(95)91691-1.

Donato, F., Tagger, A., Chiesa, R., Ribero, M. L., Tomasoni, V., Fasola, M., et al. (1997). Hepatitis B and C virus infection, alcohol drinking, and hepatocellular carcinoma: A case-control study in Italy. Brescia HCC study. *Hepatology*, *26*(3), 579–584. https://doi.org/10.1002/hep.510260308.

Donato, F., Tagger, A., Gelatti, U., Parrinello, G., Boffetta, P., Albertini, A., et al. (2002). Alcohol and hepatocellular carcinoma: The effect of lifetime intake and hepatitis virus infections in men and women. *American Journal of Epidemiology, 155*(4), 323–331. https://doi.org/10.1093/aje/155.4.323.

Duvet, S., Cocquerel, L., Pillez, A., Cacan, R., Verbert, A., Moradpour, D., et al. (1998). Hepatitis C virus glycoprotein complex localization in the endoplasmic reticulum involves a determinant for retention and not retrieval. *The Journal of Biological Chemistry, 273*(48), 32088–32095. https://doi.org/10.1074/jbc.273.48.32088.

El-Fatah Fahmy Hanno, A., Mohiedeen, K. M., Deghedy, A., & Sayed, R. (2014). Serum complements C3 and C4 in chronic HCV infection and their correlation with response to pegylated interferon and ribavirin treatment. *Arab Journal of Gastroenterology, 15*(2), 58–62. https://doi.org/10.1016/j.ajg.2014.04.005.

El-Khoueiry, A. B., Sangro, B., Yau, T., Crocenzi, T. S., Kudo, M., Hsu, C., et al. (2017). Nivolumab in patients with advanced hepatocellular carcinoma (CheckMate 040): An open-label, non-comparative, phase 1/2 dose escalation and expansion trial. *Lancet, 389*(10088), 2492–2502. https://doi.org/10.1016/S0140-6736(17)31046-2.

El-Serag, H. B., Kanwal, F., Richardson, P., & Kramer, J. (2016). Risk of hepatocellular carcinoma after sustained virological response in veterans with hepatitis C virus infection. *Hepatology, 64*(1), 130–137. https://doi.org/10.1002/hep.28535.

El-Shamy, A., Branch, A. D., Schiano, T. D., & Gorevic, P. D. (2018). The complement system and C1q in chronic hepatitis C virus infection and mixed cryoglobulinemia. *Frontiers in Immunology, 9*, 1001. https://doi.org/10.3389/fimmu.2018.01001.

Fitian, A. I., & Cabrera, R. (2017). Disease monitoring of hepatocellular carcinoma through metabolomics. *World Journal of Hepatology, 9*(1), 1–17. https://doi.org/10.4254/wjh.v9.i1.1.

Furutani, T., Hino, K., Okuda, M., Gondo, T., Nishina, S., Kitase, A., et al. (2006). Hepatic iron overload induces hepatocellular carcinoma in transgenic mice expressing the hepatitis C virus polyprotein. *Gastroenterology, 130*(7), 2087–2098. https://doi.org/10.1053/j.gastro.2006.02.060.

Gardiner, D., Lalezari, J., Lawitz, E., DiMicco, M., Ghalib, R., Reddy, K. R., et al. (2013). A randomized, double-blind, placebo-controlled assessment of BMS-936558, a fully human monoclonal antibody to programmed death-1 (PD-1), in patients with chronic hepatitis C virus infection. *PLoS One, 8*(5), e63818. https://doi.org/10.1371/journal.pone.0063818.

Gieseler, R. K., Marquitan, G., Schlattjan, M., Sowa, J. P., Bechmann, L. P., Timm, J., et al. (2011). Hepatocyte apoptotic bodies encasing nonstructural HCV proteins amplify hepatic stellate cell activation: Implications for chronic hepatitis C. *Journal of Viral Hepatitis, 18*(11), 760–767. https://doi.org/10.1111/j.1365-2893.2010.01362.x.

Gong, G., Waris, G., Tanveer, R., & Siddiqui, A. (2001). Human hepatitis C virus NS5A protein alters intracellular calcium levels, induces oxidative stress, and activates STAT-3 and NF-kappa B. *Proceedings of the National Academy of Sciences of the United States of America, 98*(17), 9599–9604. https://doi.org/10.1073/pnas.171311298.

Goossens, N., & Hoshida, Y. (2015). Hepatitis C virus-induced hepatocellular carcinoma. *Clinical and Molecular Hepatology, 21*(2), 105–114. https://doi.org/10.3350/cmh.2015.21.2.105.

Greten, T. F., Wang, X. W., & Korangy, F. (2015). Current concepts of immune based treatments for patients with HCC: From basic science to novel treatment approaches. *Gut, 64*(5), 842–848. https://doi.org/10.1136/gutjnl-2014-307990.

Guo, J., Hao, C., Wang, C., & Li, L. (2018). Long noncoding RNA PVT1 modulates hepatocellular carcinoma cell proliferation and apoptosis by recruiting EZH2. *Cancer Cell International, 18*, 98. https://doi.org/10.1186/s12935-018-0582-3.

Ha, M., & Kim, V. N. (2014). Regulation of microRNA biogenesis. *Nature Reviews. Molecular Cell Biology*, *15*(8), 509–524. https://doi.org/10.1038/nrm3838.

Hamdane, N., Juhling, F., Crouchet, E., El Saghire, H., Thumann, C., Oudot, M. A., et al. (2019). HCV-induced epigenetic changes associated with liver cancer risk persist after sustained virologic response. *Gastroenterology*, *156*(8), 2313–2329. e2317 https://doi.org/10.1053/j.gastro.2019.02.038.

Han, Y., Niu, J., Wang, D., & Li, Y. (2016). Hepatitis C virus protein interaction network analysis based on hepatocellular carcinoma. *PLoS One*, *11*(4), e0153882. https://doi.org/10.1371/journal.pone.0153882.

Hernandez-Bartolome, A., Lopez-Rodriguez, R., Garcia-Buey, L., Martin-Vilchez, S., Rodriguez-Munoz, Y., Borque, M. J., et al. (2017). Intrahepatic angiopoietin-2 correlates with chronic hepatitis C progression and is induced in hepatitis C virus replicon systems. *Liver International*, *37*(8), 1148–1156. https://doi.org/10.1111/liv.13352.

Higgs, M. R., Lerat, H., & Pawlotsky, J. M. (2013). Hepatitis C virus-induced activation of beta-catenin promotes c-Myc expression and a cascade of pro-carcinogenetic events. *Oncogene*, *32*(39), 4683–4693. https://doi.org/10.1038/onc.2012.484.

Holmgaard, R. B., Zamarin, D., Li, Y., Gasmi, B., Munn, D. H., Allison, J. P., et al. (2015). Tumor-expressed IDO recruits and activates MDSCs in a treg-dependent manner. *Cell Reports*, *13*(2), 412–424. https://doi.org/10.1016/j.celrep.2015.08.077.

Hoshida, Y., Fuchs, B. C., Bardeesy, N., Baumert, T. F., & Chung, R. T. (2014). Pathogenesis and prevention of hepatitis C virus-induced hepatocellular carcinoma. *Journal of Hepatology*, *61*(1 Suppl), S79–S90. https://doi.org/10.1016/j.jhep.2014.07.010.

Hossain, M. S., Raheem, E., Sultana, T. A., Ferdous, S., Nahar, N., Islam, S., et al. (2017). Thalassemias in South Asia: Clinical lessons learnt from Bangladesh. *Orphanet Journal of Rare Diseases*, *12*(1), 93. https://doi.org/10.1186/s13023-017-0643-z.

Hou, J., Zhang, H., Sun, B., & Karin, M. (2020). The immunobiology of hepatocellular carcinoma in humans and mice: Basic concepts and therapeutic implications. *Journal of Hepatology*, *72*(1), 167–182. https://doi.org/10.1016/j.jhep.2019.08.014.

Huang, Y. H., Liang, K. H., Chien, R. N., Hu, T. H., Lin, K. H., Hsu, C. W., et al. (2017). A circulating MicroRNA signature capable of assessing the risk of hepatocellular carcinoma in cirrhotic patients. *Scientific Reports*, 7(1), 523. https://doi.org/10.1038/s41598-017-00631-9.

Huang, T. S., Lin, C. L., Lu, M. J., Yeh, C. T., Liang, K. H., Sun, C. C., et al. (2017). Diabetes, hepatocellular carcinoma, and mortality in hepatitis C-infected patients: A population-based cohort study. *Journal of Gastroenterology and Hepatology*, *32*(7), 1355–1362. https://doi.org/10.1111/jgh.13670.

Hui, J. M., Sud, A., Farrell, G. C., Bandara, P., Byth, K., Kench, J. G., et al. (2003). Insulin resistance is associated with chronic hepatitis C virus infection and fibrosis progression [corrected]. *Gastroenterology*, *125*(6), 1695–1704. https://doi.org/10.1053/j.gastro.2003.08.032.

Ioannou, G. N., Bryson, C. L., Weiss, N. S., Miller, R., Scott, J. D., & Boyko, E. J. (2013). The prevalence of cirrhosis and hepatocellular carcinoma in patients with human immunodeficiency virus infection. *Hepatology*, *57*(1), 249–257. https://doi.org/10.1002/hep.25800.

Irshad, M., Gupta, P., & Irshad, K. (2017). Molecular basis of hepatocellular carcinoma induced by hepatitis C virus infection. *World Journal of Hepatology*, *9*(36), 1305–1314. https://doi.org/10.4254/wjh.v9.i36.1305.

Iyer, M. K., Niknafs, Y. S., Malik, R., Singhal, U., Sahu, A., Hosono, Y., et al. (2015). The landscape of long noncoding RNAs in the human transcriptome. *Nature Genetics*, *47*(3), 199–208. https://doi.org/10.1038/ng.3192.

Janjua, N. Z., Chong, M., Kuo, M., Woods, R., Wong, J., Yoshida, E. M., et al. (2017). Long-term effect of sustained virological response on hepatocellular carcinoma in patients with hepatitis C in Canada. *Journal of Hepatology*, *66*(3), 504–513. https://doi.org/10.1016/j.jhep.2016.10.028.

Jiang, Y. F., He, B., Li, N. P., Ma, J., Gong, G. Z., & Zhang, M. (2011). The oncogenic role of NS5A of hepatitis C virus is mediated by up-regulation of survivin gene expression in the hepatocellular cell through p53 and NF-kappaB pathways. *Cell Biology International*, *35*(12), 1225–1232. https://doi.org/10.1042/CBI20110102.

Jopling, C. L., Yi, M., Lancaster, A. M., Lemon, S. M., & Sarnow, P. (2005). Modulation of hepatitis C virus RNA abundance by a liver-specific microRNA. *Science*, *309*(5740), 1577–1581. https://doi.org/10.1126/science.1113329.

Kamel, M. M., Matboli, M., Sallam, M., Montasser, I. F., Saad, A. S., & El-Tawdi, A. H. F. (2016). Investigation of long noncoding RNAs expression profile as potential serum biomarkers in patients with hepatocellular carcinoma. *Translational Research*, *168*, 134–145. https://doi.org/10.1016/j.trsl.2015.10.002.

Kanda, T., Goto, T., Hirotsu, Y., Moriyama, M., & Omata, M. (2019). Molecular mechanisms driving progression of liver cirrhosis towards hepatocellular carcinoma in chronic hepatitis B and C infections: A review. *International Journal of Molecular Sciences*, *20*(6), 1358. https://doi.org/10.3390/ijms20061358.

Kanda, T., Steele, R., Ray, R., & Ray, R. B. (2008). Hepatitis C virus core protein augments androgen receptor-mediated signaling. *Journal of Virology*, *82*(22), 11066–11072. https://doi.org/10.1128/JVI.01300-08.

Kanwal, F., Kramer, J. R., Ilyas, J., Duan, Z., & El-Serag, H. B. (2014). HCV genotype 3 is associated with an increased risk of cirrhosis and hepatocellular cancer in a national sample of U.S. veterans with HCV. *Hepatology*, *60*(1), 98–105. https://doi.org/10.1002/hep.27095.

Kared, H., Fabre, T., Bedard, N., Bruneau, J., & Shoukry, N. H. (2013). Galectin-9 and IL-21 mediate cross-regulation between Th17 and Treg cells during acute hepatitis C. *PLoS Pathogens*, *9*(6), e1003422. https://doi.org/10.1371/journal.ppat.1003422.

Kasai, D., Adachi, T., Deng, L., Nagano-Fujii, M., Sada, K., Ikeda, M., et al. (2009). HCV replication suppresses cellular glucose uptake through down-regulation of cell surface expression of glucose transporters. *Journal of Hepatology*, *50*(5), 883–894. https://doi.org/10.1016/j.jhep.2008.12.029.

Keck, Z. Y., Pierce, B. G., Lau, P., Lu, J., Wang, Y., Underwood, A., et al. (2019). Broadly neutralizing antibodies from an individual that naturally cleared multiple hepatitis C virus infections uncover molecular determinants for E2 targeting and vaccine design. *PLoS Pathogens*, *15*(5), e1007772. https://doi.org/10.1371/journal.ppat.1007772.

Keoshkerian, E., Hunter, M., Cameron, B., Nguyen, N., Sugden, P., Bull, R., et al. (2016). Hepatitis C-specific effector and regulatory CD4 T-cell responses are associated with the outcomes of primary infection. *Journal of Viral Hepatitis*, *23*(12), 985–993. https://doi.org/10.1111/jvh.12576.

Kim, J. H., Lee, C. H., & Lee, S. W. (2019). Exosomal transmission of MicroRNA from HCV replicating cells stimulates transdifferentiation in hepatic stellate cells. *Molecular Therapy- -Nucleic Acids*, *14*, 483–497. https://doi.org/10.1016/j.omtn.2019.01.006.

Khatun, M., & Ray, R. B. (2019). Mechanisms underlying hepatitis C virus-associated hepatic fibrosis. *Cell*, *8*(10), 1249. https://doi.org/10.3390/cells8101249.

Kim, H., Meyer, K., Di Bisceglie, A. M., & Ray, R. (2013). Hepatitis C virus suppresses C9 complement synthesis and impairs membrane attack complex function. *Journal of Virology*, *87*(10), 5858–5867. https://doi.org/10.1128/JVI.00174-13.

Koike, K. (2007). Hepatitis C virus contributes to hepatocarcinogenesis by modulating metabolic and intracellular signaling pathways. *Journal of Gastroenterology and Hepatology*, *22*(Suppl. 1), S108–S111. https://doi.org/10.1111/j.1440-1746.2006.04669.x.

Korenaga, M., Wang, T., Li, Y., Showalter, L. A., Chan, T., Sun, J., et al. (2005). Hepatitis C virus core protein inhibits mitochondrial electron transport and increases reactive oxygen species (ROS) production. *The Journal of Biological Chemistry*, *280*(45), 37481–37488. https://doi.org/10.1074/jbc.M506412200.

Kozbial, K., Moser, S., Schwarzer, R., Laferl, H., Al-Zoairy, R., Stauber, R., et al. (2016). Unexpected high incidence of hepatocellular carcinoma in cirrhotic patients with sustained virologic response following interferon-free direct-acting antiviral treatment. *Journal of Hepatology*, *65*(4), 856–858. https://doi.org/10.1016/j.jhep.2016.06.009.

Kramer, J. R., Kowalkowski, M. A., Duan, Z., & Chiao, E. Y. (2015). The effect of HIV viral control on the incidence of hepatocellular carcinoma in veterans with hepatitis C and HIV coinfection. *Journal of Acquired Immune Deficiency Syndromes*, *68*(4), 456–462. https://doi.org/10.1097/QAI.0000000000000494.

Kruse, R. L., Kramer, J. R., Tyson, G. L., Duan, Z., Chen, L., El-Serag, H. B., et al. (2014). Clinical outcomes of hepatitis B virus coinfection in a United States cohort of hepatitis C virus-infected patients. *Hepatology*, *60*(6), 1871–1878. https://doi.org/10.1002/hep.27337.

Kumar, V., Kato, N., Urabe, Y., Takahashi, A., Muroyama, R., Hosono, N., et al. (2011). Genome-wide association study identifies a susceptibility locus for HCV-induced hepatocellular carcinoma. *Nature Genetics*, *43*(5), 455–458. https://doi.org/10.1038/ng.809.

Kwon, Y. C., Bose, S. K., Steele, R., Meyer, K., Di Bisceglie, A. M., Ray, R. B., et al. (2015). Promotion of cancer stem-like cell properties in hepatitis C virus-infected hepatocytes. *Journal of Virology*, *89*(22), 11549–11556. https://doi.org/10.1128/JVI.01946-15.

Kwon, Y. C., Kim, H., Meyer, K., Di Bisceglie, A. M., & Ray, R. (2016). Distinct CD55 isoform synthesis and inhibition of complement-dependent cytolysis by hepatitis C virus. *Journal of Immunology*, *197*(4), 1127–1136. https://doi.org/10.4049/jimmunol.1600631.

Kwon, Y. C., Ray, R. B., & Ray, R. (2014). Hepatitis C virus infection: Establishment of chronicity and liver disease progression. *EXCLI Journal*, *13*, 977–996. https://www.ncbi.nlm.nih.gov/pubmed/26417315.

Kwon, Y. C., Sasaki, R., Meyer, K., & Ray, R. (2017). Hepatitis C virus core protein modulates Endoglin (CD105) signaling pathway for Liver pathogenesis. *Journal of Virology*, *91*(21). https://doi.org/10.1128/JVI.01235-17.

Landeras-Bueno, S., & Ortin, J. (2016). Regulation of influenza virus infection by long non-coding RNAs. *Virus Research*, *212*, 78–84. https://doi.org/10.1016/j.virusres.2015.08.008.

Lauer, G. M. (2013). Immune responses to hepatitis C virus (HCV) infection and the prospects for an effective HCV vaccine or immunotherapies. *The Journal of Infectious Diseases*, *207*(Suppl. 1), S7–S12. https://doi.org/10.1093/infdis/jis762.

Lazarus, J. V., Roel, E., & Elsharkawy, A. M. (2020). Hepatitis C virus epidemiology and the impact of interferon-free hepatitis C virus therapy. *Cold Spring Harbor Perspectives in Medicine*, *10*(3), a036913. https://doi.org/10.1101/cshperspect.a036913.

Lechner, F., Wong, D. K., Dunbar, P. R., Chapman, R., Chung, R. T., Dohrenwend, P., et al. (2000). Analysis of successful immune responses in persons infected with hepatitis C virus. *The Journal of Experimental Medicine*, *191*(9), 1499–1512. https://doi.org/10.1084/jem.191.9.1499.

Lee, M. H., Hsiao, T. I., Subramaniam, S. R., Le, A. K., Vu, V. D., Trinh, H. N., et al. (2017). HCV genotype 6 increased the risk for hepatocellular carcinoma among Asian patients with Liver cirrhosis. *The American Journal of Gastroenterology*, *112*(7), 1111–1119. https://doi.org/10.1038/ajg.2017.123.

Li, K., Foy, E., Ferreon, J. C., Nakamura, M., Ferreon, A. C., Ikeda, M., et al. (2005). Immune evasion by hepatitis C virus NS3/4A protease-mediated cleavage of the toll-like receptor 3 adaptor protein TRIF. *Proceedings of the National Academy of Sciences of the United States of America*, *102*(8), 2992–2997. https://doi.org/10.1073/pnas.0408824102.

Li, S., Tan, H. Y., Wang, N., Zhang, Z. J., Lao, L., Wong, C. W., et al. (2015). The role of oxidative stress and antioxidants in Liver diseases. *International Journal of Molecular Sciences*, *16*(11), 26087–26124. https://doi.org/10.3390/ijms161125942.

Lim, H. L., Myers, B. M., Hamilton, B. A., Davis, G. L., & Lau, J. Y. (1995). Plasma glutathione concentration in patients with chronic hepatitis C virus infection. *Journal of Viral Hepatitis*, *2*(4), 211–214. https://doi.org/10.1111/j.1365-2893.1995.tb00031.x.

Lin, M. V., King, L. Y., & Chung, R. T. (2015). Hepatitis C virus-associated cancer. *Annual Review of Pathology*, *10*, 345–370. https://doi.org/10.1146/annurev-pathol-012414-040323.

Lin, C., & Yang, L. (2018). Long noncoding RNA in cancer: Wiring signaling circuitry. *Trends in Cell Biology*, *28*(4), 287–301. https://doi.org/10.1016/j.tcb.2017.11.008.

Lindenbach, B. D., & Rice, C. M. (2013). The ins and outs of hepatitis C virus entry and assembly. *Nature Reviews. Microbiology*, *11*(10), 688–700. https://doi.org/10.1038/nrmicro3098.

Liu, S., Qiu, J., He, G., Liang, Y., Wang, L., Liu, C., et al. (2019). LncRNA MALAT1 acts as a miR-125a-3p sponge to regulate FOXM1 expression and promote hepatocellular carcinoma progression. *Journal of Cancer*, *10*(26), 6649–6659. https://doi.org/10.7150/jca.29213.

Machida, K., Cheng, K. T., Lai, C. K., Jeng, K. S., Sung, V. M., & Lai, M. M. (2006). Hepatitis C virus triggers mitochondrial permeability transition with production of reactive oxygen species, leading to DNA damage and STAT3 activation. *Journal of Virology*, *80*(14), 7199–7207. https://doi.org/10.1128/JVI.00321-06.

Mahmoudvand, S., Shokri, S., Taherkhani, R., & Farshadpour, F. (2019). Hepatitis C virus core protein modulates several signaling pathways involved in hepatocellular carcinoma. *World Journal of Gastroenterology*, *25*(1), 42–58. https://doi.org/10.3748/wjg.v25.i1.42.

Majumder, M., Ghosh, A. K., Steele, R., Ray, R., & Ray, R. B. (2001). Hepatitis C virus NS5A physically associates with p53 and regulates p21/waf1 gene expression in a p53-dependent manner. *Journal of Virology*, *75*(3), 1401–1407. https://doi.org/10.1128/JVI.75.3.1401-1407.2001.

Matsuzaki, K., Murata, M., Yoshida, K., Sekimoto, G., Uemura, Y., Sakaida, N., et al. (2007). Chronic inflammation associated with hepatitis C virus infection perturbs hepatic transforming growth factor beta signaling, promoting cirrhosis and hepatocellular carcinoma. *Hepatology*, *46*(1), 48–57. https://doi.org/10.1002/hep.21672.

Mattos, A. A., Marcon Pdos, S., Araujo, F. S., Coral, G. P., & Tovo, C. V. (2015). Hepatocellular carcinoma in a non-cirrhotic patient with sustained virological response after hepatitis C treatment. *Revista do Instituto de Medicina Tropical de São Paulo*, *57*(6), 519–522. https://doi.org/10.1590/S0036-46652015000600011.

Mazumdar, B., Kim, H., Meyer, K., Bose, S. K., Di Bisceglie, A. M., Ray, R. B., et al. (2013). Hepatitis C virus infection upregulates CD55 expression on the hepatocyte surface and promotes association with virus particles. *Journal of Virology*, *87*(14), 7902–7910. https://doi.org/10.1128/JVI.00917-13.

Mazumdar, B., Kim, H., Meyer, K., Bose, S. K., Di Bisceglie, A. M., Ray, R. B., et al. (2012). Hepatitis C virus proteins inhibit C3 complement production. *Journal of Virology*, *86*(4), 2221–2228. https://doi.org/10.1128/JVI.06577-11.

Mazzanti, R., Messerini, L., Monsacchi, L., Buzzelli, G., Zignego, A. L., Foschi, M., et al. (1997). Chronic viral hepatitis induced by hepatitis C but not hepatitis B virus infection correlates with increased liver angiogenesis. *Hepatology*, *25*(1), 229–234. https://doi.org/10.1002/hep.510250142.

McGivern, D. R., & Lemon, S. M. (2009). Tumor suppressors, chromosomal instability, and hepatitis C virus-associated liver cancer. *Annual Review of Pathology*, *4*, 399–415. https://doi.org/10.1146/annurev.pathol.4.110807.092202.

McMahon, B. J., Bruden, D., Townshend-Bulson, L., Simons, B., Spradling, P., Livingston, S., et al. (2017). Infection with hepatitis C virus genotype 3 is an independent risk factor for end-stage liver disease, hepatocellular carcinoma, and liver-related death. *Clinical Gastroenterology and Hepatology*, *15*(3), 431–437, e432. https://doi.org/10.1016/j.cgh.2016.10.012.

Mehta, N., & Yao, F. Y. (2017). Con: Treating hepatitis C virus with direct-acting antivirals: Fear not the perceived threat of hepatocellular carcinoma. *Liver Transplantation*, *23*(12), 1596–1600. https://doi.org/10.1002/lt.24959.

Meringer, H., Shibolet, O., & Deutsch, L. (2019). Hepatocellular carcinoma in the post-hepatitis C virus era: Should we change the paradigm? *World Journal of Gastroenterology*, *25*(29), 3929–3940. https://doi.org/10.3748/wjg.v25.i29.3929.

Messerini, L., Novelli, L., & Comin, C. E. (2004). Microvessel density and clinicopathological characteristics in hepatitis C virus and hepatitis B virus related hepatocellular carcinoma. *Journal of Clinical Pathology*, *57*(8), 867–871. https://doi.org/10.1136/jcp.2003.015784.

Meyer, K., Basu, A., Saito, K., Ray, R. B., & Ray, R. (2005). Inhibition of hepatitis C virus core protein expression in immortalized human hepatocytes induces cytochrome c-independent increase in Apaf-1 and caspase-9 activation for cell death. *Virology*, *336*(2), 198–207. https://doi.org/10.1016/j.virol.2005.03.016.

Midgard, H., Weir, A., Palmateer, N., Lo Re, V., 3rd, Pineda, J. A., Macias, J., et al. (2016). HCV epidemiology in high-risk groups and the risk of reinfection. *Journal of Hepatology*, *65*(Suppl. 1), S33–S45. https://doi.org/10.1016/j.jhep.2016.07.012.

Miki, D., Ochi, H., Hayes, C. N., Abe, H., Yoshima, T., Aikata, H., et al. (2011). Variation in the DEPDC5 locus is associated with progression to hepatocellular carcinoma in chronic hepatitis C virus carriers. *Nature Genetics*, *43*(8), 797–800. https://doi.org/10.1038/ng.876.

Missale, G., Bertoni, R., Lamonaca, V., Valli, A., Massari, M., Mori, C., et al. (1996). Different clinical behaviors of acute hepatitis C virus infection are associated with different vigor of the anti-viral cell-mediated immune response. *The Journal of Clinical Investigation*, *98*(3), 706–714. https://doi.org/10.1172/JCI118842.

Miyakawa, H., Izumi, N., Marumo, F., & Sato, C. (1996). Roles of alcohol, hepatitis virus infection, and gender in the development of hepatocellular carcinoma in patients with liver cirrhosis. *Alcoholism, Clinical and Experimental Research*, *20*(Suppl. 1), 91A–94A. https://doi.org/10.1111/j.1530-0277.1996.tb01738.x.

Moore, M. A., Park, C. B., & Tsuda, H. (1998). Implications of the hyperinsulinaemia-diabetes-cancer link for preventive efforts. *European Journal of Cancer Prevention*, 7(2), 89–107. https://www.ncbi.nlm.nih.gov/pubmed/9818771.

Moorman, J. P., Zhang, C. L., Ni, L., Ma, C. J., Zhang, Y., Wu, X. Y., et al. (2011). Impaired hepatitis B vaccine responses during chronic hepatitis C infection: Involvement of the PD-1 pathway in regulating CD4(+) T cell responses. *Vaccine*, *29*(17), 3169–3176. https://doi.org/10.1016/j.vaccine.2011.02.052.

Moreno-Cubero, E., & Larrubia, J. R. (2016). Specific CD8(+) T cell response immunotherapy for hepatocellular carcinoma and viral hepatitis. *World Journal of Gastroenterology*, *22*(28), 6469–6483. https://doi.org/10.3748/wjg.v22.i28.6469.

Moriya, K., Yotsuyanagi, H., Shintani, Y., Fujie, H., Ishibashi, K., Matsuura, Y., et al. (1997). Hepatitis C virus core protein induces hepatic steatosis in transgenic mice. *The Journal of General Virology*, *78*(Pt 7), 1527–1531. https://doi.org/10.1099/0022-1317-78-7-1527.

Moucari, R., Asselah, T., Cazals-Hatem, D., Voitot, H., Boyer, N., Ripault, M. P., et al. (2008). Insulin resistance in chronic hepatitis C: Association with genotypes 1 and 4, serum HCV RNA level, and liver fibrosis. *Gastroenterology*, *134*(2), 416–423. https://doi.org/10.1053/j.gastro.2007.11.010.

Mukherjee, A., Di Bisceglie, A. M., & Ray, R. B. (2015). Hepatitis C virus-mediated enhancement of microRNA miR-373 impairs the JAK/STAT signaling pathway. *Journal of Virology*, *89*(6), 3356–3365. https://doi.org/10.1128/JVI.03085-14.

Mukherjee, A., Shrivastava, S., Bhanja Chowdhury, J., Ray, R., & Ray, R. B. (2014). Transcriptional suppression of miR-181c by hepatitis C virus enhances homeobox A1 expression. *Journal of Virology*, *88*(14), 7929–7940. https://doi.org/10.1128/JVI.00787-14.
Munn, D. H., & Mellor, A. L. (2016). IDO in the tumor microenvironment: Inflammation, counter-regulation, and tolerance. *Trends in Immunology*, *37*(3), 193–207. https://doi.org/10.1016/j.it.2016.01.002.
Nazzal, M., Sur, S., Steele, R., Khatun, M., Patra, T., Phillips, N., et al. (2020). Establishment of a patient-derived xenograft tumor from hepatitis C-associated liver cancer and evaluation of imatinib treatment efficacy. *Hepatology*, *72*(2), 379–388. https://doi.org/10.1002/hep.31298.
Negash, A. A., Ramos, H. J., Crochet, N., Lau, D. T., Doehle, B., Papic, N., et al. (2013). IL-1beta production through the NLRP3 inflammasome by hepatic macrophages links hepatitis C virus infection with liver inflammation and disease. *PLoS Pathogens*, *9*(4), e1003330. https://doi.org/10.1371/journal.ppat.1003330.
Neumann-Haefelin, C., & Thimme, R. (2013). Adaptive immune responses in hepatitis C virus infection. *Current Topics in Microbiology and Immunology*, *369*, 243–262. https://doi.org/10.1007/978-3-642-27340-7_10.
Osburn, W. O., Fisher, B. E., Dowd, K. A., Urban, G., Liu, L., Ray, S. C., et al. (2010). Spontaneous control of primary hepatitis C virus infection and immunity against persistent reinfection. *Gastroenterology*, *138*(1), 315–324. https://doi.org/10.1053/j.gastro.2009.09.017.
Osburn, W. O., Snider, A. E., Wells, B. L., Latanich, R., Bailey, J. R., Thomas, D. L., et al. (2014). Clearance of hepatitis C infection is associated with the early appearance of broad neutralizing antibody responses. *Hepatology*, *59*(6), 2140–2151. https://doi.org/10.1002/hep.27013.
Parvaiz, F., Manzoor, S., Iqbal, J., McRae, S., Javed, F., Ahmed, Q. L., et al. (2014). Hepatitis C virus nonstructural protein 5A favors upregulation of gluconeogenic and lipogenic gene expression leading towards insulin resistance: A metabolic syndrome. *Archives of Virology*, *159*(5), 1017–1025. https://doi.org/10.1007/s00705-013-1892-3.
Patra, T., Meyer, K., Ray, R. B., & Ray, R. (2020a). A combination of AZD5363 and FH5363 induces lethal autophagy in transformed hepatocytes. *Cell Death & Disease*, *11*(7), 540. https://doi.org/10.1038/s41419-020-02741-1.
Patra, T., Meyer, K., Ray, R. B., & Ray, R. (2020b). Hepatitis C virus mediated inhibition of miR-181c activates ATM signaling and promotes hepatocyte growth. *Hepatology*, *71*(3), 780–793. https://doi.org/10.1002/hep.30893.
Patra, T., Ray, R. B., & Ray, R. (2019). Strategies to circumvent host innate immune response by hepatitis C virus. *Cells*, *8*(3), 274. https://doi.org/10.3390/cells8030274.
Patra, T., Sasaki, R., Meyer, K., Ray, R. B., & Ray, R. (2019). Transforming growth factor beta acts as a regulatory molecule for lipogenic pathways among hepatitis C virus genotype-specific infections. *Journal of Virology*, *93*(18), e00811–19. https://doi.org/10.1128/JVI.00811-19.
Paul, D., Madan, V., & Bartenschlager, R. (2014). Hepatitis C virus RNA replication and assembly: Living on the fat of the land. *Cell Host & Microbe*, *16*(5), 569–579. https://doi.org/10.1016/j.chom.2014.10.008.
Pazienza, V., Vinciguerra, M., Andriulli, A., & Mangia, A. (2010). Hepatitis C virus core protein genotype 3a increases SOCS-7 expression through PPAR-{gamma} in Huh-7 cells. *The Journal of General Virology*, *91*(Pt 7), 1678–1686. https://doi.org/10.1099/vir.0.020644-0.
Pea, A., Jamieson, N. B., & Braconi, C. (2020). Biology and clinical application of regulatory RNAs in hepatocellular carcinoma. *Hepatology*. https://doi.org/10.1002/hep.31225.
Peng, N., He, J., Li, J., Huang, H., Huang, W., Liao, Y., et al. (2020). Long noncoding RNA MALAT1 inhibits the apoptosis and autophagy of hepatocellular carcinoma cell by targeting the microRNA-146a/PI3K/Akt/mTOR axis. *Cancer Cell International*, *20*, 165. https://doi.org/10.1186/s12935-020-01231-w.

Perez, S., Kaspi, A., Domovitz, T., Davidovich, A., Lavi-Itzkovitz, A., Meirson, T., et al. (2019). Hepatitis C virus leaves an epigenetic signature post cure of infection by direct-acting antivirals. *PLoS Genetics*, *15*(6), e1008181. https://doi.org/10.1371/journal.pgen.1008181.

Perlemuter, G., Sabile, A., Letteron, P., Vona, G., Topilco, A., Chretien, Y., et al. (2002). Hepatitis C virus core protein inhibits microsomal triglyceride transfer protein activity and very low density lipoprotein secretion: A model of viral-related steatosis. *The FASEB Journal*, *16*(2), 185–194. https://doi.org/10.1096/fj.01-0396com.

Persico, M., Russo, R., Persico, E., Svelto, M., Spano, D., Andolfo, I., et al. (2009). SOCS3 and IRS-1 gene expression differs between genotype 1 and genotype 2 hepatitis C virus-infected HepG2 cells. *Clinical Chemistry and Laboratory Medicine*, *47*(10), 1217–1225. https://doi.org/10.1515/CCLM.2009.280.

Pestka, J. M., Zeisel, M. B., Blaser, E., Schurmann, P., Bartosch, B., Cosset, F. L., et al. (2007). Rapid induction of virus-neutralizing antibodies and viral clearance in a single-source outbreak of hepatitis C. *Proceedings of the National Academy of Sciences of the United States of America*, *104*(14), 6025–6030. https://doi.org/10.1073/pnas.0607026104.

Pio, R., Ajona, D., Ortiz-Espinosa, S., Mantovani, A., & Lambris, J. D. (2019). Complementing the cancer-immunity cycle. *Frontiers in Immunology*, *10*, 774. https://doi.org/10.3389/fimmu.2019.00774.

Platten, M., von Knebel Doeberitz, N., Oezen, I., Wick, W., & Ochs, K. (2014). Cancer immunotherapy by targeting IDO1/TDO and their downstream effectors. *Frontiers in Immunology*, *5*, 673. https://doi.org/10.3389/fimmu.2014.00673.

Plissonnier, M. L., Herzog, K., Levrero, M., & Zeisel, M. B. (2018). Non-coding RNAs and hepatitis C virus-induced hepatocellular carcinoma. *Viruses*, *10*(11), 591. https://doi.org/10.3390/v10110591.

Prendergast, G. C., Smith, C., Thomas, S., Mandik-Nayak, L., Laury-Kleintop, L., Metz, R., et al. (2014). Indoleamine 2,3-dioxygenase pathways of pathogenic inflammation and immune escape in cancer. *Cancer Immunology, Immunotherapy*, *63*(7), 721–735. https://doi.org/10.1007/s00262-014-1549-4.

Raff, E. J., Kakati, D., Bloomer, J. R., Shoreibah, M., Rasheed, K., & Singal, A. K. (2015). Diabetes mellitus predicts occurrence of cirrhosis and hepatocellular cancer in alcoholic liver and non-alcoholic fatty liver diseases. *Journal of Clinical and Translational Hepatology*, *3*(1), 9–16. https://doi.org/10.14218/JCTH.2015.00001.

Ramiere, C., Rodriguez, J., Enache, L. S., Lotteau, V., Andre, P., & Diaz, O. (2014). Activity of hexokinase is increased by its interaction with hepatitis C virus protein NS5A. *Journal of Virology*, *88*(6), 3246–3254. https://doi.org/10.1128/JVI.02862-13.

Rao, C. V., Asch, A. S., & Yamada, H. Y. (2017). Frequently mutated genes/pathways and genomic instability as prevention targets in liver cancer. *Carcinogenesis*, *38*(1), 2–11. https://doi.org/10.1093/carcin/bgw118.

Raught, B., Gingras, A. C., & Sonenberg, N. (2001). The target of rapamycin (TOR) proteins. *Proceedings of the National Academy of Sciences of the United States of America*, *98*(13), 7037–7044. https://doi.org/10.1073/pnas.121145898.

Ray, R. B., & Ray, R. (2019). Hepatitis C virus manipulates humans as its favorite host for a long-term relationship. *Hepatology*, *69*(2), 889–900. https://doi.org/10.1002/hep.30214.

Rehermann, B. (2013). Pathogenesis of chronic viral hepatitis: Differential roles of T cells and NK cells. *Nature Medicine*, *19*(7), 859–868. https://doi.org/10.1038/nm.3251.

Reig, M., Marino, Z., Perello, C., Inarrairaegui, M., Ribeiro, A., Lens, S., et al. (2016). Unexpected high rate of early tumor recurrence in patients with HCV-related HCC undergoing interferon-free therapy. *Journal of Hepatology*, *65*(4), 719–726. https://doi.org/10.1016/j.jhep.2016.04.008.

Rinn, J. L., & Chang, H. Y. (2012). Genome regulation by long noncoding RNAs. *Annual Review of Biochemistry*, *81*, 145–166. https://doi.org/10.1146/annurev-biochem-051410-092902.

Roche, B., Coilly, A., Duclos-Vallee, J. C., & Samuel, D. (2018). The impact of treatment of hepatitis C with DAAs on the occurrence of HCC. *Liver International*, *38*(Suppl. 1), 139–145. https://doi.org/10.1111/liv.13659.

Ronca, R., Benkheil, M., Mitola, S., Struyf, S., & Liekens, S. (2017). Tumor angiogenesis revisited: Regulators and clinical implications. *Medicinal Research Reviews*, *37*(6), 1231–1274. https://doi.org/10.1002/med.21452.

Saab, S. (2014). Surveillance for hepatocellular carcinoma. *Gastroenterología y Hepatología*, *10*(11), 745–747. https://www.ncbi.nlm.nih.gov/pubmed/28435411.

Sadri Nahand, J., Bokharaei-Salim, F., Salmaninejad, A., Nesaei, A., Mohajeri, F., Moshtzan, A., et al. (2019). microRNAs: Key players in virus-associated hepatocellular carcinoma. *Journal of Cellular Physiology*, *234*(8), 12188–12225. https://doi.org/10.1002/jcp.27956.

Sanchez-Elsner, T., Botella, L. M., Velasco, B., Langa, C., & Bernabeu, C. (2002). Endoglin expression is regulated by transcriptional cooperation between the hypoxia and transforming growth factor-beta pathways. *The Journal of Biological Chemistry*, *277*(46), 43799–43808. https://doi.org/10.1074/jbc.M207160200.

Sasaki, R., Devhare, P. B., Steele, R., Ray, R., & Ray, R. B. (2017). Hepatitis C virus-induced CCL5 secretion from macrophages activates hepatic stellate cells. *Hepatology*, *66*(3), 746–757. https://doi.org/10.1002/hep.29170.

Sasaki, R., Kanda, T., Kato, N., Yokosuka, O., & Moriyama, M. (2018). Hepatitis C virus-associated hepatocellular carcinoma after sustained virologic response. *World Journal of Hepatology*, *10*(12), 898–906. https://doi.org/10.4254/wjh.v10.i12.898.

Sasaki, R., Sur, S., Cheng, Q., Steele, R., & Ray, R. B. (2019). Repression of MicroRNA-30e by hepatitis C virus enhances fatty acid synthesis. *Hepatology Communications*, *3*(7), 943–953. https://doi.org/10.1002/hep4.1362.

Serfaty, L., Andreani, T., Giral, P., Carbonell, N., Chazouilleres, O., & Poupon, R. (2001). Hepatitis C virus induced hypobetalipoproteinemia: A possible mechanism for steatosis in chronic hepatitis C. *Journal of Hepatology*, *34*(3), 428–434. https://doi.org/10.1016/s0168-8278(00)00036-2.

Shata, M. T., Anthony, D. D., Carlson, N. L., Andrus, L., Brotman, B., Tricoche, N., et al. (2002). Characterization of the immune response against hepatitis C infection in recovered, and chronically infected chimpanzees. *Journal of Viral Hepatitis*, *9*(6), 400–410. https://doi.org/10.1046/j.1365-2893.2002.00373.x.

Sheikh, M. Y., Choi, J., Qadri, I., Friedman, J. E., & Sanyal, A. J. (2008). Hepatitis C virus infection: Molecular pathways to metabolic syndrome. *Hepatology*, *47*(6), 2127–2133. https://doi.org/10.1002/hep.22269.

Shi, S. T., Polyak, S. J., Tu, H., Taylor, D. R., Gretch, D. R., & Lai, M. M. (2002). Hepatitis C virus NS5A colocalizes with the core protein on lipid droplets and interacts with apolipoproteins. *Virology*, *292*(2), 198–210. https://doi.org/10.1006/viro.2001.1225.

Shi, L., Wang, J. M., Ren, J. P., Cheng, Y. Q., Ying, R. S., Wu, X. Y., et al. (2014). KLRG1 impairs CD4+ T cell responses via p16ink4a and p27kip1 pathways: Role in hepatitis B vaccine failure in individuals with hepatitis C virus infection. *Journal of Immunology*, *192*(2), 649–657. https://doi.org/10.4049/jimmunol.1302069.

Shimomura, Y., Takaki, A., Wada, N., Yasunaka, T., Ikeda, F., Maruyama, T., et al. (2017). The serum oxidative/anti-oxidative stress balance becomes dysregulated in patients with non-alcoholic steatohepatitis associated with hepatocellular carcinoma. *Internal Medicine*, *56*(3), 243–251. https://doi.org/10.2169/internalmedicine.56.7002.

Shrivastava, S., Mukherjee, A., Ray, R., & Ray, R. B. (2013). Hepatitis C virus induces interleukin-1beta (IL-1beta)/IL-18 in circulatory and resident liver macrophages. *Journal of Virology*, *87*(22), 12284–12290. https://doi.org/10.1128/JVI.01962-13.

Shrivastava, S., Petrone, J., Steele, R., Lauer, G. M., Di Bisceglie, A. M., & Ray, R. B. (2013). Up-regulation of circulating miR-20a is correlated with hepatitis C virus-mediated liver disease progression. *Hepatology*, *58*(3), 863–871. https://doi.org/10.1002/hep.26296.

Shrivastava, S., Steele, R., Ray, R., & Ray, R. B. (2015). MicroRNAs: Role in hepatitis C virus pathogenesis. *Genes & Diseases*, *2*(1), 35–45. https://doi.org/10.1016/j.gendis.2015.01.001.

Simonetti, R. G., Camma, C., Fiorello, F., Cottone, M., Rapicetta, M., Marino, L., et al. (1992). Hepatitis C virus infection as a risk factor for hepatocellular carcinoma in patients with cirrhosis. A case-control study. *Annals of Internal Medicine*, *116*(2), 97–102. https://doi.org/10.7326/0003-4819-116-2-97.

Smyk-Pearson, S., Tester, I. A., Klarquist, J., Palmer, B. E., Pawlotsky, J. M., Golden-Mason, L., et al. (2008). Spontaneous recovery in acute human hepatitis C virus infection: Functional T-cell thresholds and relative importance of CD4 help. *Journal of Virology*, *82*(4), 1827–1837. https://doi.org/10.1128/JVI.01581-07.

Spizzo, R., Almeida, M. I., Colombatti, A., & Calin, G. A. (2012). Long non-coding RNAs and cancer: A new frontier of translational research? *Oncogene*, *31*(43), 4577–4587. https://doi.org/10.1038/onc.2011.621.

Stoermer, K. A., & Morrison, T. E. (2011). Complement and viral pathogenesis. *Virology*, *411*(2), 362–373. https://doi.org/10.1016/j.virol.2010.12.045.

Su, A. I., Pezacki, J. P., Wodicka, L., Brideau, A. D., Supekova, L., Thimme, R., et al. (2002). Genomic analysis of the host response to hepatitis C virus infection. *Proceedings of the National Academy of Sciences of the United States of America*, *99*(24), 15669–15674. https://doi.org/10.1073/pnas.202608199.

Sur, S., Sasaki, R., Devhare, P., Steele, R., Ray, R., & Ray, R. B. (2018). Association between MicroRNA-373 and Long noncoding RNA NORAD in hepatitis C virus-infected hepatocytes impairs Wee1 expression for growth promotion. *Journal of Virology*, *92*(20), e01215–18. https://doi.org/10.1128/JVI.01215-18.

Syed, G. H., Amako, Y., & Siddiqui, A. (2010). Hepatitis C virus hijacks host lipid metabolism. *Trends in Endocrinology and Metabolism*, *21*(1), 33–40. https://doi.org/10.1016/j.tem.2009.07.005.

Takaki, A., Wiese, M., Maertens, G., Depla, E., Seifert, U., Liebetrau, A., et al. (2000). Cellular immune responses persist and humoral responses decrease two decades after recovery from a single-source outbreak of hepatitis C. *Nature Medicine*, *6*(5), 578–582. https://doi.org/10.1038/75063.

Takaki, A., & Yamamoto, K. (2015). Control of oxidative stress in hepatocellular carcinoma: Helpful or harmful? *World Journal of Hepatology*, 7(7), 968–979. https://doi.org/10.4254/wjh.v7.i7.968.

Tardif, K. D., Mori, K., & Siddiqui, A. (2002). Hepatitis C virus subgenomic replicons induce endoplasmic reticulum stress activating an intracellular signaling pathway. *Journal of Virology*, *76*(15), 7453–7459. https://doi.org/10.1128/jvi.76.15.7453-7459.2002.

Toraih, E. A., Ellawindy, A., Fala, S. Y., Al Ageeli, E., Gouda, N. S., Fawzy, M. S., et al. (2018). Oncogenic long noncoding RNA MALAT1 and HCV-related hepatocellular carcinoma. *Biomedicine & Pharmacotherapy*, *102*, 653–669. https://doi.org/10.1016/j.biopha.2018.03.105.

Tseng, C. T., & Klimpel, G. R. (2002). Binding of the hepatitis C virus envelope protein E2 to CD81 inhibits natural killer cell functions. *The Journal of Experimental Medicine*, *195*(1), 43–49. https://doi.org/10.1084/jem.20011145.

Tsigkos, S., Zhou, Z., Kotanidou, A., Fulton, D., Zakynthinos, S., Roussos, C., et al. (2006). Regulation of Ang2 release by PTEN/PI3-kinase/Akt in lung microvascular endothelial cells. *Journal of Cellular Physiology*, *207*(2), 506–511. https://doi.org/10.1002/jcp.20592.
Urbani, S., Amadei, B., Fisicaro, P., Tola, D., Orlandini, A., Sacchelli, L., et al. (2006). Outcome of acute hepatitis C is related to virus-specific CD4 function and maturation of antiviral memory CD8 responses. *Hepatology*, *44*(1), 126–139. https://doi.org/10.1002/hep.21242.
van de Veerdonk, M. C., Kind, T., Marcus, J. T., Mauritz, G. J., Heymans, M. W., Bogaard, H. J., et al. (2011). Progressive right ventricular dysfunction in patients with pulmonary arterial hypertension responding to therapy. *Journal of the American College of Cardiology*, *58*(24), 2511–2519. https://doi.org/10.1016/j.jacc.2011.06.068.
Van Renne, N., Roca Suarez, A. A., Duong, F. H. T., Gondeau, C., Calabrese, D., Fontaine, N., et al. (2018). miR-135a-5p-mediated downregulation of protein tyrosine phosphatase receptor delta is a candidate driver of HCV-associated hepatocarcinogenesis. *Gut*, *67*(5), 953–962. https://doi.org/10.1136/gutjnl-2016-312270.
Vescovo, T., Refolo, G., Vitagliano, G., Fimia, G. M., & Piacentini, M. (2016). Molecular mechanisms of hepatitis C virus-induced hepatocellular carcinoma. *Clinical Microbiology and Infection*, *22*(10), 853–861. https://doi.org/10.1016/j.cmi.2016.07.019.
Villani, R., Vendemiale, G., & Serviddio, G. (2018). Molecular mechanisms involved in HCC recurrence after direct-acting antiviral therapy. *International Journal of Molecular Sciences*, *20*(1). https://doi.org/10.3390/ijms20010049.
Villanueva, A. (2019). Hepatocellular carcinoma. *The New England Journal of Medicine*, *380*(15), 1450–1462. https://doi.org/10.1056/NEJMra1713263.
Vrancken, K., Paeshuyse, J., & Liekens, S. (2012). Angiogenic activity of hepatitis B and C viruses. *Antiviral Chemistry & Chemotherapy*, *22*(4), 159–170. https://doi.org/10.3851/IMP1987.
Wei, L., Wang, X., Lv, L., Liu, J., Xing, H., Song, Y., et al. (2019). The emerging role of microRNAs and long noncoding RNAs in drug resistance of hepatocellular carcinoma. *Molecular Cancer*, *18*(1), 147. https://doi.org/10.1186/s12943-019-1086-z.
Wu, S., Ai, H., Zhang, K., Yun, H., & Xie, F. (2019). Long non-coding RNA EGOT promotes the malignant phenotypes of hepatocellular carcinoma cells and increases the expression of HMGA2 via down-regulating miR-33a-5p. *Oncotargets and Therapy*, *12*, 11623–11635. https://doi.org/10.2147/OTT.S218308.
Wu, S. C., Chang, S. C., Wu, H. Y., Liao, P. J., & Chang, M. F. (2008). Hepatitis C virus NS5A protein down-regulates the expression of spindle gene Aspm through PKR-p38 signaling pathway. *The Journal of Biological Chemistry*, *283*(43), 29396–29404. https://doi.org/10.1074/jbc.M802821200.
Xiao, J. N., Yan, T. H., Yu, R. M., Gao, Y., Zeng, W. L., Lu, S. W., et al. (2017). Long noncoding RNA UCA1 regulates the expression of Snail2 by miR-203 to promote hepatocellular carcinoma progression. *Journal of Cancer Research and Clinical Oncology*, *143*(6), 981–990. https://doi.org/10.1007/s00432-017-2370-1.
Yang, W., Hood, B. L., Chadwick, S. L., Liu, S., Watkins, S. C., Luo, G., et al. (2008). Fatty acid synthase is up-regulated during hepatitis C virus infection and regulates hepatitis C virus entry and production. *Hepatology*, *48*(5), 1396–1403. https://doi.org/10.1002/hep.22508.
Zavaglia, C., Okolicsanyi, S., Cesarini, L., Mazzarelli, C., Pontecorvi, V., Ciaccio, A., et al. (2017). Is the risk of neoplastic recurrence increased after prescribing direct-acting antivirals for HCV patients whose HCC was previously cured? *Journal of Hepatology*, *66*(1), 236–237. https://doi.org/10.1016/j.jhep.2016.08.016.

Zekri, A. N., Youssef, A. S., El-Desouky, E. D., Ahmed, O. S., Lotfy, M. M., Nassar, A. A., et al. (2016). Serum microRNA panels as potential biomarkers for early detection of hepatocellular carcinoma on top of HCV infection. *Tumour Biology*, *37*(9), 12273–12286. https://doi.org/10.1007/s13277-016-5097-8.

Zhang, Y., Wei, W., Cheng, N., Wang, K., Li, B., Jiang, X., et al. (2012). Hepatitis C virus-induced up-regulation of microRNA-155 promotes hepatocarcinogenesis by activating Wnt signaling. *Hepatology*, *56*(5), 1631–1640. https://doi.org/10.1002/hep.25849.

Zhang, C., Yang, X., Qi, Q., Gao, Y., Wei, Q., & Han, S. (2018). lncRNA-HEIH in serum and exosomes as a potential biomarker in the HCV-related hepatocellular carcinoma. *Cancer Biomarkers*, *21*(3), 651–659. https://doi.org/10.3233/CBM-170727.

Zhang, H., Zhu, C., Zhao, Y., Li, M., Wu, L., Yang, X., et al. (2015). Long non-coding RNA expression profiles of hepatitis C virus-related dysplasia and hepatocellular carcinoma. *Oncotarget*, *6*(41), 43770–43778. https://doi.org/10.18632/oncotarget.6087.

Zhao, B., Lu, Y., Cao, X., Zhu, W., Kong, L., Ji, H., et al. (2019). MiRNA-124 inhibits the proliferation, migration and invasion of cancer cell in hepatocellular carcinoma by down-regulating lncRNA-UCA1. *Oncotargets and Therapy*, *12*, 4509–4516. https://doi.org/10.2147/OTT.S205169.

Zheng, Y., Hlady, R. A., Joyce, B. T., Robertson, K. D., He, C., Nannini, D. R., et al. (2019). DNA methylation of individual repetitive elements in hepatitis C virus infection-induced hepatocellular carcinoma. *Clinical Epigenetics*, *11*(1), 145. https://doi.org/10.1186/s13148-019-0733-y.

Zhu, A. X., Finn, R. S., Edeline, J., Cattan, S., Ogasawara, S., Palmer, D., et al. (2018). Pembrolizumab in patients with advanced hepatocellular carcinoma previously treated with sorafenib (KEYNOTE-224): A non-randomised, open-label phase 2 trial. *The Lancet Oncology*, *19*(7), 940–952. https://doi.org/10.1016/S1470-2045(18)30351-6.

CHAPTER FOUR

NAFLD-related HCC

Bubu A. Banini[a] and Arun J. Sanyal[b,*]
[a]Section of Digestive Diseases, Yale University, New Haven, CT, United States
[b]Division of Gastroenterology, Hepatology and Nutrition, Virginia Commonwealth University, Richmond, VA, United States
*Corresponding author: e-mail address: arun.sanyal@vcuhealth.org

Contents

1. The problem of NAFLD-related HCC 144
2. Metabolic syndrome and NAFLD in driving HCC 146
   2.1 Obesity and HCC 147
   2.2 Type 2 DM and HCC 148
3. NAFLD-related HCC 149
   3.1 Patients with cirrhosis 149
   3.2 Patients without cirrhosis 150
4. Molecular mechanisms of NAFLD-related HCC 150
   4.1 Genetic mechanisms 151
   4.2 Epigenetic mechanisms 153
   4.3 Metabolic and endocrine pathways 153
   4.4 Immunologic mechanisms 154
   4.5 Gut microbiota and dietary modulation 155
5. Prevention and surveillance of NAFLD-related HCC 156
6. Management of NAFLD-related HCC 157
7. Conclusions/future directions 158
Acknowledgment 159
Disclosures 159
References 160

Abstract

Nonalcoholic fatty liver disease (NAFLD) is one of the major drivers for the rising trend in hepatocellular carcinoma (HCC). Over the past three decades, the incidence of both NAFLD and HCC have increased two- to threefold. It has been forecasted that the number of patients with NAFLD in the Unites States will reach 101 million by 2030; global increase is also foreseen. This trend will likely continue to translate into increased HCC in the Unites States and across the globe. In this chapter, we summarize the current evidence linking NAFLD, metabolic syndrome, particularly obesity and type 2 diabetes mellitus, and HCC. We describe the main molecular mechanisms connecting these metabolic perturbations and hepatocarcinogenesis.

Advances in Cancer Research, Volume 149
ISSN 0065-230X
https://doi.org/10.1016/bs.acr.2020.11.001

1. The problem of NAFLD-related HCC

Nonalcoholic fatty liver disease (NAFLD) is a rapidly growing indication for liver transplant due to hepatocellular carcinoma (HCC) (Wong et al., 2015). In the United States, the prevalence of NAFLD is forecasted to reach almost 101 million by 2030 (Estes, Razavi, Loomba, Younossi, & Sanyal, 2018). The term NAFLD encompasses a spectrum of liver diseases ranging from nonalcoholic fatty liver (NAFL) to nonalcoholic steatohepatitis (NASH) (de Alwis & Day, 2008; Farrell & Larter, 2006). NASH is the more aggressive subtype of NAFLD and has been estimated to occur in up to a third of patients with NAFLD (Estes et al., 2018). NASH is characterized by hepatocellular injury and liver inflammation and increases the risk of disease progression to fibrosis, cirrhosis and HCC (Banini & Sanyal, 2016).

From the mid-1970s to the mid-2000s, the overall incidence of HCC tripled from 1.4 per 100,000 persons per year to 4.8 per 100,000 persons per year (Altekruse, McGlynn, & Reichman, 2009; Baffy, Brunt, & Caldwell, 2012; Mittal & El-Serag, 2013), in tandem with NAFLD prevalence which increased two- to threefold from the mid-1980s to the mid-2000s (Lazo et al., 2013). From 2004 to 2013, NAFLD showed an increase of 170% as the etiology for end stage liver disease in new liver transplant waitlist registrants, compared to chronic hepatitis C (HCV) and alcohol-related liver disease which increased by 14% and 45%, respectively (Wong et al., 2015). An HCC etiological factor by itself, NAFLD can also act synergistically with other risk factors such as alcohol and chronic viral hepatitis to drive HCC progression (Petta & Craxi, 2010). The evidence supporting NAFLD as an etiological risk factor for HCC is derived from a number of studies showing clinical or biopsy-proven NAFLD, NAFLD-related cirrhosis or cryptogenic cirrhosis in patients with HCC.

Recently, a panel of experts from 22 countries proposed adopting the term metabolic dysfunction-associated fatty liver disease (MAFLD) to replace NAFLD, which not only eliminates the prominent role that alcohol exclusion played in the historical definition of the disease, but also aims to outline positive criteria with which to define the disease (Eslam et al., 2020). The panel also recommended that MAFLD be diagnosed based on the presence of hepatic steatosis in addition to overweight/obesity, type 2 diabetes mellitus, or metabolic dysregulation. The adoption of the term MAFLD is currently in its infancy but may experience more widespread use in the near future.

There are several reasons why NAFLD is a salient contributor to the burden of HCC. First, the sheer number of patients with NAFLD increases the prevalence of NAFLD-associated end stage liver disease. According to 2017 global estimates, NAFLD accounts for 60% of the burden of chronic liver disease worldwide, far surpassing the prevalence of other common etiologies of chronic liver disease (Moon, Singal, & Tapper, 2020). Second, although about one-third of NAFLD-related HCC cases occur in individuals without cirrhosis, the total number of cases of NAFLD is very high, translating into a low incidence rate for HCC that does not meet standards for surveillance of the population at risk. There are currently no accepted approaches to identify the subset of individuals with pre-cirrhotic NAFLD who will develop HCC. Third, the asymptomatic nature of NAFLD allows the disease to progress silently in many patients, with decompensated cirrhosis or HCC being the first manifestation of the disease (Fazel, Koenig, Sayiner, Goodman, & Younossi, 2016). Several studies have shown that NAFLD-related cirrhosis is diagnosed later than cirrhosis related to other causes of chronic liver disease (Ratziu et al., 2002), and that NAFLD-related HCC patients are at higher risk of being undiagnosed (Singal et al., 2012). Fourth, even when NAFLD progresses to cirrhosis and raises the radar for HCC surveillance, co-existent obesity in majority of NAFLD patients creates technical challenges in the performance of ultrasound and increases the likelihood that small or early cancers may be missed. Patients with body mass index (BMI) $>30\,\text{kg/m}^2$ were found to have ultrasound sensitivity of 0.76, compared to 0.87 for patients with BMI $<30\,\text{kg/m}^2$ (Samoylova, Mehta, Roberts, & Yao, 2018). Also, patients with NASH have decreased ultrasound sensitivity of 0.59 compared to 0.84 for other etiologies of chronic liver disease (Samoylova et al., 2018). Fifth, the length of time from NAFLD diagnosis to HCC development is relatively shorter than other etiologies of chronic liver disease predisposing to HCC, likely due to the disease being diagnosed later and in more advanced stages. Patients with NAFLD-related HCC are generally older, have shorter survival and are more likely to succumb to HCC compared to other etiologies (Younossi et al., 2015). Sixth, there are currently no medications approved to treat NAFLD and curtail disease progression. This should hopefully change in the next few years given the vast number of medications in the pipeline of testing for NAFLD. All the above-mentioned factors compound to make NAFLD a unique and formidable etiological risk factor for HCC.

2. Metabolic syndrome and NAFLD in driving HCC

Metabolic syndrome is the strongest risk factor for development of NAFLD and NASH (Adams, Waters, Knuiman, Elliott, & Olynyk, 2009; Dowman, Tomlinson, & Newsome, 2011). The syndrome is a condition of metabolic dysregulation characterized by accumulation of visceral fat, insulin resistance, and hyperlipidemia which predisposes the individual to type 2 diabetes mellitus (T2DM), hypertension and atherosclerotic cardiovascular disease (Grundy et al., 2005). An individual with metabolic syndrome has three or more of the following conditions: abdominal obesity defined as waist circumference over 40 in. (102 cm) in men and 35 in. (88 cm) in women; serum triglycerides over 150 mg/dL or being on triglyceride-lowering medication; serum high density lipoprotein (HDL) cholesterol less than 40 mg/dL in men and less than 50 mg/dL in women or being on HDL-increasing medication; blood pressure greater than 130/85 or being on anti-hypertensive medication; and fasting plasma glucose greater than 100 mg/dl or being on glucose-lowering medication. The relationship between NAFLD and components of metabolic syndrome appears to be reciprocal, especially pertaining to T2DM and cardiovascular disease. Over 70% of patients with T2DM have NAFLD (Arrese, Barrera, Triantafilo, & Arab, 2019), with resolution of fatty liver leading to a reduced risk of T2DM (Sung, Wild, & Byrne, 2013). Patients with NAFLD also have increased risk of cardiovascular disease, which is the leading cause of mortality among this patient population (Mantovani, Ballestri, Lonardo, & Targher, 2016).

Given the close association between NAFLD and metabolic syndrome, it may be difficult to determine which co-morbid condition posed the greatest risk factor for HCC development in any particular individual, as these different risk factors could also have incremental effects. It is well-known that both obesity and T2DM independently exert a carcinogenic potential, although historically, the contribution of NAFLD to HCC development in such individuals may not have been appreciated given the often quiescent nature of NAFLD. In light of the association between NAFLD, obesity and T2DM, some hepatology practice guidelines recommend individualized HCC screening for NAFLD in patients with these co-morbidities, while others suggest the use of clinical decision aids such as NAFLD fibrosis score, (NFS), fibrosis-4 index (FIB-4) or vibration controlled transient elastography (VCTE) for further risk stratification among such patients (Chalasani et al., 2018; Farrell, Chitturi, Lau, & Sollano, 2007). Excess body weight, T2DM, and NAFLD also synergize with other risk factors to increase HCC risk. Both obesity and T2DM synergize with HBV or HCV to increase HCC risk more than 100-fold when compared to patients without any risk factors (Chen et al., 2008). Currently available data linking the individual features of metabolic syndrome, including obesity and T2DM, and HCC, will be examined.

2.1 Obesity and HCC

Obesity is a complex, multifactorial disease which increases the risk of several malignancies including HCC (Vucenik & Stains, 2012). The increasing trend of obesity and metabolic syndrome will unfortunately continue to translate into increased cancer incidence (Arnold et al., 2016). Obesity is defined as a BMI $\geq 30\,\mathrm{kg/m^2}$; BMI values ≥ 25 to $< 30\,\mathrm{kg/m^2}$ are referred to as overweight. Excess accumulation of fatty tissue overwhelms the body's storage capacity, resulting in deposition of ectopic fat which alters several metabolic, inflammatory and immunologic pathways. These perturbation may affect DNA repair and function as well as alter epigenetic processes leading to malignant transformation (Park, Morley, Kim, Clegg, & Scherer, 2014).

Several large-scale studies have demonstrated that excess body weight is a risk factor for HCC. In a meta-analysis of 11 cohort studies, patients who had overweight or obesity were found to have relative risk (RR) of HCC of 1.17 and 1.89, respectively, compared to individuals with normal weight (Larsson & Wolk, 2007). The risk of dying from HCC was found to be 4.5 times higher in men with BMI $\geq 35\,\mathrm{kg/m^2}$ compared to those with normal BMI in a cohort of 900,000 adults (Calle, Rodriguez, Walker-Thurmond, & Thun, 2003). The summary RR of HCC was 17% and 89% for patients with overweight and obesity, respectively, with RR increasing by 24% for every $5\,\mathrm{kg/m^2}$ increase in BMI (Calle et al., 2003).

Despite the common use of BMI in quantitating obesity, this measure has several limitations. It is inaccurate in the elderly who may be losing height due to aging; among muscular individuals; among those of Asian descent; and is unable to differentiate between adipose tissue and lean tissue. Nonetheless, other measures of obesity, including visceral/abdominal adiposity which is estimated by waist to hip ratio, have been found to correlate with HCC risk and essentially recapitulate the HCC trends shown by BMI data (Schlesinger et al., 2013).

The mechanisms underlying the obesity-cancer link are still being elucidated. So far, available data suggests that these mechanisms include: (1) insulin resistance and insulin like growth factor signaling; (2) chronic inflammation and oxidative stress pathways; (3) sex hormone biosynthesis and pathways; (4) cytokine and adipocytokine signaling; (5) disruption of circadian rhythms and nutrient metabolism; and (6) alteration of intestinal microbiome. For a more in-depth review of the potential mechanisms underlying obesity-related carcinogenesis, the reader is referred to a recent review by Avgerinos, Spyrou, Mantzoros, and Dalamaga (2019).

2.2 Type 2 DM and HCC

Data supporting T2DM as an etiological risk factor for HCC has been mixed. Several studies indicate that T2DM contributes a two- to threefold increased risk of HCC independent of obesity (Kyrgiou et al., 2017; Park & Colditz, 2018). In a large public health center based prospective study, T2DM conferred a hazard ratio for HCC of 2.24 in males and 1.94 in females compared to patients without diabetes (Inoue et al., 2006). In a large longitudinal cohort study of 173,643 patients, HCC was increased over twofold in patients with T2DM compared to those without T2DM, with HCC risk increasing with time of follow-up (El-Serag, Tran, & Everhart, 2004). Other studies did not demonstrate a link between T2DM and HCC (Chen et al., 2013). Yet, some studies found that T2DM conferred increased HCC risk only if other risk factors such as chronic viral hepatitis B- or C- or alcohol-related cirrhosis were present (El-Serag, Richardson, & Everhart, 2001). Further studies exploring T2DM as an etiological risk factor for HCC would need to dissect the etiological role of other conditions that commonly associate with T2DM, including obesity and NAFLD in promoting HCC.

There are several mechanisms through with insulin resistance and T2DM could increase HCC risk. Hyperinsulinemia and insulin resistance in the liver, adipose tissue and muscle may lead to increased levels of insulin-like growth factor 1 (IGF-1) which stimulates proliferation of hepatocytes (Moore, Park, & Tsuda, 1998; Rosenfeld, 2003). Insulin resistance also acts on hepatic stellate cells to increase matrix protein secretion and decrease fatty acid mitochondrial beta oxidation (Mota, Banini, Cazanave, & Sanyal, 2016; Paradis et al., 2001), resulting in hepatic inflammation and fibrosis. T2DM and insulin resistance also create a state of chronic inflammation that may result in elevated inflammatory mediators including tumor necrosis factor-alpha (TNF-α) and interleukin-6 (IL-6) (Bastard et al., 2000; Fain, 2010; Goyal, Faizy, Siddiqui, & Singhai, 2012) and deregulation of apoptosis which could lead to hepatocarcinogenesis (Nakagawa et al., 2009; Park et al., 2010). In addition, insulin resistance alters epigenetic processes leading to activation of proinflammatory pathways including JAK/STAT and NFkB (Aggarwal et al., 2009).

In a meta-analysis including 334,307 patients with T2DM, metformin use reduced the incidence of HCC by 50% (Zhang, Gao, Fang, Zhao, & Yao, 2013). It has been proposed that metformin likely exerts antineoplastic effects via both insulin-dependent as well as insulin-independent pathways (Chen et al., 2013). HCC incidence in individuals with T2DM was increased in those taking sulfonylurea or insulin, and unchanged with glitazone use (Zhang et al. 2013).

3. NAFLD-related HCC

3.1 Patients with cirrhosis

NASH, the aggressive form of NAFLD, significantly increases the risk of disease progression to cirrhosis and HCC (Adams et al., 2005). In a recent meta-analysis of 86 studies, biopsy-proven NASH was found in approximately 7–30% of NAFLD patients without a clinical indication for biopsy; and in approximately 61–69% of NAFLD patients with a clinical indication for biopsy (Younossi et al., 2016). Estimates suggest that about 10–25% of patients with NASH progress to cirrhosis over a period of 10–20 years (Ekstedt et al., 2006; Matteoni et al., 1999). Among NASH patients with cirrhosis, the yearly incidence of HCC has been estimated to be 2.6–2.7% (Ascha et al., 2010; O'Leary, Landaverde, Jennings, Goldstein, & Davis, 2011). Although patients with NASH-related cirrhosis are at lower risk of HCC compared to cirrhosis related to alcohol or chronic viral hepatitis (Bhala et al., 2011), the much higher prevalence of NAFLD (Kabbany et al., 2017) and the rise in NAFLD-related HCC (Wong et al., 2015) over the past few decades offers an outlook on what is already a public health burden.

Cryptogenic cirrhosis is a term which refers to cirrhosis of unknown etiology. Over the past decade, is has become apparent that most of these cases are attributable to NASH (Bugianesi et al., 2002; Bugianesi et al., 2004; Starley, Calcagno, & Harrison, 2010). Most patients with cryptogenic cirrhosis are phenotypically similar to those with NASH cirrhosis, with features of metabolic syndrome including obesity and T2DM being a commonality.

Furthermore, about 25% and 16% of patients who were transplanted for cryptogenic cirrhosis developed NAFLD and NASH, respectively, within 2 years post-transplant, suggesting that cryptogenic cirrhosis and NASH cirrhosis are the same entity (Ong et al., 2001).

3.2 Patients without cirrhosis

Although HCC typically occurs in a background of cirrhosis, about 20–40% of NAFLD-related HCC occurs in patients without cirrhosis (Alexander, Torbenson, Wu, & Yeh, 2013; Tokushige, Hashimoto, Horie, Taniai, & Higuchi, 2011; Younes & Bugianesi, 2018). Non-cirrhotic HCC also occurs in patients with HBV and inherited liver diseases such as Wilsons disease and glycogen storage disease (Barrow & Ritter, 1988) but appears to be of greater magnitude among those with NAFLD given the number of patients at risk. In a cohort of 292 Japanese patients with NAFLD-related HCC, 38% of patients did not have cirrhosis (Tokushige et al., 2011). Similarly, an epidemiological study of 632 cases in England showed that one-third of patients with NAFLD-related HCC did not have cirrhosis (Dyson et al., 2014). In a VA study of 1500 patients with HCC, patients with non-cirrhotic NAFLD had a fivefold risk of having HCC, compared to those with non-cirrhotic HCV (Mittal et al., 2016).

Among patients with NAFLD-related HCC without cirrhosis, HCC generally presents in advanced stages compared to those with cirrhosis. Majority of such patients were found to present with symptomatic disease and large tumors, with a median survival of 7.2 months (Dyson et al., 2014). In 54 patients with NAFLD-related HCC, of whom 8 were non-cirrhotic, there were no statistically significant differences in BMI, age, level of HCC differentiation or number of risk factors; however non-cirrhotic patients had greater tumor diameter at diagnoses (Leung et al., 2015). Among patients with NAFLD-related HCC without cirrhosis, fibrosis stage has been found to correlate with tumor size and disease prognosis (Leung et al., 2015).

4. Molecular mechanisms of NAFLD-related HCC

The initial hypothesis proposed to explain the pathogenesis of NAFLD was the two-hit hypothesis, which proposed that insulin resistance and accumulation of excess fatty acids lead to steatosis, the first hit of NASH. This sensitizes the liver to a second hit involving oxidative stress, dysfunction of mitochondria and peroxidation of lipids that facilitate hepatic inflammation and fibrosis (Day & James, 1998). Over the past few decades, a large body of work supports a much more complex pathogenesis for NASH. These studies also highlight the vast heterogeneity in disease progression and in clinical manifestations among different individuals (Alonso et al., 2017). The current concept for NASH pathogenesis proposes that metabolic overload of the liver as a consequence of excess carbohydrates and fats leads to insulin resistance, increased adiposity and increased free fatty acids. This results in hepatocellular injury and cell death, with resultant chronic inflammation and fibrogenic remodeling that can predispose to cirrhosis and HCC (Cusi, 2012; Friedman, Neuschwander-Tetri, Rinella, & Sanyal, 2018; Mota et al., 2016; Sanyal, 2019) (Fig. 1). These inflammatory, fibrogenic and oncogenic pathways are likely modulated by genetic and epigenetic mechanisms, metabolic and endocrine signaling, immunologic mechanisms and gut microbiota.

4.1 Genetic mechanisms

The patatin-like phospholipase domain-containing protein 3 (PNPLA3) gene is a known determinant of hepatic fat content. The PNPLA3 rs738409 C > G single nucleotide polymorphism (SNP) results in an isoleucine to methionine substitution next to the catalytic domain, resulting in the PNPLA3 I148M variant. This SNP was first linked to hepatic fat content, steatosis and liver inflammation by two independent Genome-wide association studies (GWAS) (Romeo et al., 2008; Yuan et al., 2008). Hepatic overexpression of the I148M variant results in impaired triacylglycerol hydrolysis and leads to triacylglycerol accumulation (Smagris et al., 2015). The I148M variant has also been shown to predispose to HCC when metabolic risk factors such as T2DM and obesity are present (Ueyama et al., 2016; Valenti, Dongiovanni, Ginanni Corradini, Burza, & Romeo, 2013). The exact mechanisms linking PNPLA3 mutation to HCC are being investigated.

Another SNP implicated in increased HCC risk among patients with NAFLD is the transmembrane 6 superfamily member 2 gene (TM6SF2). The rs58542926 C > T genetic variant, resulting in a glutamic acid to lysine substitution, predisposes to increased steatosis and advanced fibrosis among patients with NASH (Kozlitina et al., 2014; Liu et al., 2014) and has been associated with NASH-related HCC (Chen, Xia, Xin, Lin, & Xuan, 2015). Other genetic abnormalities likely predisposing to HCC in patients with NAFLD include the C282Y and H63D mutations in the human hemostatic

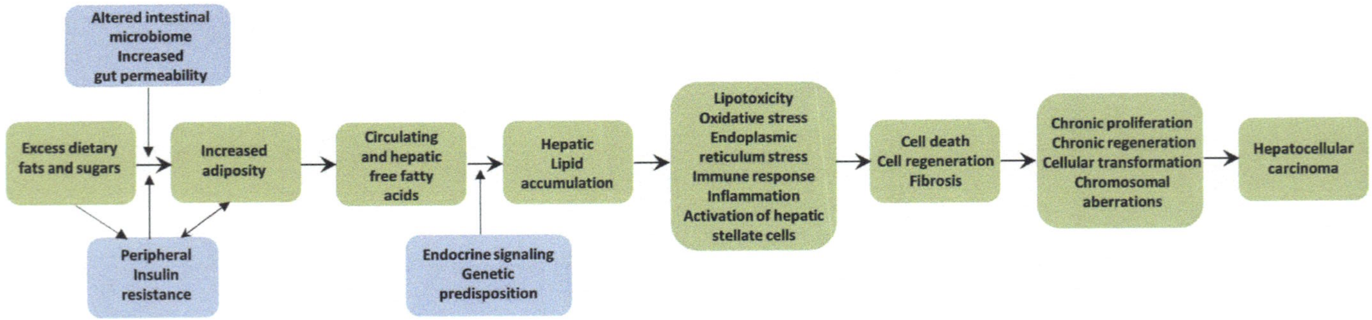

Fig. 1 Model of disease development in NAFLD-related HCC. Metabolic overload as a result of excess dietary fats and sugars leads to insulin resistance, increased adiposity and increased free fatty acids. Genetic susceptibility and endocrine signaling pathways modulate susceptibility to hepatic lipid accumulation. This triggers several processes including lipotoxic and oxidative stress, endoplasmic reticulum stress, innate immune response, inflammation and activation of hepatic stellate cells. These changes drive cell death, regeneration and fibrosis. Chronic proliferation and regeneration drive cellular transformation and chromosomal aberrations that lead to carcinogenesis.

iron regulator (HFE) gene (Bugianesi et al., 2002; Bugianesi et al., 2004; Nelson et al., 2007) and the rs641738 C>T variant of the membrane bound O-acyltransferase domain-containing 7 (MBOAT7) gene (Donati et al., 2017).

4.2 Epigenetic mechanisms

Emerging data indicate that development of HCC is associated with epigenetic modifications including differential DNA methylation, histone modification, and RNA interference. These modifications may differentially silence genes involved in DNA repair and tumor suppression, while activating those involved in fibrosis progression and oncogenesis (Tryndyak et al., 2011). Aberrant DNA methylation by the enzyme DNA methyl transferase (DNMT) is a cardinal feature of carcinogenesis. In humans, DNMT levels were found to be higher in patients with NASH compared to those with NAFL (Pirola et al., 2013). NAFLD-related HCC is associated with hypermethylation of the glycine *N*-methyltransferase (*GNMT*) promoter, resulting in reduced expression of the gene. GNMT inhibits HCC growth through unknown mechanisms (Simile et al., 2018), and has been noted to be absent in fast growing HCC and present at low levels in slow growing HCC (Simile et al., 2018).

Whole-exome sequencing of 10 patients with NAFLD-related HCC showed a pattern of recurrent mutations in 12 genes, with the most common mutations occurring in the TERT promoter (82%), CTNNB1 (45%) and TP53 (36%) (Kim et al., 2016). Chromosomal loss and gains were also noted, with all 10 NAFLD-related HCC samples showing loss of chromosome 8p (Kim et al., 2016). Genome-wide analysis of copy number variations in patients with NASH showed differences in exportin 4 (XPO4) and phosphodiesterase 1B (PDE1B) genes, known to be associated with disease progression to cirrhosis and HCC (Zain et al., 2014). In addition, several microRNA (miRNA) appear to play a role in NAFLD-related HCC, including miR-21, miR-23, miR-29, miR-93, miR-106, miR-155, miR-221, miR-222, and miR-519 (de Conti et al., 2017).

4.3 Metabolic and endocrine pathways

Caloric overload occurring in NAFLD leads to insulin resistance which alters signaling in several pathways and predispose to carcinogenesis. Interaction of IGF-1 with its receptor activates the PI3K and MAPK signaling pathways, leading to cell proliferation and inhibition of apoptosis

(Janku, Kaseb, Tsimberidou, Wolff, & Kurzrock, 2014; Yang & Liu, 2017). PI3K signaling alters cell cycle regulation by cyclin D1, apoptosis by Mdm2/p53, and regulation of cell growth by mTOR (Kudo et al., 2011). The activation of MAPK signaling results in the transcriptional upregulation of several protooncogenes including c-fos and c-jun, resulting in downstream activation of Wnt/β-catenin signaling, and upregulating hepatic fibrosis and carcinogenesis (Chettouh, Lequoy, Fartoux, Vigouroux, & Desbois-Mouthon, 2015).

Insulin resistance and mitochondrial lipotoxicity also result in oxidative and endoplasmic reticulum (ER) stress. Chronic oxidative stress leads to calcium release from ER, increased mitochondrial and lysosomal permeabilization, and activation of caspase 3 and 9-related apoptosis (Mazurkevich, Dzhurko, & Gorlachev, 1980; Novo & Parola, 2008), facilitating hepatic inflammation and carcinogenesis (Fu et al., 2011).

Sex hormone signaling has also been implicated in NAFLD-related HCC. The likelihood of developing both NAFLD and HCC is higher in males than females, although this trend is reversed after the age of 60 (Gan, Chitturi, & Farrell, 2011; Hashimoto & Tokushige, 2011). Androgen signaling appears to upregulate HCC pathways, while estrogen signaling exerts the opposite effect (Nagasue, Yu, Yukaya, Kohno, & Nakamura, 1995). The increased risk of HCC in females after 60 years of age has been attributed to loss of the protective effects of estrogen (Iyer, Kalra, Kaul, Payton, & Kaul, 2017). Activation of the androgen receptor through binding by androgen or via MAPK and PI3K signaling results in transcriptional activation of cell cycle-related kinase (CCRK) which activates signaling through β-catenin and T-cell factor, increasing HCC risk (Awuah & Monga, 2012; Kanda, Jiang, & Yokosuka, 2014).

4.4 Immunologic mechanisms

Several immunologic mechanisms appear to be involved in NASH progression to HCC. These include cytokine signaling pathways, signaling hormones including leptin and adiponectin, and immune cells.

Oxidative stress and insulin resistance stimulate signaling via nuclear factor kappa-light-chain-enhancer of activated B-cells (NF-κB) and inhibitor of kappa light polypeptide gene enhancer in B-cells, kinase beta (IKKβ), resulting in increased hepatocyte survival (He & Karin, 2011). IKKβ can also be activated by lipids to promote insulin resistance (Cai et al., 2005) likely through TLR pathways. Along those lines, TLR-deficient mice exhibited decreased steatosis and inflammation despite high fat diet (Jia et al., 2014). Regulation of TLR-mediated inflammatory response has been found to be protective against HCC (Li et al., 2015). Oxidative stress also alters the level of several cytokines and adipokines, including tumor necrosis factor-alpha (TNF-α) and interleukin-6 (IL-6) (Park et al., 2010).

Signaling hormones such as leptin and adiponectin can activate or inhibit cytokine signaling (Carbone, La Rocca, & Matarese, 2012). Obesity and NAFLD are typically characterized by high leptin and low adiponectin. Leptin facilitates insulin resistance and hepatic steatosis, and upregulates fibrosis and angiogenesis through activation of TNF-α and IL-6 signaling. Binding of leptin to the Ob receptor results in activation of several oncogenic signaling pathways including JAK2/STAT, MAPK, and PI3K/Akt (Jiang, Sun, & Sun, 2014; Villanueva et al., 2008), favoring cell proliferation and inhibiting apoptosis. Leptin also upregulates TERT, resulting in tumor cell immortalization (Stefanou, Papanikolaou, Furukawa, Nakamura, & Tsezou, 2010). Adiponectin, on the other hand, downregulates inflammatory cytokines and inhibits angiogenesis (Xing et al., 2015). Adiponectin also reduces activation of STAT3 and Akt (Sharma et al., 2010). Under conditions of insulin resistance, adiponectin levels are reduced, thus abrogating its anti-inflammatory and anti-carcinogenic effects (Shen et al., 2016).

The role of adaptive immune response in NASH-related HCC is also being examined. Linoleic acid-induced disruption of mitochondrial function results in higher levels of oxidative damage compared to other free fatty acids, leading to loss of CD4(+) T lymphocytes and promoting carcinogenesis (Ma et al., 2016). Induction of intrahepatic CD8(+) T cells, NKT cells, and inflammatory cytokines by choline-deficient high-fat diet was associated with increased liver damage with upregulation of NASH and HCC (Wolf et al., 2014). Cell injury also triggers the Hedgehog signaling pathway which is involved in cell repair and regeneration (Zheng et al., 2013). Dysregulated hedgehog signaling and excessive activation of the hedgehog pathway resulting in aberrant cellular repair and malignant transformation has been shown (Sircana, Paschetta, Saba, Molinaro, & Musso, 2019; Zheng et al., 2013).

4.5 Gut microbiota and dietary modulation

The gut flora has been shown to influence the development of NAFLD, obesity, insulin resistance and other HCC etiological risk factors. Microbiota transplantation studies show that gut microbiota from patients with obesity extract energy more efficiency compared to those from non-obese individuals (Turnbaugh et al., 2006; Turnbaugh et al., 2009). As such, antibiotic treatment in mice has been shown to abrogate high fat diet-induced NAFLD (Jiang et al., 2015). Moreover, introduction of gut microflora from conventionally raised mice into germ-free mice results in insulin resistance and increased body fat in the recipient mice (Backhed et al., 2004; Wostmann, Larkin, Moriarty, & Bruckner-Kardoss, 1983). Insulin resistance can also occur through gut microbiota-associated decrease in levels of anorectic hormones such as GLP-1 and PYY (Bliss & Whiteside, 2018).

Gut microbiota-induced alterations in intestinal permeability can affect inflammation and immune response. Pathogen-associated molecular patterns (PAMPs) such as lipopolysaccharides (LPS) in the outer membrane of gram-negative bacteria can be recognized by pattern recognition receptors (PRRs) such as TLR, resulting in activation of innate immune responses (Valentini et al., 2014). Murine studies showed that high fat diet increased intestinal permeability and resulted in a two- to threefold increase in serum LPS levels (Cani et al., 2007), upregulating plasma triglyceride and inflammatory markers. Other PAMPs, including flagellin, peptidoglycans, nucleic acid variants or unmethylated CpG motifs can also activate immune response through recognized by TLRs (Valentini et al., 2014). TLR4 and intestinal microbiota were found to be crucial for progression of HCC in mice through upregulation of cell proliferation and inhibition of apoptosis (Dapito et al., 2012).

Induction of senescence-associated secretory phenotype (SASP) in mice by gut bacteria promotes obesity-associated HCC (Yoshimoto et al., 2013) and may also have a role in NAFLD-related HCC. Obese mice were found to express increased levels of deoxycholic acid (DCA), a metabolite that causes DNA damage. DCA-mediated activation of hepatic stellate cells (HSCs) promoted HCC development after exposure of mice to chemical carcinogen. Similar HSC activation was also noted in patients with NAFLD. Notably, HCC development in mice was abrogated by blocking DCA production or reducing gut bacteria (Yoshimoto et al., 2013).

5. Prevention and surveillance of NAFLD-related HCC

HCC surveillance in patients with NAFLD has been problematic due to several factors. Most patients with NAFLD are asymptomatic and may present only when the disease is in its advanced stages at which point they may have already developed HCC. Even in patients diagnosed with non-cirrhotic NAFLD and at risk for HCC, surveillance is not cost-effective in the absence of cirrhosis and thus not recommended. Given the increased risk of HCC in patients with NAFLD co-occurring with obesity and T2DM, some guidelines suggest individualized HCC surveillance regimens or further clinical risk stratification in such patients (Chalasani et al., 2018; Farrell et al., 2007). However, there are no specific surveillance recommendations in patients with non-cirrhotic NAFLD.

Addressing obesity is crucial to decreasing the risk of cardiometabolic complications and NAFLD. Measures to prevent or reverse obesity include dietary intervention, physical activity, pharmacotherapy and bariatric surgery. Weight loss has been found to decrease NASH fibrosis, a key determinant of progression to HCC. NAFLD patients with obesity should be counseled on the beneficial effect of weight loss, and referred for behavioral, medical or surgical interventions as feasible in order to promote weight loss and decrease or reverse attendant metabolic complications.

Metformin use has been found to reduce the risk of HCC by about 50% in patients with T2DM (Zhang et al. 2013), hence NAFLD patients with concomitant T2DM should be placed on metformin if feasible. Aside from decreased HCC risk, metformin use can also lead to weight loss, leading to potential improvement in metabolic syndrome which commonly occurs among patients with NAFLD. Statin use has also been found to be beneficial in patients with NAFLD. Statins may decrease carcinogenesis through beneficial effects on immunomodulatory, antiproliferative, antiangiogenic, and proapoptotic pathways. A meta-analysis including 1.5 million patients and 4298 cases of HCC showed a 37% reduction in HCC incidence among patients on statin (Bitterman, Lefer, & Lefer, 1988). Thus, patients with NAFLD should be placed on statin if feasible in order to decrease both HCC risk as well as cardiovascular risk.

6. Management of NAFLD-related HCC

Treatment options for NAFLD-related HCC are the same as for HCC related to other etiologies (Galle et al., 2018; Marrero et al., 2018) (Fig. 2). In patients with NAFLD-related HCC, several factors, including older age, cardiometabolic comorbidities, late diagnosis, and advanced HCC at the time of diagnosis may preclude potentially curative treatment options from being offered, with several of these patients receiving palliative chemotherapy or TACE (Weinmann et al., 2015). Even so, given the large

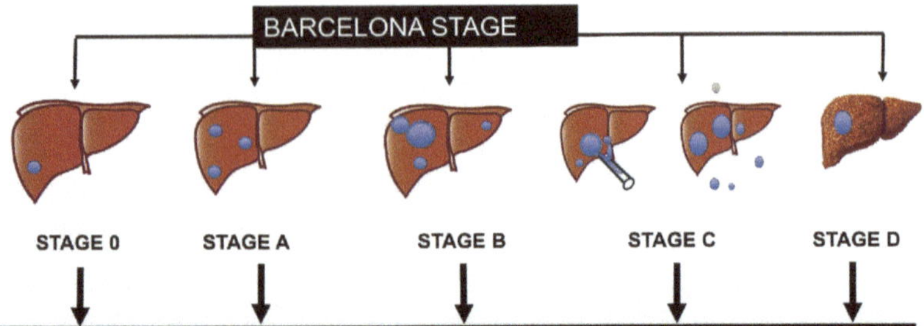

Level of Evidence	Stage 0	Stage A	Stage B	Stage C	Stage D
1	Resection		TACE	Sorafenib (1L) Lenvatinib (1L) Regorafenib (2L) Cabozantinib (2L)	
2	RFA MWA	Resection OLT RFA MWA TARE TACE SBRT	TARE Downsize OLT	Nivolumab (2L)	OLT BSC
3				TARE	

Fig. 2 Recommendations for treatment of hepatocellular carcinoma according to Barcelona Clinic Liver Cancer (BCLC) stage. Recommendations are categorized based on level of evidence. For further details, please refer to Marrero et al. (2018). Abbreviations: BSC, best supportive care; MWA, microwave ablation; OLT, orthotopic liver transplant; RFA, radiofrequency ablation; SBRT, stereotactic body radiation therapy; TACE, transarterial chemoembolization; TARE; transarterial radioembolization; 1 L, first-line therapy; 2 L, second-line therapy *Reproduced with permission from Marrero, J. A., Kulik, L. M., Sirlin, C. B., Zhu, A. X., Finn, R. S., Abecassis, M. M., et al. (2018). Diagnosis, staging, and Management of Hepatocellular Carcinoma: 2018 practice guidance by the American Association for the Study of Liver Diseases.* Hepatology, 68*(2), 723–750. doi:10.1002/hep.29913.*

number of patients with NAFLD, NASH-related HCC is a rapidly increasing indication for liver transplant among patients with HCC (Wong, Cheung, & Ahmed, 2014). At a molecular level, most NAFLD-related HCC are in the S1 subtype of HCC (Cazanave et al., 2017; Hoshida et al., 2009), with activation of WNT, beta-catenin and TGF-β signaling. Subclassification of HCC and further molecular characterization may show utility in the development of biomarkers to predict response to therapy.

7. Conclusions/future directions

NAFLD has become one of the most common causes of HCC. NAFLD-related HCC can occur in both cirrhotic and non-cirrhotic patients. This upward trend in NAFLD-related HCC will likely continue over the next several decades as the incidence of both obesity and metabolic syndrome increase. The development of NAFLD appears to be triggered by metabolic overload as a consequence of excess carbohydrates and fats that leads to insulin resistance, increased adiposity and fatty acid accumulation. These processes upregulate various pathways, including lipotoxic and oxidative stress, endoplasmic reticulum stress and innate immune response. Chronic inflammation, regeneration, and fibrogenic remodeling predispose to cellular transformation and chromosomal aberrations that can lead to hepatocarcinogenesis. The vast heterogeneity in NAFLD progression and in clinical manifestations among different individuals is likely due to modifiers from genetic, epigenetic, metabolic, endocrine, immunologic, and gut microbial mechanisms.

The complex interaction of various mechanisms involved in development and progression of NAFLD and NAFLD-related HCC raise several themes for the future. The development of predictive biomarkers would help identify NAFLD patients at risk for disease progression. These biomarkers would also assist in personalized therapy by predicting patients who are likely to respond to a particular therapy, whether pharmacologic, immunologic, or otherwise. Given the multiple mechanisms involved in disease development and progression, treatments that would show superior efficacy for NAFLD as well as NAFLD-related HCC are likely to be those that address multiple pathways. It is our hope that on-going and future studies aimed at improving our understanding of the molecular processes involved in NAFLD-related HCC will help toward addressing these needs.

Acknowledgment

Dr. Arun Sanyal receives funding from NIH project RO1 DK 105961.

Disclosures

Dr. Sanyal is President of Sanyal Bio. He has stock options in Genfit, Galmed, Exhalenz, Durect, Tiziana, Algernon and Indalo. He has served as a consultant to Intercept, Gilead, Bristol Myers Squibb, Novartis, Pfizer, Lilly, Novo Nordisk, Astra Zeneca, Medimmune, Merck, Allergan, Albireo, Boehringer Ingelhiem, Celgene, NGM, Glympse, Conatus, Genentech, Tern, Takeda, Hemoshear, Immuron, Surrozen, Poxel, Path AI, Second Genome, Zydus, Chiasma, Surrozen, Poxel, Blade, Pliant, Liposcience, Cymabay, Salix, Ferring and Teva. His institution has received grants from Intercept, Gilead, Novartis, Merck, Astra Zeneca, Malinckrodt, Pfizer, Lilly, Salix and Bristol Myers Squibb. VCU has ownership interests in Sanyal Bio.

Dr. Banini serves on the NASH advisory board for Boehringer Ingelhiem.

References

Adams, L. A., Lymp, J. F., St Sauver, J., Sanderson, S. O., Lindor, K. D., Feldstein, A., et al. (2005). The natural history of nonalcoholic fatty liver disease: A population-based cohort study. *Gastroenterology*, *129*(1), 113–121. https://doi.org/10.1053/j.gastro.2005.04.014.

Adams, L. A., Waters, O. R., Knuiman, M. W., Elliott, R. R., & Olynyk, J. K. (2009). NAFLD as a risk factor for the development of diabetes and the metabolic syndrome: An eleven-year follow-up study. *The American Journal of Gastroenterology*, *104*(4), 861–867. https://doi.org/10.1038/ajg.2009.67.

Aggarwal, B. B., Kunnumakkara, A. B., Harikumar, K. B., Gupta, S. R., Tharakan, S. T., Koca, C., et al. (2009). Signal transducer and activator of transcription-3, inflammation, and cancer: How intimate is the relationship? *Annals of the New York Academy of Sciences*, *1171*, 59–76. https://doi.org/10.1111/j.1749-6632.2009.04911.x.

Alexander, J., Torbenson, M., Wu, T. T., & Yeh, M. M. (2013). Non-alcoholic fatty liver disease contributes to hepatocarcinogenesis in non-cirrhotic liver: A clinical and pathological study. *Journal of Gastroenterology and Hepatology*, *28*(5), 848–854. https://doi.org/10.1111/jgh.12116.

Alonso, C., Fernandez-Ramos, D., Varela-Rey, M., Martinez-Arranz, I., Navasa, N., Van Liempd, S. M., et al. (2017). Metabolomic identification of subtypes of nonalcoholic steatohepatitis. *Gastroenterology*, *152*(6), 1449–1461. e1447. https://doi.org/10.1053/j.gastro.2017.01.015.

Altekruse, S. F., McGlynn, K. A., & Reichman, M. E. (2009). Hepatocellular carcinoma incidence, mortality, and survival trends in the United States from 1975 to 2005. *Journal of Clinical Oncology*, *27*(9), 1485–1491. https://doi.org/10.1200/JCO.2008.20.7753.

Arnold, M., Leitzmann, M., Freisling, H., Bray, F., Romieu, I., Renehan, A., et al. (2016). Obesity and cancer: An update of the global impact. *Cancer Epidemiology*, *41*, 8–15. https://doi.org/10.1016/j.canep.2016.01.003.

Arrese, M., Barrera, F., Triantafilo, N., & Arab, J. P. (2019). Concurrent nonalcoholic fatty liver disease and type 2 diabetes: Diagnostic and therapeutic considerations. *Expert Review of Gastroenterology & Hepatology*, *13*(9), 849–866. https://doi.org/10.1080/17474124.2019.1649981.

Ascha, M. S., Hanouneh, I. A., Lopez, R., Tamimi, T. A., Feldstein, A. F., & Zein, N. N. (2010). The incidence and risk factors of hepatocellular carcinoma in patients with nonalcoholic steatohepatitis. *Hepatology*, *51*(6), 1972–1978. https://doi.org/10.1002/hep.23527.

Avgerinos, K. I., Spyrou, N., Mantzoros, C. S., & Dalamaga, M. (2019). Obesity and cancer risk: Emerging biological mechanisms and perspectives. *Metabolism*, *92*, 121–135. https://doi.org/10.1016/j.metabol.2018.11.001.

Awuah, P. K., & Monga, S. P. (2012). Cell cycle-related kinase links androgen receptor and beta-catenin signaling in hepatocellular carcinoma: Why are men at a loss? *Hepatology*, *55*(3), 970–973. https://doi.org/10.1002/hep.24774.

Backhed, F., Ding, H., Wang, T., Hooper, L. V., Koh, G. Y., Nagy, A., et al. (2004). The gut microbiota as an environmental factor that regulates fat storage. *Proceedings of the National Academy of Sciences of the United States of America*, *101*(44), 15718–15723. https://doi.org/10.1073/pnas.0407076101.

Baffy, G., Brunt, E. M., & Caldwell, S. H. (2012). Hepatocellular carcinoma in non-alcoholic fatty liver disease: An emerging menace. *Journal of Hepatology*, *56*(6), 1384–1391. https://doi.org/10.1016/j.jhep.2011.10.027.

Banini, B. A., & Sanyal, A. J. (2016). Nonalcoholic fatty liver disease: Epidemiology, pathogenesis, natural history, diagnosis, and current treatment options. *Clinical Medicine Insights. Therapeutics*, *8*, 75–84. https://doi.org/10.4137/cmt.s18885.

Barrow, S. E., & Ritter, J. M. (1988). Prostacyclin biosynthesis in vivo: Measurements in plasma and urine. *Prostaglandins, Leukotrienes, and Essential Fatty Acids*, *33*(3), 213–220. https://doi.org/10.1016/0952-3278(88)90033-6.

Bastard, J. P., Jardel, C., Bruckert, E., Blondy, P., Capeau, J., Laville, M., et al. (2000). Elevated levels of interleukin 6 are reduced in serum and subcutaneous adipose tissue of obese women after weight loss. *The Journal of Clinical Endocrinology and Metabolism*, *85*(9), 3338–3342. https://doi.org/10.1210/jcem.85.9.6839.

Bhala, N., Angulo, P., van der Poorten, D., Lee, E., Hui, J. M., Saracco, G., et al. (2011). The natural history of nonalcoholic fatty liver disease with advanced fibrosis or cirrhosis: An international collaborative study. *Hepatology*, *54*(4), 1208–1216. https://doi.org/10.1002/hep.24491.

Bitterman, H., Lefer, D. J., & Lefer, A. M. (1988). Novel mechanism of action of a prostacyclin enhancing agent in haemorrhagic shock. *Naunyn-Schmiedeberg's Archives of Pharmacology*, *337*(6), 679–686. https://doi.org/10.1007/BF00175796.

Bliss, E. S., & Whiteside, E. (2018). The gut-brain Axis, the human gut microbiota and their integration in the development of obesity. *Frontiers in Physiology*, *9*, 900. https://doi.org/10.3389/fphys.2018.00900.

Bugianesi, E., Leone, N., Vanni, E., Marchesini, G., Brunello, F., Carucci, P., et al. (2002). Expanding the natural history of nonalcoholic steatohepatitis: From cryptogenic cirrhosis to hepatocellular carcinoma. *Gastroenterology*, *123*(1), 134–140. https://doi.org/10.1053/gast.2002.34168.

Bugianesi, E., Manzini, P., D'Antico, S., Vanni, E., Longo, F., Leone, N., et al. (2004). Relative contribution of iron burden, HFE mutations, and insulin resistance to fibrosis in nonalcoholic fatty liver. *Hepatology*, *39*(1), 179–187. https://doi.org/10.1002/hep.20023.

Cai, D., Yuan, M., Frantz, D. F., Melendez, P. A., Hansen, L., Lee, J., et al. (2005). Local and systemic insulin resistance resulting from hepatic activation of IKK-beta and NF-kappaB. *Nature Medicine*, *11*(2), 183–190. https://doi.org/10.1038/nm1166.

Calle, E. E., Rodriguez, C., Walker-Thurmond, K., & Thun, M. J. (2003). Overweight, obesity, and mortality from cancer in a prospectively studied cohort of U.S. adults. *The New England Journal of Medicine*, *348*(17), 1625–1638. https://doi.org/10.1056/NEJMoa021423.

Cani, P. D., Amar, J., Iglesias, M. A., Poggi, M., Knauf, C., Bastelica, D., et al. (2007). Metabolic endotoxemia initiates obesity and insulin resistance. *Diabetes*, *56*(7), 1761–1772. https://doi.org/10.2337/db06-1491.

Carbone, F., La Rocca, C., & Matarese, G. (2012). Immunological functions of leptin and adiponectin. *Biochimie*, *94*(10), 2082–2088. https://doi.org/10.1016/j.biochi.2012.05.018.

Cazanave, S., Podtelezhnikov, A., Jensen, K., Seneshaw, M., Kumar, D. P., Min, H. K., et al. (2017). The Transcriptomic signature of disease development and progression of nonalcoholic fatty liver disease. *Scientific Reports*, 7(1), 17193. https://doi.org/10.1038/s41598-017-17370-6.

Chalasani, N., Younossi, Z., Lavine, J. E., Charlton, M., Cusi, K., Rinella, M., et al. (2018). The diagnosis and management of nonalcoholic fatty liver disease: Practice guidance from the American Association for the Study of Liver Diseases. *Hepatology*, *67*(1), 328–357. https://doi.org/10.1002/hep.29367.

Chen, C. T., Chen, J. Y., Wang, J. H., Chang, K. C., Tseng, P. L., Kee, K. M., et al. (2013). Diabetes mellitus, metabolic syndrome and obesity are not significant risk factors for hepatocellular carcinoma in an HBV- and HCV-endemic area of southern Taiwan. *The Kaohsiung Journal of Medical Sciences*, *29*(8), 451–459. https://doi.org/10.1016/j.kjms.2012.12.006.

Chen, H. P., Shieh, J. J., Chang, C. C., Chen, T. T., Lin, J. T., Wu, M. S., et al. (2013). Metformin decreases hepatocellular carcinoma risk in a dose-dependent manner: Population-based and in vitro studies. *Gut*, *62*(4), 606–615. https://doi.org/10.1136/gutjnl-2011-301708.

Chen, L. Z., Xia, H. H., Xin, Y. N., Lin, Z. H., & Xuan, S. Y. (2015). TM6SF2 E167K variant, a novel genetic susceptibility variant, contributing to nonalcoholic fatty liver disease. *Journal of Clinical and Translational Hepatology*, *3*(4), 265–270. https://doi.org/10.14218/JCTH.2015.00023.

Chen, C. L., Yang, H. I., Yang, W. S., Liu, C. J., Chen, P. J., You, S. L., et al. (2008). Metabolic factors and risk of hepatocellular carcinoma by chronic hepatitis B/C infection: A follow-up study in Taiwan. *Gastroenterology*, *135*(1), 111–121. https://doi.org/10.1053/j.gastro.2008.03.073.

Chettouh, H., Lequoy, M., Fartoux, L., Vigouroux, C., & Desbois-Mouthon, C. (2015). Hyperinsulinaemia and insulin signalling in the pathogenesis and the clinical course of hepatocellular carcinoma. *Liver International*, *35*(10), 2203–2217. https://doi.org/10.1111/liv.12903.

Cusi, K. (2012). Role of obesity and lipotoxicity in the development of nonalcoholic steatohepatitis: Pathophysiology and clinical implications. *Gastroenterology*, *142*(4), 711–725. e716. https://doi.org/10.1053/j.gastro.2012.02.003.

Dapito, D. H., Mencin, A., Gwak, G. Y., Pradere, J. P., Jang, M. K., Mederacke, I., et al. (2012). Promotion of hepatocellular carcinoma by the intestinal microbiota and TLR4. *Cancer Cell*, *21*(4), 504–516. https://doi.org/10.1016/j.ccr.2012.02.007.

Day, C. P., & James, O. F. (1998). Steatohepatitis: A tale of two "hits"? *Gastroenterology*, *114*(4), 842–845. https://doi.org/10.1016/s0016-5085(98)70599-2.

de Alwis, N. M., & Day, C. P. (2008). Non-alcoholic fatty liver disease: The mist gradually clears. *Journal of Hepatology*, *48*(Suppl. 1), S104–S112. https://doi.org/10.1016/j.jhep.2008.01.009.

de Conti, A., Ortega, J. F., Tryndyak, V., Dreval, K., Moreno, F. S., Rusyn, I., et al. (2017). MicroRNA deregulation in nonalcoholic steatohepatitis-associated liver carcinogenesis. *Oncotarget*, *8*(51), 88517–88528. https://doi.org/10.18632/oncotarget.19774.

Donati, B., Dongiovanni, P., Romeo, S., Meroni, M., McCain, M., Miele, L., et al. (2017). MBOAT7 rs641738 variant and hepatocellular carcinoma in non-cirrhotic individuals. *Scientific Reports*, 7(1), 4492. https://doi.org/10.1038/s41598-017-04991-0.

Dowman, J. K., Tomlinson, J. W., & Newsome, P. N. (2011). Systematic review: The diagnosis and staging of non-alcoholic fatty liver disease and non-alcoholic steatohepatitis. *Alimentary Pharmacology & Therapeutics*, *33*(5), 525–540. https://doi.org/10.1111/j.1365-2036.2010.04556.x.

Dyson, J., Jaques, B., Chattopadyhay, D., Lochan, R., Graham, J., Das, D., et al. (2014). Hepatocellular cancer: The impact of obesity, type 2 diabetes and a multidisciplinary team. *Journal of Hepatology*, *60*(1), 110–117. https://doi.org/10.1016/j.jhep.2013.08.011.

Ekstedt, M., Franzen, L. E., Mathiesen, U. L., Thorelius, L., Holmqvist, M., Bodemar, G., et al. (2006). Long-term follow-up of patients with NAFLD and elevated liver enzymes. *Hepatology*, *44*(4), 865–873. https://doi.org/10.1002/hep.21327.

El-Serag, H. B., Richardson, P. A., & Everhart, J. E. (2001). The role of diabetes in hepatocellular carcinoma: A case-control study among United States veterans. *The American Journal of Gastroenterology*, *96*(8), 2462–2467. https://doi.org/10.1111/j.1572-0241.2001.04054.x.

El-Serag, H. B., Tran, T., & Everhart, J. E. (2004). Diabetes increases the risk of chronic liver disease and hepatocellular carcinoma. *Gastroenterology*, *126*(2), 460–468. https://doi.org/10.1053/j.gastro.2003.10.065.

Eslam, M., Newsome, P. N., Sarin, S. K., Anstee, Q. M., Targher, G., Romero-Gomez, M., et al. (2020). A new definition for metabolic dysfunction-associated fatty liver disease: An international expert consensus statement. *Journal of Hepatology, 73*(1), 202–209. https://doi.org/10.1016/j.jhep.2020.03.039.

Estes, C., Razavi, H., Loomba, R., Younossi, Z., & Sanyal, A. J. (2018). Modeling the epidemic of nonalcoholic fatty liver disease demonstrates an exponential increase in burden of disease. *Hepatology, 67*(1), 123–133. https://doi.org/10.1002/hep.29466.

Fain, J. N. (2010). Release of inflammatory mediators by human adipose tissue is enhanced in obesity and primarily by the nonfat cells: A review. *Mediators of Inflammation, 2010*, 513948. https://doi.org/10.1155/2010/513948.

Farrell, G. C., Chitturi, S., Lau, G. K., & Sollano, J. D. (2007). Guidelines for the assessment and management of non-alcoholic fatty liver disease in the Asia-Pacific region: Executive summary. *Journal of Gastroenterology and Hepatology, 22*(6), 775–777. https://doi.org/10.1111/j.1440-1746.2007.05002.x.

Farrell, G. C., & Larter, C. Z. (2006). Nonalcoholic fatty liver disease: From steatosis to cirrhosis. *Hepatology, 43*(2 Suppl. 1), S99–S112. https://doi.org/10.1002/hep.20973.

Fazel, Y., Koenig, A. B., Sayiner, M., Goodman, Z. D., & Younossi, Z. M. (2016). Epidemiology and natural history of non-alcoholic fatty liver disease. *Metabolism, 65*(8), 1017–1025. https://doi.org/10.1016/j.metabol.2016.01.012.

Friedman, S. L., Neuschwander-Tetri, B. A., Rinella, M., & Sanyal, A. J. (2018). Mechanisms of NAFLD development and therapeutic strategies. *Nature Medicine, 24*(7), 908–922. https://doi.org/10.1038/s41591-018-0104-9.

Fu, S., Yang, L., Li, P., Hofmann, O., Dicker, L., Hide, W., et al. (2011). Aberrant lipid metabolism disrupts calcium homeostasis causing liver endoplasmic reticulum stress in obesity. *Nature, 473*(7348), 528–531. https://doi.org/10.1038/nature09968.

Galle, P. R., Forner, A., Llovet, J. M., Mazzaferro, V., Piscaglia, F., Raoul, J. L., et al. (2018). EASL clinical practice guidelines: Management of hepatocellular carcinoma. *Journal of Hepatology, 69*(1), 182–236. https://doi.org/10.1016/j.jhep.2018.03.019.

Gan, L., Chitturi, S., & Farrell, G. C. (2011). Mechanisms and implications of age-related changes in the liver: Nonalcoholic fatty liver disease in the elderly. *Current Gerontology and Geriatrics Research, 2011*, 831536. https://doi.org/10.1155/2011/831536.

Goyal, R., Faizy, A. F., Siddiqui, S. S., & Singhai, M. (2012). Evaluation of TNF-alpha and IL-6 levels in obese and non-obese diabetics: Pre- and postinsulin effects. *North American Journal of Medical Sciences, 4*(4), 180–184. https://doi.org/10.4103/1947-2714.94944.

Grundy, S. M., Cleeman, J. I., Daniels, S. R., Donato, K. A., Eckel, R. H., Franklin, B. A., et al. (2005). Diagnosis and management of the metabolic syndrome: An American Heart Association/National Heart, Lung, and Blood Institute Scientific Statement. *Circulation, 112*(17), 2735–2752.

Hashimoto, E., & Tokushige, K. (2011). Prevalence, gender, ethnic variations, and prognosis of NASH. *Journal of Gastroenterology, 46*(Suppl. 1), 63–69. https://doi.org/10.1007/s00535-010-0311-8.

He, G., & Karin, M. (2011). NF-kappaB and STAT3—Key players in liver inflammation and cancer. *Cell Research, 21*(1), 159–168. https://doi.org/10.1038/cr.2010.183.

Hoshida, Y., Nijman, S. M., Kobayashi, M., Chan, J. A., Brunet, J. P., Chiang, D. Y., et al. (2009). Integrative transcriptome analysis reveals common molecular subclasses of human hepatocellular carcinoma. *Cancer Research, 69*(18), 7385–7392. https://doi.org/10.1158/0008-5472.CAN-09-1089.

Inoue, M., Iwasaki, M., Otani, T., Sasazuki, S., Noda, M., & Tsugane, S. (2006). Diabetes mellitus and the risk of cancer: Results from a large-scale population-based cohort study in Japan. *Archives of Internal Medicine, 166*(17), 1871–1877. https://doi.org/10.1001/archinte.166.17.1871.

Iyer, J. K., Kalra, M., Kaul, A., Payton, M. E., & Kaul, R. (2017). Estrogen receptor expression in chronic hepatitis C and hepatocellular carcinoma pathogenesis. *World Journal of Gastroenterology, 23*(37), 6802–6816. https://doi.org/10.3748/wjg.v23.i37.6802.

Janku, F., Kaseb, A. O., Tsimberidou, A. M., Wolff, R. A., & Kurzrock, R. (2014). Identification of novel therapeutic targets in the PI3K/AKT/mTOR pathway in hepatocellular carcinoma using targeted next generation sequencing. *Oncotarget*, *5*(10), 3012–3022. https://doi.org/10.18632/oncotarget.1687.

Jia, L., Vianna, C. R., Fukuda, M., Berglund, E. D., Liu, C., Tao, C., et al. (2014). Hepatocyte toll-like receptor 4 regulates obesity-induced inflammation and insulin resistance. *Nature Communications*, *5*, 3878. https://doi.org/10.1038/ncomms4878.

Jiang, N., Sun, R., & Sun, Q. (2014). Leptin signaling molecular actions and drug target in hepatocellular carcinoma. *Drug Design, Development and Therapy*, *8*, 2295–2302. https://doi.org/10.2147/DDDT.S69004.

Jiang, C., Xie, C., Li, F., Zhang, L., Nichols, R. G., Krausz, K. W., et al. (2015). Intestinal farnesoid X receptor signaling promotes nonalcoholic fatty liver disease. *The Journal of Clinical Investigation*, *125*(1), 386–402. https://doi.org/10.1172/JCI76738.

Kabbany, M. N., Conjeevaram Selvakumar, P. K., Watt, K., Lopez, R., Akras, Z., Zein, N., et al. (2017). Prevalence of nonalcoholic Steatohepatitis-associated cirrhosis in the United States: An analysis of National Health and nutrition examination survey data. *The American Journal of Gastroenterology*, *112*(4), 581–587. https://doi.org/10.1038/ajg.2017.5.

Kanda, T., Jiang, X., & Yokosuka, O. (2014). Androgen receptor signaling in hepatocellular carcinoma and pancreatic cancers. *World Journal of Gastroenterology*, *20*(28), 9229–9236. https://doi.org/10.3748/wjg.v20.i28.9229.

Kim, S. K., Ueda, Y., Hatano, E., Kakiuchi, N., Takeda, H., Goto, T., et al. (2016). TERT promoter mutations and chromosome 8p loss are characteristic of nonalcoholic fatty liver disease-related hepatocellular carcinoma. *International Journal of Cancer*, *139*(11), 2512–2518. https://doi.org/10.1002/ijc.30379.

Kozlitina, J., Smagris, E., Stender, S., Nordestgaard, B. G., Zhou, H. H., Tybjaerg-Hansen, A., et al. (2014). Exome-wide association study identifies a TM6SF2 variant that confers susceptibility to nonalcoholic fatty liver disease. *Nature Genetics*, *46*(4), 352–356. https://doi.org/10.1038/ng.2901.

Kudo, Y., Tanaka, Y., Tateishi, K., Yamamoto, K., Yamamoto, S., Mohri, D., et al. (2011). Altered composition of fatty acids exacerbates hepatotumorigenesis during activation of the phosphatidylinositol 3-kinase pathway. *Journal of Hepatology*, *55*(6), 1400–1408. https://doi.org/10.1016/j.jhep.2011.03.025.

Kyrgiou, M., Kalliala, I., Markozannes, G., Gunter, M. J., Paraskevaidis, E., Gabra, H., et al. (2017). Adiposity and cancer at major anatomical sites: Umbrella review of the literature. *BMJ*, *356*, j477. https://doi.org/10.1136/bmj.j477.

Larsson, S. C., & Wolk, A. (2007). Overweight, obesity and risk of liver cancer: A meta-analysis of cohort studies. *British Journal of Cancer*, *97*(7), 1005–1008. https://doi.org/10.1038/sj.bjc.6603932.

Lazo, M., Hernaez, R., Eberhardt, M. S., Bonekamp, S., Kamel, I., Guallar, E., et al. (2013). Prevalence of nonalcoholic fatty liver disease in the United States: The third national health and nutrition examination survey, 1988–1994. *American Journal of Epidemiology*, *178*(1), 38–45. https://doi.org/10.1093/aje/kws448.

Leung, C., Yeoh, S. W., Patrick, D., Ket, S., Marion, K., Gow, P., et al. (2015). Characteristics of hepatocellular carcinoma in cirrhotic and non-cirrhotic non-alcoholic fatty liver disease. *World Journal of Gastroenterology*, *21*(4), 1189–1196. https://doi.org/10.3748/wjg.v21.i4.1189.

Li, W., Xiao, J., Zhou, X., Xu, M., Hu, C., Xu, X., et al. (2015). STK4 regulates TLR pathways and protects against chronic inflammation-related hepatocellular carcinoma. *The Journal of Clinical Investigation*, *125*(11), 4239–4254. https://doi.org/10.1172/JCI81203.

Liu, Y. L., Reeves, H. L., Burt, A. D., Tiniakos, D., McPherson, S., Leathart, J. B., et al. (2014). TM6SF2 rs58542926 influences hepatic fibrosis progression in patients with non-alcoholic fatty liver disease. *Nature Communications*, *5*, 4309. https://doi.org/10.1038/ncomms5309.

Ma, C., Kesarwala, A. H., Eggert, T., Medina-Echeverz, J., Kleiner, D. E., Jin, P., et al. (2016). NAFLD causes selective CD4(+) T lymphocyte loss and promotes hepatocarcinogenesis. *Nature*, *531*(7593), 253–257. https://doi.org/10.1038/nature16969.

Mantovani, A., Ballestri, S., Lonardo, A., & Targher, G. (2016). Cardiovascular disease and myocardial abnormalities in nonalcoholic fatty liver disease. *Digestive Diseases and Sciences*, *61*(5), 1246–1267. https://doi.org/10.1007/s10620-016-4040-6.

Marrero, J. A., Kulik, L. M., Sirlin, C. B., Zhu, A. X., Finn, R. S., Abecassis, M. M., et al. (2018). Diagnosis, staging, and Management of Hepatocellular Carcinoma: 2018 practice guidance by the American Association for the Study of Liver Diseases. *Hepatology*, *68*(2), 723–750. https://doi.org/10.1002/hep.29913.

Matteoni, C. A., Younossi, Z. M., Gramlich, T., Boparai, N., Liu, Y. C., & McCullough, A. J. (1999). Nonalcoholic fatty liver disease: A spectrum of clinical and pathological severity. *Gastroenterology*, *116*(6), 1413–1419. https://doi.org/10.1016/s0016-5085(99)70506-8.

Mazurkevich, G. S., Dzhurko, B. I., & Gorlachev, V. I. (1980). Technic of determining cardiac output by the thermodilution method means of making corrections for changes in body temperature. *Patologicheskaia Fiziologiia i Èksperimental'naia Terapiia*, (2), 74–75. Retrieved from https://www.ncbi.nlm.nih.gov/pubmed/7393639.

Mittal, S., & El-Serag, H. B. (2013). Epidemiology of hepatocellular carcinoma: Consider the population. *Journal of Clinical Gastroenterology*, *47*(Suppl), S2–S6. https://doi.org/10.1097/MCG.0b013e3182872f29.

Mittal, S., El-Serag, H. B., Sada, Y. H., Kanwal, F., Duan, Z., Temple, S., et al. (2016). Hepatocellular carcinoma in the absence of cirrhosis in United States veterans is associated with nonalcoholic fatty liver disease. *Clinical Gastroenterology and Hepatology*, *14*(1), 124–131. e121 https://doi.org/10.1016/j.cgh.2015.07.019.

Moon, A. M., Singal, A. G., & Tapper, E. B. (2020). Contemporary epidemiology of chronic liver disease and cirrhosis. *Clinical Gastroenterology and Hepatology*, *18*, 2650–2666. https://doi.org/10.1016/j.cgh.2019.07.060.

Moore, M. A., Park, C. B., & Tsuda, H. (1998). Implications of the hyperinsulinaemia-diabetes-cancer link for preventive efforts. *European Journal of Cancer Prevention*, 7(2), 89–107. Retrieved from https://www.ncbi.nlm.nih.gov/pubmed/9818771.

Mota, M., Banini, B. A., Cazanave, S. C., & Sanyal, A. J. (2016). Molecular mechanisms of lipotoxicity and glucotoxicity in nonalcoholic fatty liver disease. *Metabolism*, *65*(8), 1049–1061. https://doi.org/10.1016/j.metabol.2016.02.014.

Nagasue, N., Yu, L., Yukaya, H., Kohno, H., & Nakamura, T. (1995). Androgen and oestrogen receptors in hepatocellular carcinoma and surrounding liver parenchyma: Impact on intrahepatic recurrence after hepatic resection. *The British Journal of Surgery*, *82*(4), 542–547. https://doi.org/10.1002/bjs.1800820435.

Nakagawa, H., Maeda, S., Yoshida, H., Tateishi, R., Masuzaki, R., Ohki, T., et al. (2009). Serum IL-6 levels and the risk for hepatocarcinogenesis in chronic hepatitis C patients: An analysis based on gender differences. *International Journal of Cancer*, *125*(10), 2264–2269. https://doi.org/10.1002/ijc.24720.

Nelson, J. E., Bhattacharya, R., Lindor, K. D., Chalasani, N., Raaka, S., Heathcote, E. J., et al. (2007). HFE C282Y mutations are associated with advanced hepatic fibrosis in Caucasians with nonalcoholic steatohepatitis. *Hepatology*, *46*(3), 723–729. https://doi.org/10.1002/hep.21742.

Novo, E., & Parola, M. (2008). Redox mechanisms in hepatic chronic wound healing and fibrogenesis. *Fibrogenesis & Tissue Repair, 1*(1), 5. https://doi.org/10.1186/1755-1536-1-5.

O'Leary, J. G., Landaverde, C., Jennings, L., Goldstein, R. M., & Davis, G. L. (2011). Patients with NASH and cryptogenic cirrhosis are less likely than those with hepatitis C to receive liver transplants. *Clinical Gastroenterology and Hepatology, 9*(8), 700–704. e701. https://doi.org/10.1016/j.cgh.2011.04.007.

Ong, J., Younossi, Z. M., Reddy, V., Price, L. L., Gramlich, T., Mayes, J., et al. (2001). Cryptogenic cirrhosis and posttransplantation nonalcoholic fatty liver disease. *Liver Transplantation, 7*(9), 797–801. https://doi.org/10.1053/jlts.2001.24644.

Paradis, V., Perlemuter, G., Bonvoust, F., Dargere, D., Parfait, B., Vidaud, M., et al. (2001). High glucose and hyperinsulinemia stimulate connective tissue growth factor expression: A potential mechanism involved in progression to fibrosis in nonalcoholic steatohepatitis. *Hepatology, 34*(4 Pt. 1), 738–744. https://doi.org/10.1053/jhep.2001.28055.

Park, Y., & Colditz, G. A. (2018). Diabetes and adiposity: A heavy load for cancer. *The Lancet Diabetes and Endocrinology, 6*(2), 82–83. https://doi.org/10.1016/S2213-8587(17)30396-0.

Park, E. J., Lee, J. H., Yu, G. Y., He, G., Ali, S. R., Holzer, R. G., et al. (2010). Dietary and genetic obesity promote liver inflammation and tumorigenesis by enhancing IL-6 and TNF expression. *Cell, 140*(2), 197–208. https://doi.org/10.1016/j.cell.2009.12.052.

Park, J., Morley, T. S., Kim, M., Clegg, D. J., & Scherer, P. E. (2014). Obesity and cancer—Mechanisms underlying tumour progression and recurrence. *Nature Reviews. Endocrinology, 10*(8), 455–465. https://doi.org/10.1038/nrendo.2014.94.

Petta, S., & Craxi, A. (2010). Hepatocellular carcinoma and non-alcoholic fatty liver disease: From a clinical to a molecular association. *Current Pharmaceutical Design, 16*(6), 741–752. https://doi.org/10.2174/138161210790883787.

Pirola, C. J., Gianotti, T. F., Burgueno, A. L., Rey-Funes, M., Loidl, C. F., Mallardi, P., et al. (2013). Epigenetic modification of liver mitochondrial DNA is associated with histological severity of nonalcoholic fatty liver disease. *Gut, 62*(9), 1356–1363. https://doi.org/10.1136/gutjnl-2012-302962.

Ratziu, V., Bonyhay, L., Di Martino, V., Charlotte, F., Cavallaro, L., Sayegh-Tainturier, M. H., et al. (2002). Survival, liver failure, and hepatocellular carcinoma in obesity-related cryptogenic cirrhosis. *Hepatology, 35*(6), 1485–1493. https://doi.org/10.1053/jhep.2002.33324.

Romeo, S., Kozlitina, J., Xing, C., Pertsemlidis, A., Cox, D., Pennacchio, L. A., et al. (2008). Genetic variation in PNPLA3 confers susceptibility to nonalcoholic fatty liver disease. *Nature Genetics, 40*(12), 1461–1465. https://doi.org/10.1038/ng.257.

Rosenfeld, R. G. (2003). Insulin-like growth factors and the basis of growth. *The New England Journal of Medicine, 349*(23), 2184–2186. https://doi.org/10.1056/NEJMp038156.

Samoylova, M. L., Mehta, N., Roberts, J. P., & Yao, F. Y. (2018). Predictors of ultrasound failure to detect hepatocellular carcinoma. *Liver Transplantation, 24*(9), 1171–1177. https://doi.org/10.1002/lt.25202.

Sanyal, A. J. (2019). Past, present and future perspectives in nonalcoholic fatty liver disease. *Nature Reviews. Gastroenterology & Hepatology, 16*(6), 377–386. https://doi.org/10.1038/s41575-019-0144-8.

Schlesinger, S., Aleksandrova, K., Pischon, T., Fedirko, V., Jenab, M., Trepo, E., et al. (2013). Abdominal obesity, weight gain during adulthood and risk of liver and biliary tract cancer in a European cohort. *International Journal of Cancer, 132*(3), 645–657. https://doi.org/10.1002/ijc.27645.

Sharma, D., Wang, J., Fu, P. P., Sharma, S., Nagalingam, A., Mells, J., et al. (2010). Adiponectin antagonizes the oncogenic actions of leptin in hepatocellular carcinogenesis. *Hepatology, 52*(5), 1713–1722. https://doi.org/10.1002/hep.23892.

Shen, J., Yeh, C. C., Wang, Q., Gurvich, I., Siegel, A. B., & Santella, R. M. (2016). Plasma adiponectin and hepatocellular carcinoma survival among patients without liver transplantation. *Anticancer Research, 36*(10), 5307–5314. https://doi.org/10.21873/anticanres.11103.

Simile, M. M., Latte, G., Feo, C. F., Feo, F., Calvisi, D. F., & Pascale, R. M. (2018). Alterations of methionine metabolism in hepatocarcinogenesis: The emergent role of glycine N-methyltransferase in liver injury. *Annals of Gastroenterology, 31*(5), 552–560. https://doi.org/10.20524/aog.2018.0288.

Singal, A. G., Yopp, A. C., Gupta, S., Skinner, C. S., Halm, E. A., Okolo, E., et al. (2012). Failure rates in the hepatocellular carcinoma surveillance process. *Cancer Prevention Research (Philadelphia, Pa.), 5*(9), 1124–1130. https://doi.org/10.1158/1940-6207.CAPR-12-0046.

Sircana, A., Paschetta, E., Saba, F., Molinaro, F., & Musso, G. (2019). Recent insight into the role of fibrosis in nonalcoholic Steatohepatitis-related hepatocellular carcinoma. *International Journal of Molecular Sciences, 20*(7), 1745. https://doi.org/10.3390/ijms20071745.

Smagris, E., BasuRay, S., Li, J., Huang, Y., Lai, K. M., Gromada, J., et al. (2015). Pnpla3I148M knockin mice accumulate PNPLA3 on lipid droplets and develop hepatic steatosis. *Hepatology, 61*(1), 108–118. https://doi.org/10.1002/hep.27242.

Starley, B. Q., Calcagno, C. J., & Harrison, S. A. (2010). Nonalcoholic fatty liver disease and hepatocellular carcinoma: A weighty connection. *Hepatology, 51*(5), 1820–1832. https://doi.org/10.1002/hep.23594.

Stefanou, N., Papanikolaou, V., Furukawa, Y., Nakamura, Y., & Tsezou, A. (2010). Leptin as a critical regulator of hepatocellular carcinoma development through modulation of human telomerase reverse transcriptase. *BMC Cancer, 10*, 442. https://doi.org/10.1186/1471-2407-10-442.

Sung, K. C., Wild, S. H., & Byrne, C. D. (2013). Resolution of fatty liver and risk of incident diabetes. *The Journal of Clinical Endocrinology and Metabolism, 98*(9), 3637–3643. https://doi.org/10.1210/jc.2013-1519.

Tokushige, K., Hashimoto, E., Horie, Y., Taniai, M., & Higuchi, S. (2011). Hepatocellular carcinoma in Japanese patients with nonalcoholic fatty liver disease, alcoholic liver disease, and chronic liver disease of unknown etiology: Report of the nationwide survey. *Journal of Gastroenterology, 46*(10), 1230–1237. https://doi.org/10.1007/s00535-011-0431-9.

Tryndyak, V. P., Han, T., Muskhelishvili, L., Fuscoe, J. C., Ross, S. A., Beland, F. A., et al. (2011). Coupling global methylation and gene expression profiles reveal key pathophysiological events in liver injury induced by a methyl-deficient diet. *Molecular Nutrition & Food Research, 55*(3), 411–418. https://doi.org/10.1002/mnfr.201000300.

Turnbaugh, P. J., Hamady, M., Yatsunenko, T., Cantarel, B. L., Duncan, A., Ley, R. E., et al. (2009). A core gut microbiome in obese and lean twins. *Nature, 457*(7228), 480–484. https://doi.org/10.1038/nature07540.

Turnbaugh, P. J., Ley, R. E., Mahowald, M. A., Magrini, V., Mardis, E. R., & Gordon, J. I. (2006). An obesity-associated gut microbiome with increased capacity for energy harvest. *Nature, 444*(7122), 1027–1031. https://doi.org/10.1038/nature05414.

Ueyama, M., Nishida, N., Korenaga, M., Korenaga, K., Kumagai, E., Yanai, H., et al. (2016). The impact of PNPLA3 and JAZF1 on hepatocellular carcinoma in non-viral hepatitis patients with type 2 diabetes mellitus. *Journal of Gastroenterology, 51*(4), 370–379. https://doi.org/10.1007/s00535-015-1116-6.

Valenti, L., Dongiovanni, P., Ginanni Corradini, S., Burza, M. A., & Romeo, S. (2013). PNPLA3 I148M variant and hepatocellular carcinoma: A common genetic variant for a rare disease. *Digestive and Liver Disease, 45*(8), 619–624. https://doi.org/10.1016/j.dld.2012.12.006.

Valentini, M., Piermattei, A., Di Sante, G., Migliara, G., Delogu, G., & Ria, F. (2014). Immunomodulation by gut microbiota: Role of toll-like receptor expressed by T cells. *Journal of Immunology Research*, *2014*, 586939. https://doi.org/10.1155/2014/586939.

Villanueva, A., Chiang, D. Y., Newell, P., Peix, J., Thung, S., Alsinet, C., et al. (2008). Pivotal role of mTOR signaling in hepatocellular carcinoma. *Gastroenterology*, *135*(6), 1972–1983. 1983 e1971–1911. https://doi.org/10.1053/j.gastro.2008.08.008.

Vucenik, I., & Stains, J. P. (2012). Obesity and cancer risk: Evidence, mechanisms, and recommendations. *Annals of the New York Academy of Sciences*, *1271*, 37–43. https://doi.org/10.1111/j.1749-6632.2012.06750.x.

Weinmann, A., Alt, Y., Koch, S., Nelles, C., Duber, C., Lang, H., et al. (2015). Treatment and survival of non-alcoholic steatohepatitis associated hepatocellular carcinoma. *BMC Cancer*, *15*, 210. https://doi.org/10.1186/s12885-015-1197-x.

Wolf, M. J., Adili, A., Piotrowitz, K., Abdullah, Z., Boege, Y., Stemmer, K., et al. (2014). Metabolic activation of intrahepatic CD8+ T cells and NKT cells causes nonalcoholic steatohepatitis and liver cancer via cross-talk with hepatocytes. *Cancer Cell*, *26*(4), 549–564. https://doi.org/10.1016/j.ccell.2014.09.003.

Wong, R. J., Aguilar, M., Cheung, R., Perumpail, R. B., Harrison, S. A., Younossi, Z. M., et al. (2015). Nonalcoholic steatohepatitis is the second leading etiology of liver disease among adults awaiting liver transplantation in the United States. *Gastroenterology*, *148*(3), 547–555. https://doi.org/10.1053/j.gastro.2014.11.039.

Wong, R. J., Cheung, R., & Ahmed, A. (2014). Nonalcoholic steatohepatitis is the most rapidly growing indication for liver transplantation in patients with hepatocellular carcinoma in the U.S. *Hepatology*, *59*(6), 2188–2195. https://doi.org/10.1002/hep.26986.

Wostmann, B. S., Larkin, C., Moriarty, A., & Bruckner-Kardoss, E. (1983). Dietary intake, energy metabolism, and excretory losses of adult male germfree Wistar rats. *Laboratory Animal Science*, *33*(1), 46–50. Retrieved from https://www.ncbi.nlm.nih.gov/pubmed/6834773.

Xing, S. Q., Zhang, C. G., Yuan, J. F., Yang, H. M., Zhao, S. D., & Zhang, H. (2015). Adiponectin induces apoptosis in hepatocellular carcinoma through differential modulation of thioredoxin proteins. *Biochemical Pharmacology*, *93*(2), 221–231. https://doi.org/10.1016/j.bcp.2014.12.001.

Yang, S., & Liu, G. (2017). Targeting the Ras/Raf/MEK/ERK pathway in hepatocellular carcinoma. *Oncology Letters*, *13*(3), 1041–1047. https://doi.org/10.3892/ol.2017.5557.

Yoshimoto, S., Loo, T. M., Atarashi, K., Kanda, H., Sato, S., Oyadomari, S., et al. (2013). Obesity-induced gut microbial metabolite promotes liver cancer through senescence secretome. *Nature*, *499*(7456), 97–101. https://doi.org/10.1038/nature12347.

Younes, R., & Bugianesi, E. (2018). Should we undertake surveillance for HCC in patients with NAFLD? *Journal of Hepatology*, *68*(2), 326–334. https://doi.org/10.1016/j.jhep.2017.10.006.

Younossi, Z. M., Koenig, A. B., Abdelatif, D., Fazel, Y., Henry, L., & Wymer, M. (2016). Global epidemiology of nonalcoholic fatty liver disease-meta-analytic assessment of prevalence, incidence, and outcomes. *Hepatology*, *64*(1), 73–84. https://doi.org/10.1002/hep.28431.

Younossi, Z. M., Otgonsuren, M., Henry, L., Venkatesan, C., Mishra, A., Erario, M., et al. (2015). Association of nonalcoholic fatty liver disease (NAFLD) with hepatocellular carcinoma (HCC) in the United States from 2004 to 2009. *Hepatology*, *62*(6), 1723–1730. https://doi.org/10.1002/hep.28123.

Yuan, X., Waterworth, D., Perry, J. R., Lim, N., Song, K., Chambers, J. C., et al. (2008). Population-based genome-wide association studies reveal six loci influencing plasma levels of liver enzymes. *American Journal of Human Genetics*, *83*(4), 520–528. https://doi.org/10.1016/j.ajhg.2008.09.012.

Zain, S. M., Mohamed, R., Cooper, D. N., Razali, R., Rampal, S., Mahadeva, S., et al. (2014). Genome-wide analysis of copy number variation identifies candidate gene loci associated with the progression of non-alcoholic fatty liver disease. *PLoS One*, *9*(4), e95604. https://doi.org/10.1371/journal.pone.0095604.
Zhang, H., Gao, C., Fang, L., Zhao, H. C., & Yao, S. K. (2013). Metformin and reduced risk of hepatocellular carcinoma in diabetic patients: A meta-analysis. *Scandinavian Journal of Gastroenterology*, *48*(1), 78–87. https://doi.org/10.3109/00365521.2012.719926.
Zheng, X., Zeng, W., Gai, X., Xu, Q., Li, C., Liang, Z., et al. (2013). Role of the hedgehog pathway in hepatocellular carcinoma (review). *Oncology Reports*, *30*(5), 2020–2026. https://doi.org/10.3892/or.2013.2690.

CHAPTER FIVE

Atypical immunometabolism and metabolic reprogramming in liver cancer: Deciphering the role of gut microbiome

Rachel M. Golonka and Matam Vijay-Kumar*
Department of Physiology and Pharmacology, University of Toledo College of Medicine and Life Sciences, Toledo, OH, United States
*Corresponding author: e-mail address: matamvijay.kumar@utoledo.edu

Contents

1. Hepatocellular carcinoma (HCC): A heterogenous and complex liver disease 174
2. Gut microbiota dysbiosis as an identifier of HCC 177
3. HCC microenvironment favors immunosuppression: Is gut microbiota a driver of this immunologic signature? 179
   3.1 Cellular senescence 179
   3.2 Tumor associated myeloid cells 185
   3.3 Tumor infiltrating lymphocytes (TILs) 192
4. Hijacking energy sources: Does gut microbiota have an influence on metabolic reprogramming in HCC? 201
   4.1 Warburg effect 201
   4.2 Aerobic glycolysis 202
   4.3 Fructolysis 205
   4.4 Rewiring of the tricarboxylic acid cycle 207
   4.5 Glutamine addiction 208
5. Immunological and gut microbiota therapeutics for HCC 209
   5.1 Immune checkpoint inhibitors 209
   5.2 Adoptive cellular immunotherapy 211
   5.3 Probiotics 214
   5.4 Prebiotics 215
   5.5 Fecal microbiota transplantation (FMT) 216
6. Summary and future perspectives 217
Acknowledgment 218
References 218

Abstract

Hepatocellular carcinoma (HCC) is the fourth leading cause of cancer-related mortality worldwide. Much recent research has delved into understanding the underlying molecular mechanisms of HCC pathogenesis, which has revealed to be heterogenous and

Advances in Cancer Research, Volume 149
ISSN 0065-230X
https://doi.org/10.1016/bs.acr.2020.10.004

complex. Two major hallmarks of HCC include: (i) a hijacked immunometabolism and (ii) a reprogramming in metabolic processes. We posit that the gut microbiota is a third component in an entanglement triangle contributing to HCC progression. Besides metagenomic studies highlighting the diagnostic potential in the gut microbiota profile, recent research is pinpointing the gut microbiota as an instigator, not just a mere bystander, in HCC. In this chapter, we discuss mechanistic insights on atypical immunometabolism and metabolic reprogramming in HCC, including the examination of tumor-associated macrophages and neutrophils, tumor-infiltrating lymphocytes (e.g., T-cell exhaustion, regulatory T-cells, natural killer T-cells), the Warburg effect, rewiring of the tricarboxylic acid cycle, and glutamine addiction. We further discuss the potential involvement of the gut microbiota in these characteristics of hepatocarcinogenesis. An immediate highlight is that microbiota metabolites (e.g., short chain fatty acids, secondary bile acids) can impair anti-tumor responses, which aggravates HCC. Lastly, we describe the rising 'new era' of immunotherapies (e.g., immune checkpoint inhibitors, adoptive T-cell transfer) and discuss for the potential incorporation of gut microbiota targeted therapeutics (e.g., probiotics, fecal microbiota transplantation) to alleviate HCC. Altogether, this chapter invigorates for continuous research to decipher the role of gut microbiome in HCC from its influence on immunometabolism and metabolic reprogramming.

Abbreviations

ACT	adoptive T-cell transfer
AFP	α-fetoprotein
AKT	protein kinase B
ATP	adenosine triphosphate
B_{reg}	regulatory B-cell
CAR	chimeric antigen receptor
CCL	C-C motif ligand
CCl_4	carbon tetrachloride
CCR	C-C chemokine receptor
CD	cluster of differentiation
CIK	cytokine-induced killer cell
COX2	cyclooxygenase 2
CTLA-4	cytotoxic T-lymphocyte-associated antigen 4
CX3CR1	C-X3-C motif chemokine receptor 1
CXCL	C-C motif chemokine ligand
DC	dendritic cell
DEN	diethylnitrosamine
DHAP	dihydroxyacetone phosphate
EMT	epithelial-mesenchymal transition
ERK	extracellular signal regulated kinase
FBP1	fructose-1,6-bisphosphatase 1
FMT	fecal microbiota transplantation
FOS	fructo-oligosaccharides
FOXP3	forkhead box P3

GAP	glyceraldehyde 3-phosphate
GLS1	glutaminase 1
GLS2	glutaminase 2
GM-CSF	granulocyte-macrophage colony-stimulating factor
GPC3	glypican-3
GPR	G-protein coupled receptor
HBV	hepatitis B virus
HCC	hepatocellular carcinoma
HCV	hepatitis C virus
HDAC	histone deacetylase
HDACi	histone deacetylase inhibitor
HIF-1α	hypoxia-inducible factor-1α
HLA	human leukocyte antigen
HMGB1	high-mobility group B1
HSC	hepatic stellate cell
IDH	isocitrate dehydrogenase
IFN	interferon
Ig	immunoglobulin
IL	interleukin
iNKT	invariant natural killer T-cell
ISAPP	International Scientific Association for Probiotics and Prebiotics
LAG3	lymphocyte-activation gene 3
LPS	lipopolysaccharide
M1	pro-inflammatory macrophage
M2	anti-inflammatory macrophage
MAPK	mitogen activated protein kinase
MDDC	monocyte-derived dendritic cell
MDSC	myeloid-derived suppressor cell
M-MDSC	monocytic myeloid-derived suppressor cell
MyD88	myeloid differentiation primary response 88
N1	tumor suppressive neutrophil
N2	oncogenic neutrophil
NAFLD	non-alcoholic fatty liver disease
NASH	non-alcoholic steatohepatitis
NETs	neutrophil extracellular traps
NF-κB	nuclear factor kappa-light-chain-enhance of activated B cells
NK	natural killer cell
NKG2D	NK group 2 member D
NKG2DL	NK group 2 member D ligand
NKT	natural killer T-cell
Nrf2	nuclear factor (erythroid-derived 2)-like 2
PCK/PEPCK	phosphoenolpyruvate carboxykinase
PD-1	programmed cell death protein-1
PD-L1	programmed cell death ligand-1
PI3K	phosphatidylinositol 3-kinase
PMN-MDSC	polymorphonuclear myeloid-derived suppressor cells

RFA	radiofrequency
ROS	reactive oxygen species
SASP	senescence-associated secretory phenotype
SCFA	short chain fatty acid
SDH	succinate dehydrogenase
SIRT	selective internal radiation
STAT3	signal transducer and activator of transcription 3
TAA	tumor associated antigen
TACE	transcatheter arterial chemoembolization
TAM	tumor associated macrophage
TAN	tumor associated neutrophils
TBF-β1	transforming growth factor-β1
TCA	tricarboxylic acid cyle
TCR	T-cell receptor
$\mathbf{T_{FH}}$	T-follicular helper cells
TIGIT	T-cell immunoglobulin and ITIM domain (TIGIT)
TIL	tumor infiltrating lymphocyte
TIM3	T-cell immunoglobulin and mucin domain-3
TLR	Toll-like receptor
$\mathbf{T_{reg}}$	regulatory T-cells

1. Hepatocellular carcinoma (HCC): A heterogenous and complex liver disease

Old Germanic and Romance languages defined the liver as "life" (Riva et al., 2011), which is attributed to the several moonlighting functions it possesses. For one, the liver is responsible for preserving appropriate macro and micronutrient homeostasis, which involves sustaining bile and xenobiotic metabolisms (Trefts, Gannon, & Wasserman, 2017). Strategic anatomical placement makes the liver a "first-hit" organ for exposure to host molecules, noxious diet-derived metabolites, viable bacteria, microbial components, and xenobiotic compounds transported from the gastrointestinal tract via the portal vein. In response, phagocytic Kupffer cells (resident hepatic macrophages) are activated as a "firewall" to promote the clearance of commensals, opportunistic gut bacteria, pathogenic microbes, and toxins (Balmer et al., 2014; Macpherson, Heikenwalder, & Ganal-Vonarburg, 2016). This is essential in cases of intestinal barrier dysfunction, referred to as a "leaky gut"

(Konturek et al., 2018). To prevent unintended organ damage from these self-defense mechanisms, the liver encompasses immunotolerant operations to balance inflammatory signals, which involves co-inhibitory signaling and immunosuppressive responses (Kubes & Jenne, 2018). Importantly, damage to the hepatic barricade makes the liver vulnerable to various microbiota-associated conditions, including non-alcoholic fatty liver disease (NAFLD) and non-alcoholic steatohepatitis (NASH) (Campo, Eiseler, Apfel, & Pyrsopoulos, 2019). Further, prolonged immunotolerance can progress liver dysfunction to create an ideal environment for tumorigenesis.

For this chapter, we are focused on hepatocellular carcinoma (HCC) because it is the most frequent form of primary liver cancer and the fourth most common cause of cancer-related deaths worldwide (Villanueva, 2019; Yang et al., 2019). The highest liver cancer incidences from 1990 to 2017 were found with male dominance, in the high-risk elderly population, and a strong prevalence in developed countries (Liu et al., 2020). In 2015, there were 854,000 new liver cancer cases and 810,000 deaths globally (Global Burden of Disease Liver Cancer Collaboration et al., 2017). For 2020, it is projected that there will be 42,810 new HCC cases with a mortality of 30,160 deaths in the United States alone (Siegel, Miller, & Jemal, 2020). Further projections for 2030 predict the number of liver cancer incidences to increase by threefold (since 2005) in the United States (Valery et al., 2018). Detrimentally, the highly heterogenous and asymptomatic features of HCC makes early therapeutic intervention almost negligible, which is reflected on the 2- and 5-year survival rates being less than 50% and only 10%, respectively (Golabi et al., 2017).

The oral regimen sorafenib (a multikinase inhibitor) is the front-line pharmaceutical approved for advanced HCC management (Llovet et al., 2008), but HCC cases with sorafenib resistance require co-therapeutic agents (Jiang et al., 2018). In addition, sorafenib treatment can cause a "paradoxical" upregulation of mitogen-activated protein kinases (MAPK) to sustain HCC proliferation (Chen et al., 2017). Also, sorafenib can become reactivated in the gut by microbial β-glucuronidase enzymes, which can (i) vary the pharmacokinetics of sorafenib between patients and (ii) lead to drug toxicity, as indicated by adverse side effects like hypertension, hand-foot skin reactions, and gastrointestinal mucositis (Boudou-Rouquette et al., 2012; Brose, Frenette, Keefe, & Stein, 2014; Ervin et al., 2019; Vasilyeva et al., 2015). Accordingly, a rising "new era" of promising immunotherapies are being developed against HCC, such as immune checkpoint inhibitors (Pinato et al., 2020).

Designing effective immunotherapies requires in-depth understanding of how cancer cells hijack immunometabolism and reprogram host metabolic function to sustain HCC growth and metastasis (Ngwa, Edwards, Philip, & Chen, 2019). There are multiple models of murine HCC, such as overexpression of oncogenes by genetic engineering, viral infection (e.g., hepatitis virus B/C), hepato- and chemo-toxic agents (e.g., carbon tetrachloride/CCl_4, diethylnitrosamine/DEN), aflatoxin administration, alcohol overconsumption, and choline-deficient dietary intervention (De Minicis et al., 2013; Vandamme, 2014). This allows researchers to study the diverse etiological risk factors and heterogeneous complexity in HCC initiation/progression (Yang, Hainaut, et al., 2019). One new environmental risk factor is the gut microbiota, but until recently there had not been an appropriate model to study gut microbiota dependent HCC. It is noteworthy that our group has discovered that feeding an enriched prebiotic fiber diet results in HCC development in a subset ($\sim$40%) of gut dysbiotic Toll-like receptor 5 deficient mice (TLR5, receptor for bacterial flagellin) (Singh et al., 2018).

Other murine HCC models have been developed to also account for understanding the role of immunometabolism in liver tumorigenesis. To highlight, Liu et al. have designed a spontaneous HCC model in mice that recapitulates T-cell exhaustion in the tumor microenvironment (Liu et al., 2018). There are also many human databases that are important sources for sequencing and modeling analyses, which have been imperative for identifying distinct immune cell populations, defining the metabolic phenotype, and classifying the gut microbiota in HCC patients (Bolouri et al., 2020; Ren et al., 2019; Rohr-Udilova et al., 2018; Zheng et al., 2017). To summarize, there is a monumental crosstalk between all types of immune cell populations that can support HCC evasion by favoring immunosuppression. HCC is also correlated with the reprogramming of metabolic functions to favor the Warburg effect (i.e., aerobic glycolysis) for tumor proliferation. Further, HCC patients exhibit a dysbiotic gut microbiome. Throughout this chapter, we will review the atypical immunometabolism and metabolic reprogramming denoted in HCC, in coordination with linking the role of gut microbiota (as previewed in Fig. 1).

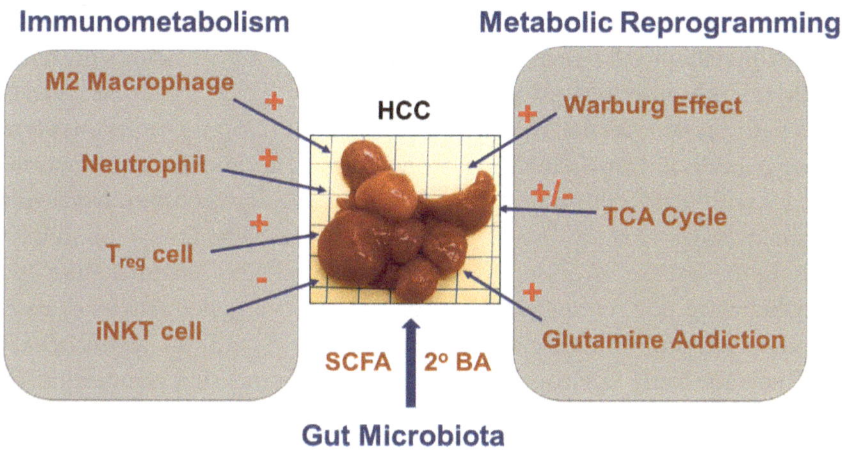

Fig. 1 Immunometabolism, metabolic reprogramming, and the gut microbiota collectively contribute to hepatocellular carcinoma (HCC). Literature indicates that HCC cells modify the tumor microenvironment and immune cell profile to increase M2 macrophages, tumor-associated neutrophils, and regulatory T-cell (T_{reg}) recruitment and function but lessen the abundance and activity of invariant natural killer T-cells (iNKT). In addition, metabolically active HCC cells require initiation of the Warburg effect for lactate production and glutamine addiction to sustain proliferation. Depending on nutrient rich vs deprived conditions, the tricarboxylic acid (TCA) cycle will either be up- or downregulated. The gut microbiota is another component in HCC pathogenesis, where the accumulation of metabolites like short chain fatty acids (SCFA) and secondary bile acids (2° BA) promote liver tumorigenesis by impairing anti-tumor responses. *Source: Credit: Rachel M. Golonka.*

2. Gut microbiota dysbiosis as an identifier of HCC

The gut microbiome comprises trillions of indigenous bacteria, archaea, bacteriophages, eukaryotes, viruses, and fungi that can tip the host fitness scale toward either healthy or at risk for disease (Lynch & Pedersen, 2016). Excitingly, the Culturable Genome Reference comprises an extensive collection of "healthy" human gut microbiota encompassing all major bacteria phyla, which provides a reference microbiome and gut bacterial genome (Zou et al., 2019). Next generation Illumina-based metagenomic sequencing finds that the bacterial genome outweighs the human genome by 150-fold (Qin et al., 2010), but mathematical modeling highlights that the human to bacterial cell number ratio is 1:1 (Sender, Fuchs, & Milo, 2016). Importantly, ongoing global sample collection, advancement in bacterial culturing, and continuous metagenomic analyses are cooperatively identifying the phylogenetic diversity between each populations individualized microbiome blueprint (Almeida et al., 2019; Li et al., 2014). Metagenomic studies have also found that transitioning from in utero, infancy, childhood, and adulthood correlates with taxonomic and functional changes within the gut microbiome (Collado, Rautava, Aakko, Isolauri, & Salminen, 2016; Derrien, Alvarez, & de Vos, 2019; Rampelli et al., 2020). As such, these differences between an "immature" and "mature/aged" gut microbiota are uncovering potential mechanisms to aging-related diseases, like inflammaging (Franceschi, Garagnani, Vitale, Capri, & Salvioli, 2017).

There are many environmental and genetic factors that can perturb the gut microbiota into a state of dysbiosis or imbalance (Rothschild et al., 2018). For instance, the genetic differences between identical and fraternal twins were found to correlate with distinct gut microbiota compositions (Goodrich et al., 2014). Other reports have shown that the gut microbiota profile is distinguishable when comparing vaginal to C-section born infants (Dominguez-Bello et al., 2016; Shao et al., 2019), breastfed to formula fed children (Fan et al., 2013; Timmerman et al., 2017), and omnivore vs vegetarian fed individuals (De Angelis et al., 2020). Another layer of complexity entails that there is a crosstalk between epigenetics and the gut microbiota, which can influence toward either health or disease states (Qin & Wade, 2018). Additionally, the intestinal environment in either aerobic or hypoxic conditions can consequently mitigate the colonization of obligate anaerobic gut bacteria (Litvak, Byndloss, & Baumler, 2018). Regardless of the trigger, an altered gut microbiota profile is becoming a common associative indicator of pathogenesis for multiple diseases (Carding, Verbeke, Vipond, Corfe, & Owen, 2015). Generally, gut microbiota dysbiosis is denoted with an increase in the Firmicutes to Bacteroidetes ratio (the two dominant phyla of the microbiota), in coordination with the bloom of opportunistic pathogens like γ-Proteobacteria and reduced abundance of beneficial commensals like *Akkermansia* spp.

Much recent research is emphasizing that the gut microbiota and their metabolites are pathological instigators and not mere bystanders in HCC through the gut-liver axis (Tripathi et al., 2018). The "catch all" term gut microbiota dysbiosis has become widely recognized as a profound biomarker for HCC (Ni et al., 2019; Zheng et al., 2020; Zhou et al., 2020), where fecal and blood metagenomic signatures of the microbiome have been proposed as potential diagnostic tools for liver cancer (Cho et al., 2019; Ren et al., 2019). Specifically, both the Firmicutes and γ-Proteobacteria were elevated in the blood microbiome signature for HCC patients. The fecal microbiota in NAFLD-HCC patients showed a similar microbiota profile, including a pathological bloom for *Enterobacteriaceae* spp. and *Streptococcus* spp., but mitigation of *Akkermansia* spp., which was collectively associated with an inflammatory milieu (Ponziani et al., 2019). Other analyses depict a strong correlation of elevated *Bacteroides* and *Clostridia* species in NASH-HCC patients, in addition to increased systemic circulation of the microbial endotoxin lipopolysaccharide (LPS) (Xie et al., 2016). Translocation of viable bacteria and toxic substances like LPS (*alias* endotoxemia) is a hallmark of chronic liver diseases, which can contribute to inflammation, fibrosis, and HCC by activation of pattern recognition receptors (e.g., Toll-like receptors/TLR) (Backhed, 2011; Dapito et al., 2012; Mamedova et al., 2013). In addition, the translocation/leakage of other gut metabolites, like short chain fatty acids (SCFA) and secondary bile acids, can contribute to HCC by modulating immunometabolism (Ohno, 2020). Importantly, sterilizing the gut microbiota through antibiotic treatment can abate hepatocarcinogenesis (Mamedova et al., 2013; Yamada et al., 2018).

Considering the strong cross-talk between the gut microbiome and host immunity (Zheng, Liwinski, & Elinav, 2020), the next sections of this chapter will be describing the influence of the gut microbiota and intestinal metabolites on atypical immunometabolism in HCC (as summarized in Table 1).

3. HCC microenvironment favors immunosuppression: Is gut microbiota a driver of this immunologic signature?

3.1 Cellular senescence

Hepatocyte senescence and immune surveillance are two extrinsic gatekeepers in the "senescence surveillance" program to abate HCC. Normally, premalignant senescent hepatocytes create a "secretory phenotype" of chemokine and cytokine signals for immune-dependent clearance, but impairment of $CD4^+$ T-cell mediated adaptive immune responses can drive liver tumorigenesis (Kang et al., 2011). To maintain this oncogenic phenotype, the maturation of recruited C-C chemokine receptor (CCR) type 2 $(CCR2)^+$ myeloid precursors required for senescence is inhibited, which drives murine HCC and worsens the prognosis for HCC patients (Eggert et al., 2016). In coordination, tumor cells switch the transcriptome profile toward replicative senescence or "immortality" (Ozturk et al., 2006; Yildiz et al., 2013) and alter chromatin status to lessen DNA damage responses (Wu et al., 2017), both of which bypass senescence surveillance. Further, chronic liver injury can direct hepatocyte proliferation and HCC initiation rather than senescence, whereas senescence surveillance would be functional to impede HCC under severe acute hepatic injury (Wang, Chen, et al., 2018).

Table 1 Microbiota-immunity axis in HCC.

Chapter section	Cell type or immunity response	Immune signature in HCC	Gut microbe or intestinal metabolite targets	Microbiota-immunity interaction in HCC (known and potential mechanisms)	Reference(s)
3.1	Cellular senescence	Cellular senescence is impaired during HCC initiation. However, senescence under established HCC causes inflammation and the replenishment of hepatocytes that foster HCC growth	Lipoteichoic acid (LTA; Gram-positive bacterial cell wall component) and Deoxycholate (DCA; secondary bile acid produced by Gram-positive bacteria) promote senescence-associated secretory phenotype (SASP)	LTA and DCA induces SASP factors (e.g., IL-1β, Groα/CXCL1, and IL-6). Further, LTA activates Toll-like receptor 2, which follows with cyclooxygenase 2 (COX2) mediated prostaglandin E_2 (PGE2) production and PGE2 receptor 4 (PTGER4) signaling to suppress anti-tumor immunity	Eggert et al. (2016), Kang et al. (2011), Loo et al. (2017), Ho et al. (2020), Yoshimoto et al. (2013)
3.2.1	Tumor-associated macrophages (TAMs)	Progression toward M2 polarization and impediment for M1 polarization. This is from activation of multiple signaling pathways (e.g., IL-6/STAT3) and alteration of the tumor microenvironment (e.g., hypoxia). M2 macrophages support HCC by multiple mechanisms, including immunosuppression	Imbalance of enriched Gram-negative bacteria (e.g., Proteobacteria, Bacteroidetes) and diminished Gram-positive bacteria (e.g. Firmicutes) support M2 polarization. Further, elevated translocation of the Gram-negative gut microbial component, lipopolysaccharide (LPS) is noted	Absence of Gram-positive bacteria (via vancomycin treatment) induced IL-25 secretion and IL-25 dependent M2 activation, where subsequent CXCL10 production facilitated HCC invasion. Alongside, LPS-TLR4 signaling on macrophages crosstalk with T-cells for secretion of M2 polarization factors (e.g., IL-21, IFNγ)	Chen et al. (2016), Ke et al. (2019), Kong et al. (2016), Li et al. (2019), Miura et al. (2016), Yao, Li, Zhang, Chen, and Wang (2018), Yao et al. (2017)

3.2.2	Tumor-associated neutrophils (TANs)	Favor N2 immunosuppressive subtype compared to the N1 anti-tumor subset. TANs promote HCC growth and metastasis by recruiting immunosuppressive immune cells and secreting neutrophil extracellular traps (NETs)	Gut microbiota, in general, influence the circulating neutrophil population, but association with TANs in HCC is unclear. Yet, presence of gut microbiota is related to modulating the sensitivity for NETs formation in HCC-independent conditions. There are no current direct links between TANs and the gut microbiota in HCC	Potential, underexplored mechanisms include the gut microbiota translocation of unmethylated CpG islands from bacterial DNA activates TLR9-dependent NETs formation, which may stimulate HCC metastasis	Fridlender et al. (2009), Mishalian et al. (2013), Seo, Gu, Jung, Kim, and Kim (2019), Singh et al. (2018), Zhou et al. (2016), Yang et al. (2020), Yang et al. (2020), Yoon et al. (2018)
3.2.3	Myeloid-derived tumor suppressor cells (MDSCs)	Compared to monocytic (M)-MDSCs, polymorphonuclear (PMN)-MDSCs are more prominent in HCC. MDSCs support HCC by modulating host immunity to favor immunosuppression	Gut microbiota, in general, influence circulating MDSC populations. Absence of gut microbiota is related to increased levels of M- and PMN-MDSCs in the bone marrow, spleen, and mesenteric lymph nodes. However, there are no current direct links between MDSCs and the gut microbiota in HCC	Potential, underexplored mechanisms include a diverse microbiota could be beneficial to reduce PMN-MDSCs and protect from HCC	Chiu et al. (2017), Hathaway-Schrader et al. (2019), Li et al. (2019), Li et al. (2020), Nan et al. (2018)

Continued

Table 1 Microbiota-immunity axis in HCC.—cont'd

Chapter section	Cell type or immunity response	Immune signature in HCC	Gut microbe or intestinal metabolite targets	Microbiota-immunity interaction in HCC (known and potential mechanisms)	Reference(s)
3.3.1	Cytotoxic $CD8^+$ T-cells	$CD8^+$ T-cells are in an "exhausted" phenotype with limited cytotoxic function from immune checkpoint signaling and crosstalk from other immune populations	There is a consortium of 11 bacteria that can stimulate $CD8^+$ T-cells with anti-tumor immunity against colorectal cancer, which needs exploration in HCC. Consortium: *Parabacteroides distasonis*, *Parabacteroides gordonii*, *Alistipes senegalensis*, *Parabacteroides johnsonii*, *Paraprevotella xylaniphila*, *Bacteroides dorei*, *Bacteroides uniformis JCM 5828*, *Eubacterium limosum*, *Ruminococcaceae bacterium* cv2, *Phascolarctobacterium faecium*, *Fusobacterium ulcerans*. The gut metabolite inosine stimulates IFNγ producing $CD8^+$ T-cells, which protects a spectrum of cancers, but has not yet been explored in HCC. CD8+ T-cells from HCC patients are more reactive toward commensals with anti-tumor effects (e.g., *Escherichia coli*, *Enterococcus faecium*, *Bifidobacterium longum*, *Bacteroides fragilis*, and *Enterococcus hirae*)	$CD103^+$ dendritic cells induce IFNγ $CD8^+$ T-cells in a major histocompatibility (MHC) class IA-dependent manner in HCC patients	Flecken et al. (2014), Li et al. (2018), Liu et al. (2019), Liu, Tseng, et al. (2018), Mager et al. (2020), Rong et al. (2017), Shi et al. (2011), Parry et al. (2005), Patsoukis et al. (2012), Tanoue et al. (2019), Wherry (2011), Yu et al. (2020)

3.3.2	Regulatory T cells (Tregs)	Tregs negatively regulate cytotoxic $CD8^+$ T-cells and promote an immunosuppressive tumor microenvironment	*Bacteroides* spp. induce Tregs in the gut by TLR2 dependent dendritic cell activation. Also, *Clostridia* spp. (e.g., clusters IV and XIVa) induce Tregs via TGF-β1 stimulation of FOXP3 expression. Additionally, SCFA (e.g., butyrate, acetate, and propionate) induce colonic Tregs by G-protein coupled receptor signaling and HDAC inhibitory activity. Further, secondary bile acids can promote Treg generation. Despite the strong evidence that the gut microbiota and intestinal metabolites regulate both colonic and peripheral Treg differentiation, there are no current direct links between Tregs and the gut microbiota in HCC	Potential, underexplored mechanisms include gut microbiota dysbiosis in HCC diminishes SCFA-producing bacteria, where the lack of colonic Tregs followed with increased gut permeability allows for the translocation of products (e.g., secondary bile acids, LPS) that promote HCC precursors, like inflammation and fibrosis, and Treg mediated suppression of anti-tumor responses	Atarashi et al. (2013), Atarashi et al. (2011), Dasgupta, Erturk-Hasdemir, Ochoa-Reparaz, Reinecker, and Kasper (2014), Duscha et al. (2020), Hu et al. (2019), Langhans et al. (2019), Nakajima et al. (2017), Narushima et al. (2014), Sefik et al. (2015), Smith et al. (2013), Tu et al. (2016), Wu, Lin, et al. (2020), Wu, Singer, et al., 2020, Yan et al. (2013)
3.3.3	γδ T cells	γδ T-cells have the plasticity to be either tumor promoting or tumor suppressive	Activation of γδ T-cells by microbial lipid antigens is reported, but whether this promotes HCC is unknown	A potential, underexplored mechanism includes lipid antigens from the gut microbiota stimulates hepatic γδ T-cells followed with IL-17A production, which promotes HCC growth	Jiang et al. (2019), Kouakanou et al. (2020), Ma et al. (2014)

Continued

Table 1 Microbiota-immunity axis in HCC.—cont'd

Chapter section	Cell type or immunity response	Immune signature in HCC	Gut microbe or intestinal metabolite targets	Microbiota-immunity interaction in HCC (known and potential mechanisms)	Reference(s)
3.3.4	Natural Killer T (NKT) cells	Invariant NKT (iNKT) cells are generally hindered in HCC, but there is caution for a potential double edge sword	Gut microbial glycolipids activate hepatic iNKT cells. The HDAC inhibitory activity of butyrate can inhibit iNKT function. Secondary bile acids can also counteract iNKT responses and aggravate HCC	Secondary bile acids negatively regulate iNKT cell recruitment and activation through modulation of CXCL16 levels, which worsens HCC. Whether glycolipids or SCFA from the gut microbiota protect or worsen HCC, respectively, through iNKT activation is unclear	Bandyopadhyay, Marrero, and Kumar (2016), Chen et al. (2014), Fu et al. (2020), Lee, Koh, Chang, Kim, and Chung (2019) Ma et al. (2018), Wei et al. (2016), Wolf et al. (2014)
3.3.5	Tumor infiltrating B-cells	Newly defined to promote HCC by regulatory B-cell (Breg) dependent mitigation of cytotoxic T-cell function	Potentially LPS and LTA, but there are no current direct links between Bregs and the gut microbiota in HCC	Potential, underlying mechanisms include LPS-TLR4 dependent upregulation of Bregs or LPS-mediated promotion of IgA^+ cells, which worsen HCC. In line with this, gut microbiota activation of TLR2 (potentially through LTA) promotes IL-10 producing B-cells	Mishima et al. (2019), Nagatake et al. (2019), Rosser et al. (2014), Shalapour et al. (2017), Xiao et al. (2016)

Source: Credit: Rachel M. Golonka.

Despite the evidence that pharmaceutical restoration of senescence shows promise to prevent HCC (Baell et al., 2018; Qu et al., 2014, 2013; Wang, Vegna, Jin, Benedict, et al., 2019; Wang, Wang, Wang, Qin, et al., 2019; Wang, Huang, et al., 2020; Wu et al., 2017), instigating senescence when HCC is fully established becomes problematic. This is because senescence can induce an inflammatory milieu that stimulates the expansion of periportal hepatic progenitor cells to replenish hepatocytes and thus, foster HCC growth (Eggert et al., 2016; Ho et al., 2020). It is noteworthy that the gut microbial compounds lipoteichoic acid (Gram-positive bacterial cell wall component) and deoxycholate (secondary bile acid) can also collaboratively prompt the senescence-associated secretory phenotype (SASP) and cyclooxygenase 2 (COX2)-mediated suppression of anti-tumor immunity, which collectively promotes obesity-induced HCC (Loo et al., 2017; Yoshimoto et al., 2013). Inhibition of SASP by depleting deoxycholate generating gut bacteria (via dietary difructose anhydride III, which decreases 7α-dehydroxylation) gave protection against obesity-associated HCC development (Yoshimoto et al., 2013). This pinpoints the gut microbiota as an alternative target for modulating cellular senescence to halt HCC.

3.2 Tumor associated myeloid cells

The complexity of targeting hepatocyte senescence calls for more understanding on how the immunologic signature can be reprogrammed to prevent and alleviate HCC. Accordingly, much immunogenomic research has unveiled the heterogeneity of immune cell populations in the liver tumor microenvironment, which has provided promising prognostic tools and immunotherapy targets (Dong et al., 2020; Liu, Wu, Qi, Qin, & Zhu, 2020; Wu et al., 2020). This section of the review will focus on the role of tumor associated myeloid cells and lymphocytes in HCC. Further, we incorporate evidence to describe a potential role of the gut microbiota in regulating the underlying immune signature during HCC pathogenesis (as summarized in Table 1).

3.2.1 Tumor associated monocytes/macrophages (TAMs)

Monocytes are distinctively identified between rodents and humans based on the presence or absence of certain chemokine receptor (e.g., C-X3-C Motif Chemokine Receptor 1/CX3CR1, CCR2), inflammatory (e.g., Ly6C), and other surface markers (e.g. cluster of differentiation 14/CD14, CD16). Comparatively, mouse monocytes are classified with two main subtypes, $CX3CR1^{low}CCR2^{high}Ly6C^{+}$ and $CX3CR1^{high}CCR2^{low}Ly6C^{-}$, whereas human monocyte subsets are set as classical ($CD14^{++}CD16^{-}$), intermediate ($CD14^{++}CD16^{+}$), and non-classical monocytes ($CD14^{+}CD16^{++}$) (Yang,

Zhang, Yu, Yang, & Wang, 2014). After monocytes differentiate and mature into these subtypes, chemokines will signal for cell migration from the bone marrow to systemic circulation for another stage of differentiation. Accordingly, mouse $Ly6C^{+}$ monocytes differentiate into the steady $Ly6C^{-}$ subset (Yona et al., 2013) and human classical monocytes can take a two-step differentiation sequence to generate non-classical monocytes (Zawada et al., 2012). During early inflammation, mouse $CCR2^{high}Ly6C^{+}$ inflammatory and $CCR2^{low}Ly6C^{-}$ resident monocytes are believed to be precursors for the polarization into M1 inflammatory and M2 anti-inflammatory macrophages, respectively (Auffray et al., 2007). Human $CD14^{+}CD16^{-}$ and $CD14^{low}CD16^{+}$ resemble the mouse inflammatory and resident monocytes, respectively (Geissmann, Jung, & Littman, 2003). Besides differences in surface markers, machine learning algorithms have created an image-based approach to identify macrophage subsets from morphological distinctions of the M1 and M2 phenotypes (Rostam, Reynolds, Alexander, Gadegaard, & Ghaemmaghami, 2017).

Kupffer cells are the liver resident macrophages that are imperative to maintain immunotolerance by phagocytosis of toxins derived from bacteria and xenobiotics (Krenkel & Tacke, 2017). It is noteworthy that circulating monocytes can adopt the transcriptional profile of Kupffer cells when the latter is depleted (Scott et al., 2016). In the Kupffer cell niche, interactions with hepatic stellate cells (HSCs), hepatocytes, and endothelial cells can also initiate monocyte recruitment. This is followed with a cascade of Notch, liver X receptor-α, and transforming growth factor-β1 (TGF-β1) interactive signaling, which programs monocytes to repopulate the liver macrophage population as Kupffer cells (Bonnardel et al., 2019; Sakai et al., 2019). Analogous to circulating monocytes/macrophages, Kupffer cells can differentiate into M1 and M2 macrophages during chronic liver inflammation, but these subtypes can act as a double-edged sword on hepatic health. In terms of NAFLD and NASH, M2-polarized macrophages exert protective effects by being resistant to apoptosis and preventing lipid accumulation, whereas M1 macrophage dominance worsened these liver diseases (Bai et al., 2017; Luo, Xu, Wang, Wu, & Hua, 2017). In contrast, in vitro studies with well-established HCC cell lines and clinical HCC specimens (e.g., blood and liver tissue) delineate that HCC favors the M2 phenotype because this supports tumor growth (impedes apoptosis), cell migration, epithelial-mesenchymal transition (EMT)-dependent metastasis, pharmaceutical resistance and, most importantly, creates an immunosuppressive tumor microenvironment (Aras & Zaidi, 2017; Dong et al., 2019; Guo, Li, et al., 2017; Yao et al., 2018; Yeung et al., 2015; Zhang, Zhang, et al., 2018; Zhang, Wang, et al., 2018).

There are many mechanisms that HCC tumors instigate to drive tumor associated macrophage (TAM) M2-polarization (Tian, Hou, Liu, Han, & Wei, 2019). In vitro studies and in situ hepatoma mouse models indicate that hypoxia (Ke et al., 2019), autophagy (Shiau et al., 2020), $CD8^+$ T-cell suppression (Peranzoni et al., 2018), and fatty acid oxidation (Zhang, Wang, et al., 2018) are HCC mediated techniques for sustaining growth. Additionally, multiple pathways are activated such as Hedgehog (Petty et al., 2019), Hippo (Guo, Zhao, et al., 2017; Zhou, Zhou, et al., 2018; Zhou et al., 2019), neurotensin/interleukin (IL)-8 (Xiao et al., 2018), and Notch/Wnt/β-catenin signaling (Yang, Ye, et al., 2018; Ye et al., 2019). Also, HCC cells favor a cytokine profile for upregulation of IL-6 and subsequent signal transducer and activator of transcription 3 (STAT3) signaling, where inhibition of IL-6 via IL-12 or IL-37 abrogates M2 polarization and HCC growth (Kong et al., 2016; Lanaya et al., 2014; Wang et al., 2016; Yin et al., 2018; Zhang, Zhang, He, Han, & Sun, 2020). Further, methylation of the colony-stimulating factor-1 receptor regulates macrophage differentiation, where inhibition of this receptor reverts to M1 polarization (Ao et al., 2017). In line with this, there are multiple reported mechanisms to how TAMs support HCC pathogenesis, including shifting from apoptosis to autophagy (Guo, Li, et al., 2017), promoting angiogenesis (Bartneck et al., 2019), supporting liver cancer stem cells (Chen et al., 2019; Fan et al., 2014; Wan et al., 2014), causing resistance to anti-cancer therapies (Fu et al., 2019), and creating an immunosuppressive microenvironment (Yao, Ba, et al., 2017). As such, M2 polarization and the monocyte/granulocyte to lymphocyte ratio gives poor prognosis for HCC patients (Guo, Li, et al., 2017; Yeung et al., 2015; Zhou et al., 2015).

It must be emphasized that there is a potential for the gut microbiota to facilitate macrophage polarization in HCC. To date, there is only one article that directly links the gut microbiota and M2 polarization during HCC. In 2019, Li et al. found that vancomycin (Gram-positive specific bacterial antibiotic) intervention to an orthotopic-transplanted liver tumor model in BALB/c nude mice caused gut microbiota dysbiosis, intestinal architecture dysfunction, and IL-25 secretion (Li, Ma, et al., 2019). IL-25 supported M2 polarization by alternatively activating macrophages that subsequently secreted C-C motif chemokine ligand (CXCL)-10, which further stimulated HCC growth and invasion (Li, Ma, et al., 2019). Besides this one piece of evidence, there could be many other mechanisms that have not yet been explored. For example, orthotopic HCC induction in BALB/c mice demonstrates that the tumor microenvironment will stress for mitochondrial DNA release to activate Toll-like receptor 9 (TLR9) signaling, followed by NF-κB activation and C-C motif ligand (CCL)2-CCR2 signaling, which induces TAM recruitment and M2 polarization (Bao et al., 2019; Li et al., 2017). Considering that TLR9 recognizes unmethylated DNA as CpG islands, which is a parallel feature between mitochondrial and bacterial DNA (Bhagirath, Dwivedi, & Liaw, 2015; Dalpke, Frank, Peter, & Heeg, 2006; Fang, Wei, & Wei, 2016), it is plausible that translocation of bacterial DNA from the gut microbiota could instigate TLR9-dependent signaling and support M2 polarization in HCC.

In vitro studies by Feng et al. show that HCC cells can also implement M2 polarization by switching from cellular respiration to aerobic glycolysis for lactate production (Feng et al., 2018). Following, lactate induces reactive oxygen species (ROS) generation, which activates nuclear factor (erythroid-derived 2)-like 2 (Nrf2) and subsequent changes in the cytokine profile to initiate M2 polarization (Feng et al., 2018). Intriguingly, probiotic administration of *Clostridium butyricum* to CCl_4-treated mice caused a positive inversion for more Nrf2 activation and less NF-κB signaling, which was associated with a rebalanced gut microbiota, increased anti-oxidant responses, and alleviation of liver injury (Liu et al., 2017). Comparatively, gut microbiota ablation by antibiotics protects from high-fat diet induced NASH-HCC in mice, and this was correlated with upregulated Nrf2, increased levels of anti-oxidant metabolites [e.g., glutathione, nicotinamide adenine dinucleotide $(NAD)^+$ and nicotinamide adenine dinucleotide phosphate $(NADP)^+$], and reduced lipid accumulation (Yamada, Kimura, Saito, & Saito, 2018). This piece of evidence suggests that the gut microbiota could promote M2 polarization by modulating oxidative stress and lipid metabolism in a Nrf2-dependent manner. Considering that high fat diet-induced dysbiosis was already found to cause M2 polarization in intestinal adenoma-adenocarcinoma (Liu, Guo, et al., 2020), there might be a parallel role of the gut microbiota in TAM differentiation during HCC pathogenesis.

From studying primary HCC cell lines and the NASH-HCC murine model, it was uncovered that TLR4 signaling induces M2 polarization through a crosstalk between macrophages and T-cells (Miura et al., 2016; Yao et al., 2018). Accordingly, TLR4 activation on macrophages instigates STAT3 signaling on T-cells and subsequent induction of pro-tumorigenic $IL\text{-}21^+$ T-follicular helper (T_{FH})-like cells (Chen, Xiao, et al., 2016). Following, T_{FH}-like cells employ IL-21 and interferon-γ (IFNγ) signaling that develop conditions for M2 polarization (Chen, Xiao, et al., 2016). It is noteworthy that gut-bacterial derived LPS is one of the most potent ligands for TLR4 activation (Park & Lee, 2013). Importantly, gut microbiota dysbiosis can coincide with intestinal barrier dysfunction, resulting in metabolic endotoxemia or the excessive leakage of microbial products, like LPS, into systemic and enterohepatic circulation (Cani et al., 2007). The hepatic LPS-TLR4 axis triggers inflammatory responses from Kupffer cells, which has been implicated in the pathogenesis of obesity, insulin resistance, NAFLD, and NASH (Cani et al., 2007; Rivera et al., 2007; Xue et al., 2017). Additionally, an imbalance of the gut microbiota was found to stimulate cathepsin K, which was found to aggravate colorectal cancer by inducing TLR4-mediated M2 polarization in a mammalian target of rapamycin-dependent manner (Li et al., 2019). Determining whether TLR4 activation by LPS and/or cathepsin K induces M2 polarization in HCC could provide a novel mechanistic link in how gut microbiota dysbiosis contributes to liver tumorigenesis.

3.2.2 Tumor associated neutrophils (TANs)

The neutrophil to lymphocyte ratio has been well established as a diagnostic, prognostic, overall survival, preoperative, postoperative, and recurrence biomarker for HCC (Chu et al., 2018; He et al., 2015; He, Peng, et al., 2016; Li, Wen, et al., 2015; Li, Xing, et al., 2017; Mano et al., 2013; Margetts et al., 2018; Qi et al., 2016; Tajiri et al., 2016; Tsuda et al., 2012; Wang, Yao, Zhang, Xie, et al., 2019). The cytokine profile encompassing the tumor microenvironment can determine whether TANs differentiate into the N1 anti-tumor or N2 immunosuppressive subtype. In the early stages of HCC initiation, TANs are located within the periphery and exhibit a N1 phenotype that are cytotoxic toward tumor cells (Mishalian et al., 2013). Comparatively, neutrophils differentiate into the N2 phenotype at late stages of HCC, which is due to the cytokine profile alteration to favor CCL2 and CCL17, along with the upregulation of TGF-β (Fridlender et al., 2009; Zhou, Zhou, et al., 2016). As such, neutrophil depletion and TGF-β blockade have been found to abate HCC growth, where inhibition of TGF-β in particular could switch the TAN population from the N2 to N1 phenotype (Fridlender et al., 2009; Mishalian et al., 2013; Wilson et al., 2015).

In line with TAMs, there are mechanisms that HCC cells implement for TAN recruitment, differentiation, and longevity. For example, tumors can modify the chemokine environment to limit signals for apoptosis and thus, prolong the life span of TANs (Haider et al., 2019; Li, Xu, et al., 2015; Peng et al., 2020; Zhou et al., 2012). To highlight, chemokine secretion from tumor associated fibroblasts prime TANs to resist apoptosis and impair T-cell function via the programmed cell death protein-1 (PD-1)/ programmed cell death protein ligand-1 (PD-L1) pathway (Cheng et al., 2018). Moreover, tumors can secrete hyaluronan fragments that support anti-apoptosis of TANs via TLR4 signaling (Wu et al., 2011). TANs also have their own internal methods to sustain longevity, which includes promoting autophagy and downregulating apoptosis (Li, Chen, et al., 2015; Wu et al., 2011). Moreover, TANs support HCC progression by recruiting TAMs and regulatory T-cells (Zhou, Zhou, et al., 2016), supporting angiogenesis (Kuang et al., 2011), producing metastatic factors (Peng et al., 2020), and excreting neutrophil extracellular traps (NETs) (Seo et al., 2019;van der Windt et al., 2018 ; Yang, Luo, et al., 2020). Interestingly, internalization of NETs into HCC cells activates TLR4/9-COX2 signaling to support metastasis (Yang, Luo, et al., 2020). Accordingly, ablation of NETs via deoxyribonuclease 1 (NETs degrader) or genetic deficiency of peptidyl deaminase 4 (rate limiting enzyme for NETs formation) protects mice from HCC growth and metastasis (van der Windt et al., 2018; Yang, Luo, et al., 2020).

There are no current direct links on the gut microbiota influencing TAN pathology in HCC. However, the diversity and composition of the gut microbiota can influence circulating and hepatic neutrophil levels. Accordingly, the greater the diversity of the gut microbiota, the lower the blood neutrophil to lymphocyte ratio (Yoon et al., 2018). The absence of the gut microbiota with germ-free rats lowers systemic neutrophil count (Yang, Chakraborty, et al., 2020). In line with this, gut microbiota ablation by broad-spectrum antibiotics in wild-type mice decreased LPS-TLR4 signaling from HSCs, which correlated with lowered levels of the chemoattractant CXCL1 and diminished neutrophil migration (Bigorgne et al., 2016). This further holds true under inflammatory conditions, where the intestinal flora in mice mediate neutrophil migration to the peritoneal cavity by myeloid differentiation primary response 88 (MyD88) signaling after being challenged with macrophage inflammatory protein 2 (Karmarkar & Rock, 2013).

The gut microbiota has also been found to influence NETs formation, but this has not been delineated in the context of HCC. In our previous study, we found that gut microbiota-dependent HCC through dietary feeding of the purified, fermentable fiber inulin to mice results in elevated systemic and hepatic neutrophil levels (Singh et al., 2018). Additionally, we observed increased abundance of NETs associated proteins, including lipocalin 2, elastase, and myeloperoxidase (Singh et al., 2018), which suggests that there could be more NETs formation. Under conditions of mesenteric I/R injury, neutrophils from germ-free mice were found to be more sensitive to LPS-induced NETs generation, which had explained the increased severity of I/R injury in the germ-free mice (Ascher et al., 2020). This study emphasized that the gut microbiota is required to set a physiological standard of NETs formation, which would then inhibit hyperreactivity under disease conditions. Furthermore, similar to M2 polarization, mitochondrial DNA can induce NETs formation through TLR9-dependent signaling in mice (Mallavia et al., 2020), which emphasizes the potential that leaked bacterial DNA could contribute to the increase NETs formation in HCC.

3.2.3 Myeloid-derived suppressor cells (MDSCs)

There are two subclasses of MDSCs: Ly6G^{+} polymorphonuclear (PMN)-MDSCs and Ly6G^{-} monocytic (M)-MDSCs, where each population has differentially expressed chemokine profiles (Zhao et al., 2015). As a whole, MDSCs exert immunosuppression (Chiu et al., 2017), reduce T-cell immunity (Hsieh, Hung, Chiang, Tsai, & He, 2019; Nan et al., 2018; Zhou et al., 2018), and instigate drug resistance (Xu et al., 2017; Yu et al., 2019), which has made this immune population a poor prognosis and low survival marker for HCC patients (Iwata et al., 2016; Mizukoshi et al., 2016; Zhang, Fu, Li, & Yan, 2019). There are multiple independent mechanisms by which HCC tumor cells activate MDSC recruitment and differentiation, such as creating a hypoxic environment for chemokine secretion (Chiu et al., 2017, 2016), inducing tumor activated fibroblast- and tumor activated liver stromal cell-derived cytokines (Deng et al., 2017; Zhang et al., 2017), and triggering HSC activation and signaling (Hsieh et al., 2019; Xu, Zhao, et al., 2016). Diminishing the accumulation of M-MDSCs and PMN-MDSCs have been found to beneficially abate HCC growth in vitro, in vivo, and ex vivo (Ho et al., 2015; Li, Jiang, et al., 2020; Liu, Zhou, et al., 2020; Nan et al., 2018; Pang, Song, Zhang, Tu, & Niu, 2016). Comparatively, PMN-MDSCs are the dominant subtype in hepatocarcinogenesis because, in part, this subtype is mobilized from the spleen to the liver for facilitating tumorigenesis (Li, Jiang, et al., 2020; Li, Zhang, et al., 2019). It must be emphasized that PMN-MDSCs are not synonymous to normal neutrophils nor TANs. Transcriptomics revealed that normal neutrophils and PMN-MDSCs have a similar overlapping mRNA profile that is distinctive when compared to TANs (Fridlender et al., 2012), but all three immune cell populations are functionally and phenotypically contrast (Ostrand-Rosenberg & Fenselau, 2018).

There are no current publications that has found an interrelationship between MDSCs and the gut microbiota in the context of HCC. However, there are lines of evidence that the gut microbiota does modulate MDSC levels in general. For example, perturbation of the gut microbiota via antibiotics increased both M- and PMN-MDSCs in the bone marrow, spleen, and mesenteric lymph nodes in mice (Hathaway-Schrader et al., 2019). In line with this, ethanol-induced alcohol liver disease in mice is normally associated with low MDSCs, but berberine (a plant alkaloid) administration changed the gut bacteria to favor the expansion of PMN-MDSCs (Li et al., 2020). Further, gut microbiota dysbiosis was found to drive PMN-MDSC upregulation during murine colitis-associated tumorigenesis (Harusato et al., 2019). These observations should invigorate future studies to explore the potential of PMN-MDSC regulation from the gut microbiota in HCC.

3.3 Tumor infiltrating lymphocytes (TILs)

There is a natural T-cell repertoire in the tumor microenvironment that serve to survey and eliminate tumor associated antigens (TAA) and to maintain immunotolerance. This immune population includes cytotoxic CD8$^+$ T-cells, regulatory T-cells (Treg), γδ T-cells, and natural killer T cells (NKT), which contribute to the immunosuppressive gradient found between tumor and non-tumor regions (Chew et al., 2017). HCC tumor cells modify the surroundings to antagonize the anti-tumor functions of these lymphocytes, which includes veering the immunoregulatory network toward immunosuppression. Recent evidence demonstrates that metabolites generated from the gut microbiota can direct to favor immunosuppression and dampen anti-tumor responses. In this section, we will describe the role of each TIL in HCC and discuss how the gut microbiota influences each TIL's anti-tumor function (as summarized in Table 1).

3.3.1 Effector and exhausted T-cells: CD8$^+$ cytotoxic T lymphocyte dysfunction

Short-term interaction between TAA and T-cell receptors initiates differentiation into cytotoxic, effector CD8$^+$ T-cells. This immune population can inhibit HCC by secreting cytokines, like IFN-γ and IL-33, which provides a positive feedback loop to enhance T-cell cytotoxicity (Bhat, Leggatt,

Waterhouse, & Frazer, 2017; Brunner et al., 2015; Jin et al., 2018). After consistent and repetitive exposure to the TAA, the expansion of effector $CD8^+$ T-cells passes a peak that triggers a transition toward memory T-cell differentiation (Chang, Wherry, & Goldrath, 2014). During HCC, prolonged TAA interaction diminishes effector $CD8^+$ T-cells levels by switching to an exhausted, dysfunctional $CD8^+$ T-cell phenotype (Wherry, 2011). Importantly, this phenotypical switch is reversible during early stages of liver tumorigenesis but becomes permanent when HCC is fully established (Schietinger et al., 2016).

T-cell exhaustion is hallmarked with sustained expression of inhibitory receptors that are more popularly known for being immune checkpoint inhibitor targets, including programmed cell death protein-1 (PD-1)/ programmed cell death ligand-1 (PD-L1, *alias* B7-H1), cytotoxic T-lymphocyte-associated antigen 4 (CTLA-4), lymphocyte-activation gene 3 (LAG3), T-cell immunoglobulin and mucin domain-3 (TIM3), and T-cell immunoglobulin and ITIM domain (TIGIT) (Di Blasi et al., 2020; Grywalska, Pasiarski, Gozdz, & Rolinski, 2018; Li, Zhang, Jin, Yao, & Wu, 2013; Liu, Liu, & Chen, 2018; Long et al., 2018; Onuma et al., 2020; Pinato et al., 2020; Wherry & Kurachi, 2015; Wu, Kryczek, Chen, Zou, & Welling, 2009; Zong et al., 2019). Exhausted $CD8^+$ T-cells also acquire surface makers, such as forkhead box P3 (FOXP3) and CD25, which resemble immunosuppressive Tregs (Li et al., 2015). This is further associated with reduced cell proliferation and depleted production of effector cytokines (e.g., IFN-γ, IL-2, TNF-α) (Wang et al., 2018; Wang, He, Shen, Lu, & Sun, 2019). Accordingly, low effector $CD8^+$ TILs but high exhausted $CD8^+$ TILs are poor prognosis indicators for HCC patients (Ma et al., 2019; Shi et al., 2011; Sideras et al., 2017; Yao et al., 2017). Intriguingly, there are many TAA-expressing T-cell populations whose antigen specificity still requires identification (Di Blasi et al., 2020).

Besides being diagnostic surface markers of T-cell exhaustion, immune checkpoints have direct mechanisms for suppressing the cytotoxic function of $CD8^+$ T-cells. The PD-1/PD-L1 pathway and double positive $PD1^+TIGIT^+CD8^+$ T-cells, for example, can promote $CD8^+$ T-cell exhaustion, instigate apoptosis, and deplete effector cytokine production (Li et al., 2018; Liu, Li, Wang, et al., 2019; Shi et al., 2011). In addition, PD-1 and CTLA-4 signaling inhibit $CD8^+$ T-cell activation by suppressing the phosphatidylinositol 3-kinase (PI3K)-protein kinase B (*alias* AKT) axis through independent mechanisms (Parry et al., 2005). Ablation of PI3K-AKT signaling and MAPK/extracellular signal regulated kinase (ERK) signaling by PD-1 also abates the cell cycle and proliferation of T-cells (Patsoukis et al., 2012). Survival of exhausted T-cells is mediated through TIM-3/p53 dependent upregulation of anti-apoptosis genes (Ji et al., 2018).

The tumor microenvironment also provides a positive feedback loop to support immune checkpoints, which is mostly by altering the cytokine profile. For example, IL-15 and IL-21 are diminished, which impairs T-cell survival, function, and survival (Batra et al., 2020). Also, classical IL-22-producing T helper cells (Th22 cells) are skewed toward IL-17 production rather than IFN-γ due to PD-1 signaling, which suppresses $CD8^+$ T-cell cytotoxicity, increases angiogenesis, and aggravates cancer growth (Zhao et al., 2011). As such, immune checkpoint inhibitor therapy has become a rising force to abate HCC (Inada et al., 2019; Tang et al., 2019; Zhou et al., 2017). In line with this, the leading HCC therapeutic sorafenib has been demonstrated to lower exhausted $CD8^+$ T-cell levels and restore IFN-γ responses by inhibiting immune checkpoints (Chen et al., 2014; Kalathil, Hutson, Barbi, Iyer, & Thanavala, 2019). Besides immune checkpoint therapy, targeting transcription factors (e.g., thymocyte selection-associated high mobility group box protein, TOX) that protect PD-1 from degradation and promote PD-1 surface translocation could serve as potential adjuvants (Wang et al., 2019). Other options to reactivate T-cell cytotoxicity and IFN-γ responses is by priming with dendritic-cell pulsing (Chieochansin, Thepmalee, Grainok, Junking, & Yenchitsomanus, 2019; Zhou, Ma, Li, Cui, & Song, 2016) and upregulating IL-7 levels (Teng, Ding, Cai, Luo, & Wang, 2019).

Crosstalk between immune cell populations delineate another causation for impairment of activated effector $CD8^+$ T-cells (Kalathil, Lugade, Miller, Iyer, & Thanavala, 2013). As described, TAMs and MDSCs can crosstalk with TILs to suppress cytotoxic $CD8^+$ T-cell activity (Liu, Tseng, et al., 2018), where antagonizing these immune populations can restore $CD8^+$ T-cell function (Yao, Ba, et al., 2017). Immunosuppressive Tregs, dendritic cells, and HSCs have also been found to block $CD8^+$ T-cell activation (Harimoto et al., 2013; Inada et al., 2019; Zhao et al., 2012), where depletion of Tregs can restore cytotoxic $CD8^+$ T-cell levels but IFN-γ production is still impaired (Flecken et al., 2014). Comparatively, immune regulation by sorafenib can reduce Treg-mediated repression of $CD8^+$ T-cell function and restore IFN-γ production (Chen, Yan, et al., 2014; Kalathil et al., 2019; Shibolet et al., 2004). Another population includes liver-resident immunoglobulin-A (IgA^+) producing cells that are upregulated under inflammatory conditions, which contribute to the overexpression of immunosuppressive PD-L1 and anti-inflammatory IL-10, followed by suppression of $CD8^+$ T lymphocytes and HCC progression (Shalapour et al., 2017).

The gut microbiota is an additional central player in $CD8^+$ T-cell function, especially at early age. When compared to conventionally raised mice, germ-free conditions lower $CD8^+$ T cell populations (Semenkovich et al., 2016). Alongside, introducing the gut microbiota from conventionally raised mice to reared adult germ-free mice negatively impacts $CD8^+$ T-cell levels in the same manner as in pure germ-free conditions (Semenkovich et al., 2016), which emphasizes that obtaining the gut microbiota at birth heavily influences immune responses. Accordingly, a mouse model of maternal antibiotic treatment causes microbiota dysbiosis in the gastrointestinal tract of the offspring, which is highlighted by the bloom of *Enterococcus faecalis* (Gonzalez-Perez et al., 2016). Importantly, infants from antibiotic-treated mothers have basal dysfunction of $CD8^+$ T-cell function as indicated with reduced IFN-γ production (Gonzalez-Perez et al., 2016). Introducing microbial products, like LPS, to stimulate Toll-like receptors could rescue effector $CD8^+$ T-cell activity (Gonzalez-Perez & Lamouse-Smith, 2017). Early exposure of the gut microbiome was also found to stimulate TNF-α secretion for differentiation and maturation of dendritic cells (Kohler, Delbauve, Smout, Torres, & Flamand, 2020), but whether this response promotes T-cell differentiation and maturation is unclear.

There is much evidence that regulation of cytotoxic $CD8^+$ T-cells by the gut microbiome influences anti-tumor responses and tumorigenesis, *albeit* this has not been thoroughly investigated in terms of HCC except for one article. Rong et al. show that circulating $CD8^+$ T-cells from HCC patients are more reactive toward a series of commensals and other bacteria previously associated with anti-tumor effects, including *Escherichia coli*, *Enterococcus faecium*, *Bifidobacterium longum*, *Bacteroides fragilis*, and *Enterococcus hirae* (Rong et al., 2017). Importantly, *Enterococcus hirae*-reactive $CD8^+$ T-cells had an inverse relationship with $PD\text{-}1^+CD8^+$ T cells. Also, both *Bifidobacterium longum*-reactive and *Enterococcus hirae*-reactive $CD8^+$ T cells were associated with better disease-free survival time for HCC patients (Rong et al., 2017). The identification of gut bacteria that stimulate IFN-γ-producing $CD8^+$ T-cells with anti-tumor immunity and the knowledge that gut microbiota dysbiosis impairs general tumor suppressive responses from cytotoxic T cells (Tanoue et al., 2019; Yu et al., 2020) emphasizes that more research is required to delineate how the gut microbiota impacts HCC pathogenesis by modulating $CD8^+$ T cell levels and function. In line with this, certain gut metabolites (e.g., butyrate) have been positively associated with improved $CD8^+$ T cell function (Luu et al., 2018), but whether continuous exposure could lead to hyperreactivity and T-cell exhaustion in HCC needs further exploration.

3.3.2 Immunosuppressive regulatory T cells (Tregs)

Double positivity of $CD4^+CD25^+$ cells define classic Tregs, but tumor infiltrating Tregs consist of multiple subsets with an important common denominator of forkhead box P3 expression (FOXP3; transcription factor and master regulator of Tregs). Accordingly, mutations of the *FOXP3* gene causes delocalization from the nucleus to cytoplasm and subsequent transcriptional dysregulation of FOXP3 target genes, including anti-tumor responses, which could contribute to HCC pathogenesis (Ren et al., 2020). Compared to naïve $CD4^+CD25^+FOXP3^-$ Tregs, the acquisition of FOXP3 expression reflects diminished T-cell effector function, including IFN-γ production (Kalathil et al., 2013; Li, Peng, et al., 2015; Yan et al., 2013). Additionally, $FOXP3^+$ Tregs contribute to the immunosuppressive microenvironment in HCC by inhibiting cytotoxic $CD8^+$ T-cells, which is correlated with worse prognosis (Langhans et al., 2019; Tu et al., 2016; Yan et al., 2013). This inhibitory crosstalk could be because Tregs also acquire the immune checkpoints, CTLA-4 and TIM-3 (S. Kalathil et al., 2013; Li, Peng, et al., 2015; Yan et al., 2013). In line with $CD4^+FOXP3^+$ Tregs, $CD8^+FOXP3^+$ Tregs are found to be very abundant in the late stages of HCC, which is further associated with metastasis (Yang et al., 2010). As such, epigenetic control of FOXP3 or depletion of Tregs upregulates anti-tumor responses (Liu et al., 2019; Zhou et al., 2013), which could be useful in abating HCC but this needs investigation.

There is a consortium of gut microbes and intestinal antigens that regulate the Treg cell lineage. For one, symbiotic members of the *Bacteroides* can induce Tregs in the gut by TLR2-dependent dendritic cell activation (Dasgupta et al., 2014; Sefik et al., 2015). In addition, a collection of *Clostridia* species, particularly clusters IV and XIVa, can induce colonic Tregs by sustaining a TGF-β1 enriched environment and stimulating FOXP3 expression (Atarashi et al., 2011; Atarashi et al., 2013; Narushima et al., 2014). Comparatively, commensals and microbial antigens can switch conventional, helper T-cells to reactive Tregs, which may be independent of TGF-β1 signaling (Nutsch et al., 2016; Zhu et al., 2019). Importantly, defining the "Treg induction score" by the gut microbiota is suggested as an important tool for bacteriotherapeutics in humans (Stein et al., 2018). This is because Tregs have a bidirectional relationship with the gut microbiota, where FOXP3$^+$ Tregs contribute to bacterial diversity, inflammatory responses, and immune-modulating effects of the gut microbiota, which creates a positive feedback loop for Treg induction (Kawamoto et al., 2014; Kehrmann et al., 2020). Collectively, the gut microbiota plays an essential role in priming the immune system, which is extra critical in early life stages (Hansen et al., 2014).

The gut microbiota generates a diverse chemical library of metabolites. Short chain fatty acids (SCFA) like acetate, propionate, and butyrate are fermented products of indigestible carbohydrates (i.e., dietary fiber). Both acetate and propionate have been found to induce the differentiation of anti-inflammatory Treg cells through activation of the G-protein coupled SCFA receptors, GPR41 and GPR43, respectively (Duscha et al., 2020; Hu et al., 2019; Nakajima et al., 2017; Smith et al., 2013; Wu et al., 2020). Additionally, propionate and butyrate can boost the peripheral and colonic Treg cell populations in mice by their function as a histone deacetylase inhibitor (HDACi), which is a property that is absent for acetate (Alrafas, Busbee, Chitrala, Nagarkatti, & Nagarkatti, 2020; Arpaia et al., 2013; Furusawa et al., 2013; Ho et al., 2017; Kaisar, Pelgrom, van der Ham, Yazdanbakhsh, & Everts, 2017; Park et al., 2015; Tao et al., 2007; Zhou et al., 2018). Simultaneously, butyrate-induced Tregs was dependent on GPR109A activation (Furusawa et al., 2013; Kaisar et al., 2017; Singh et al., 2014; Smith et al., 2013). Intriguingly, butyrate produced by the Firmicutes and Fusobacteria was found to increase TGF-β1 expression independent of activating GPR41, GPR43, and GPR109a but dependent on its HDACi activity (Martin-Gallausiaux et al., 2018), which would subsequently increase Treg production. In line with SCFA, much research demonstrates that gut microbiota-derived secondary bile acids and its derivatives, produced primarily by *Clostridia* species, can instigate immunosuppression by inhibiting and promoting Th17 and Treg cell differentiation, respectively (Campbell et al., 2020; Hang et al., 2019; Song et al., 2020). Mechanistically, secondary bile acids can bind directly to the retinoid-related orphan receptor γT (RORγT) to inhibit Th17 production, but increase mitochondrial ROS production and subsequently upregulate FOXP3 expression to enhance Treg cell differentiation (Campbell et al., 2020; Hang et al., 2019; Song et al., 2020). As such, depletion of the gut bile acid pool significantly diminishes the Treg population and thus, causes an immunological imbalance (Song et al., 2020).

While these gut metabolites are mostly reported to induce Tregs in the mucosa, it must be acknowledged that this relationship is imperative for intestinal homeostasis and subsequent liver health. For example, loss of bacterial-derived SCFA, followed with diminished colonic Tregs, impairs gut barrier function and increases translocation of microbial products that can cause cirrhosis (Juanola et al., 2018). This includes secondary bile acids themselves who act as inflammagens and cause HSCs to transdifferentiate into myofibroblasts, which are contributing factors in HCC pathogenesis (Yang & Duan, 2016). Further, excessive leakage of endotoxins like LPS can stimulate TLR4 signaling that facilitates Treg recruitment to the HCC microenvironment and promotes metastasis by inhibiting apoptosis (Dapito et al., 2012; Yang et al., 2012). Future studies are required to delineate whether an impaired gut metabolite profile causes gut permeability from a lack in Tregs and persists an endotoxemia-associated immunosuppressive environment during HCC.

3.3.3 γδ T cells

T-cells are defined by their expression of specific heterodimeric T-cell receptors (TCR). $CD4^+$ helper and $CD8^+$ cytotoxic T-cells are classified with αβ TCRs, in which these receptors recognize specific human leukocyte antigens (HLA) (homolog of the major histocompatibility complex) and subsequently fine-tune the adaptive immune system (Sebestyen, Prinz, Dechanet-Merville, Silva-Santos, & Kuball, 2020). Comparatively, there is another subset identified as γδ T cells that do not rely on the HLA system, where the broad spectrum recognition makes γδ T cells more easily transferable between individuals, especially in the context of cancer immunotherapy (Legut, Cole, & Sewell, 2015). The cancer suppressive actions of γδ T cells include tumor lysis and IFN-γ production independent of TCR signaling (via IL-12 and IL-18 co-stimulation), which has made adoptive transfer of γδ T cells clinically relevant (Schilbach et al., 2020).

In HCC, the abundance of γδ T cells and $PD-1^+$ γδ T cells was significantly lower and higher, respectively, where it is postulated that PD-1 suppression of IL-2 and IL-21 production causes impairment in the cytotoxic effects of γδ T cells (Jiang et al., 2019). In line with this, infiltrating Tregs repress IFN-γ secretion from γδ T cells in TGF-β and IL-10 dependent manners (Yi et al., 2013), which could be further amplified from vitamin C enhancement of FOXP3 expression on Tregs (Kouakanou et al., 2020). Therefore, increasing γδ T-cell recruitment and cytotoxic effects can serve as a potential, effective therapeutic for HCC without expanding immunosuppression (Tian et al., 2018; Zhao et al., 2020). However, γδ T cells have the plasticity to be either tumor suppressive or oncogenic, which holds true for HCC. For example, γδ T-cells promote tumor growth by IL-17A mediated recruitment of immunosuppressive MDSCs, where MDSCs can further stimulate IL-17A from γδ T cells creating a positive feedback system (Ma et al., 2014). This emphasizes that caution is required for γδ T cell-dependent immunotherapy in HCC patients.

The gut microbiota has been found to influence both intestinal and liver resident γδ T-cells. In conditions of intestinal dysbiosis, γδ T-cells are diminished following upregulation of Tregs and alteration of dendritic cell activity (Benakis et al., 2016). Analogous to Tregs, γδ T-cells are essential for gut barrier function via IL-22 and calprotectin secretion (Tyler et al., 2017), and thus, impairment in intestinal γδ T-cells could cause gut permeability issues. Comparatively, gut commensals and microbial-derived lipid antigens can induce hepatic γδ T-cells activation, survival, and proliferation, which includes subsequent IL-17A production (Li et al., 2017). Whether these microbial glycolipids can promote HCC through the γδ T cell-IL-17A axis is unclear. Interestingly, exposure to *Lactobacillus gasseri* can induce hepatic fibrosis and inflammation via γδ T-cell activation and IL-17 production (Tedesco et al., 2018). Future research is needed to further explore the immunological axis between the gut microbiota and γδ T-cells in HCC pathogenesis.

3.3.4 Natural killer T (NKT) cells

The dual expression of natural killer (NK) receptors and a TCR (α or β chain) composes the hybrid composition of $CD3^+CD56^+$ NKT cells. There are four subgroups of NKT cells, but only types I and II recognize CD1d-restricted antigens (Kronenberg & Gapin, 2002), whereas types III and IV are stimulated by inflammatory cytokines secreted from antigen presenting cells (Tupin, Kinjo, & Kronenberg, 2007). Comparatively, CD1d-restricted invariant NKT (iNKT) cells are the most well studied tumor suppressive NKT subtype in many lines of cancer (Vivier, Ugolini, Blaise, Chabannon, & Brossay, 2012). In terms of HCC, iNKT cells are regarded to have tumor suppressive abilities, such as removal of senescent hepatocytes, anti-fibrotic effects, cholesterol-dependent IFN-γ secretion, and stimulating $CD8^+$ cytotoxic T-cells (Fu et al., 2020; Lin et al., 2016; Miyagi et al., 2010; Mossanen et al., 2019; Nakano et al., 2012).

Ex vivo studies reveal that HCC tumor cells can counteract the benefits of iNKT cells by metabolically switching toward lactate production, which subsequently reduces peroxisome proliferator-activated receptor-γ signaling, cholesterol synthesis, and IFN-γ generation (Fu et al., 2020). In addition, the initial inflammatory milieu exhibited in β-catenin-induced murine HCC can reduce iNKT cell recruitment, but this immune population still has the capability to produce high levels of Th1/2 cytokines (Anson et al., 2012). Further, the suppression of iNKT cells could cause the upregulation of compensatory pathways to enhance TANs recruitment (Chen et al., 2019), where neutrophils themselves can further reduce iNKT responses (Wingender et al., 2012). As such, low iNKT and IFN-γ levels are promising predictors of increased hepatocarcinogenesis risk and diminished overall survival (Kawarabayashi et al., 2000; Xiao et al., 2013).

Intriguingly, a mouse NASH-HCC model uncovered that NKT cells caused steatosis and cooperatively induced liver injury with $CD8^{+}$ T-cells (Wolf et al., 2014), which emphasizes the potential double-edge sword of NKT cells in HCC that needs further exploration. In support of this, gut bacteria have been found as a source of glycolipids, which activate hepatic iNKT cells that contribute to immune-mediated liver injury (Chen, Wei, et al., 2014; Wei et al., 2016). In particular, there is a positive association between *Bacteroides* spp. and iNKT cell abundance in the gastrointestinal-associated lymphoid tissue (Paquin-Proulx et al., 2017). Comparatively, there have been recent reports of gut metabolites inhibiting iNKT cell function. For example, the HDACi activity from butyrate can suppress the cytokine production of iNKT cells (Bandyopadhyay et al., 2016; Lee, Koh, et al., 2019). Alongside, overabundance of secondary bile acids was recently found to worsen HCC by negatively regulating iNKT cell recruitment and activation through modulation of CXCL16 levels (Ma et al., 2018). Collectively, the crosstalk between the gut microbiota and T-cell repertoire is a strong mediator of HCC.

3.3.5 Tumor infiltrating B-cells

Compared to T-cell mediated prevention of initial tumor formation, tumor infiltrating B-cells have been regarded to work against established tumors (Schneider et al., 2012). Accordingly, the collection of $CD20^{+}$ B-cells, naïve B-cells, IgM^{+} memory B-cells, $CD27^{-}$ isotype switched memory B cells, and plasma cells were found to correlate with improved HCC survival (Garnelo et al., 2017; Shi et al., 2013; Zhang et al., 2019). However, much recent research is redefining the multiple B-cell subsets for their unveiling pathological role in HCC. For example, B-cell depletion improved senescence-mediated fibrosis resolution, increased T-cell anti-tumor responses, suppressed M2 polarization, and abated HCC tumorigenesis in mice (Faggioli et al., 2018; Liu et al., 2015; Wei et al., 2019). Infiltrating B-cells are now associated with tumor aggressiveness and lower survival in HCC patients (Faggioli et al., 2018). Additionally, there is a novel immunotolerant PD-1^{high}IL-10^{+} expressing B-cell subset classified as regulatory B-cells (Bregs), which mitigate T-cell function (Ouyang et al., 2016; Xiao et al., 2016; Zhou, Sprengers, et al., 2017) and cause subsequent defects in B-cell help during advanced HCC (Zhou et al., 2016). Intriguingly, the role of Bregs to advance tumor growth and invasiveness is independent of Tregs, but actually via the CD40/CD154 pathway (Shao et al., 2014; Xue et al., 2016). Further, the hypoxic HCC microenvironment can instigate high-mobility group B1 (HMGB1)-expressing tumor cell-released autophagosomes, which can subsequently induce Bregs, suppress $CD8^{+}$ T-cell anti-tumor responses, and activate HMGB1-TLR2/4-MAPK dependent HCC immune evasion (Ye et al., 2018; Zhang et al., 2019; Zhou et al., 2016).

Importantly, TLR4 mediates upregulation of B-cell lymphoma 6, which is required for PD-1^{high}IL-10^{+} Breg induction (Xiao et al., 2016), but whether the gut microbiota LPS-TLR4 axis contributes to this phenotype is unknown. Other studies have shown that gut microbiota activation of TLR2, MyD88, and PI3K pathways, in addition to upregulating IL-1β and IL-6 expressions, can induce IL-10 producing B-cells (Mishima et al., 2019; Nagatake et al., 2019; Rosser et al., 2014), but whether this supports HCC is also unclear. Interestingly, minimal levels of LPS can cause a class-switch DNA recombination for B-cells to become IgA producers, which is further stimulated by TGF-β1 (Coffman, Lebman, & Shrader, 1989). Liver injury also causes a substantial increase in IgA production from the gut-associated lymphoid tissues, which then reacts toward commensals for clearance of gut bacterial-derived antigens (Moro-Sibilot et al., 2016). As mentioned prior, PD-$L1^{+}$IL-$10^{+}$$IgA^{+}$ liver resident cells can stimulate HCC pathogenesis by interfering with cytotoxic $CD8^{+}$ T-cell anti-tumor responses (Shalapour et al., 2017). It would be intriguing to determine whether gut microbiota regulation of B-cell differentiation to IgA^{+} cells would contribute to HCC.

4. Hijacking energy sources: Does gut microbiota have an influence on metabolic reprogramming in HCC?

4.1 Warburg effect

In the roaring 1920s, the pioneer Otto Warburg discovered that cancer cells shift their energy metabolism to favor rapid uptake and fermentation of glucose for lactate production rather than oxidative phosphorylation

(Warburg, 1925). This phenomenon of aerobic glycolysis was termed the Warburg effect. Since uncovering this fundamental concept of cancer cell metabolism, further research has revealed that the metabolic status in tumors can vary based on the microenvironment. For instance, cancer cells have the dual metabolic flexibility to switch between a glycolytic and non-glycolytic phenotype depending on lactate levels. This means that lactic acidosis caused from the Warburg effect can initiate tumor cells to increase oxygen consumption rate and slow glycolytic rate, to minimize lactate production, and to incorporate glucose carbon(s) for cellular mass (Xie et al., 2014). In addition, a new two-compartment model named the "reverse Warburg effect" details that cancer cells can hijack the energy storage from fibroblasts within the tumor microenvironment. This involves cancer cells inducing oxidative stress and subsequent aerobic glycolysis in the fibroblasts, where the lactate is then shuffled to support tumor metabolism (Fu et al., 2017). Furthermore, metabolic reprogramming from cancer cells indicate that metabolites themselves can be oncogenic by not only serving as an important energy source, but also by altering cell signaling (Ward & Thompson, 2012). In this section, we discuss the underexplored, potential role of gut microbiota as a potential source and gatekeeper of various metabolites that could support the Warburg effect and other metabolic reprograms during HCC (as summarized in Fig. 2).

4.2 Aerobic glycolysis

The liver is a central hub for glycolysis and glycogen metabolism. When glycogen storages become saturated, the remaining glucose is catabolized by a series of enzymatic reactions for the production of pyruvate, which enters the tricarboxylic acid cycle (TCA; *alias* Krebs cycle) (Han, Kang, Kim, Choi, & Koo, 2016). Mathematical modeling has demonstrated that there is a spectrum of glucose metabolism rates depending on whether the cancer is dependent on glycolytic or non-glycolytic tumor growth (Shamsi, Saghafian, Dejam, & Sanati-Nezhad, 2018). Comparatively, HCC is generally associated with glucose deprivation due to excessive glycolysis but suppressed gluconeogenesis (Bian et al., 2017). This Warburg effect of active glucose uptake followed with glucose intolerance is essential for HCC progression on multiple levels, including (i) ATP and lactate production for cellular growth (de la Cruz-Lopez, Castro-Munoz, Reyes-Hernandez, Garcia-Carranca, & Manzo-Merino, 2019), (ii) EMT for metastasis (Jo et al., 2020), (iii) immunosuppressive lymphocyte infiltration (X. Li et al., 2020),

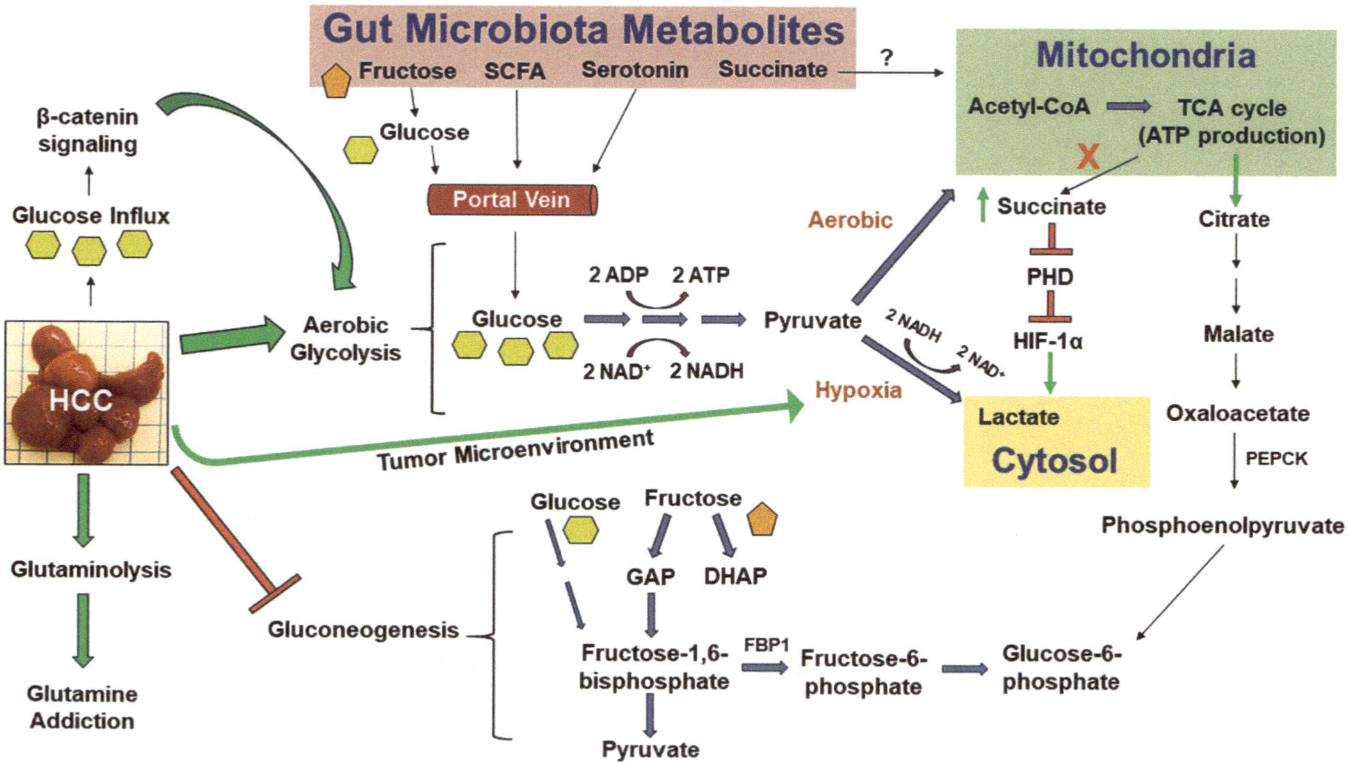

Fig. 2 See figure legend on opposite page.

and (iv) glucose-influx induced β-catenin signaling activation (Hung et al., 2019), which instigates a positive feedback cycle to support glycolysis (Liu et al., 2019). As such, disrupted expression of glucose metabolism related genes have been associated with poor prognosis for HCC patients (Ahn et al., 2009; Amann et al., 2009; Chen et al., 2015; Daskalow et al., 2009; Gong et al., 2012; Paudyal et al., 2008; Shi et al., 2018).

The influence of gut microbiota on glucose metabolism could provide further mechanistic insight to HCC pathogenesis, but this has been under-explored. Therefore, understanding how the gut microbiota impacts glucose metabolism in obesity, type II diabetes, and NAFLD could provide some clues because these metabolic and liver diseases are three top risk factors for HCC. Ironically, one of the most well used animal models to induce these conditions includes high-fat diet administration, where dietary lifestyle is also one of the most potent factors for modulating the gut microbiota. It is well established from animal studies that perturbation of the gut bacterial ecosystem by high-fat diet intervention results in lower species richness (*alias* diversity) but higher species evenness (*alias* abundance), which is associated with hyperglycemia (Acharya, Gao, Bless, Chen, & Tetel, 2019; Daniel et al., 2014; Gu et al., 2019; Madsen et al., 2019). Human studies with obese individuals also found similar correlations (Ridaura et al., 2013; Turnbaugh et al., 2009). Metabolomic studies further demonstrate that an obesity-associated gut microbiota reflects differences in circulating metabolite levels that are involved in glycolysis and the TCA cycle, in addition to bile acid,

Fig. 2 Potential implications of the gut microbiota in favoring the Warburg effect in HCC. The hypoxic tumor microenvironment favors aerobic glycolysis, which encompasses pyruvate fermentation into lactate within the cytosol. Alongside the positive feedback loop initiated by β-catenin signaling to induce aerobic glycolysis, enterohepatic circulation of various gut microbiota metabolites, such as fructose, short chain fatty acids (SCFA), serotonin, and succinate could also, perhaps, fuel the Warburg effect. In coordination, impairment in the tricarboxylic acid (TCA) cycle can over accumulate succinate levels, where succinate is a potent inhibitor of prolyl hydrolases (PHD), which frees hypoxia-inducible factor 1-α (HIF-1α) to promote aerobic glycolysis. Even under conditions of working TCA function, the reduced expression of rate limiting enzymes (e.g., PEPCK, FBP1) diminishes gluconeogenesis, which is a favorable outcome for HCC. In collection with a "glutamine" addiction phenotype, intense metabolic reprogramming collectively supports HCC proliferation. ADP, adenine di-phosphate (ADP); ATP, adenine triphosphate; NAD, nicotinamide adenine dinucleotide; NADH, nicotinamide adenine dinucleotide + hydrogen (H); GAP, glyceraldehyde 3-phosphate; DHAP, dihydroxyacetone phosphate; FBP1, fructose bisphophastase-1; PEPCK, phosphoenolpyruvate carboxykinase. *Source: Credit: Rachel M. Golonka.*

carbohydrate, lipid, and amino acid metabolisms (Aron-Wisnewsky et al., 2019; Calvani et al., 2014; Daniel et al., 2014; Fujisaka et al., 2018; Gu et al., 2019; Hoyles et al., 2018; Lin, An, Tang, & Wang, 2019; Mayengbam et al., 2019; Ottosson et al., 2018).

There are other characteristics of the gut microbiota that could be supporting HCC by modulating glucose metabolism independent from contributing to metabolic diseases. A study by Kim et al. demonstrated that gut microbiota generation of SCFA improves antibody production by increasing glycolysis, oxidative phosphorylation, and fatty acid synthesis (Kim, Qie, Park, & Kim, 2016). This may raise a hypothesis that enterohepatic cycling of SCFA from the gut microbiota to the liver could potentially drive HCC through stimulating glycolysis. In the same line, it is plausible for the gut microbiota to be pathological on glucose metabolism during HCC when considering that (i) gut microbes manufacture around 95% of serotonin production (Martin, Osadchiy, Kalani, & Mayer, 2018; Yano et al., 2015), (ii) gut microbiome regulates glucose homeostasis through serotonin (Martin et al., 2019), and (iii) serotonin activates glycolysis in tumor cells (Sola-Penna et al., 2020). Additionally, serotonin can sustain HCC proliferation under nutrient-deprived conditions (Soll et al., 2010), which could be thought as an alternative energy source for HCC cells without having to either induce macroautophagy processes (Niture et al., 2020) or metabolically re-activate ketolysis and subsequent ketone body production (Huang et al., 2016). Future studies are certainly needed to understand the interrelationship between gut microbiota metabolites and glucose metabolism in the context of HCC development.

4.3 Fructolysis

After glucose is metabolized in the liver, there are two reaction pathways that pyruvate can undertake depending on oxygen bioavailability. In aerobic conditions, pyruvate would enter the mitochondria and serve as a source of acetyl-CoA for the TCA cycle. Under anaerobic conditions, such as the hypoxic tumor microenvironment for HCC cells, pyruvate will remain in the cytosol and be further fermented into the end product of aerobic glycolysis, lactate. Despite the gut microbiota being also capable to synthesize lactate, it is unlikely that it would reach the liver before being further metabolized into other products, such as SCFA (Scheiman et al., 2019). Intriguingly, there might be another means for which the gut microbiota indirectly supports the Warburg effect in HCC. A recent study by Jang et al. demonstrates that when fructose clearance becomes saturated in the small intestine, the remaining fructose is metabolized into glucose and other organic acids like acetate by the gut microbiota (Jang et al., 2018). It is important to note that the probability for enterohepatic circulation of these metabolites is high because the fructose that escapes intestinal catabolism can not only impair gut colonization of beneficial commensal bacteria (e.g., *Lactobacillus, Bacteroides thetaiotaomicron, Eubacterium*) (Astbury et al., 2018; Jones et al., 2019; Townsend et al., 2019), but can also cause gut barrier dysfunction (Song et al., 2018). Therefore, one potential mechanism to the previously reported liver carcinogenic effects of high fructose levels (Dewdney, Roberts, Qiao, George, & Hebbard, 2020; Kumamoto et al., 2013; Laguna, Alegret, & Roglans, 2014; Ozawa, Maehara, Kai, Arai, & Miyazaki, 2016) could be that the additional intestinal glucose supply fuels hepatic glycolysis. Together, both fructose and acetate are also potent contributors of hepatic de novo lipogenesis (Jang et al., 2020; Zhao, Jang, et al., 2020), which is considered an important precursor and driver of HCC (Li, Che, et al., 2016).

There is an irony that HCC cells can markedly reduce the rate of hepatic fructose metabolism for multiple reasons, including to retain ATP levels (Softic, Cohen, & Kahn, 2016) and to limit ROS production (Li, Qian, et al., 2016; Softic et al., 2016). Additionally, halting fructose metabolism could later limit the activity of gluconeogenesis. To quickly summarize hepatic fructolysis, the two main products are dihydroxyacetone phosphate (DHAP) and glyceraldehyde 3-phosphate (GAP). Following, GAP can be reversibly transformed into fructose-1,6-bisphosphate, which is then irreversibly dephosphorylated by fructose-1,6-bisphosphatase 1 (FBP1) into fructose-6-phosphate (Lee & Cha, 2018). FBP1 is one of the key intestinal gluconeogenesis enzymes to antagonize aerobic glycolysis (Ghanem, Kitanovic, Holzwarth, & Wolfl, 2017; Wang & Dong, 2019), but its expression is either downregulated or repressed during HCC (Barcena-Varela et al., 2019; Chen, Li, Zhou, Liu, & Huang, 2017; Chen, Yu, et al., 2016; Fan et al., 2020; Hirata et al., 2016; Jin et al., 2017; Krause et al., 2017; Liu, Jin, et al., 2018; Yang, Jin, et al., 2017; Yang, Wang, et al., 2017; Yang, Ning, Wang, Yan, & Meng, 2019). Many epigenetic mechanisms have been uncovered to cause FBP1 silence in HCC, including (i) tripartite motif-containing protein 28 dependent ubiquitination and degradation (Jin et al., 2017), (ii) hypoxia-inducible factor-1 (HIF-1)α signaling in coordination with Snail dependent transcriptional repression (Fan et al., 2020), (iii) heat shock factors (Yang, Ning, et al., 2019), (iv) HDAC activity (Yang, Jin, et al., 2017), and (v) microRNA interactions (Zhang, Li, et al., 2018). Accordingly, restoration of FBP1 inhibited HCC tumorigenesis by impeding glucose uptake, glycolysis, and lactate production (Chen, Li, Zhou, et al., 2017; Hirata et al., 2016; Yang, Jin, et al., 2017), while also promoting cell apoptosis via ROS (Krause et al., 2017; Lu, Yu, & Zhu, 2014) and suppressing Snail dependent EMT (Liu et al., 2018). It would be interesting to determine if gut microbiota metabolites that are capable of epigenetic modifications (e.g., HDACi activity of SCFA) can silence FBP1 and thus, impair gluconeogenesis and support HCC.

4.4 Rewiring of the tricarboxylic acid cycle

In line with FBP1, another rate limiting enzyme for gluconeogenesis is phosphoenolpyruvate carboxykinase (PEPCK or PCK), but this is also downregulated in primary HCC (Liu, Jin, et al., 2018). Intriguingly, forced expression of cytosolic PCK1 and mitochondrial PCK2 was sufficient to impede HCC progression under glucose-deprived conditions by causing TCA cataplerosis, where the scarcity of bioenergetic TCA metabolites induced oxidative stress and apoptosis (Liu, Jin, et al., 2018). As such, HCC cells can rewire for elevated expression of TCA related genes (Nwosu et al., 2017) and pronounce TCA activity (Fukuda et al., 2002), but this may be more directed for nutrient-deprived conditions because HCC has been associated with downregulation of the TCA cycle as well (Huang et al., 2013). This is further supported with the identification of single nucleotide polymorphisms in TCA related enzymes, such as the succinate dehydrogenase (SDH) complex and isocitrate dehydrogenase (IDH), which can predict poor prognosis and early recurrence of HCC (Li et al., 2019; Wan et al., 2015).

Accordingly, deficiency of just one component in the SDH complex can favor the Warburg effect in vitro, as evident in the switch to aerobic glycolysis from cellular respiration, followed with aggressive HCC and metastasis (Li, Liang, et al., 2019; Tseng et al., 2018). In coordination, impaired SDH activity causes an elevated repository of succinate with subsequent inhibition of prolyl hydroxylases and stabilization of HIF-1α, which, in turn, promotes aerobic glycolysis and HCC (Xiao et al., 2012; Zhang et al., 2015). Similar to this, downregulation of IDH accumulates the TCA intermediate itaconate, which acts as an endogenous succinate dehydrogenase inhibitor and exerts anti-inflammatory effects through modulation of succinate levels (Cordes et al., 2016; Lampropoulou et al., 2016; Tannahill et al., 2013), but whether this contributes to the Warburg effect and HCC pathogenesis is unknown.

It is noteworthy that certain microbes within the gut bacterial community have an evolutionary starch utilization system and/or alternative TCA cycle to synthesize succinate (Kwong, Zheng, & Moran, 2017). Compared to mammalian SDH oxidizing succinate into fumarate, obligate anaerobic commensal such as *Bacteroides* spp., *Prevotella* spp., and *Lactobacllius* spp. ferment or reduce fumarate into succinate, which generates levels ranging between 0.5 and 5 mM in the human luminal gut (Curtis et al., 2014; De Vadder et al., 2016; Hwang et al., 2017; Spiga et al., 2017). Intriguingly, intestinal succinate was found to improve glucose homeostasis in mice by serving as a substrate for gluconeogenesis activation (De Vadder et al., 2016). Therefore, it is plausible that perturbation of the gut microbiota, like with high-fat or fructose diet, could be contributing to HCC tumorigenesis by diminishing succinate-producing gut bacteria, which would further impede intestinal gluconeogenesis. If this is not the case, then possibly gut barrier dysfunction caused by the same interventions could potentially leak gut bacterial-derived succinate that reaches the liver and supports the Warburg effect and HCC. Future studies should explore the potential relation of gut microbiota derived TCA metabolites in HCC pathogenesis.

4.5 Glutamine addiction

In addition to favoring aerobic glycolysis and rewiring the TCA cycle, HCC cells metabolically reprogram to induce glutaminolysis, which is the catabolism of glutamine to glutamate. This phenotype of "glutamine addiction" is directed to support uncontrolled cell proliferation, replenish TCA intermediates, increase ATP generation, advance nucleotide synthesis, and promote glutathione production, which are essential during glycolytic deprivation conditions (Dai et al., 2020; Nilsson et al., 2020; Shiratori et al., 2019; Tompkins et al., 2019). Coordination of repressing glutaminases (e.g., GLS1/GLS2), upregulating glutamine transporters (e.g., ASCT2), and increasing exportation of glutamate are required to maintain glutamine bioavailability and its benefits for HCC cells (Kuo et al., 2016; Nilsson et al., 2020; Sun et al., 2016; Xu et al., 2016). As such, restoration of the rate limiting enzyme GLS2 has been prompted as a therapeutic option against HCC based on its tumor suppressive effects to arrest the G2/M phase in the cell cycle (Lopez de la Oliva et al., 2020) and to inhibit Snail-dependent metastasis during HCC (Kuo et al., 2016). Opposingly, inhibition of GLS1 is suggested as a targeted therapeutic based on its bioenergetic effects to support liver tumorigenesis (Huang, Gan, Wang, Xu, & Xie, 2019; The PLOS Genetics Staff, 2016; Xiang et al., 2015).

Glutamine metabolism in the intestine has been extensively studied (Kim & Kim, 2017), but whether the gut microbiota plays a role is unclear. Comparatively, there is evidence that glutamine modulates the gut microbiota and the intestinal-immune axis. Oral supplementation of glutamine was found to favorably decrease the Firmicutes to Bacteroidetes ratio in obese individuals (de Souza et al., 2015). Secretory immunoglobulin A (sIgA; major microbiota homeostasis factor) concentrations were found to be elevated following glutamine administration in mice (Wu et al., 2016). Additionally, glutamine demonstrated to protect against intestinal permeability and translocation of toxins in rodent models of bacterial overgrowth and bile duct ligation induced jaundice (Azuma et al., 2009; dos Santos et al., 2010; Karatepe et al., 2010; Margaritis et al., 2005). Evidence from Wells et al. postulates that glutamine and bile inhibits gut bacterial adhesion/encroachment to the intestinal mucosa, which subsequently prevents bacterial translocation (Wells, Jechorek, & Erlandsen, 1995). Considering that HCC cells can overly consume glutamine, the limited availability of this amino acid could disrupt gut microbiota and intestinal homeostasis, which would result in the leak of hazardous compounds that circulate to the liver. Future studies should investigate whether the potential competition of glutamine in the gut-liver axis reflects with alterations in the gut microbiota associated with HCC pathogenesis.

5. Immunological and gut microbiota therapeutics for HCC

The current front-line medical treatments for HCC include: (i) pharmaceutics (e.g., sorafenib), (ii) surgical resection, (iii) chemoradiation, (iv) selective internal radiation (SIRT), (v) transcatheter arterial chemoembolization (TACE), (vi) radiofrequency ablation (RFA), and most recently (vii) immunotherapy (Daher, Massarwa, Benson, & Khoury, 2018). In this final section, we discuss the multitude of different adjuvant immunotherapies that halt HCC and how the gut microbiota can regulate the efficacy of these approaches. Further, we describe how interventions to the gut microbiota, such as probiotics, prebiotics, and fecal microbiota transplantation could also serve as potential adjuvant therapeutics to thwart HCC.

5.1 Immune checkpoint inhibitors

As mentioned earlier, an indicator of atypical immunometabolism includes innate and adaptive immune cells acquiring immune checkpoints, which impair anti-tumor responses. Accordingly, immune checkpoints have become one of the main targets for immunotherapy in all types of cancers, including HCC [as reviewed in (Pinato et al., 2020)]. This has expanded to use a combinatorial approach with multiple immune checkpoint inhibitors for a synergistic effect (Wolchok et al., 2013). Interestingly, recent reports depict cautionary evidence that the diversity of the gut microbiota can dictate the efficacy of immune checkpoint inhibitors, *albeit* underexplored in HCC. The one study by Li et al. show that anti-PD-1 treated HCC patients have a more diverse gut microbiota, where elevated *Faecalibacterium* abundance was associated with better progression free survival (Li & Ye, 2020). In line with this, there is a strong correlation between anti-PD-1 immunotherapy and gut microbiota diversity in lung cancer, which is further associated with increased T-cell and NKT cell levels (Jin et al., 2019).

Importantly, meta-analyses among other cancer types (e.g., melanoma, lung cancer) reveal that reduction of gut microbiota diversity by antibiotics treatment diminishes the clinical benefits from immune checkpoint inhibitors (Derosa et al., 2018; Huang et al., 2019; Schett et al., 2020). What is more interesting is that the negative effects of antibiotics seem to occur when administered as a pre-treatment to checkpoint inhibitors, but not while given concurrently (Elkrief et al., 2019; Pinato et al., 2019). Excitingly, fecal microbiota transplantation from responder cancer patients into germ-free or antibiotic-treated mice restored the tumor suppressive effects of PD-1 inhibition (Routy et al., 2018).

Other studies demonstrate that there is a bidirectional relationship between immune checkpoint inhibitors and the gut microbiome composition. PD-1 blockade had a corresponding increase in gut microbiota diversity, including higher abundance of *Ruminococcaceae*, which was suggested to enhance anti-PD-1 immunotherapy (Gopalakrishnan et al., 2018). Ironically, *Ruminococcaceae* is one of the main bacterial groups to generate SCFA, where a recent study by Coutzac et al. uncovers that butyrate and propionate limit the efficacy of CTLA-4 inhibitors, which is associated with higher Tregs and lower survival (Coutzac et al., 2020). Comparatively, a recent study by Mager et al. delineates that the gut metabolite, inosine, which is produced by *Bifidobacterium pseudolongum* and *Akkermansia muciniphila*, can enhance anti-CTLA-4 and anti-PD-L1 therapy by activating anti-tumor responding T-cells (Mager et al., 2020). Future studies should explore (i) whether the gut microbiota is necessary for maximum efficacy of checkpoint inhibitor treatment in HCC patients and (ii) discern which gut metabolites have promoting or demoting effects on immune checkpoint inhibitor therapy.

5.2 Adoptive cellular immunotherapy

5.2.1 Engineering adoptive T-cell transfer (ACT)

ACT is one pillar of immunotherapy involving infusion of tumor-specific cytotoxic T-cells that can recognize, target, and destroy cancer cells (Perica, Varela, Oelke, & Schneck, 2015). There have been many ACT formulas designed to suppress HCC. For example, a TCR-engineered T-cell to recognize HCV has been found to efficiently treat HCV-induced HCC (Spear et al., 2016). In line with this, generating a GPC3367-specific T-cell receptor can identify and kill the glypican-3 positive ($GPC3^{+}$) HCC subset (Dargel et al., 2015). Another type of ACT includes α-fetoprotein (AFP)-specific murine TCR genes that recognize and kill human HCC cancer cells (Zhu et al., 2018). Importantly, introducing other cofactors, like hyper-IL-15, can beneficially increase ACT efficacy, as reflected by the expansion of $CD8^{+}$ T-cells and subsequent IFN-γ production to cytotoxically inhibit HCC (Cheng et al., 2014).

It is noteworthy that the gut microbiota composition contributes to the efficacy of ACT therapy. For instance, ACT + Gram-positive specific antibiotic, vancomycin treatment inhibits cancer growth, but this tumor suppression is not evident when ACT is paired with either neomycin or metronidazole intervention (Uribe-Herranz et al., 2018). Accordingly, vancomycin complemented ACT by expanding cytotoxic $CD8^{+}$ T-cells and dendritic cells (Uribe-Herranz et al., 2018), which suggests that Gram-positive bacteria may hinder ACT efficiency. Similar to this, oral administration of the Gram-negative bacterium *Bacteroides fragilis* or adoptive transfer of *B. fragilis*-reactive T-cells could rescue ACT therapy in germ-free mice (Vetizou et al., 2015). Future studies with gnotobiotic mouse models are needed to further determine which gut bacteria support or demote ACT therapy.

5.2.2 Autologous cytokine-induced killer (CIK) cells

CIK cells are ex vivo products of expanded polyclonal T lymphocytes comprising $CD3^{+}CD56^{+}$ cells, $CD3^{+}CD56^{-}$ cytotoxic T-cells, and $CD3^{-}CD56^{+}$ NK cells (Garofano et al., 2019; Meng et al., 2017; Yoon et al., 2019). Adjuvant CIK therapy has been monumental in providing better prognosis, less recurrence, and long-term overall survival for HCC patients (Lee et al., 2015; Lee, Lee, et al., 2019; Yoon et al., 2019). These positive effects of CIK cells are further delineated with the synergistic effects of TACE and RFA therapies (Huang, Li, et al., 2013; Jia et al., 2019;

Wu, Leboeuf, et al., 2016). Not to mention, CIK immunotherapy has been a very important alternative strategy with minimal side effects for those who are unable to undergo surgery (Yu et al., 2014).

The tumor suppressive property of CIK cells are attributed to the subsequent increase in NKT-mediated cytotoxicity (Leboeuf et al., 2014; Weng et al., 2008), potentially from NKG2D-ligand activation (Rong et al., 2016). However, CIK therapy is recognized to be mostly sufficient to treat early stage tumorigenesis rather than advanced HCC. This could be due to a few reasons. For one, MDSCs elevate post CIK treatment, which could inhibit the tumor suppressive effects of CIK cells (Yu et al., 2019). Another reason could be from the acidic pH environment of cancer cells, where sodium bicarbonate could be used as a solvent to neutralize the acidity in the tumor microenvironment and restore CIK function (Yuan et al., 2016). Intriguingly, co-treatment with IL-6 can advance CIK anti-tumor effects by decreasing Treg levels (Lin et al., 2012). Surprisingly, there are no studies to date that have explored whether the gut microbiota effects the efficiency rate of CIK or if CIK therapy alters the gut microbiota.

5.2.3 Vaccination with pulsed dendritic cells

As antigen-presenting cells, dendritic cells (DCs) are mediators between the innate and adaptive immune systems. This has generated the exploitation concept to prime or "pulse" isolated DCs with tumor-associated antigens and administer them as a vaccine. There are four main extraction sources in obtaining DCs for vaccine purposes: (i) monocyte-derived dendritic cells (MDDCs) isolated from peripheral blood mononuclear cells, (ii) $CD34^+$ precursors in the bone marrow, (iii) whole blood, which is a mixture of DCs, B-cells, monocytes, and NK cells, and (iv) $CD1c^+$ DCs and pDCs (Sabado, Balan, & Bhardwaj, 2017). Comparatively, MDDCs are the most common vaccine immunization approach for cancer immunotherapy. Accordingly, isolated $CD14^+$ human monocytes are first stimulated with IL-4 and GM-CSF to generate naïve DCs, which are later exposed to a mature stimulus while simultaneously being loaded with tumor-associated antigens (Sabado et al., 2017). Loading includes incubation with either peptides, proteins, RNA, tumor lysates, or multiple tumor cell lines, where the last two approaches provides a large spectrum of tumor antigens and expands the breadth for $CD8^+$ T-cell anti-tumor immunity (Alaniz, Rizzo, & Mazzolini, 2014; Carreno et al., 2015; Chieochansin et al., 2019; O'Neill & Bhardwaj, 2005).

Importantly, this field is expanding to deliver pulsed-DCs via lentivirus (Liu et al., 2011; Wan et al., 2015) and exosome methods (Lu et al., 2017), which have provided more potent efficacy to suppress HCC cells. Pulsed-DCs have been also inputted as a combination therapy with CIKs (Cao, Kong, Liu, & Wang, 2019; Jung et al., 2016; Wang et al., 2015; Yang et al., 2018), T-cell transfer (Sun et al., 2016), PD-1 checkpoint inhibition (Zhang et al., 2019), and TACE (He et al., 2016), which provides better outcomes for HCC patients. While the gut microbiota is known to generally impact DC function and immunization of vaccines (Hagan et al., 2019; Reis et al., 2018), it is unclear as to whether the gut microbiota influences the efficacy of pulsed-DC vaccines.

5.2.4 Chimeric antigen receptor (CAR) therapy

The construction of a CAR mimics TCRs but it is fashioned to have tumor-antigen recognition and intracellular activation. The design comprises of an extracellular immunoglobulin target-binding domain to recognize antigens independent of HLA, a hinge region, a transmembrane domain that anchors CAR to the cell membrane, and intracellular domains for signaling (Guedan, Calderon, Posey, & Maus, 2019; June & Sadelain, 2018). Later generation models have incorporated co-stimulatory receptors to enhance the efficiency of CAR therapy (Guedan et al., 2019). In 2018, CAR immunotherapy was named an "Advance of the Year" and received Federal Drug and Administration approval (Lindner, Johnson, Brown, & Wang, 2020).

There are two categories of CAR therapy: CAR-T cells and CAR-NK cells. Much more attention has been given to CAR-T cell manufacturing, but one common design engineered for both models include GPC3-specific CAR, which has been potent in treating $GPC3^+$ HCC (Liu et al., 2019; Yu et al., 2018). Further research on GPC3-engineered CAR-T cells revealed that the tumor suppressive effects correlate with induced effector memory T-cells (Wang et al., 2020). Importantly, adding dual armory such as transgenic IL-12, IL-15, or IL-21 expression enhanced GPC3-CAR T-cell therapy (Batra et al., 2020; Liu, Di, Shi, et al., 2019). In line with this, designing T-cells that carry a complementary receptor in addition to GPC3 maintains anti-HCC responses without causing off-tumor toxicity (Chen et al., 2017). Intriguingly, Liu et al. developed a split anti-GPC3 CAR-T cell model to treat HCC and to solve the issue of rigid CAR-T cell construction (Liu et al., 2020).

While CAR-T cell therapy has provided imperative success, this therapy has limited efficacy against solid tumors like HCC. Accordingly, designing a

CAR-T cell that targets a protein overexpressed on malignant cells, such as NK group 2 member D ligands (NKG2DL), serves as a potent CAR-T cell model to abate HCC (Sun et al., 2019). Impressively, the measles virus vaccine strain Edmonston enhanced the anti-HCC effects of adoptive transfer of $CD8^+$ $NKG2D^+$ cells (Chen et al., 2017), which ponders to wonder if this vaccine could synergistically improve NKG2DL-specific CAR-T cell therapy. Future studies are also needed to determine whether the gut microbiota can influence CAR therapy, when considering the impacts it has on ACT and immune checkpoint therapy (Abid, Shah, Maatman, & Hari, 2019).

5.3 Probiotics

In 2013, the International Scientific Association for Probiotics and Prebiotics (ISAPP) had a consensus agreement regarding the Food and Agriculture Organization and the World Health Organization definition of a probiotic: "live microorganisms which when administered in adequate amounts confer a health benefit on the host" (Hill et al., 2014). Various strains of *Lactobacilli* and *Bifidobacteria*, *E. coli* strain Nissle 1917, VSL#3, and Prohop are the most common types of probiotics that are incorporated as food supplements (Pandey, Naik, & Vakil, 2015). One popular use of probiotics is to improve gut motility, which is further associated with a balanced gut microbiota and alleviation of constipation (Dimidi, Christodoulides, Scott, & Whelan, 2017). These probiotics have been also highly advertised as potent dietary therapeutics against NAFLD [as reviewed in (Meng, Li, Li, Gan, & Li, 2018)].

Many combinations of probiotic bacterial strains have been utilized as a dietary approach to inhibit HCC development [as reviewed in (Wan & El-Nezami, 2018)]. For example, administration of VSL#3 (combination of several *Lactobacilli* and *Bifidobacteria* strains and a single strain of *Streptococcus*) to DEN-treated rats restored the lost *Lactobacilli* and *Bifidobacteria*, where the rebalance in gut microbiota was further associated with alleviation of inflammation, reduced gut permeability, and suppression of HCC (Zhang et al., 2012). Prohop, which is a probiotic mixture of heat-inactivated VSL#3, in combination with *L. rhamnosus* GG and viable *E. coli* Nissle 1917, can also protect against gut microbiota-associated HCC development (Li et al., 2016). The positive effects of Prohop included shifting the microbiota toward an anti-inflammatory bacterial community that encompasses *Bacteroidetes*, *Prevotella* and *Oscillibacter*, which corresponded with increased intestinal Treg cells and reduced Th17 cells (Li, Sung, et al., 2016).

Despite these beneficial effects, there are conflicting reports as to the full effectiveness and potential downfalls of probiotics on host health. For one, Zmora et al. demonstrates that probiotics have a marked colonization resistance, as indicated with region and strain-specific mucosal localization patterns in the human gut microbiome (Zmora et al., 2018). In addition, probiotics were found to delay the reconstitution of the gut microbiome after antibiotic treatment when compared to spontaneous/regular recovery (Suez et al., 2018). Another concern is whether the substantial increase of SCFA from probiotic administration can hinder anti-tumor iNKT responses and advance immunosuppressive Tregs. Further investigation is certainly required to understand the benefits vs costs of probiotics in HCC.

5.4 Prebiotics

In 2016, ISAPP made a consensus on the appropriate definition for a prebiotic: "a substrate that is selectively utilized by host microorganisms conferring a health benefit" (Gibson et al., 2017). Prebiotics are complex carbohydrates, such as fructans, fructo-oligosaccharides (FOS), and galacto-oligosaccharides, which provide essential nutrients for probiotic bacteria, including *Bifidobacteria* and *Lactobacilli* (Gibson et al., 2017). However, there is a chance of cross-feeding, where fermented products generated from the commensal bacteria could promote opportunistic pathogens (Davani-Davari et al., 2019). Other evidence also pinpoints that prebiotics can have variable and negative impacts on host health, including HCC development.

Inulin (β 2$\rightarrow$1 linkages) is the most widely studied and utilized plant fructan. Reports have stated that inulin protects against drug- (Kalantari, Asadmasjedi, Abyaz, Mahdavinia, & Mohammadtaghvaei, 2019) and chemical-induced liver injury (Correa-Ferreira et al., 2017; Liu, Lu, Wen, Kan, & Jin, 2015). Additionally, inulin was found to improve gut barrier integrity, thereby reducing pro-inflammatory LPS-TLR4 signaling in macrophages during alcoholic liver disease (Yang et al., 2019). Accordingly, extraction and purification of inulin from its natural source has become a highly touted commercial product for processed foods (Mensink, Frijlink, van der Voort Maarschalk, & Hinrichs, 2015).

Alarmingly, our group discovered that prolonged feeding of refined inulin for 24 weeks causes HCC in a subset of mice with preexisting gut dysbiosis (Singh et al., 2018). This phenotype of gut microbiota-dependent HCC held true when the mice were fed refined FOS and pectin, but HCC was not as severe when compared to inulin (Singh et al., 2018). Importantly, germ-free conditions and broad-spectrum antibiotics could halt inulin-induced HCC, including specific ablation of Gram-positive bacteria from vancomycin treatment (Golonka, Yeoh, & Vijay-Kumar, 2019; Singh et al., 2018; Singh et al., 2020). One striking observation in this new murine HCC model was the elevation of circulating bile acids (i.e., cholemia), including the secondary bile acid deoxycholate generated from Gram-positive *Clostridia* species. Recalling that deoxycholate provokes SASP in hepatic stellate cells, this could explain, at least in part, the promotion of pro-inflammatory and tumorigenic factors in the liver following inulin feeding.

As further discussed in more detail from our previous review (Golonka et al., 2019), current research highlights prebiotic fibers to have both beneficial and detrimental effects. Our research suggests that purified ingredients incorporated into foods may negatively affect host hepatic health in individuals with gut microbiota dysbiosis, but more confirmatory studies are required.

5.5 Fecal microbiota transplantation (FMT)

As another means to rebalance the gut microbiota composition, FMT involves transferring gut bacteria from an individual with a "healthy gut microbiota" to another person with dysbiosis. An immediate necessity for FMT would be toward bacterial recolonization post antibiotic treatment, which has shown to be more effective when compared to probiotics (Suez et al., 2018). FMT therapy has been mostly studied in the context of *Clostridium difficile* infection with favorable outcomes (Cammarota et al., 2017), but now this has expanded as a potential therapeutic for other conditions such as metabolic syndrome and chronic liver diseases (Ferrere et al., 2017; Lechner, Yee, Limketkai, & Pham, 2020; Sung, Kim, Hong, & Suk, 2016; Zhou et al., 2017). Besides exhibiting therapeutic potential, FMT transfer from disease models to germ-free rodents is becoming a novel method to study whether the gut microbiota is a cause or consequence of disease pathogenesis (de Groot, Frissen, de Clercq, & Nieuwdorp, 2017; Murakami et al., 2016). This includes stool transplant from "healthy" and "dysbiotic" human-associated microbiota to a mouse model of disease and germ-free mice, respectively, (Britton et al., 2020; Wong et al., 2017). Two important pieces of information can be uncovered from FMT: (i) if dysbiosis human-associated microbiota recapitulates the condition in germ-free mice, then this confirms the disease is gut microbiota-dependent, and (ii) if intervention with a "healthy" human-associated gut microbiota alleviates disease conditions in mice, then microbiota sequencing can help classify what comprises a "healthy" gut microbiota for clinical implications.

In terms of HCC, there is the promising thought that FMT could provide anti-tumor effects based on its alleviation for NAFLD/NASH, but more research is required to discern the makeup of a "unhealthy" gut microbiome in HCC (Delaune et al., 2018). From our fermentable fiber-induced, gut microbiota-dependent HCC model, we highlight that tumorigenesis was associated with elevated levels of Firmicutes, γ-Proteobacteria, and *Clostridia* species (Singh et al., 2018). Another concern to raise is the impact of FMT on the immune system. It is already established that FMT can improve early-life innate and adaptive immunity development (Xiang et al., 2020) and regulate intestinal inflammation (Burrello et al., 2018), but whether FMT could cause secondary advantages or consequences on anti-tumor immune responses needs to be explored. Future studies should also determine whether FMT transfer from human HCC patients to germ-free animals can recapitulate liver tumorigenesis.

6. Summary and future perspectives

In this chapter, we discussed multiple molecular and immunological components that are dysregulated to favor HCC. This included atypical immunometabolism that stimulates M2 polarization, TAN recruitment, and MDSCs differentiation. Additionally, there is an imbalance of more immunosuppressive Tregs and Bregs but less anti-tumor $CD8^+$ T-cells and NKT cells, which contributes to HCC pathogenesis. Metabolic reprogramming toward an environment comprised of the Warburg effect (i.e., lactate production) and a glutamine addiction phenotype, in conjunction with repressed TCA activity, also supports HCC development. We further discussed the underexplored role of the gut microbiota in modulating these systems in HCC, where bacterial-derived metabolites have a major negative impact on immunoregulatory operations that would normally suppress HCC. Whether the gut microbiota influences metabolic reprogramming is less clear, which emphasizes that future work is necessary. Lastly, we outlined that some of the front-line immunological therapies work best as a combinatorial treatment. We propose that manipulation of the gut microbiota could also prove potent in abating HCC initiation, where probiotics and FMT seem more fruitful than the potential harmful effects from prebiotics. Altogether, this chapter begins the journey to decipher the role of gut microbiome in HCC from its influence on immunometabolism and metabolic reprogramming.

Acknowledgment

M. Vijay-Kumar is supported by R01 grant from the National Institutes of Health (NIH) [grant number CA219144].

References

Abid, M. B., Shah, N. N., Maatman, T. C., & Hari, P. N. (2019). Gut microbiome and CAR-T therapy. *Experimental Hematology & Oncology*, *8*, 31. https://doi.org/10.1186/s40164-019-0155-8.

Acharya, K. D., Gao, X., Bless, E. P., Chen, J., & Tetel, M. J. (2019). Estradiol and high fat diet associate with changes in gut microbiota in female ob/ob mice. *Scientific Reports*, *9*(1), 20192. https://doi.org/10.1038/s41598-019-56723-1.

Ahn, K. J., Hwang, H. S., Park, J. H., Bang, S. H., Kang, W. J., Yun, M., et al. (2009). Evaluation of the role of hexokinase type II in cellular proliferation and apoptosis using human hepatocellular carcinoma cell lines. *Journal of Nuclear Medicine*, *50*(9), 1525–1532. https://doi.org/10.2967/jnumed.108.060780.

Alaniz, L., Rizzo, M. M., & Mazzolini, G. (2014). Pulsing dendritic cells with whole tumor cell lysates. *Methods in Molecular Biology*, *1139*, 27–31. https://doi.org/10.1007/978-1-4939-0345-0_3.

Almeida, A., Mitchell, A. L., Boland, M., Forster, S. C., Gloor, G. B., Tarkowska, A., et al. (2019). A new genomic blueprint of the human gut microbiota. *Nature*, *568*(7753), 499–504. https://doi.org/10.1038/s41586-019-0965-1.

Alrafas, H. R., Busbee, P. B., Chitrala, K. N., Nagarkatti, M., & Nagarkatti, P. (2020). Alterations in the gut microbiome and suppression of histone deacetylases by resveratrol are associated with attenuation of colonic inflammation and protection against colorectal cancer. *Journal of Clinical Medicine*, *9*(6), 1796. https://doi.org/10.3390/jcm9061796.

Amann, T., Maegdefrau, U., Hartmann, A., Agaimy, A., Marienhagen, J., Weiss, T. S., et al. (2009). GLUT1 expression is increased in hepatocellular carcinoma and promotes tumorigenesis. *The American Journal of Pathology*, *174*(4), 1544–1552. https://doi.org/10.2353/ajpath.2009.080596.

Anson, M., Crain-Denoyelle, A. M., Baud, V., Chereau, F., Gougelet, A., Terris, B., et al. (2012). Oncogenic beta-catenin triggers an inflammatory response that determines the aggressiveness of hepatocellular carcinoma in mice. *The Journal of Clinical Investigation*, *122*(2), 586–599. https://doi.org/10.1172/JCI43937.

Ao, J. Y., Zhu, X. D., Chai, Z. T., Cai, H., Zhang, Y. Y., Zhang, K. Z., et al. (2017). Colony-stimulating factor 1 receptor blockade inhibits tumor growth by altering the polarization of tumor-associated macrophages in hepatocellular carcinoma. *Molecular Cancer Therapeutics*, *16*(8), 1544–1554. https://doi.org/10.1158/1535-7163.MCT-16-0866.

Aras, S., & Zaidi, M. R. (2017). TAMeless traitors: Macrophages in cancer progression and metastasis. *British Journal of Cancer*, *117*(11), 1583–1591. https://doi.org/10.1038/bjc.2017.356.

Aron-Wisnewsky, J., Prifti, E., Belda, E., Ichou, F., Kayser, B. D., Dao, M. C., et al. (2019). Major microbiota dysbiosis in severe obesity: Fate after bariatric surgery. *Gut*, *68*(1), 70–82. https://doi.org/10.1136/gutjnl-2018-316103.

Arpaia, N., Campbell, C., Fan, X., Dikiy, S., van der Veeken, J., deRoos, P., et al. (2013). Metabolites produced by commensal bacteria promote peripheral regulatory T-cell generation. *Nature*, *504*(7480), 451–455. https://doi.org/10.1038/nature12726.

Ascher, S., Wilms, E., Pontarollo, G., Formes, H., Bayer, F., Muller, M., et al. (2020). Gut microbiota restricts NETosis in acute mesenteric ischemia-reperfusion injury. *Arteriosclerosis, Thrombosis, and Vascular Biology*, *40*(9), 2279–2292. https://doi.org/10.1161/ATVBAHA.120.314491.

Astbury, S., Song, A., Zhou, M., Nielsen, B., Hoedl, A., Willing, B. P., et al. (2018). High fructose intake during pregnancy in rats influences the maternal microbiome and gut development in the offspring. *Frontiers in Genetics*, *9*, 203. https://doi.org/10.3389/fgene.2018.00203.

Atarashi, K., Tanoue, T., Oshima, K., Suda, W., Nagano, Y., Nishikawa, H., et al. (2013). Treg induction by a rationally selected mixture of Clostridia strains from the human microbiota. *Nature*, *500*(7461), 232–236. https://doi.org/10.1038/nature12331.

Atarashi, K., Tanoue, T., Shima, T., Imaoka, A., Kuwahara, T., Momose, Y., et al. (2011). Induction of colonic regulatory T cells by indigenous Clostridium species. *Science*, *331*(6015), 337–341. https://doi.org/10.1126/science.1198469.

Auffray, C., Fogg, D., Garfa, M., Elain, G., Join-Lambert, O., Kayal, S., et al. (2007). Monitoring of blood vessels and tissues by a population of monocytes with patrolling behavior. *Science*, *317*(5838), 666–670. https://doi.org/10.1126/science.1142883.

Azuma, H., Mishima, S., Oda, J., Homma, H., Sasaki, H., Hisamura, M., et al. (2009). Enteral supplementation enriched with glutamine, fiber, and oligosaccharide prevents gut translocation in a bacterial overgrowth model. *The Journal of Trauma*, *66*(1), 110–114. https://doi.org/10.1097/TA.0b013e318193109b.

Backhed, F. (2011). Programming of host metabolism by the gut microbiota. *Annals of Nutrition & Metabolism*, *58*(Suppl. 2), 44–52. https://doi.org/10.1159/000328042.

Baell, J. B., Leaver, D. J., Hermans, S. J., Kelly, G. L., Brennan, M. S., Downer, N. L., et al. (2018). Inhibitors of histone acetyltransferases KAT6A/B induce senescence and arrest tumour growth. *Nature*, *560*(7717), 253–257. https://doi.org/10.1038/s41586-018-0387-5.

Bai, L., Liu, X., Zheng, Q., Kong, M., Zhang, X., Hu, R., et al. (2017). M2-like macrophages in the fibrotic liver protect mice against lethal insults through conferring apoptosis resistance to hepatocytes. *Scientific Reports*, 7(1). https://doi.org/10.1038/s41598-017-11303-z, 10518.

Balmer, M. L., Slack, E., de Gottardi, A., Lawson, M. A., Hapfelmeier, S., Miele, L., et al. (2014). The liver may act as a firewall mediating mutualism between the host and its gut commensal microbiota. *Science Translational Medicine*, *6*(237), 237ra266. https://doi.org/10.1126/scitranslmed.3008618.

Bandyopadhyay, K., Marrero, I., & Kumar, V. (2016). NKT cell subsets as key participants in liver physiology and pathology. *Cellular & Molecular Immunology*, *13*(3), 337–346. https://doi.org/10.1038/cmi.2015.115.

Bao, D., Zhao, J., Zhou, X., Yang, Q., Chen, Y., Zhu, J., et al. (2019). Mitochondrial fission-induced mtDNA stress promotes tumor-associated macrophage infiltration and HCC progression. *Oncogene*, *38*(25), 5007–5020. https://doi.org/10.1038/s41388-019-0772-z.

Barcena-Varela, M., Caruso, S., Llerena, S., Alvarez-Sola, G., Uriarte, I., Latasa, M. U., et al. (2019). Dual targeting of histone methyltransferase G9a and DNA-methyltransferase 1 for the treatment of experimental hepatocellular carcinoma. *Hepatology*, *69*(2), 587–603. https://doi.org/10.1002/hep.30168.

Bartneck, M., Schrammen, P. L., Mockel, D., Govaere, O., Liepelt, A., Krenkel, O., et al. (2019). The CCR2(+) macrophage subset promotes pathogenic angiogenesis for tumor vascularization in fibrotic livers. *Cellular and Molecular Gastroenterology and Hepatology*, 7(2), 371–390. https://doi.org/10.1016/j.jcmgh.2018.10.007.

Batra, S. A., Rathi, P., Guo, L., Courtney, A. N., Fleurence, J., Balzeau, J., et al. (2020). Glypican-3-specific CAR T cells coexpressing IL15 and IL21 have superior expansion and antitumor activity against hepatocellular carcinoma. *Cancer Immunology Research*, *8*(3), 309–320. https://doi.org/10.1158/2326-6066.CIR-19-0293.

Benakis, C., Brea, D., Caballero, S., Faraco, G., Moore, J., Murphy, M., et al. (2016). Commensal microbiota affects ischemic stroke outcome by regulating intestinal gammadelta T cells. *Nature Medicine*, *22*(5), 516–523. https://doi.org/10.1038/nm.4068.

Bhagirath, V. C., Dwivedi, D. J., & Liaw, P. C. (2015). Comparison of the proinflammatory and procoagulant properties of nuclear, mitochondrial, and bacterial DNA. *Shock*, *44*(3), 265–271. https://doi.org/10.1097/SHK.0000000000000397.

Bhat, P., Leggatt, G., Waterhouse, N., & Frazer, I. H. (2017). Interferon-gamma derived from cytotoxic lymphocytes directly enhances their motility and cytotoxicity. *Cell Death & Disease*, *8*(6), e2836. https://doi.org/10.1038/cddis.2017.67.

Bian, X. L., Chen, H. Z., Yang, P. B., Li, Y. P., Zhang, F. N., Zhang, J. Y., et al. (2017). Nur77 suppresses hepatocellular carcinoma via switching glucose metabolism toward gluconeogenesis through attenuating phosphoenolpyruvate carboxykinase sumoylation. *Nature Communications*, *8*. https://doi.org/10.1038/ncomms14420, 14420.

Bigorgne, A. E., John, B., Ebrahimkhani, M. R., Shimizu-Albergine, M., Campbell, J. S., & Crispe, I. N. (2016). TLR4-dependent secretion by hepatic stellate cells of the neutrophil-chemoattractant CXCL1 mediates liver response to gut microbiota. *PLoS One*, *11*(3). https://doi.org/10.1371/journal.pone.0151063, e0151063.

Bolouri, H., Young, M., Beilke, J., Johnson, R., Fox, B., Huang, L., et al. (2020). Integrative network modeling reveals mechanisms underlying T cell exhaustion. *Scientific Reports*, *10*(1), 1915. https://doi.org/10.1038/s41598-020-58600-8.

Bonnardel, J., T'Jonck, W., Gaublomme, D., Browaeys, R., Scott, C. L., Martens, L., et al. (2019). Stellate cells, hepatocytes, and endothelial cells imprint the kupffer cell identity on monocytes colonizing the liver macrophage niche. *Immunity*, *51*(4), 638–654. e639 https://doi.org/10.1016/j.immuni.2019.08.017.

Boudou-Rouquette, P., Ropert, S., Mir, O., Coriat, R., Billemont, B., Tod, M., et al. (2012). Variability of sorafenib toxicity and exposure over time: A pharmacokinetic/pharmacodynamic analysis. *The Oncologist*, *17*(9), 1204–1212. https://doi.org/10.1634/theoncologist.2011-0439.

Britton, G. J., Contijoch, E. J., Spindler, M. P., Aggarwala, V., Dogan, B., Bongers, G., et al. (2020). Defined microbiota transplant restores Th17/RORgammat(+) regulatory T cell balance in mice colonized with inflammatory bowel disease microbiotas. *Proceedings of the National Academy of Sciences of the United States of America*, *117*(35), 21536–21545. https://doi.org/10.1073/pnas.1922189117.

Brose, M. S., Frenette, C. T., Keefe, S. M., & Stein, S. M. (2014). Management of sorafenib-related adverse events: A clinician's perspective. *Seminars in Oncology*, *41*(Suppl. 2), S1–S16. https://doi.org/10.1053/j.seminoncol.2014.01.001.

Brunner, S. M., Rubner, C., Kesselring, R., Martin, M., Griesshammer, E., Ruemmele, P., et al. (2015). Tumor-infiltrating, interleukin-33-producing effector-memory CD8(+) T cells in resected hepatocellular carcinoma prolong patient survival. *Hepatology*, *61*(6), 1957–1967. https://doi.org/10.1002/hep.27728.

Burrello, C., Garavaglia, F., Cribiu, F. M., Ercoli, G., Lopez, G., Troisi, J., et al. (2018). Therapeutic faecal microbiota transplantation controls intestinal inflammation through IL10 secretion by immune cells. *Nature Communications*, *9*(1), 5184. https://doi.org/10.1038/s41467-018-07359-8.

Calvani, R., Brasili, E., Pratico, G., Sciubba, F., Roselli, M., Finamore, A., et al. (2014). Application of NMR-based metabolomics to the study of gut microbiota in obesity. *Journal of Clinical Gastroenterology*, *48*(Suppl. 1), S5–S7. https://doi.org/10.1097/MCG.0000000000000236.

Cammarota, G., Ianiro, G., Tilg, H., Rajilic-Stojanovic, M., Kump, P., Satokari, R., et al. (2017). European consensus conference on faecal microbiota transplantation in clinical practice. *Gut*, *66*(4), 569–580. https://doi.org/10.1136/gutjnl-2016-313017.

Campbell, C., McKenney, P. T., Konstantinovsky, D., Isaeva, O. I., Schizas, M., Verter, J., et al. (2020). Bacterial metabolism of bile acids promotes generation of peripheral regulatory T cells. *Nature*, *581*(7809), 475–479. https://doi.org/10.1038/s41586-020-2193-0.

Campo, L., Eiseler, S., Apfel, T., & Pyrsopoulos, N. (2019). Fatty liver disease and gut microbiota: A comprehensive update. *Journal of Clinical and Translational Hepatology*, 7(1), 56–60. https://doi.org/10.14218/JCTH.2018.00008.

Cani, P. D., Amar, J., Iglesias, M. A., Poggi, M., Knauf, C., Bastelica, D., et al. (2007). Metabolic endotoxemia initiates obesity and insulin resistance. *Diabetes*, *56*(7), 1761–1772. https://doi.org/10.2337/db06-1491.

Cao, J., Kong, F. H., Liu, X., & Wang, X. B. (2019). Immunotherapy with dendritic cells and cytokine-induced killer cells for hepatocellular carcinoma: A meta-analysis. *World Journal of Gastroenterology*, *25*(27), 3649–3663. https://doi.org/10.3748/wjg.v25.i27.3649.

Carding, S., Verbeke, K., Vipond, D. T., Corfe, B. M., & Owen, L. J. (2015). Dysbiosis of the gut microbiota in disease. *Microbial Ecology in Health and Disease*, *26*. https://doi.org/10.3402/mehd.v26.26191, 26191.

Carreno, B. M., Magrini, V., Becker-Hapak, M., Kaabinejadian, S., Hundal, J., Petti, A. A., et al. (2015). Cancer immunotherapy. A dendritic cell vaccine increases the breadth and diversity of melanoma neoantigen-specific T cells. *Science*, *348*(6236), 803–808. https://doi.org/10.1126/science.aaa3828.

Chang, J. T., Wherry, E. J., & Goldrath, A. W. (2014). Molecular regulation of effector and memory T cell differentiation. *Nature Immunology*, *15*(12), 1104–1115. https://doi.org/10.1038/ni.3031.

Chen, L., Gu, J., Qian, Y., Li, M., Qian, Y., Xu, M., et al. (2019). Deletion of C-C motif chemokine ligand 5 worsens invariant natural killer T-cell-mediated hepatitis via compensatory up-regulation of CXCR2-related chemokine activity. *Cellular and Molecular Gastroenterology and Hepatology*, 7(3), 623–639. https://doi.org/10.1016/j.jcmgh.2018.12.009.

Chen, C., Li, K., Jiang, H., Song, F., Gao, H., Pan, X., et al. (2017). Development of T cells carrying two complementary chimeric antigen receptors against glypican-3 and asialoglycoprotein receptor 1 for the treatment of hepatocellular carcinoma. *Cancer Immunology, Immunotherapy*, *66*(4), 475–489. https://doi.org/10.1007/s00262-016-1949-8.

Chen, R., Li, J., Zhou, X., Liu, J., & Huang, G. (2017). Fructose-1,6-bisphosphatase 1 reduces (18)F FDG uptake in hepatocellular carcinoma. *Radiology*, *284*(3), 844–853. https://doi.org/10.1148/radiol.2017161607.

Chen, Y., Liu, Y. C., Sung, Y. C., Ramjiawan, R. R., Lin, T. T., Chang, C. C., et al. (2017). Overcoming sorafenib evasion in hepatocellular carcinoma using CXCR4-targeted nanoparticles to co-deliver MEK-inhibitors. *Scientific Reports*, 7. https://doi.org/10.1038/srep44123, 44123.

Chen, Z., Lu, X., Wang, Z., Jin, G., Wang, Q., Chen, D., et al. (2015). Co-expression of PKM2 and TRIM35 predicts survival and recurrence in hepatocellular carcinoma. *Oncotarget*, *6*(4), 2538–2548. https://doi.org/10.18632/oncotarget.2991.

Chen, J., Wei, Y., He, J., Cui, G., Zhu, Y., Lu, C., et al. (2014). Natural killer T cells play a necessary role in modulating of immune-mediated liver injury by gut microbiota. *Scientific Reports*, *4*, 7259. https://doi.org/10.1038/srep07259.

Chen, Y., Wen, H., Zhou, C., Su, Q., Lin, Y., Xie, Y., et al. (2019). TNF-alpha derived from M2 tumor-associated macrophages promotes epithelial-mesenchymal transition and cancer stemness through the Wnt/beta-catenin pathway in SMMC-7721

hepatocellular carcinoma cells. *Experimental Cell Research*, *378*(1), 41–50. https://doi.org/10.1016/j.yexcr.2019.03.005.
Chen, M. M., Xiao, X., Lao, X. M., Wei, Y., Liu, R. X., Zeng, Q. H., et al. (2016). Polarization of tissue-resident TFH-like cells in human hepatoma bridges innate monocyte inflammation and M2b macrophage polarization. *Cancer Discovery*, *6*(10), 1182–1195. https://doi.org/10.1158/2159-8290.CD-16-0329.
Chen, M. L., Yan, B. S., Lu, W. C., Chen, M. H., Yu, S. L., Yang, P. C., et al. (2014). Sorafenib relieves cell-intrinsic and cell-extrinsic inhibitions of effector T cells in tumor microenvironment to augment antitumor immunity. *International Journal of Cancer*, *134*(2), 319–331. https://doi.org/10.1002/ijc.28362.
Chen, S. S., Yu, K. K., Ling, Q. X., Huang, C., Li, N., Zheng, J. M., et al. (2016). The combination of three molecular markers can be a valuable predictive tool for the prognosis of hepatocellular carcinoma patients. *Scientific Reports*, *6*. https://doi.org/10.1038/srep24582, 24582.
Chen, A., Zhang, Y., Meng, G., Jiang, D., Zhang, H., Zheng, M., et al. (2017). Oncolytic measles virus enhances antitumour responses of adoptive CD8(+)NKG2D(+) cells in hepatocellular carcinoma treatment. *Scientific Reports*, 7(1), 5170. https://doi.org/10.1038/s41598-017-05500-z.
Cheng, L., Du, X., Wang, Z., Ju, J., Jia, M., Huang, Q., et al. (2014). Hyper-IL-15 suppresses metastatic and autochthonous liver cancer by promoting tumour-specific CD8+ T cell responses. *Journal of Hepatology*, *61*(6), 1297–1303. https://doi.org/10.1016/j.jhep.2014.07.004.
Cheng, Y., Li, H., Deng, Y., Tai, Y., Zeng, K., Zhang, Y., et al. (2018). Cancer-associated fibroblasts induce PDL1+ neutrophils through the IL6-STAT3 pathway that foster immune suppression in hepatocellular carcinoma. *Cell Death & Disease*, *9*(4), 422. https://doi.org/10.1038/s41419-018-0458-4.
Chew, V., Lai, L., Pan, L., Lim, C. J., Li, J., Ong, R., et al. (2017). Delineation of an immunosuppressive gradient in hepatocellular carcinoma using high-dimensional proteomic and transcriptomic analyses. *Proceedings of the National Academy of Sciences of the United States of America*, *114*(29), E5900–E5909. https://doi.org/10.1073/pnas.1706559114.
Chieochansin, T., Thepmalee, C., Grainok, J., Junking, M., & Yenchitsomanus, P. T. (2019). Cytolytic activity of effector T-lymphocytes against hepatocellular carcinoma is improved by dendritic cells pulsed with pooled tumor antigens. *Scientific Reports*, *9*(1). https://doi.org/10.1038/s41598-019-54087-0, 17668.
Chiu, D. K., Tse, A. P., Xu, I. M., Di Cui, J., Lai, R. K., Li, L. L., et al. (2017). Hypoxia inducible factor HIF-1 promotes myeloid-derived suppressor cells accumulation through ENTPD2/CD39L1 in hepatocellular carcinoma. *Nature Communications*, *8*(1), 517. https://doi.org/10.1038/s41467-017-00530-7.
Chiu, D. K., Xu, I. M., Lai, R. K., Tse, A. P., Wei, L. L., Koh, H. Y., et al. (2016). Hypoxia induces myeloid-derived suppressor cell recruitment to hepatocellular carcinoma through chemokine (C-C motif) ligand 26. *Hepatology*, *64*(3), 797–813. https://doi.org/10.1002/hep.28655.
Cho, E. J., Leem, S., Kim, S. A., Yang, J., Lee, Y. B., Kim, S. S., et al. (2019). Circulating microbiota-based metagenomic signature for detection of hepatocellular carcinoma. *Scientific Reports*, *9*(1), 7536. https://doi.org/10.1038/s41598-019-44012-w.
Chu, M. O., Shen, C. H., Chang, T. S., Xu, H. W., Yen, C. W., Lu, S. N., et al. (2018). Pretreatment inflammation-based markers predict survival outcomes in patients with early stage hepatocellular carcinoma after radiofrequency ablation. *Scientific Reports*, *8*(1). https://doi.org/10.1038/s41598-018-34543-z, 16611.
Coffman, R. L., Lebman, D. A., & Shrader, B. (1989). Transforming growth factor beta specifically enhances IgA production by lipopolysaccharide-stimulated murine B lymphocytes. *The Journal of Experimental Medicine*, *170*(3), 1039–1044. https://doi.org/10.1084/jem.170.3.1039.

Collado, M. C., Rautava, S., Aakko, J., Isolauri, E., & Salminen, S. (2016). Human gut colonisation may be initiated in utero by distinct microbial communities in the placenta and amniotic fluid. *Scientific Reports*, *6*. https://doi.org/10.1038/srep23129, 23129.

Cordes, T., Wallace, M., Michelucci, A., Divakaruni, A. S., Sapcariu, S. C., Sousa, C., et al. (2016). Immunoresponsive gene 1 and itaconate inhibit succinate dehydrogenase to modulate intracellular succinate levels. *The Journal of Biological Chemistry*, *291*(27), 14274–14284. https://doi.org/10.1074/jbc.M115.685792.

Correa-Ferreira, M. L., Verdan, M. H., Dos Reis Livero, F. A., Galuppo, L. F., Telles, J. E., Alves Stefanello, M. E., et al. (2017). Inulin-type fructan and infusion of Artemisia vulgaris protect the liver against carbon tetrachloride-induced liver injury. *Phytomedicine*, *24*, 68–76. https://doi.org/10.1016/j.phymed.2016.11.017.

Coutzac, C., Jouniaux, J. M., Paci, A., Schmidt, J., Mallardo, D., Seck, A., et al. (2020). Systemic short chain fatty acids limit antitumor effect of CTLA-4 blockade in hosts with cancer. *Nature Communications*, *11*(1), 2168. https://doi.org/10.1038/s41467-020-16079-x.

Curtis, M. M., Hu, Z., Klimko, C., Narayanan, S., Deberardinis, R., & Sperandio, V. (2014). The gut commensal Bacteroides thetaiotaomicron exacerbates enteric infection through modification of the metabolic landscape. *Cell Host & Microbe*, *16*(6), 759–769. https://doi.org/10.1016/j.chom.2014.11.005.

Daher, S., Massarwa, M., Benson, A. A., & Khoury, T. (2018). Current and future treatment of hepatocellular carcinoma: An updated comprehensive review. *Journal of Clinical and Translational Hepatology*, *6*(1), 69–78. https://doi.org/10.14218/JCTH.2017.00031.

Dai, W., Xu, L., Yu, X., Zhang, G., Guo, H., Liu, H., et al. (2020). OGDHL silencing promotes hepatocellular carcinoma by reprogramming glutamine metabolism. *Journal of Hepatology*, *72*(5), 909–923. https://doi.org/10.1016/j.jhep.2019.12.015.

Dalpke, A., Frank, J., Peter, M., & Heeg, K. (2006). Activation of toll-like receptor 9 by DNA from different bacterial species. *Infection and Immunity*, *74*(2), 940–946. https://doi.org/10.1128/IAI.74.2.940-946.2006.

Daniel, H., Gholami, A. M., Berry, D., Desmarchelier, C., Hahne, H., Loh, G., et al. (2014). High-fat diet alters gut microbiota physiology in mice. *The ISME Journal*, *8*(2), 295–308. https://doi.org/10.1038/ismej.2013.155.

Dapito, D. H., Mencin, A., Gwak, G. Y., Pradere, J. P., Jang, M. K., Mederacke, I., et al. (2012). Promotion of hepatocellular carcinoma by the intestinal microbiota and TLR4. *Cancer Cell*, *21*(4), 504–516. https://doi.org/10.1016/j.ccr.2012.02.007.

Dargel, C., Bassani-Sternberg, M., Hasreiter, J., Zani, F., Bockmann, J. H., Thiele, F., et al. (2015). T cells engineered to express a T-cell receptor specific for glypican-3 to recognize and kill hepatoma cells in vitro and in mice. *Gastroenterology*, *149*(4), 1042–1052. https://doi.org/10.1053/j.gastro.2015.05.055.

Dasgupta, S., Erturk-Hasdemir, D., Ochoa-Reparaz, J., Reinecker, H. C., & Kasper, D. L. (2014). Plasmacytoid dendritic cells mediate anti-inflammatory responses to a gut commensal molecule via both innate and adaptive mechanisms. *Cell Host & Microbe*, *15*(4), 413–423. https://doi.org/10.1016/j.chom.2014.03.006.

Daskalow, K., Pfander, D., Weichert, W., Rohwer, N., Thelen, A., Neuhaus, P., et al. (2009). Distinct temporospatial expression patterns of glycolysis-related proteins in human hepatocellular carcinoma. *Histochemistry and Cell Biology*, *132*(1), 21–31. https://doi.org/10.1007/s00418-009-0590-4.

Davani-Davari, D., Negahdaripour, M., Karimzadeh, I., Seifan, M., Mohkam, M., Masoumi, S. J., et al. (2019). Prebiotics: Definition, types, sources, mechanisms, and clinical applications. *Food*, *8*(3), 92. https://doi.org/10.3390/foods8030092.

De Angelis, M., Ferrocino, I., Calabrese, F. M., De Filippis, F., Cavallo, N., Siragusa, S., et al. (2020). Diet influences the functions of the human intestinal microbiome. *Scientific Reports*, *10*(1), 4247. https://doi.org/10.1038/s41598-020-61192-y.

de Groot, P. F., Frissen, M. N., de Clercq, N. C., & Nieuwdorp, M. (2017). Fecal microbiota transplantation in metabolic syndrome: History, present and future. *Gut Microbes*, *8*(3), 253–267. https://doi.org/10.1080/19490976.2017.1293224.

de la Cruz-Lopez, K. G., Castro-Munoz, L. J., Reyes-Hernandez, D. O., Garcia-Carranca, A., & Manzo-Merino, J. (2019). Lactate in the regulation of tumor microenvironment and therapeutic approaches. *Frontiers in Oncology*, *9*, 1143. https://doi.org/10.3389/fonc.2019.01143.

De Minicis, S., Kisseleva, T., Francis, H., Baroni, G. S., Benedetti, A., Brenner, D., et al. (2013). Liver carcinogenesis: Rodent models of hepatocarcinoma and cholangiocarcinoma. *Digestive and Liver Disease*, *45*(6), 450–459. https://doi.org/10.1016/j.dld.2012.10.008.

de Souza, A. Z., Zambom, A. Z., Abboud, K. Y., Reis, S. K., Tannihao, F., Guadagnini, D., et al. (2015). Oral supplementation with L-glutamine alters gut microbiota of obese and overweight adults: A pilot study. *Nutrition*, *31*(6), 884–889. https://doi.org/10.1016/j.nut.2015.01.004.

De Vadder, F., Kovatcheva-Datchary, P., Zitoun, C., Duchampt, A., Backhed, F., & Mithieux, G. (2016). Microbiota-produced succinate improves glucose homeostasis via intestinal gluconeogenesis. *Cell Metabolism*, *24*(1), 151–157. https://doi.org/10.1016/j.cmet.2016.06.013.

Delaune, V., Orci, L. A., Lacotte, S., Peloso, A., Schrenzel, J., Lazarevic, V., et al. (2018). Fecal microbiota transplantation: A promising strategy in preventing the progression of non-alcoholic steatohepatitis and improving the anti-cancer immune response. *Expert Opinion on Biological Therapy*, *18*(10), 1061–1071. https://doi.org/10.1080/14712598.2018.1518424.

Deng, Y., Cheng, J., Fu, B., Liu, W., Chen, G., Zhang, Q., et al. (2017). Hepatic carcinoma-associated fibroblasts enhance immune suppression by facilitating the generation of myeloid-derived suppressor cells. *Oncogene*, *36*(8), 1090–1101. https://doi.org/10.1038/onc.2016.273.

Derosa, L., Hellmann, M. D., Spaziano, M., Halpenny, D., Fidelle, M., Rizvi, H., et al. (2018). Negative association of antibiotics on clinical activity of immune checkpoint inhibitors in patients with advanced renal cell and non-small-cell lung cancer. *Annals of Oncology*, *29*(6), 1437–1444. https://doi.org/10.1093/annonc/mdy103.

Derrien, M., Alvarez, A. S., & de Vos, W. M. (2019). The gut microbiota in the first decade of life. *Trends in Microbiology*, *27*(12), 997–1010. https://doi.org/10.1016/j.tim.2019.08.001.

Dewdney, B., Roberts, A., Qiao, L., George, J., & Hebbard, L. (2020). A sweet connection? Fructose's role in hepatocellular carcinoma. *Biomolecules*, *10*(4), 496. https://doi.org/10.3390/biom10040496.

Di Blasi, D., Boldanova, T., Mori, L., Terracciano, L., Heim, M. H., & De Libero, G. (2020). Unique T-cell populations define immune-inflamed hepatocellular carcinoma. *Cellular and Molecular Gastroenterology and Hepatology*, *9*(2), 195–218. https://doi.org/10.1016/j.jcmgh.2019.08.004.

Dimidi, E., Christodoulides, S., Scott, S. M., & Whelan, K. (2017). Mechanisms of action of probiotics and the gastrointestinal microbiota on gut motility and constipation. *Advances in Nutrition*, *8*(3), 484–494. https://doi.org/10.3945/an.116.014407.

Dominguez-Bello, M. G., De Jesus-Laboy, K. M., Shen, N., Cox, L. M., Amir, A., Gonzalez, A., et al. (2016). Partial restoration of the microbiota of cesarean-born infants via vaginal microbial transfer. *Nature Medicine*, *22*(3), 250–253. https://doi.org/10.1038/nm.4039.

Dong, L. Q., Peng, L. H., Ma, L. J., Liu, D. B., Zhang, S., Luo, S. Z., et al. (2020). Heterogeneous immunogenomic features and distinct escape mechanisms in multifocal hepatocellular carcinoma. *Journal of Hepatology*, *72*(5), 896–908. https://doi.org/10.1016/j.jhep.2019.12.014.

Dong, N., Shi, X., Wang, S., Gao, Y., Kuang, Z., Xie, Q., et al. (2019). M2 macrophages mediate sorafenib resistance by secreting HGF in a feed-forward manner in hepatocellular carcinoma. *British Journal of Cancer, 121*(1), 22–33. https://doi.org/10.1038/s41416-019-0482-x.

dos Santos, R., Viana, M. L., Generoso, S. V., Arantes, R. E., Davisson Correia, M. I., & Cardoso, V. N. (2010). Glutamine supplementation decreases intestinal permeability and preserves gut mucosa integrity in an experimental mouse model. *JPEN Journal of Parenteral and Enteral Nutrition, 34*(4), 408–413. https://doi.org/10.1177/0148607110362530.

Duscha, A., Gisevius, B., Hirschberg, S., Yissachar, N., Stangl, G. I., Eilers, E., et al. (2020). Propionic acid shapes the multiple sclerosis disease course by an immunomodulatory mechanism. *Cell, 180*(6), 1067–1080, e1016. https://doi.org/10.1016/j.cell.2020.02.035.

Eggert, T., Wolter, K., Ji, J., Ma, C., Yevsa, T., Klotz, S., et al. (2016). Distinct functions of senescence-associated immune responses in liver tumor surveillance and tumor progression. *Cancer Cell, 30*(4), 533–547. https://doi.org/10.1016/j.ccell.2016.09.003.

Elkrief, A., El Raichani, L., Richard, C., Messaoudene, M., Belkaid, W., Malo, J., et al. (2019). Antibiotics are associated with decreased progression-free survival of advanced melanoma patients treated with immune checkpoint inhibitors. *Oncoimmunology, 8*(4). https://doi.org/10.1080/2162402X.2019.1568812, e1568812.

Ervin, S. M., Hanley, R. P., Lim, L., Walton, W. G., Pearce, K. H., Bhatt, A. P., et al. (2019). Targeting regorafenib-induced toxicity through inhibition of gut microbial beta-glucuronidases. *ACS Chemical Biology, 14*(12), 2737–2744. https://doi.org/10.1021/acschembio.9b00663.

Faggioli, F., Palagano, E., Di Tommaso, L., Donadon, M., Marrella, V., Recordati, C., et al. (2018). B lymphocytes limit senescence-driven fibrosis resolution and favor hepatocarcinogenesis in mouse liver injury. *Hepatology, 67*(5), 1970–1985. https://doi.org/10.1002/hep.29636.

Fan, W., Huo, G., Li, X., Yang, L., Duan, C., Wang, T., et al. (2013). Diversity of the intestinal microbiota in different patterns of feeding infants by Illumina high-throughput sequencing. *World Journal of Microbiology and Biotechnology, 29*(12), 2365–2372. https://doi.org/10.1007/s11274-013-1404-3.

Fan, Q. M., Jing, Y. Y., Yu, G. F., Kou, X. R., Ye, F., Gao, L., et al. (2014). Tumor-associated macrophages promote cancer stem cell-like properties via transforming growth factor-beta1-induced epithelial-mesenchymal transition in hepatocellular carcinoma. *Cancer Letters, 352*(2), 160–168. https://doi.org/10.1016/j.canlet.2014.05.008.

Fan, Z., Zheng, W., Li, H., Wu, W., Liu, X., Sun, Z., et al. (2020). LOXL2 upregulates hypoxiainducible factor1alpha signaling through SnailFBP1 axis in hepatocellular carcinoma cells. *Oncology Reports, 43*(5), 1641–1649. https://doi.org/10.3892/or.2020.7541.

Fang, C., Wei, X., & Wei, Y. (2016). Mitochondrial DNA in the regulation of innate immune responses. *Protein & Cell,* 7(1), 11–16. https://doi.org/10.1007/s13238-015-0222-9.

Feng, R., Morine, Y., Ikemoto, T., Imura, S., Iwahashi, S., Saito, Y., et al. (2018). Nrf2 activation drive macrophages polarization and cancer cell epithelial-mesenchymal transition during interaction. *Cell Communication and Signaling: CCS, 16*(1), 54. https://doi.org/10.1186/s12964-018-0262-x.

Ferrere, G., Wrzosek, L., Cailleux, F., Turpin, W., Puchois, V., Spatz, M., et al. (2017). Fecal microbiota manipulation prevents dysbiosis and alcohol-induced liver injury in mice. *Journal of Hepatology, 66*(4), 806–815. https://doi.org/10.1016/j.jhep.2016.11.008.

Flecken, T., Schmidt, N., Hild, S., Gostick, E., Drognitz, O., Zeiser, R., et al. (2014). Immunodominance and functional alterations of tumor-associated antigen-specific CD8+ T-cell responses in hepatocellular carcinoma. *Hepatology, 59*(4), 1415–1426. https://doi.org/10.1002/hep.26731.

Franceschi, C., Garagnani, P., Vitale, G., Capri, M., & Salvioli, S. (2017). Inflammaging and 'Garb-aging'. *Trends in Endocrinology and Metabolism, 28*(3), 199–212. https://doi.org/10.1016/j.tem.2016.09.005.

Fridlender, Z. G., Sun, J., Kim, S., Kapoor, V., Cheng, G., Ling, L., et al. (2009). Polarization of tumor-associated neutrophil phenotype by TGF-beta: "N1" versus "N2" TAN. *Cancer Cell, 16*(3), 183–194. https://doi.org/10.1016/j.ccr.2009.06.017.

Fridlender, Z. G., Sun, J., Mishalian, I., Singhal, S., Cheng, G., Kapoor, V., et al. (2012). Transcriptomic analysis comparing tumor-associated neutrophils with granulocytic myeloid-derived suppressor cells and normal neutrophils. *PLoS One,* 7(2), e31524. https://doi.org/10.1371/journal.pone.0031524.

Fu, S., He, K., Tian, C., Sun, H., Zhu, C., Bai, S., et al. (2020). Impaired lipid biosynthesis hinders anti-tumor efficacy of intratumoral iNKT cells. *Nature Communications, 11*(1), 438. https://doi.org/10.1038/s41467-020-14332-x.

Fu, Y., Liu, S., Yin, S., Niu, W., Xiong, W., Tan, M., et al. (2017). The reverse Warburg effect is likely to be an Achilles' heel of cancer that can be exploited for cancer therapy. *Oncotarget, 8*(34), 57813–57825. https://doi.org/10.18632/oncotarget.18175.

Fu, X. T., Song, K., Zhou, J., Shi, Y. H., Liu, W. R., Shi, G. M., et al. (2019). Tumor-associated macrophages modulate resistance to oxaliplatin via inducing autophagy in hepatocellular carcinoma. *Cancer Cell International, 19,* 71. https://doi.org/10.1186/s12935-019-0771-8.

Fujisaka, S., Avila-Pacheco, J., Soto, M., Kostic, A., Dreyfuss, J. M., Pan, H., et al. (2018). Diet, genetics, and the gut microbiome drive dynamic changes in plasma metabolites. *Cell Reports, 22*(11), 3072–3086. https://doi.org/10.1016/j.celrep.2018.02.060.

Fukuda, H., Ebara, M., Okuyama, M., Sugiura, N., Yoshikawa, M., Saisho, H., et al. (2002). Increased metabolizing activities of the tricarboxylic acid cycle and decreased drug metabolism in hepatocellular carcinoma. *Carcinogenesis, 23*(12), 2019–2023. https://doi.org/10.1093/carcin/23.12.2019.

Furusawa, Y., Obata, Y., Fukuda, S., Endo, T. A., Nakato, G., Takahashi, D., et al. (2013). Commensal microbe-derived butyrate induces the differentiation of colonic regulatory T cells. *Nature, 504*(7480), 446–450. https://doi.org/10.1038/nature12721.

Garnelo, M., Tan, A., Her, Z., Yeong, J., Lim, C. J., Chen, J., et al. (2017). Interaction between tumour-infiltrating B cells and T cells controls the progression of hepatocellular carcinoma. *Gut, 66*(2), 342–351. https://doi.org/10.1136/gutjnl-2015-310814.

Garofano, F., Gonzalez-Carmona, M. A., Skowasch, D., Schmidt-Wolf, R., Abramian, A., Hauser, S., et al. (2019). Clinical trials with combination of cytokine-induced killer cells and dendritic cells for cancer therapy. *International Journal of Molecular Sciences, 20*(17), 4307. https://doi.org/10.3390/ijms20174307.

Geissmann, F., Jung, S., & Littman, D. R. (2003). Blood monocytes consist of two principal subsets with distinct migratory properties. *Immunity, 19*(1), 71–82. https://doi.org/10.1016/s1074-7613(03)00174-2.

Ghanem, A., Kitanovic, A., Holzwarth, J., & Wolfl, S. (2017). Mutational analysis of fructose-1,6-bis-phosphatase FBP1 indicates partially independent functions in gluconeogenesis and sensitivity to genotoxic stress. *Microbial Cell, 4*(2), 52–63. https://doi.org/10.15698/mic2017.02.557.

Gibson, G. R., Hutkins, R., Sanders, M. E., Prescott, S. L., Reimer, R. A., Salminen, S. J., et al. (2017). Expert consensus document: The International Scientific Association for Probiotics and Prebiotics (ISAPP) consensus statement on the definition and scope of prebiotics. *Nature Reviews. Gastroenterology & Hepatology, 14*(8), 491–502. https://doi.org/10.1038/nrgastro.2017.75.

Global Burden of Disease Liver Cancer Collaboration, Akinyemiju, T., Abera, S., Ahmed, M., Alam, N., Alemayohu, M. A., et al. (2017). The burden of primary liver cancer and underlying etiologies from 1990 to 2015 at the global, regional, and national level: Results from the global burden of disease study 2015. *JAMA Oncology*, *3*(12), 1683–1691. https://doi.org/10.1001/jamaoncol.2017.3055.

Golabi, P., Fazel, S., Otgonsuren, M., Sayiner, M., Locklear, C. T., & Younossi, Z. M. (2017). Mortality assessment of patients with hepatocellular carcinoma according to underlying disease and treatment modalities. *Medicine (Baltimore)*, *96*(9). https://doi.org/10.1097/MD.0000000000005904, e5904.

Golonka, R., Yeoh, B. S., & Vijay-Kumar, M. (2019). Dietary additives and supplements revisited: The fewer, the safer for liver and gut health. *Current Pharmacology Reports*, *5*(4), 303–316. https://doi.org/10.1007/s40495-019-00187-4.

Gong, L., Cui, Z., Chen, P., Han, H., Peng, J., & Leng, X. (2012). Reduced survival of patients with hepatocellular carcinoma expressing hexokinase II. *Medical Oncology*, *29*(2), 909–914. https://doi.org/10.1007/s12032-011-9841-z.

Gonzalez-Perez, G., Hicks, A. L., Tekieli, T. M., Radens, C. M., Williams, B. L., & Lamouse-Smith, E. S. (2016). Maternal antibiotic treatment impacts development of the neonatal intestinal microbiome and antiviral immunity. *Journal of Immunology*, *196*(9), 3768–3779. https://doi.org/10.4049/jimmunol.1502322.

Gonzalez-Perez, G., & Lamouse-Smith, E. S. (2017). Gastrointestinal microbiome dysbiosis in infant mice alters peripheral CD8(+) T cell receptor signaling. *Frontiers in Immunology*, *8*, 265. https://doi.org/10.3389/fimmu.2017.00265.

Goodrich, J. K., Waters, J. L., Poole, A. C., Sutter, J. L., Koren, O., Blekhman, R., et al. (2014). Human genetics shape the gut microbiome. *Cell*, *159*(4), 789–799. https://doi.org/10.1016/j.cell.2014.09.053.

Gopalakrishnan, V., Spencer, C. N., Nezi, L., Reuben, A., Andrews, M. C., Karpinets, T. V., et al. (2018). Gut microbiome modulates response to anti-PD-1 immunotherapy in melanoma patients. *Science*, *359*(6371), 97–103. https://doi.org/10.1126/science.aan4236.

Grywalska, E., Pasiarski, M., Gozdz, S., & Rolinski, J. (2018). Immune-checkpoint inhibitors for combating T-cell dysfunction in cancer. *Oncotargets and Therapy*, *11*, 6505–6524. https://doi.org/10.2147/OTT.S150817.

Gu, Y., Liu, C., Zheng, N., Jia, W., Zhang, W., & Li, H. (2019). Metabolic and gut microbial characterization of obesity-prone mice under a high-fat diet. *Journal of Proteome Research*, *18*(4), 1703–1714. https://doi.org/10.1021/acs.jproteome.8b00945.

Guedan, S., Calderon, H., Posey, A. D., Jr., & Maus, M. V. (2019). Engineering and design of chimeric antigen receptors. *Molecular Therapy—Methods & Clinical Development*, *12*, 145–156. https://doi.org/10.1016/j.omtm.2018.12.009.

Guo, B., Li, L., Guo, J., Liu, A., Wu, J., Wang, H., et al. (2017). M2 tumor-associated macrophages produce interleukin-17 to suppress oxaliplatin-induced apoptosis in hepatocellular carcinoma. *Oncotarget*, *8*(27), 44465–44476. https://doi.org/10.18632/oncotarget.17973.

Guo, X., Zhao, Y., Yan, H., Yang, Y., Shen, S., Dai, X., et al. (2017). Single tumor-initiating cells evade immune clearance by recruiting type II macrophages. *Genes & Development*, *31*(3), 247–259. https://doi.org/10.1101/gad.294348.116.

Hagan, T., Cortese, M., Rouphael, N., Boudreau, C., Linde, C., Maddur, M. S., et al. (2019). Antibiotics-driven gut microbiome perturbation alters immunity to vaccines in humans. *Cell*, *178*(6), 1313–1328 e1313. https://doi.org/10.1016/j.cell.2019.08.010.

Haider, C., Hnat, J., Wagner, R., Huber, H., Timelthaler, G., Grubinger, M., et al. (2019). Transforming growth factor-beta and Axl induce CXCL5 and neutrophil recruitment in hepatocellular carcinoma. *Hepatology*, *69*(1), 222–236. https://doi.org/10.1002/hep.30166.

Han, H. S., Kang, G., Kim, J. S., Choi, B. H., & Koo, S. H. (2016). Regulation of glucose metabolism from a liver-centric perspective. *Experimental & Molecular Medicine*, *48*, e218. https://doi.org/10.1038/emm.2015.122.

Hang, S., Paik, D., Yao, L., Kim, E., Trinath, J., Lu, J., et al. (2019). Bile acid metabolites control TH17 and Treg cell differentiation. *Nature, 576*(7785), 143–148. https://doi.org/10.1038/s41586-019-1785-z.

Hansen, C. H., Andersen, L. S., Krych, L., Metzdorff, S. B., Hasselby, J. P., Skov, S., et al. (2014). Mode of delivery shapes gut colonization pattern and modulates regulatory immunity in mice. *Journal of Immunology, 193*(3), 1213–1222. https://doi.org/10.4049/jimmunol.1400085.

Harimoto, H., Shimizu, M., Nakagawa, Y., Nakatsuka, K., Wakabayashi, A., Sakamoto, C., et al. (2013). Inactivation of tumor-specific CD8(+) CTLs by tumor-infiltrating tolerogenic dendritic cells. *Immunology and Cell Biology, 91*(9), 545–555. https://doi.org/10.1038/icb.2013.38.

Harusato, A., Viennois, E., Etienne-Mesmin, L., Matsuyama, S., Abo, H., Osuka, S., et al. (2019). Early-life microbiota exposure restricts myeloid-derived suppressor cell-driven colonic tumorigenesis. *Cancer Immunology Research*, 7(4), 544–551. https://doi.org/10.1158/2326-6066.CIR-18-0444.

Hathaway-Schrader, J. D., Steinkamp, H. M., Chavez, M. B., Poulides, N. A., Kirkpatrick, J. E., Chew, M. E., et al. (2019). Antibiotic perturbation of gut microbiota dysregulates osteoimmune cross talk in postpubertal skeletal development. *The American Journal of Pathology, 189*(2), 370–390. https://doi.org/10.1016/j.ajpath.2018.10.017.

He, M., Peng, A., Huang, X. Z., Shi, D. C., Wang, J. C., Zhao, Q., et al. (2016). Peritumoral stromal neutrophils are essential for c-Met-elicited metastasis in human hepatocellular carcinoma. *Oncoimmunology, 5*(10). https://doi.org/10.1080/2162402X.2016.1219828, e1219828.

He, G., Zhang, H., Zhou, J., Wang, B., Chen, Y., Kong, Y., et al. (2015). Peritumoural neutrophils negatively regulate adaptive immunity via the PD-L1/PD-1 signalling pathway in hepatocellular carcinoma. *Journal of Experimental & Clinical Cancer Research, 34*, 141. https://doi.org/10.1186/s13046-015-0256-0.

He, G., Zheng, C., Huo, H., Zhang, H., Zhu, Z., Li, J., et al. (2016). TACE combined with dendritic cells and cytokine-induced killer cells in the treatment of hepatocellular carcinoma: A meta-analysis. *International Immunopharmacology, 40*, 436–442. https://doi.org/10.1016/j.intimp.2016.09.015.

Hill, C., Guarner, F., Reid, G., Gibson, G. R., Merenstein, D. J., Pot, B., et al. (2014). Expert consensus document. The International Scientific Association for Probiotics and Prebiotics consensus statement on the scope and appropriate use of the term probiotic. *Nature Reviews. Gastroenterology & Hepatology, 11*(8), 506–514. https://doi.org/10.1038/nrgastro.2014.66.

Hirata, H., Sugimachi, K., Komatsu, H., Ueda, M., Masuda, T., Uchi, R., et al. (2016). Decreased expression of fructose-1,6-bisphosphatase associates with glucose metabolism and tumor progression in hepatocellular carcinoma. *Cancer Research, 76*(11), 3265–3276. https://doi.org/10.1158/0008-5472.CAN-15-2601.

Ho, R. H., Chan, J. C. Y., Fan, H., Kioh, D. Y. Q., Lee, B. W., & Chan, E. C. Y. (2017). In silico and in vitro interactions between short chain fatty acids and human histone deacetylases. *Biochemistry, 56*(36), 4871–4878. https://doi.org/10.1021/acs.biochem.7b00508.

Ho, V., Lim, T. S., Lee, J., Steinberg, J., Szmyd, R., Tham, M., et al. (2015). TLR3 agonist and Sorafenib combinatorial therapy promotes immune activation and controls hepatocellular carcinoma progression. *Oncotarget, 6*(29), 27252–27266. https://doi.org/10.18632/oncotarget.4583.

Ho, T. C., Wang, E. Y., Yeh, K. H., Jeng, Y. M., Horng, J. H., Wu, L. L., et al. (2020). Complement C1q mediates the expansion of periportal hepatic progenitor cells in senescence-associated inflammatory liver. *Proceedings of the National Academy of Sciences of the United States of America, 117*(12), 6717–6725. https://doi.org/10.1073/pnas.1918028117.

Hoyles, L., Fernandez-Real, J. M., Federici, M., Serino, M., Abbott, J., Charpentier, J., et al. (2018). Molecular phenomics and metagenomics of hepatic steatosis in non-diabetic obese women. *Nature Medicine*, *24*(7), 1070–1080. https://doi.org/10.1038/s41591-018-0061-3.

Hsieh, C. C., Hung, C. H., Chiang, M., Tsai, Y. C., & He, J. T. (2019). Hepatic stellate cells enhance liver cancer progression by inducing myeloid-derived suppressor cells through interleukin-6 signaling. *International Journal of Molecular Sciences*, *20*(20), 5079. https://doi.org/10.3390/ijms20205079.

Hu, M., Eviston, D., Hsu, P., Marino, E., Chidgey, A., Santner-Nanan, B., et al. (2019). Decreased maternal serum acetate and impaired fetal thymic and regulatory T cell development in preeclampsia. *Nature Communications*, *10*(1), 3031. https://doi.org/10.1038/s41467-019-10703-1.

Huang, X., Gan, G., Wang, X., Xu, T., & Xie, W. (2019). The HGF-MET axis coordinates liver cancer metabolism and autophagy for chemotherapeutic resistance. *Autophagy*, *15*(7), 1258–1279. https://doi.org/10.1080/15548627.2019.1580105.

Huang, X. Z., Gao, P., Song, Y. X., Xu, Y., Sun, J. X., Chen, X. W., et al. (2019). Antibiotic use and the efficacy of immune checkpoint inhibitors in cancer patients: A pooled analysis of 2740 cancer patients. *Oncoimmunology*, *8*(12). https://doi.org/10.1080/2162402X.2019.1665973, e1665973.

Huang, Z. M., Li, W., Li, S., Gao, F., Zhou, Q. M., Wu, F. M., et al. (2013). Cytokine-induced killer cells in combination with transcatheter arterial chemoembolization and radiofrequency ablation for hepatocellular carcinoma patients. *Journal of Immunotherapy*, *36*(5), 287–293. https://doi.org/10.1097/CJI.0b013e3182948452.

Huang, D., Li, T., Wang, L., Zhang, L., Yan, R., Li, K., et al. (2016). Hepatocellular carcinoma redirects to ketolysis for progression under nutrition deprivation stress. *Cell Research*, *26*(10), 1112–1130. https://doi.org/10.1038/cr.2016.109.

Huang, Q., Tan, Y., Yin, P., Ye, G., Gao, P., Lu, X., et al. (2013). Metabolic characterization of hepatocellular carcinoma using nontargeted tissue metabolomics. *Cancer Research*, *73*(16), 4992–5002. https://doi.org/10.1158/0008-5472.CAN-13-0308.

Hung, M. H., Chen, Y. L., Chen, L. J., Chu, P. Y., Hsieh, F. S., Tsai, M. H., et al. (2019). Canagliflozin inhibits growth of hepatocellular carcinoma via blocking glucose-influx-induced beta-catenin activation. *Cell Death & Disease*, *10*(6), 420. https://doi.org/10.1038/s41419-019-1646-6.

Hwang, N., Eom, T., Gupta, S. K., Jeong, S. Y., Jeong, D. Y., Kim, Y. S., et al. (2017). Genes and gut bacteria involved in luminal butyrate reduction caused by diet and loperamide. *Genes (Basel)*, *8*(12), 350. https://doi.org/10.3390/genes8120350.

Inada, Y., Mizukoshi, E., Seike, T., Tamai, T., Iida, N., Kitahara, M., et al. (2019). Characteristics of immune response to tumor-associated antigens and immune cell profile in patients with hepatocellular carcinoma. *Hepatology*, *69*(2), 653–665. https://doi.org/10.1002/hep.30212.

Iwata, T., Kondo, Y., Kimura, O., Morosawa, T., Fujisaka, Y., Umetsu, T., et al. (2016). PD-L1(+)MDSCs are increased in HCC patients and induced by soluble factor in the tumor microenvironment. *Scientific Reports*, *6*, 39296. https://doi.org/10.1038/srep39296.

Jang, C., Hui, S., Lu, W., Cowan, A. J., Morscher, R. J., Lee, G., et al. (2018). The small intestine converts dietary fructose into glucose and organic acids. *Cell Metabolism*, *27*(2), 351–361 e353. https://doi.org/10.1016/j.cmet.2017.12.016.

Jang, C., Wada, S., Yang, S., Gosis, B., Zeng, X., Zhang, Z., et al. (2020). The small intestine shields the liver from fructose-induced steatosis. *Nature Metabolism*, *2*(7), 586–593. https://doi.org/10.1038/s42255-020-0222-9.

Ji, J., Yin, Y., Ju, H., Xu, X., Liu, W., Fu, Q., et al. (2018). Long non-coding RNA Lnc-Tim3 exacerbates CD8 T cell exhaustion via binding to Tim-3 and inducing nuclear translocation of Bat3 in HCC. *Cell Death & Disease*, *9*(5), 478. https://doi.org/10.1038/s41419-018-0528-7.

Jia, C. C., Chen, Y. H., Cai, X. R., Li, Y., Zheng, X. F., Yao, Z. C., et al. (2019). Efficacy of cytokine-induced killer cell-based immunotherapy for hepatocellular carcinoma. *American Journal of Cancer Research*, *9*(6), 1254–1265. Retrieved from https://www.ncbi.nlm.nih.gov/pubmed/31285957.

Jiang, W., Li, G., Li, W., Wang, P., Xiu, P., Jiang, X., et al. (2018). Sodium orthovanadate overcomes sorafenib resistance of hepatocellular carcinoma cells by inhibiting Na(+)/K (+)-ATPase activity and hypoxia-inducible pathways. *Scientific Reports*, *8*(1), 9706. https://doi.org/10.1038/s41598-018-28010-y.

Jiang, H., Yang, Z., Song, Z., Green, M., Song, H., & Shao, Q. (2019). gammadelta T cells in hepatocellular carcinoma patients present cytotoxic activity but are reduced in potency due to IL-2 and IL-21 pathways. *International Immunopharmacology*, *70*, 167–173. https://doi.org/10.1016/j.intimp.2019.02.019.

Jin, Y., Dong, H., Xia, L., Yang, Y., Zhu, Y., Shen, Y., et al. (2019). The diversity of gut microbiome is associated with favorable responses to anti-programmed death 1 immunotherapy in chinese patients with NSCLC. *Journal of Thoracic Oncology*, *14*(8), 1378–1389. https://doi.org/10.1016/j.jtho.2019.04.007.

Jin, Z., Lei, L., Lin, D., Liu, Y., Song, Y., Gong, H., et al. (2018). IL-33 released in the liver inhibits tumor growth via promotion of CD4(+) and CD8(+) T cell responses in hepatocellular carcinoma. *Journal of Immunology*, *201*(12), 3770–3779. https://doi.org/10.4049/jimmunol.1800627.

Jin, X., Pan, Y., Wang, L., Zhang, L., Ravichandran, R., Potts, P. R., et al. (2017). MAGE-TRIM28 complex promotes the Warburg effect and hepatocellular carcinoma progression by targeting FBP1 for degradation. *Oncogene*, *6*(4), e312. https://doi.org/10.1038/oncsis.2017.21.

Jo, H., Lee, J., Jeon, J., Kim, S. Y., Chung, J. I., Ko, H. Y., et al. (2020). The critical role of glucose deprivation in epithelial-mesenchymal transition in hepatocellular carcinoma under hypoxia. *Scientific Reports*, *10*(1), 1538. https://doi.org/10.1038/s41598-020-58124-1.

Jones, R. B., Alderete, T. L., Kim, J. S., Millstein, J., Gilliland, F. D., & Goran, M. I. (2019). High intake of dietary fructose in overweight/obese teenagers associated with depletion of Eubacterium and Streptococcus in gut microbiome. *Gut Microbes*, *10*(6), 712–719. https://doi.org/10.1080/19490976.2019.1592420.

Juanola, O., Pinero, P., Gomez-Hurtado, I., Caparros, E., Garcia-Villalba, R., Marin, A., et al. (2018). Regulatory T cells restrict permeability to bacterial antigen translocation and preserve short-chain fatty acids in experimental cirrhosis. *Hepatology Communications*, *2*(12), 1610–1623. https://doi.org/10.1002/hep4.1268.

June, C. H., & Sadelain, M. (2018). Chimeric antigen receptor therapy. *The New England Journal of Medicine*, *379*(1), 64–73. https://doi.org/10.1056/NEJMra1706169.

Jung, N. C., Lee, J. H., Choi, H. J., Hwang, S. U., Song, J. Y., Seo, H. G., et al. (2016). Dendritic cell immunotherapy combined with cytokine-induced killer cells effectively suppresses established hepatocellular carcinomas in mice. *Immunological Investigations*, *45*(6), 553–565. https://doi.org/10.1080/08820139.2016.1183025.

Kaisar, M. M. M., Pelgrom, L. R., van der Ham, A. J., Yazdanbakhsh, M., & Everts, B. (2017). Butyrate conditions human dendritic cells to prime type 1 regulatory T cells via both histone deacetylase inhibition and G protein-coupled receptor 109A signaling. *Frontiers in Immunology*, *8*, 1429. https://doi.org/10.3389/fimmu.2017.01429.

Kalantari, H., Asadmasjedi, N., Abyaz, M. R., Mahdavinia, M., & Mohammadtaghvaei, N. (2019). Protective effect of inulin on methotrexate-induced liver toxicity in mice. *Biomedicine & Pharmacotherapy*, *110*, 943–950. https://doi.org/10.1016/j.biopha.2018.11.144.

Kalathil, S. G., Hutson, A., Barbi, J., Iyer, R., & Thanavala, Y. (2019). Augmentation of IFN-gamma+ CD8+ T cell responses correlates with survival of HCC patients on sorafenib therapy. *JCI Insight*, *4*(15). https://doi.org/10.1172/jci.insight.130116.

Kalathil, S., Lugade, A. A., Miller, A., Iyer, R., & Thanavala, Y. (2013). Higher frequencies of GARP(+)CTLA-4(+)Foxp3(+) T regulatory cells and myeloid-derived suppressor cells in hepatocellular carcinoma patients are associated with impaired T-cell functionality. *Cancer Research*, *73*(8), 2435–2444. https://doi.org/10.1158/0008-5472.CAN-12-3381.

Kang, T. W., Yevsa, T., Woller, N., Hoenicke, L., Wuestefeld, T., Dauch, D., et al. (2011). Senescence surveillance of pre-malignant hepatocytes limits liver cancer development. *Nature*, *479*(7374), 547–551. https://doi.org/10.1038/nature10599.

Karatepe, O., Acet, E., Battal, M., Adas, G., Kemik, A., Altiok, M., et al. (2010). Effects of glutamine and curcumin on bacterial translocation in jaundiced rats. *World Journal of Gastroenterology*, *16*(34), 4313–4320. https://doi.org/10.3748/wjg.v16.i34.4313.

Karmarkar, D., & Rock, K. L. (2013). Microbiota signalling through MyD88 is necessary for a systemic neutrophilic inflammatory response. *Immunology*, *140*(4), 483–492. https://doi.org/10.1111/imm.12159.

Kawamoto, S., Maruya, M., Kato, L. M., Suda, W., Atarashi, K., Doi, Y., et al. (2014). Foxp3 (+) T cells regulate immunoglobulin a selection and facilitate diversification of bacterial species responsible for immune homeostasis. *Immunity*, *41*(1), 152–165. https://doi.org/10.1016/j.immuni.2014.05.016.

Kawarabayashi, N., Seki, S., Hatsuse, K., Ohkawa, T., Koike, Y., Aihara, T., et al. (2000). Decrease of CD56(+)T cells and natural killer cells in cirrhotic livers with hepatitis C may be involved in their susceptibility to hepatocellular carcinoma. *Hepatology*, *32*(5), 962–969. https://doi.org/10.1053/jhep.2000.19362.

Ke, X., Chen, C., Song, Y., Cai, Q., Li, J., Tang, Y., et al. (2019). Hypoxia modifies the polarization of macrophages and their inflammatory microenvironment, and inhibits malignant behavior in cancer cells. *Oncology Letters*, *18*(6), 5871–5878. https://doi.org/10.3892/ol.2019.10956.

Kehrmann, J., Effenberg, L., Wilk, C., Schoemer, D., Ngo Thi Phuong, N., Adamczyk, A., et al. (2020). Depletion of Foxp3(+) regulatory T cells is accompanied by an increase in the relative abundance of Firmicutes in the murine gut microbiome. *Immunology*, *159*(3), 344–353. https://doi.org/10.1111/imm.13158.

Kim, M. H., & Kim, H. (2017). The roles of glutamine in the intestine and its implication in intestinal diseases. *International Journal of Molecular Sciences*, *18*(5), 1051. https://doi.org/10.3390/ijms18051051.

Kim, M., Qie, Y., Park, J., & Kim, C. H. (2016). Gut microbial metabolites fuel host antibody responses. *Cell Host & Microbe*, *20*(2), 202–214. https://doi.org/10.1016/j.chom.2016.07.001.

Kohler, A., Delbauve, S., Smout, J., Torres, D., & Flamand, V. (2020). Very early-life exposure to microbiota-induced TNF drives the maturation of neonatal pre-cDC1. *Gut*. https://doi.org/10.1136/gutjnl-2019-319700.

Kong, L., Zhou, Y., Bu, H., Lv, T., Shi, Y., & Yang, J. (2016). Deletion of interleukin-6 in monocytes/macrophages suppresses the initiation of hepatocellular carcinoma in mice. *Journal of Experimental & Clinical Cancer Research*, *35*(1), 131. https://doi.org/10.1186/s13046-016-0412-1.

Konturek, P. C., Harsch, I. A., Konturek, K., Schink, M., Konturek, T., Neurath, M. F., et al. (2018). Gut(-)liver axis: How do gut bacteria influence the liver? *Medical Science*, *6*(3), 79. https://doi.org/10.3390/medsci6030079.

Kouakanou, L., Peters, C., Sun, Q., Floess, S., Bhat, J., Huehn, J., et al. (2020). Vitamin C supports conversion of human gammadelta T cells into FOXP3-expressing regulatory cells by epigenetic regulation. *Scientific Reports*, *10*(1), 6550. https://doi.org/10.1038/s41598-020-63572-w.

Krause, G. C., Lima, K. G., Haute, G. V., Schuster, A. D., Dias, H. B., Mesquita, F. C., et al. (2017). Fructose-1,6-bisphosphate decreases IL-8 levels and increases the activity of pro-apoptotic proteins in HepG2 cells. *Biomedicine & Pharmacotherapy*, *89*, 358–365. https://doi.org/10.1016/j.biopha.2017.01.178.

Krenkel, O., & Tacke, F. (2017). Liver macrophages in tissue homeostasis and disease. *Nature Reviews. Immunology*, *17*(5), 306–321. https://doi.org/10.1038/nri.2017.11.

Kronenberg, M., & Gapin, L. (2002). The unconventional lifestyle of NKT cells. *Nature Reviews. Immunology*, *2*(8), 557–568. https://doi.org/10.1038/nri854.

Kuang, D. M., Zhao, Q., Wu, Y., Peng, C., Wang, J., Xu, Z., et al. (2011). Peritumoral neutrophils link inflammatory response to disease progression by fostering angiogenesis in hepatocellular carcinoma. *Journal of Hepatology*, *54*(5), 948–955. https://doi.org/10.1016/j.jhep.2010.08.041.

Kubes, P., & Jenne, C. (2018). Immune responses in the liver. *Annual Review of Immunology*, *36*, 247–277. https://doi.org/10.1146/annurev-immunol-051116-052415.

Kumamoto, R., Uto, H., Oda, K., Ibusuki, R., Tanoue, S., Arima, S., et al. (2013). Dietary fructose enhances the incidence of precancerous hepatocytes induced by administration of diethylnitrosamine in rat. *European Journal of Medical Research*, *18*, 54. https://doi.org/10.1186/2047-783X-18-54.

Kuo, T. C., Chen, C. K., Hua, K. T., Yu, P., Lee, W. J., Chen, M. W., et al. (2016). Glutaminase 2 stabilizes Dicer to repress Snail and metastasis in hepatocellular carcinoma cells. *Cancer Letters*, *383*(2), 282–294. https://doi.org/10.1016/j.canlet.2016.10.012.

Kwong, W. K., Zheng, H., & Moran, N. A. (2017). Convergent evolution of a modified, acetate-driven TCA cycle in bacteria. *Nature Microbiology*, *2*. https://doi.org/10.1038/nmicrobiol.2017.67, 17067.

Laguna, J. C., Alegret, M., & Roglans, N. (2014). Simple sugar intake and hepatocellular carcinoma: Epidemiological and mechanistic insight. *Nutrients*, *6*(12), 5933–5954. https://doi.org/10.3390/nu6125933.

Lampropoulou, V., Sergushichev, A., Bambouskova, M., Nair, S., Vincent, E. E., Loginicheva, E., et al. (2016). Itaconate links inhibition of succinate dehydrogenase with macrophage metabolic remodeling and regulation of inflammation. *Cell Metabolism*, *24*(1), 158–166. https://doi.org/10.1016/j.cmet.2016.06.004.

Lanaya, H., Natarajan, A., Komposch, K., Li, L., Amberg, N., Chen, L., et al. (2014). EGFR has a tumour-promoting role in liver macrophages during hepatocellular carcinoma formation. *Nature Cell Biology*, *16*(10), 972–977. https://doi.org/10.1038/ncb3031.

Langhans, B., Nischalke, H. D., Kramer, B., Dold, L., Lutz, P., Mohr, R., et al. (2019). Role of regulatory T cells and checkpoint inhibition in hepatocellular carcinoma. *Cancer Immunology, Immunotherapy*, *68*(12), 2055–2066. https://doi.org/10.1007/s00262-019-02427-4.

Leboeuf, C., Mailly, L., Wu, T., Bour, G., Durand, S., Brignon, N., et al. (2014). In vivo proof of concept of adoptive immunotherapy for hepatocellular carcinoma using allogeneic suicide gene-modified killer cells. *Molecular Therapy*, *22*(3), 634–644. https://doi.org/10.1038/mt.2013.277.

Lechner, S., Yee, M., Limketkai, B. N., & Pham, E. A. (2020). Fecal microbiota transplantation for chronic liver diseases: Current understanding and future direction. *Digestive Diseases and Sciences*, *65*(3), 897–905. https://doi.org/10.1007/s10620-020-06100-0.

Lee, H. J., & Cha, J. Y. (2018). Recent insights into the role of ChREBP in intestinal fructose absorption and metabolism. *BMB Reports*, *51*(9), 429–436. Retrieved from https://www.ncbi.nlm.nih.gov/pubmed/30158026.

Lee, S., Koh, J., Chang, Y., Kim, H. Y., & Chung, D. H. (2019). Invariant NKT Cells functionally link microbiota-induced butyrate production and joint inflammation. *Journal of Immunology*, *203*(12), 3199–3208. https://doi.org/10.4049/jimmunol.1801314.

Lee, J. H., Lee, J. H., Lim, Y. S., Yeon, J. E., Song, T. J., Yu, S. J., et al. (2015). Adjuvant immunotherapy with autologous cytokine-induced killer cells for hepatocellular carcinoma. *Gastroenterology*, *148*(7), 1383–1391 e1386. https://doi.org/10.1053/j.gastro.2015.02.055.

Lee, J. H., Lee, J. H., Lim, Y. S., Yeon, J. E., Song, T. J., Yu, S. J., et al. (2019). Sustained efficacy of adjuvant immunotherapy with cytokine-induced killer cells for hepatocellular carcinoma: An extended 5-year follow-up. *Cancer Immunology, Immunotherapy, 68*(1), 23–32. https://doi.org/10.1007/s00262-018-2247-4.

Legut, M., Cole, D. K., & Sewell, A. K. (2015). The promise of gammadelta T cells and the gammadelta T cell receptor for cancer immunotherapy. *Cellular & Molecular Immunology, 12*(6), 656–668. https://doi.org/10.1038/cmi.2015.28.

Li, L., Che, L., Tharp, K. M., Park, H. M., Pilo, M. G., Cao, D., et al. (2016). Differential requirement for de novo lipogenesis in cholangiocarcinoma and hepatocellular carcinoma of mice and humans. *Hepatology, 63*(6), 1900–1913. https://doi.org/10.1002/hep.28508.

Li, X. F., Chen, D. P., Ouyang, F. Z., Chen, M. M., Wu, Y., Kuang, D. M., et al. (2015). Increased autophagy sustains the survival and pro-tumourigenic effects of neutrophils in human hepatocellular carcinoma. *Journal of Hepatology, 62*(1), 131–139. https://doi.org/10.1016/j.jhep.2014.08.023.

Li, F., Hao, X., Chen, Y., Bai, L., Gao, X., Lian, Z., et al. (2017). The microbiota maintain homeostasis of liver-resident gammadeltaT-17 cells in a lipid antigen/CD1d-dependent manner. *Nature Communications*, 7. https://doi.org/10.1038/ncomms13839, 13839.

Li, J., Jia, H., Cai, X., Zhong, H., Feng, Q., Sunagawa, S., et al. (2014). An integrated catalog of reference genes in the human gut microbiome. *Nature Biotechnology, 32*(8), 834–841. https://doi.org/10.1038/nbt.2942.

Li, B. H., Jiang, W., Zhang, S., Huang, N., Sun, J., Yang, J., et al. (2020). The spleen contributes to the increase in PMN-MDSCs in orthotopic H22 hepatoma mice. *Molecular Immunology, 125*, 95–103. https://doi.org/10.1016/j.molimm.2020.07.002.

Li, J., Lee, Y., Li, Y., Jiang, Y., Lu, H., Zang, W., et al. (2018). Co-inhibitory molecule B7 superfamily member 1 expressed by tumor-infiltrating myeloid cells induces dysfunction of anti-tumor CD8(+) T cells. *Immunity, 48*(4), 773–786 e775. https://doi.org/10.1016/j.immuni.2018.03.018.

Li, J., Liang, N., Long, X., Zhao, J., Yang, J., Du, X., et al. (2019). SDHC-related deficiency of SDH complex activity promotes growth and metastasis of hepatocellular carcinoma via ROS/NFkappaB signaling. *Cancer Letters, 461*, 44–55. https://doi.org/10.1016/j.canlet.2019.07.001.

Li, Q., Ma, L., Shen, S., Guo, Y., Cao, Q., Cai, X., et al. (2019). Intestinal dysbacteriosis-induced IL-25 promotes development of HCC via alternative activation of macrophages in tumor microenvironment. *Journal of Experimental & Clinical Cancer Research, 38*(1), 303. https://doi.org/10.1186/s13046-019-1271-3.

Li, X., Peng, J., Pang, Y., Yu, S., Yu, X., Chen, P., et al. (2015). Identification of a FOXP3 (+)CD3(+)CD56(+) population with immunosuppressive function in cancer tissues of human hepatocellular carcinoma. *Scientific Reports, 5*, 14757. https://doi.org/10.1038/srep14757.

Li, X., Qian, X., Peng, L. X., Jiang, Y., Hawke, D. H., Zheng, Y., et al. (2016). A splicing switch from ketohexokinase-C to ketohexokinase-A drives hepatocellular carcinoma formation. *Nature Cell Biology, 18*(5), 561–571. https://doi.org/10.1038/ncb3338.

Li, J., Sung, C. Y., Lee, N., Ni, Y., Pihlajamaki, J., Panagiotou, G., et al. (2016). Probiotics modulated gut microbiota suppresses hepatocellular carcinoma growth in mice. *Proceedings of the National Academy of Sciences of the United States of America, 113*(9), E1306–E1315. https://doi.org/10.1073/pnas.1518189113.

Li, S., Wang, N., Tan, H. Y., Chueng, F., Zhang, Z. J., Yuen, M. F., et al. (2020). Modulation of gut microbiota mediates berberine-induced expansion of immuno-suppressive cells to against alcoholic liver disease. *Clinical and Translational Medicine, 10*(4), e112. https://doi.org/10.1002/ctm2.112.

Li, C., Wen, T. F., Yan, L. N., Li, B., Wang, W. T., Yang, J. Y., et al. (2015). Postoperative neutrophil-to-lymphocyte ratio plus platelet-to-lymphocyte ratio predicts the outcomes of hepatocellular carcinoma. *The Journal of Surgical Research, 198*(1), 73–79. https://doi.org/10.1016/j.jss.2015.05.003.

Li, X., Xing, Y. F., Lei, A. H., Xiao, Q., Lin, Z. H., Hong, Y. F., et al. (2017). Neutrophil count is associated with myeloid derived suppressor cell level and presents prognostic value of for hepatocellular carcinoma patients. *Oncotarget, 8*(15), 24380–24388. https://doi.org/10.18632/oncotarget.15456.

Li, L., Xu, L., Yan, J., Zhen, Z. J., Ji, Y., Liu, C. Q., et al. (2015). CXCR2-CXCL1 axis is correlated with neutrophil infiltration and predicts a poor prognosis in hepatocellular carcinoma. *Journal of Experimental & Clinical Cancer Research, 34*, 129. https://doi.org/10.1186/s13046-015-0247-1.

Li, X., Yao, W., Yuan, Y., Chen, P., Li, B., Li, J., et al. (2017). Targeting of tumour-infiltrating macrophages via CCL2/CCR2 signalling as a therapeutic strategy against hepatocellular carcinoma. *Gut, 66*(1), 157–167. https://doi.org/10.1136/gutjnl-2015-310514.

Li, L., & Ye, J. (2020). Characterization of gut microbiota in patients with primary hepatocellular carcinoma received immune checkpoint inhibitors: A Chinese population-based study. *Medicine (Baltimore), 99*(37), e21788. https://doi.org/10.1097/MD.0000000000021788.

Li, B., Zhang, S., Huang, N., Chen, H., Wang, P., Yang, J., et al. (2019). CCL9/CCR1 induces myeloidderived suppressor cell recruitment to the spleen in a murine H22 orthotopic hepatoma model. *Oncology Reports, 41*(1), 608–618. https://doi.org/10.3892/or.2018.6809.

Li, F. J., Zhang, Y., Jin, G. X., Yao, L., & Wu, D. Q. (2013). Expression of LAG-3 is coincident with the impaired effector function of HBV-specific CD8(+) T cell in HCC patients. *Immunology Letters, 150*(1–2), 116–122. https://doi.org/10.1016/j.imlet.2012.12.004.

Li, X., Zhang, Y., Ma, W., Fu, Q., Liu, J., Yin, G., et al. (2020). Enhanced glucose metabolism mediated by CD147 contributes to immunosuppression in hepatocellular carcinoma. *Cancer Immunology, Immunotherapy, 69*(4), 535–548. https://doi.org/10.1007/s00262-019-02457-y.

Li, R., Zhou, R., Wang, H., Li, W., Pan, M., Yao, X., et al. (2019). Gut microbiota-stimulated cathepsin K secretion mediates TLR4-dependent M2 macrophage polarization and promotes tumor metastasis in colorectal cancer. *Cell Death and Differentiation, 26*(11), 2447–2463. https://doi.org/10.1038/s41418-019-0312-y.

Lin, H., An, Y., Tang, H., & Wang, Y. (2019). Alterations of bile acids and gut microbiota in obesity induced by high fat diet in rat model. *Journal of Agricultural and Food Chemistry, 67*(13), 3624–3632. https://doi.org/10.1021/acs.jafc.9b00249.

Lin, D., Lei, L., Liu, Y., Zhang, Y., Hu, B., Bao, G., et al. (2016). Membrane IL1α inhibits the development of hepatocellular carcinoma via promoting T- and NK-cell activation. *Cancer Research, 76*(11), 3179–3188. https://doi.org/10.1158/0008-5472.CAN-15-2658.

Lin, G., Wang, J., Lao, X., Wang, J., Li, L., Li, S., et al. (2012). Interleukin-6 inhibits regulatory T cells and improves the proliferation and cytotoxic activity of cytokine-induced killer cells. *Journal of Immunotherapy, 35*(4), 337–343. https://doi.org/10.1097/CJI.0b013e318255ada3.

Lindner, S. E., Johnson, S. M., Brown, C. E., & Wang, L. D. (2020). Chimeric antigen receptor signaling: Functional consequences and design implications. *Science Advances, 6*(21), eaaz3223. https://doi.org/10.1126/sciadv.aaz3223.

Liu, Y., Butterfield, L. H., Fu, X., Song, Z., Zhang, X., Lu, C., et al. (2011). Lentivirally engineered dendritic cells activate AFP-specific T cells which inhibit hepatocellular carcinoma growth in vitro and in vivo. *International Journal of Oncology, 39*(1), 245–253. https://doi.org/10.3892/ijo.2011.1004.

Liu, Y., Di, S., Shi, B., Zhang, H., Wang, Y., Wu, X., et al. (2019). Armored inducible expression of IL-12 enhances antitumor activity of glypican-3-targeted chimeric antigen receptor-engineered t cells in hepatocellular carcinoma. *Journal of Immunology, 203*(1), 198–207. https://doi.org/10.4049/jimmunol.1800033.

Liu, Q., Du, F., Huang, W., Ding, X., Wang, Z., Yan, F., et al. (2019). Epigenetic control of Foxp3 in intratumoral T-cells regulates growth of hepatocellular carcinoma. *Aging (Albany NY), 11*(8), 2343–2351. https://doi.org/10.18632/aging.101918.

Liu, J., Fu, Y., Zhang, H., Wang, J., Zhu, J., Wang, Y., et al. (2017). The hepatoprotective effect of the probiotic Clostridium butyricum against carbon tetrachloride-induced acute liver damage in mice. *Food & Function, 8*(11), 4042–4052. https://doi.org/10.1039/c7fo00355b.

Liu, T., Guo, Z., Song, X., Liu, L., Dong, W., Wang, S., et al. (2020). High-fat diet-induced dysbiosis mediates MCP-1/CCR2 axis-dependent M2 macrophage polarization and promotes intestinal adenoma-adenocarcinoma sequence. *Journal of Cellular and Molecular Medicine, 24*(4), 2648–2662. https://doi.org/10.1111/jcmm.14984.

Liu, M. X., Jin, L., Sun, S. J., Liu, P., Feng, X., Cheng, Z. L., et al. (2018). Metabolic reprogramming by PCK1 promotes TCA cataplerosis, oxidative stress and apoptosis in liver cancer cells and suppresses hepatocellular carcinoma. *Oncogene, 37*(12), 1637–1653. https://doi.org/10.1038/s41388-017-0070-6.

Liu, R., Li, Y., Tian, L., Shi, H., Wang, J., Liang, Y., et al. (2019). Gankyrin drives metabolic reprogramming to promote tumorigenesis, metastasis and drug resistance through activating beta-catenin/c-Myc signaling in human hepatocellular carcinoma. *Cancer Letters, 443*, 34–46. https://doi.org/10.1016/j.canlet.2018.11.030.

Liu, X., Li, M., Wang, X., Dang, Z., Jiang, Y., Wang, X., et al. (2019). PD-1(+) TIGIT(+) CD8(+) T cells are associated with pathogenesis and progression of patients with hepatitis B virus-related hepatocellular carcinoma. *Cancer Immunology, Immunotherapy, 68*(12), 2041–2054. https://doi.org/10.1007/s00262-019-02426-5.

Liu, G. M., Li, Q., Zhang, P. F., Shen, S. L., Xie, W. X., Chen, B., et al. (2018). Restoration of FBP1 suppressed Snail-induced epithelial to mesenchymal transition in hepatocellular carcinoma. *Cell Death & Disease, 9*(11), 1132. https://doi.org/10.1038/s41419-018-1165-x.

Litvak, Y., Byndloss, M. X., & Baumler, A. J. (2018). Colonocyte metabolism shapes the gut microbiota. *Science, 362*(6418), eaat9076. https://doi.org/10.1126/science.aat9076.

Liu, F., Liu, Y., & Chen, Z. (2018). Tim-3 expression and its role in hepatocellular carcinoma. *Journal of Hematology & Oncology, 11*(1), 126. https://doi.org/10.1186/s13045-018-0667-4.

Liu, J., Lu, J. F., Wen, X. Y., Kan, J., & Jin, C. H. (2015). Antioxidant and protective effect of inulin and catechin grafted inulin against CCl4-induced liver injury. *International Journal of Biological Macromolecules, 72*, 1479–1484. https://doi.org/10.1016/j.ijbiomac.2014.09.066.

Liu, Z., Suo, C., Mao, X., Jiang, Y., Jin, L., Zhang, T., et al. (2020). Global incidence trends in primary liver cancer by age at diagnosis, sex, region, and etiology, 1990-2017. *Cancer, 126*(10), 2267–2278. https://doi.org/10.1002/cncr.32789.

Liu, Y. T., Tseng, T. C., Soong, R. S., Peng, C. Y., Cheng, Y. H., Huang, S. F., et al. (2018). A novel spontaneous hepatocellular carcinoma mouse model for studying T-cell exhaustion in the tumor microenvironment. *Journal for Immunotherapy of Cancer, 6*(1), 144. https://doi.org/10.1186/s40425-018-0462-3.

Liu, R. X., Wei, Y., Zeng, Q. H., Chan, K. W., Xiao, X., Zhao, X. Y., et al. (2015). Chemokine (C-X-C motif) receptor 3-positive B cells link interleukin-17 inflammation to protumorigenic macrophage polarization in human hepatocellular carcinoma. *Hepatology, 62*(6), 1779–1790. https://doi.org/10.1002/hep.28020.

Liu, X., Wen, J., Yi, H., Hou, X., Yin, Y., Ye, G., et al. (2020). Split chimeric antigen receptor-modified T cells targeting glypican-3 suppress hepatocellular carcinoma growth with reduced cytokine release. *Therapeutic Advances in Medical Oncology*, *12*. https://doi.org/10.1177/1758835920910347, 1758835920910347.

Liu, T., Wu, H., Qi, J., Qin, C., & Zhu, Q. (2020). Seven immune-related genes prognostic power and correlation with tumor-infiltrating immune cells in hepatocellular carcinoma. *Cancer Medicine*, *9*, 7440–7452. https://doi.org/10.1002/cam4.3406.

Liu, M., Zhou, J., Liu, X., Feng, Y., Yang, W., Wu, F., et al. (2020). Targeting monocyte-intrinsic enhancer reprogramming improves immunotherapy efficacy in hepatocellular carcinoma. *Gut*, *69*(2), 365–379. https://doi.org/10.1136/gutjnl-2018-317257.

Llovet, J. M., Ricci, S., Mazzaferro, V., Hilgard, P., Gane, E., Blanc, J. F., et al. (2008). Sorafenib in advanced hepatocellular carcinoma. *The New England Journal of Medicine*, *359*(4), 378–390. https://doi.org/10.1056/NEJMoa0708857.

Long, L., Zhang, X., Chen, F., Pan, Q., Phiphatwatchara, P., Zeng, Y., et al. (2018). The promising immune checkpoint LAG-3: From tumor microenvironment to cancer immunotherapy. *Genes & Cancer*, *9*(5–6), 176–189. https://doi.org/10.18632/genesandcancer.180.

Loo, T. M., Kamachi, F., Watanabe, Y., Yoshimoto, S., Kanda, H., Arai, Y., et al. (2017). Gut microbiota promotes obesity-associated liver cancer through PGE2-mediated suppression of antitumor immunity. *Cancer Discovery*, 7(5), 522–538. https://doi.org/10.1158/2159-8290.CD-16-0932.

Lopez de la Oliva, A. R., Campos-Sandoval, J. A., Gomez-Garcia, M. C., Cardona, C., Martin-Rufian, M., Sialana, F. J., et al. (2020). Nuclear translocation of glutaminase GLS2 in human cancer cells associates with proliferation arrest and differentiation. *Scientific Reports*, *10*(1), 2259. https://doi.org/10.1038/s41598-020-58264-4.

Lu, Y. X., Yu, X. C., & Zhu, M. Y. (2014). Antitumor effect of fructose 1,6-bisphosphate and its mechanism in hepatocellular carcinoma cells. *Tumour Biology*, *35*(2), 1679–1685. https://doi.org/10.1007/s13277-013-1231-z.

Lu, Z., Zuo, B., Jing, R., Gao, X., Rao, Q., Liu, Z., et al. (2017). Dendritic cell-derived exosomes elicit tumor regression in autochthonous hepatocellular carcinoma mouse models. *Journal of Hepatology*, *67*(4), 739–748. https://doi.org/10.1016/j.jhep.2017.05.019.

Luo, W., Xu, Q., Wang, Q., Wu, H., & Hua, J. (2017). Effect of modulation of PPAR-gamma activity on Kupffer cells M1/M2 polarization in the development of non-alcoholic fatty liver disease. *Scientific Reports*, 7. https://doi.org/10.1038/srep44612, 44612.

Luu, M., Weigand, K., Wedi, F., Breidenbend, C., Leister, H., Pautz, S., et al. (2018). Regulation of the effector function of CD8(+) T cells by gut microbiota-derived metabolite butyrate. *Scientific Reports*, *8*(1). https://doi.org/10.1038/s41598-018-32860-x, 14430.

Lynch, S. V., & Pedersen, O. (2016). The human intestinal microbiome in health and disease. *The New England Journal of Medicine*, *375*(24), 2369–2379. https://doi.org/10.1056/NEJMra1600266.

Ma, S., Cheng, Q., Cai, Y., Gong, H., Wu, Y., Yu, X., et al. (2014). IL-17A produced by gammadelta T cells promotes tumor growth in hepatocellular carcinoma. *Cancer Research*, *74*(7), 1969–1982. https://doi.org/10.1158/0008-5472.CAN-13-2534.

Ma, C., Han, M., Heinrich, B., Fu, Q., Zhang, Q., Sandhu, M., et al. (2018). Gut microbiome-mediated bile acid metabolism regulates liver cancer via NKT cells. *Science*, *360*(6391), eaan5931. https://doi.org/10.1126/science.aan5931.

Ma, J., Zheng, B., Goswami, S., Meng, L., Zhang, D., Cao, C., et al. (2019). PD1(Hi) CD8 (+) T cells correlate with exhausted signature and poor clinical outcome in hepatocellular carcinoma. *Journal for Immunotherapy of Cancer*, 7(1), 331. https://doi.org/10.1186/s40425-019-0814-7.

Macpherson, A. J., Heikenwalder, M., & Ganal-Vonarburg, S. C. (2016). The liver at the nexus of host-microbial interactions. *Cell Host & Microbe*, *20*(5), 561–571. https://doi.org/10.1016/j.chom.2016.10.016.

Madsen, M. S. A., Holm, J. B., Palleja, A., Wismann, P., Fabricius, K., Rigbolt, K., et al. (2019). Metabolic and gut microbiome changes following GLP-1 or dual GLP-1/GLP-2 receptor agonist treatment in diet-induced obese mice. *Scientific Reports*, *9*(1). https://doi.org/10.1038/s41598-019-52103-x, 15582.

Mager, L. F., Burkhard, R., Pett, N., Cooke, N. C. A., Brown, K., Ramay, H., et al. (2020). Microbiome-derived inosine modulates response to checkpoint inhibitor immunotherapy. *Science*, *369*(6510), 1481–1489. https://doi.org/10.1126/science.abc3421.

Mallavia, B., Liu, F., Lefrancais, E., Cleary, S. J., Kwaan, N., Tian, J. J., et al. (2020). Mitochondrial DNA stimulates TLR9-dependent neutrophil extracellular trap formation in primary graft dysfunction. *American Journal of Respiratory Cell and Molecular Biology*, *62*(3), 364–372. https://doi.org/10.1165/rcmb.2019-0140OC.

Mamedova, L. K., Yuan, K., Laudick, A. N., Fleming, S. D., Mashek, D. G., & Bradford, B. J. (2013). Toll-like receptor 4 signaling is required for induction of gluconeogenic gene expression by palmitate in human hepatic carcinoma cells. *The Journal of Nutritional Biochemistry*, *24*(8), 1499–1507. https://doi.org/10.1016/j.jnutbio.2012.12.009.

Mano, Y., Shirabe, K., Yamashita, Y., Harimoto, N., Tsujita, E., Takeishi, K., et al. (2013). Preoperative neutrophil-to-lymphocyte ratio is a predictor of survival after hepatectomy for hepatocellular carcinoma: A retrospective analysis. *Annals of Surgery*, *258*(2), 301–305. https://doi.org/10.1097/SLA.0b013e318297ad6b.

Margaritis, V. G., Filos, K. S., Michalaki, M. A., Scopa, C. D., Spiliopoulou, I., Nikolopoulou, V. N., et al. (2005). Effect of oral glutamine administration on bacterial tanslocation, endotoxemia, liver and ileal morphology, and apoptosis in rats with obstructive jaundice. *World Journal of Surgery*, *29*(10), 1329–1334. https://doi.org/10.1007/s00268-005-7721-4.

Margetts, J., Ogle, L. F., Chan, S. L., Chan, A. W. H., Chan, K. C. A., Jamieson, D., et al. (2018). Neutrophils: Driving progression and poor prognosis in hepatocellular carcinoma? *British Journal of Cancer*, *118*(2), 248–257. https://doi.org/10.1038/bjc.2017.386.

Martin, C. R., Osadchiy, V., Kalani, A., & Mayer, E. A. (2018). The brain-gut-microbiome axis. *Cellular and Molecular Gastroenterology and Hepatology*, *6*(2), 133–148. https://doi.org/10.1016/j.jcmgh.2018.04.003.

Martin, A. M., Yabut, J. M., Choo, J. M., Page, A. J., Sun, E. W., Jessup, C. F., et al. (2019). The gut microbiome regulates host glucose homeostasis via peripheral serotonin. *Proceedings of the National Academy of Sciences of the United States of America*, *116*(40), 19802–19804. https://doi.org/10.1073/pnas.1909311116.

Martin-Gallausiaux, C., Beguet-Crespel, F., Marinelli, L., Jamet, A., Ledue, F., Blottiere, H. M., et al. (2018). Butyrate produced by gut commensal bacteria activates TGF-beta1 expression through the transcription factor SP1 in human intestinal epithelial cells. *Scientific Reports*, *8*(1), 9742. https://doi.org/10.1038/s41598-018-28048-y.

Mayengbam, S., Lambert, J. E., Parnell, J. A., Tunnicliffe, J. M., Nicolucci, A. C., Han, J., et al. (2019). Impact of dietary fiber supplementation on modulating microbiota-host-metabolic axes in obesity. *The Journal of Nutritional Biochemistry*, *64*, 228–236. https://doi.org/10.1016/j.jnutbio.2018.11.003.

Meng, X., Li, S., Li, Y., Gan, R. Y., & Li, H. B. (2018). Gut microbiota's relationship with liver disease and role in hepatoprotection by dietary natural products and probiotics. *Nutrients*, *10*(10), 1457. https://doi.org/10.3390/nu10101457.

Meng, Y., Yu, Z., Wu, Y., Du, T., Chen, S., Meng, F., et al. (2017). Cell-based immunotherapy with cytokine-induced killer (CIK) cells: From preparation and testing to clinical application. *Human Vaccines & Immunotherapeutics*, *13*(6), 1–9. https://doi.org/10.1080/21645515.2017.1285987.

Mensink, M. A., Frijlink, H. W., van der Voort Maarschalk, K., & Hinrichs, W. L. (2015). Inulin, a flexible oligosaccharide I: Review of its physicochemical characteristics. *Carbohydrate Polymers*, *130*, 405–419. https://doi.org/10.1016/j.carbpol.2015.05.026.

Mishalian, I., Bayuh, R., Levy, L., Zolotarov, L., Michaeli, J., & Fridlender, Z. G. (2013). Tumor-associated neutrophils (TAN) develop pro-tumorigenic properties during tumor progression. *Cancer Immunology, Immunotherapy*, *62*(11), 1745–1756. https://doi.org/10.1007/s00262-013-1476-9.

Mishima, Y., Oka, A., Liu, B., Herzog, J. W., Eun, C. S., Fan, T. J., et al. (2019). Microbiota maintain colonic homeostasis by activating TLR2/MyD88/PI3K signaling in IL-10-producing regulatory B cells. *The Journal of Clinical Investigation*, *129*(9), 3702–3716. https://doi.org/10.1172/JCI93820.

Miura, K., Ishioka, M., Minami, S., Horie, Y., Ohshima, S., Goto, T., et al. (2016). Toll-like receptor 4 on macrophage promotes the development of steatohepatitis-related hepatocellular carcinoma in mice. *The Journal of Biological Chemistry*, *291*(22), 11504–11517. https://doi.org/10.1074/jbc.M115.709048.

Miyagi, T., Takehara, T., Uemura, A., Nishio, K., Shimizu, S., Kodama, T., et al. (2010). Absence of invariant natural killer T cells deteriorates liver inflammation and fibrosis in mice fed high-fat diet. *Journal of Gastroenterology*, *45*(12), 1247–1254. https://doi.org/10.1007/s00535-010-0272-y.

Mizukoshi, E., Yamashita, T., Arai, K., Terashima, T., Kitahara, M., Nakagawa, H., et al. (2016). Myeloid-derived suppressor cells correlate with patient outcomes in hepatic arterial infusion chemotherapy for hepatocellular carcinoma. *Cancer Immunology, Immunotherapy*, *65*(6), 715–725. https://doi.org/10.1007/s00262-016-1837-2.

Moro-Sibilot, L., Blanc, P., Taillardet, M., Bardel, E., Couillault, C., Boschetti, G., et al. (2016). Mouse and human liver contain immunoglobulin a-secreting cells originating from peyer's patches and directed against intestinal antigens. *Gastroenterology*, *151*(2), 311–323. https://doi.org/10.1053/j.gastro.2016.04.014.

Mossanen, J. C., Kohlhepp, M., Wehr, A., Krenkel, O., Liepelt, A., Roeth, A. A., et al. (2019). CXCR6 inhibits hepatocarcinogenesis by promoting natural killer T- and CD4(+) T-cell-dependent control of senescence. *Gastroenterology*, *156*(6), 1877–1889, e1874. https://doi.org/10.1053/j.gastro.2019.01.247.

Murakami, M., Tognini, P., Liu, Y., Eckel-Mahan, K. L., Baldi, P., & Sassone-Corsi, P. (2016). Gut microbiota directs PPARgamma-driven reprogramming of the liver circadian clock by nutritional challenge. *EMBO Reports*, *17*(9), 1292–1303. https://doi.org/10.15252/embr.201642463.

Nagatake, T., Hirata, S. I., Koga, T., Kuroda, E., Kobari, S., Suzuki, H., et al. (2019). BLT1 mediates commensal bacteria-dependent innate immune signals to enhance antigen-specific intestinal IgA responses. *Mucosal Immunology*, *12*(5), 1082–1091. https://doi.org/10.1038/s41385-019-0175-z.

Nakajima, A., Kaga, N., Nakanishi, Y., Ohno, H., Miyamoto, J., Kimura, I., et al. (2017). Maternal high fiber diet during pregnancy and lactation influences regulatory T cell differentiation in offspring in mice. *Journal of Immunology*, *199*(10), 3516–3524. https://doi.org/10.4049/jimmunol.1700248.

Nakano, M., Saeki, C., Takahashi, H., Homma, S., Tajiri, H., & Zeniya, M. (2012). Activated natural killer T cells producing interferon-gamma elicit promoting activity to murine dendritic cell-based autoimmune hepatic inflammation. *Clinical and Experimental Immunology*, *170*(3), 274–282. https://doi.org/10.1111/j.1365-2249.2012.04664.x.

Nan, J., Xing, Y. F., Hu, B., Tang, J. X., Dong, H. M., He, Y. M., et al. (2018). Endoplasmic reticulum stress induced LOX-1(+) CD15(+) polymorphonuclear myeloid-derived suppressor cells in hepatocellular carcinoma. *Immunology*, *154*(1), 144–155. https://doi.org/10.1111/imm.12876.

Narushima, S., Sugiura, Y., Oshima, K., Atarashi, K., Hattori, M., Suematsu, M., et al. (2014). Characterization of the 17 strains of regulatory T cell-inducing human-derived Clostridia. *Gut Microbes*, *5*(3), 333–339. https://doi.org/10.4161/gmic.28572.

Ngwa, V. M., Edwards, D. N., Philip, M., & Chen, J. (2019). Microenvironmental metabolism regulates antitumor immunity. *Cancer Research*, *79*(16), 4003–4008. https://doi.org/10.1158/0008-5472.CAN-19-0617.

Ni, J., Huang, R., Zhou, H., Xu, X., Li, Y., Cao, P., et al. (2019). Analysis of the relationship between the degree of dysbiosis in gut microbiota and prognosis at different stages of primary hepatocellular carcinoma. *Frontiers in Microbiology*, *10*, 1458. https://doi.org/10.3389/fmicb.2019.01458.

Nilsson, A., Haanstra, J. R., Engqvist, M., Gerding, A., Bakker, B. M., Klingmuller, U., et al. (2020). Quantitative analysis of amino acid metabolism in liver cancer links glutamate excretion to nucleotide synthesis. *Proceedings of the National Academy of Sciences of the United States of America*, *117*(19), 10294–10304. https://doi.org/10.1073/pnas.1919250117.

Niture, S., Gyamfi, M. A., Lin, M., Chimeh, U., Dong, X., Zheng, W., et al. (2020). TNFAIP8 regulates autophagy, cell steatosis, and promotes hepatocellular carcinoma cell proliferation. *Cell Death & Disease*, *11*(3), 178. https://doi.org/10.1038/s41419-020-2369-4.

Nutsch, K., Chai, J. N., Ai, T. L., Russler-Germain, E., Feehley, T., Nagler, C. R., et al. (2016). Rapid and efficient generation of regulatory T cells to commensal antigens in the periphery. *Cell Reports*, *17*(1), 206–220. https://doi.org/10.1016/j.celrep.2016.08.092.

Nwosu, Z. C., Megger, D. A., Hammad, S., Sitek, B., Roessler, S., Ebert, M. P., et al. (2017). Identification of the consistently altered metabolic targets in human hepatocellular carcinoma. *Cellular and Molecular Gastroenterology and Hepatology*, *4*(2), 303–323. e301. https://doi.org/10.1016/j.jcmgh.2017.05.004.

Ohno, H. (2020). The impact of metabolites derived from the gut microbiota on immune regulation and diseases. *International Immunology*, *32*, 629–636. https://doi.org/10.1093/intimm/dxaa041.

O'Neill, D., & Bhardwaj, N. (2005). Generation of autologous peptide- and protein-pulsed dendritic cells for patient-specific immunotherapy. *Methods in Molecular Medicine*, *109*, 97–112. Retrieved from https://www.ncbi.nlm.nih.gov/pubmed/15585916.

Onuma, A. E., Zhang, H., Huang, H., Williams, T. M., Noonan, A., & Tsung, A. (2020). Immune checkpoint inhibitors in hepatocellular cancer: Current understanding on mechanisms of resistance and biomarkers of response to treatment. *Gene Expression*, *20*(1), 53–65. https://doi.org/10.3727/105221620X15880179864121.

Ostrand-Rosenberg, S., & Fenselau, C. (2018). Myeloid-derived suppressor cells: Immune-suppressive cells that impair antitumor immunity and are sculpted by their environment. *Journal of Immunology*, *200*(2), 422–431. https://doi.org/10.4049/jimmunol.1701019.

Ottosson, F., Brunkwall, L., Ericson, U., Nilsson, P. M., Almgren, P., Fernandez, C., et al. (2018). Connection between bmi-related plasma metabolite profile and gut microbiota. *The Journal of Clinical Endocrinology and Metabolism*, *103*(4), 1491–1501. https://doi.org/10.1210/jc.2017-02114.

Ouyang, F. Z., Wu, R. Q., Wei, Y., Liu, R. X., Yang, D., Xiao, X., et al. (2016). Dendritic cell-elicited B-cell activation fosters immune privilege via IL-10 signals in hepatocellular carcinoma. *Nature Communications*, 7. https://doi.org/10.1038/ncomms13453, 13453.

Ozawa, T., Maehara, N., Kai, T., Arai, S., & Miyazaki, T. (2016). Dietary fructose-induced hepatocellular carcinoma development manifested in mice lacking apoptosis inhibitor of macrophage (AIM). *Genes to Cells*, *21*(12), 1320–1332. https://doi.org/10.1111/gtc.12446.

Ozturk, N., Erdal, E., Mumcuoglu, M., Akcali, K. C., Yalcin, O., Senturk, S., et al. (2006). Reprogramming of replicative senescence in hepatocellular carcinoma-derived cells.

Proceedings of the National Academy of Sciences of the United States of America, *103*(7), 2178–2183. https://doi.org/10.1073/pnas.0510877103.

Pandey, K. R., Naik, S. R., & Vakil, B. V. (2015). Probiotics, prebiotics and synbiotics—A review. *Journal of Food Science and Technology*, *52*(12), 7577–7587. https://doi.org/10.1007/s13197-015-1921-1.

Pang, X., Song, H., Zhang, Q., Tu, Z., & Niu, J. (2016). Hepatitis C virus regulates the production of monocytic myeloid-derived suppressor cells from peripheral blood mononuclear cells through PI3K pathway and autocrine signaling. *Clinical Immunology*, *164*, 57–64. https://doi.org/10.1016/j.clim.2016.01.014.

Paquin-Proulx, D., Ching, C., Vujkovic-Cvijin, I., Fadrosh, D., Loh, L., Huang, Y., et al. (2017). Bacteroides are associated with GALT iNKT cell function and reduction of microbial translocation in HIV-1 infection. *Mucosal Immunology*, *10*(1), 69–78. https://doi.org/10.1038/mi.2016.34.

Park, J., Kim, M., Kang, S. G., Jannasch, A. H., Cooper, B., Patterson, J., et al. (2015). Short-chain fatty acids induce both effector and regulatory T cells by suppression of histone deacetylases and regulation of the mTOR-S6K pathway. *Mucosal Immunology*, *8*(1), 80–93. https://doi.org/10.1038/mi.2014.44.

Park, B. S., & Lee, J. O. (2013). Recognition of lipopolysaccharide pattern by TLR4 complexes. *Experimental & Molecular Medicine*, *45*, e66. https://doi.org/10.1038/emm.2013.97.

Parry, R. V., Chemnitz, J. M., Frauwirth, K. A., Lanfranco, A. R., Braunstein, I., Kobayashi, S. V., et al. (2005). CTLA-4 and PD-1 receptors inhibit T-cell activation by distinct mechanisms. *Molecular and Cellular Biology*, *25*(21), 9543–9553. https://doi.org/10.1128/MCB.25.21.9543-9553.2005.

Patsoukis, N., Brown, J., Petkova, V., Liu, F., Li, L., & Boussiotis, V. A. (2012). Selective effects of PD-1 on Akt and Ras pathways regulate molecular components of the cell cycle and inhibit T cell proliferation. *Science Signaling*, *5*(230), ra46. https://doi.org/10.1126/scisignal.2002796.

Paudyal, B., Paudyal, P., Oriuchi, N., Tsushima, Y., Nakajima, T., & Endo, K. (2008). Clinical implication of glucose transport and metabolism evaluated by 18F-FDG PET in hepatocellular carcinoma. *International Journal of Oncology*, *33*(5), 1047–1054. Retrieved from https://www.ncbi.nlm.nih.gov/pubmed/18949368.

Peng, Z. P., Jiang, Z. Z., Guo, H. F., Zhou, M. M., Huang, Y. F., Ning, W. R., et al. (2020). Glycolytic activation of monocytes regulates the accumulation and function of neutrophils in human hepatocellular carcinoma. *Journal of Hepatology*, *73*, 906–917. https://doi.org/10.1016/j.jhep.2020.05.004.

Peranzoni, E., Lemoine, J., Vimeux, L., Feuillet, V., Barrin, S., Kantari-Mimoun, C., et al. (2018). Macrophages impede CD8 T cells from reaching tumor cells and limit the efficacy of anti-PD-1 treatment. *Proceedings of the National Academy of Sciences of the United States of America*, *115*(17), E4041–E4050. https://doi.org/10.1073/pnas.1720948115.

Perica, K., Varela, J. C., Oelke, M., & Schneck, J. (2015). Adoptive T cell immunotherapy for cancer. *Rambam Maimonides Medical Journal*, *6*(1), e0004. https://doi.org/10.5041/RMMJ.10179.

Petty, A. J., Li, A., Wang, X., Dai, R., Heyman, B., Hsu, D., et al. (2019). Hedgehog signaling promotes tumor-associated macrophage polarization to suppress intratumoral CD8+ T cell recruitment. *The Journal of Clinical Investigation*, *129*(12), 5151–5162. https://doi.org/10.1172/JCI128644.

Pinato, D. J., Guerra, N., Fessas, P., Murphy, R., Mineo, T., Mauri, F. A., et al. (2020). Immune-based therapies for hepatocellular carcinoma. *Oncogene*, *39*(18), 3620–3637. https://doi.org/10.1038/s41388-020-1249-9.

Pinato, D. J., Howlett, S., Ottaviani, D., Urus, H., Patel, A., Mineo, T., et al. (2019). Association of prior antibiotic treatment with survival and response to immune checkpoint inhibitor therapy in patients with cancer. *JAMA Oncology*, *5*(12), 1774–1778. https://doi.org/10.1001/jamaoncol.2019.2785.

Ponziani, F. R., Bhoori, S., Castelli, C., Putignani, L., Rivoltini, L., Del Chierico, F., et al. (2019). Hepatocellular carcinoma is associated with gut microbiota profile and inflammation in nonalcoholic fatty liver disease. *Hepatology*, *69*(1), 107–120. https://doi.org/10.1002/hep.30036.

Qi, X., Li, J., Deng, H., Li, H., Su, C., & Guo, X. (2016). Neutrophil-to-lymphocyte ratio for the prognostic assessment of hepatocellular carcinoma: A systematic review and meta-analysis of observational studies. *Oncotarget*, 7(29), 45283–45301. https://doi.org/10.18632/oncotarget.9942.

Qin, J., Li, R., Raes, J., Arumugam, M., Burgdorf, K. S., Manichanh, C., et al. (2010). A human gut microbial gene catalogue established by metagenomic sequencing. *Nature*, *464*(7285), 59–65. https://doi.org/10.1038/nature08821.

Qin, Y., & Wade, P. A. (2018). Crosstalk between the microbiome and epigenome: Messages from bugs. *Journal of Biochemistry*, *163*(2), 105–112. https://doi.org/10.1093/jb/mvx080.

Qu, K., Lin, T., Wang, Z., Liu, S., Chang, H., Xu, X., et al. (2014). Reactive oxygen species generation is essential for cisplatin-induced accelerated senescence in hepatocellular carcinoma. *Frontiers in Medicine*, *8*(2), 227–235. https://doi.org/10.1007/s11684-014-0327-1.

Qu, K., Xu, X., Liu, C., Wu, Q., Wei, J., Meng, F., et al. (2013). Negative regulation of transcription factor FoxM1 by p53 enhances oxaliplatin-induced senescence in hepatocellular carcinoma. *Cancer Letters*, *331*(1), 105–114. https://doi.org/10.1016/j.canlet.2012.12.008.

Rampelli, S., Soverini, M., D'Amico, F., Barone, M., Tavella, T., Monti, D., et al. (2020). Shotgun metagenomics of gut microbiota in humans with up to extreme longevity and the increasing role of xenobiotic degradation. *mSystems*, *5*(2), e00124–20. https://doi.org/10.1128/mSystems.00124-20.

Reis, E. C., da Silva, L. T., da Silva, W. C., Rios, A., Duarte, A. J., Oshiro, T. M., et al. (2018). Host genetics contributes to the effectiveness of dendritic cell-based HIV immunotherapy. *Human Vaccines & Immunotherapeutics*, *14*(8), 1995–2002. https://doi.org/10.1080/21645515.2018.1463942.

Ren, Z., Li, A., Jiang, J., Zhou, L., Yu, Z., Lu, H., et al. (2019). Gut microbiome analysis as a tool towards targeted non-invasive biomarkers for early hepatocellular carcinoma. *Gut*, *68*(6), 1014–1023. https://doi.org/10.1136/gutjnl-2017-315084.

Ren, J., Liu, Y., Wang, S., Wang, Y., Li, W., Chen, S., et al. (2020). The FKH domain in FOXP3 mRNA frequently contains mutations in hepatocellular carcinoma that influence the subcellular localization and functions of FOXP3. *The Journal of Biological Chemistry*, *295*(16), 5484–5495. https://doi.org/10.1074/jbc.RA120.012518.

Ridaura, V. K., Faith, J. J., Rey, F. E., Cheng, J., Duncan, A. E., Kau, A. L., et al. (2013). Gut microbiota from twins discordant for obesity modulate metabolism in mice. *Science*, *341*(6150). https://doi.org/10.1126/science.1241214, 1241214.

Riva, M. A., Riva, E., Spicci, M., Strazzabosco, M., Giovannini, M., & Cesana, G. (2011). "The city of Hepar": Rituals, gastronomy, and politics at the origins of the modern names for the liver. *Journal of Hepatology*, *55*(5), 1132–1136. https://doi.org/10.1016/j.jhep.2011.05.011.

Rivera, C. A., Adegboyega, P., van Rooijen, N., Tagalicud, A., Allman, M., & Wallace, M. (2007). Toll-like receptor-4 signaling and Kupffer cells play pivotal roles in the pathogenesis of non-alcoholic steatohepatitis. *Journal of Hepatology*, *47*(4), 571–579. https://doi.org/10.1016/j.jhep.2007.04.019.

Rohr-Udilova, N., Klinglmuller, F., Schulte-Hermann, R., Stift, J., Herac, M., Salzmann, M., et al. (2018). Deviations of the immune cell landscape between healthy liver and hepatocellular carcinoma. *Scientific Reports*, *8*(1), 6220. https://doi.org/10.1038/s41598-018-24437-5.

Rong, Y., Dong, Z., Hong, Z., Jin, Y., Zhang, W., Zhang, B., et al. (2017). Reactivity toward Bifidobacterium longum and Enterococcus hirae demonstrate robust CD8(+) T cell response and better prognosis in HBV-related hepatocellular carcinoma. *Experimental Cell Research*, *358*(2), 352–359. https://doi.org/10.1016/j.yexcr.2017.07.009.

Rong, X. X., Wei, F., Lin, X. L., Qin, Y. J., Chen, L., Wang, H. Y., et al. (2016). Recognition and killing of cancer stem-like cell population in hepatocellular carcinoma cells by cytokine-induced killer cells via NKG2d-ligands recognition. *Oncoimmunology*, *5*(3), e1086060. https://doi.org/10.1080/2162402X.2015.1086060.

Rosser, E. C., Oleinika, K., Tonon, S., Doyle, R., Bosma, A., Carter, N. A., et al. (2014). Regulatory B cells are induced by gut microbiota-driven interleukin-1beta and interleukin-6 production. *Nature Medicine*, *20*(11), 1334–1339. https://doi.org/10.1038/nm.3680.

Rostam, H. M., Reynolds, P. M., Alexander, M. R., Gadegaard, N., & Ghaemmaghami, A. M. (2017). Image based Machine Learning for identification of macrophage subsets. *Scientific Reports*, 7(1), 3521. https://doi.org/10.1038/s41598-017-03780-z.

Rothschild, D., Weissbrod, O., Barkan, E., Kurilshikov, A., Korem, T., Zeevi, D., et al. (2018). Environment dominates over host genetics in shaping human gut microbiota. *Nature*, *555*(7695), 210–215. https://doi.org/10.1038/nature25973.

Routy, B., Le Chatelier, E., Derosa, L., Duong, C. P. M., Alou, M. T., Daillere, R., et al. (2018). Gut microbiome influences efficacy of PD-1-based immunotherapy against epithelial tumors. *Science*, *359*(6371), 91–97. https://doi.org/10.1126/science.aan3706.

Sabado, R. L., Balan, S., & Bhardwaj, N. (2017). Dendritic cell-based immunotherapy. *Cell Research*, *27*(1), 74–95. https://doi.org/10.1038/cr.2016.157.

Sakai, M., Troutman, T. D., Seidman, J. S., Ouyang, Z., Spann, N. J., Abe, Y., et al. (2019). Liver-derived signals sequentially reprogram myeloid enhancers to initiate and maintain kupffer cell identity. *Immunity*, *51*(4), 655–670 e658. https://doi.org/10.1016/j.immuni.2019.09.002.

Scheiman, J., Luber, J. M., Chavkin, T. A., MacDonald, T., Tung, A., Pham, L. D., et al. (2019). Meta-omics analysis of elite athletes identifies a performance-enhancing microbe that functions via lactate metabolism. *Nature Medicine*, *25*(7), 1104–1109. https://doi.org/10.1038/s41591-019-0485-4.

Schett, A., Rothschild, S. I., Curioni-Fontecedro, A., Krahenbuhl, S., Fruh, M., Schmid, S., et al. (2020). Predictive impact of antibiotics in patients with advanced non small-cell lung cancer receiving immune checkpoint inhibitors: Antibiotics immune checkpoint inhibitors in advanced NSCLC. *Cancer Chemotherapy and Pharmacology*, *85*(1), 121–131. https://doi.org/10.1007/s00280-019-03993-1.

Schietinger, A., Philip, M., Krisnawan, V. E., Chiu, E. Y., Delrow, J. J., Basom, R. S., et al. (2016). Tumor-specific T Cell dysfunction is a dynamic antigen-driven differentiation program initiated early during tumorigenesis. *Immunity*, *45*(2), 389–401. https://doi.org/10.1016/j.immuni.2016.07.011.

Schilbach, K., Welker, C., Krickeberg, N., Kaisser, C., Schleicher, S., & Hashimoto, H. (2020). In the absence of a TCR signal IL-2/IL-12/18-stimulated gammadelta T cells demonstrate potent anti-tumoral function through direct killing and senescence induction in cancer cells. *Cancers (Basel)*, *12*(1), 130. https://doi.org/10.3390/cancers12010130.

Schneider, C., Teufel, A., Yevsa, T., Staib, F., Hohmeyer, A., Walenda, G., et al. (2012). Adaptive immunity suppresses formation and progression of diethylnitrosamine-induced liver cancer. *Gut, 61*(12), 1733–1743. https://doi.org/10.1136/gutjnl-2011-301116.

Scott, C. L., Zheng, F., De Baetselier, P., Martens, L., Saeys, Y., De Prijck, S., et al. (2016). Bone marrow-derived monocytes give rise to self-renewing and fully differentiated Kupffer cells. *Nature Communications*, 7. https://doi.org/10.1038/ncomms10321, 10321.

Sebestyen, Z., Prinz, I., Dechanet-Merville, J., Silva-Santos, B., & Kuball, J. (2020). Translating gammadelta (gammadelta) T cells and their receptors into cancer cell therapies. *Nature Reviews. Drug Discovery, 19*(3), 169–184. https://doi.org/10.1038/s41573-019-0038-z.

Sefik, E., Geva-Zatorsky, N., Oh, S., Konnikova, L., Zemmour, D., McGuire, A. M., et al. (2015). MUCOSAL IMMUNOLOGY. Individual intestinal symbionts induce a distinct population of RORgamma(+) regulatory T cells. *Science, 349*(6251), 993–997. https://doi.org/10.1126/science.aaa9420.

Semenkovich, N. P., Planer, J. D., Ahern, P. P., Griffin, N. W., Lin, C. Y., & Gordon, J. I. (2016). Impact of the gut microbiota on enhancer accessibility in gut intraepithelial lymphocytes. *Proceedings of the National Academy of Sciences of the United States of America, 113*(51), 14805–14810. https://doi.org/10.1073/pnas.1617793113.

Sender, R., Fuchs, S., & Milo, R. (2016). Are we really vastly outnumbered? Revisiting the ratio of bacterial to host cells in humans. *Cell, 164*(3), 337–340. https://doi.org/10.1016/j.cell.2016.01.013.

Seo, J. D., Gu, J. Y., Jung, H. S., Kim, Y. J., & Kim, H. K. (2019). Contact system activation and neutrophil extracellular trap markers: Risk factors for portal vein thrombosis in patients with hepatocellular carcinoma. *Clinical and Applied Thrombosis/Hemostasis, 25*. https://doi.org/10.1177/1076029618825310, 1076029618825310.

Shalapour, S., Lin, X. J., Bastian, I. N., Brain, J., Burt, A. D., Aksenov, A. A., et al. (2017). Inflammation-induced IgA+ cells dismantle anti-liver cancer immunity. *Nature, 551*(7680), 340–345. https://doi.org/10.1038/nature24302.

Shamsi, M., Saghafian, M., Dejam, M., & Sanati-Nezhad, A. (2018). Mathematical modeling of the function of warburg effect in tumor microenvironment. *Scientific Reports, 8*(1), 8903. https://doi.org/10.1038/s41598-018-27303-6.

Shao, Y., Forster, S. C., Tsaliki, E., Vervier, K., Strang, A., Simpson, N., et al. (2019). Stunted microbiota and opportunistic pathogen colonization in caesarean-section birth. *Nature, 574*(7776), 117–121. https://doi.org/10.1038/s41586-019-1560-1.

Shao, Y., Lo, C. M., Ling, C. C., Liu, X. B., Ng, K. T., Chu, A. C., et al. (2014). Regulatory B cells accelerate hepatocellular carcinoma progression via CD40/CD154 signaling pathway. *Cancer Letters, 355*(2), 264–272. https://doi.org/10.1016/j.canlet.2014.09.026.

Shi, J. Y., Gao, Q., Wang, Z. C., Zhou, J., Wang, X. Y., Min, Z. H., et al. (2013). Margin-infiltrating CD20(+) B cells display an atypical memory phenotype and correlate with favorable prognosis in hepatocellular carcinoma. *Clinical Cancer Research, 19*(21), 5994–6005. https://doi.org/10.1158/1078-0432.CCR-12-3497.

Shi, F., Shi, M., Zeng, Z., Qi, R. Z., Liu, Z. W., Zhang, J. Y., et al. (2011). PD-1 and PD-L1 upregulation promotes CD8(+) T-cell apoptosis and postoperative recurrence in hepatocellular carcinoma patients. *International Journal of Cancer, 128*(4), 887–896. https://doi.org/10.1002/ijc.25397.

Shi, W. K., Zhu, X. D., Wang, C. H., Zhang, Y. Y., Cai, H., Li, X. L., et al. (2018). PFKFB3 blockade inhibits hepatocellular carcinoma growth by impairing DNA repair through AKT. *Cell Death & Disease, 9*(4), 428. https://doi.org/10.1038/s41419-018-0435-y.

Shiau, D. J., Kuo, W. T., Davuluri, G. V. N., Shieh, C. C., Tsai, P. J., Chen, C. C., et al. (2020). Hepatocellular carcinoma-derived high mobility group box 1 triggers M2 macrophage polarization via a TLR2/NOX2/autophagy axis. *Scientific Reports, 10*(1). https://doi.org/10.1038/s41598-020-70137-4, 13582.

Shibolet, O., Alper, R., Zlotogarov, L., Thalenfeld, B., Engelhardt, D., Rabbani, E., et al. (2004). Suppression of hepatocellular carcinoma growth via oral immune regulation towards tumor-associated antigens is associated with increased NKT and CD8+ lymphocytes. *Oncology*, *66*(4), 323–330. https://doi.org/10.1159/000078334.

Shiratori, R., Furuichi, K., Yamaguchi, M., Miyazaki, N., Aoki, H., Chibana, H., et al. (2019). Glycolytic suppression dramatically changes the intracellular metabolic profile of multiple cancer cell lines in a mitochondrial metabolism-dependent manner. *Scientific Reports*, *9*(1). https://doi.org/10.1038/s41598-019-55296-3, 18699.

Sideras, K., Biermann, K., Verheij, J., Takkenberg, B. R., Mancham, S., Hansen, B. E., et al. (2017). PD-L1, Galectin-9 and CD8(+) tumor-infiltrating lymphocytes are associated with survival in hepatocellular carcinoma. *Oncoimmunology*, *6*(2). https://doi.org/10.1080/2162402X.2016.1273309, e1273309.

Siegel, R. L., Miller, K. D., & Jemal, A. (2020). Cancer statistics, 2020. *CA: a Cancer Journal for Clinicians*, *70*(1), 7–30. https://doi.org/10.3322/caac.21590.

Singh, N., Gurav, A., Sivaprakasam, S., Brady, E., Padia, R., Shi, H., et al. (2014). Activation of Gpr109a, receptor for niacin and the commensal metabolite butyrate, suppresses colonic inflammation and carcinogenesis. *Immunity*, *40*(1), 128–139. https://doi.org/10.1016/j.immuni.2013.12.007.

Singh, V., Yeoh, B. S., Abokor, A. A., Golonka, R. M., Tian, Y., Patterson, A. D., et al. (2020). Vancomycin prevents fermentable fiber-induced liver cancer in mice with dysbiotic gut microbiota. *Gut Microbes*, *11*(4), 1077–1091. https://doi.org/10.1080/19490976.2020.1743492.

Singh, V., Yeoh, B. S., Chassaing, B., Xiao, X., Saha, P., Aguilera Olvera, R., et al. (2018). Dysregulated microbial fermentation of soluble fiber induces cholestatic liver cancer. *Cell*, *175*(3), 679–694 e622. https://doi.org/10.1016/j.cell.2018.09.004.

Smith, P. M., Howitt, M. R., Panikov, N., Michaud, M., Gallini, C. A., Bohlooly, Y. M., et al. (2013). The microbial metabolites, short-chain fatty acids, regulate colonic Treg cell homeostasis. *Science*, *341*(6145), 569–573. https://doi.org/10.1126/science.1241165.

Softic, S., Cohen, D. E., & Kahn, C. R. (2016). Role of dietary fructose and hepatic de novo lipogenesis in fatty liver disease. *Digestive Diseases and Sciences*, *61*(5), 1282–1293. https://doi.org/10.1007/s10620-016-4054-0.

Sola-Penna, M., Paixao, L. P., Branco, J. R., Ochioni, A. C., Albanese, J. M., Mundim, D. M., et al. (2020). Serotonin activates glycolysis and mitochondria biogenesis in human breast cancer cells through activation of the Jak1/STAT3/ERK1/2 and adenylate cyclase/PKA, respectively. *British Journal of Cancer*, *122*(2), 194–208. https://doi.org/10.1038/s41416-019-0640-1.

Soll, C., Jang, J. H., Riener, M. O., Moritz, W., Wild, P. J., Graf, R., et al. (2010). Serotonin promotes tumor growth in human hepatocellular cancer. *Hepatology*, *51*(4), 1244–1254. https://doi.org/10.1002/hep.23441.

Song, M., Li, X., Zhang, X., Shi, H., Vos, M. B., Wei, X., et al. (2018). Dietary copper-fructose interactions alter gut microbial activity in male rats. *American Journal of Physiology. Gastrointestinal and Liver Physiology*, *314*(1), G119–G130. https://doi.org/10.1152/ajpgi.00378.2016.

Song, X., Sun, X., Oh, S. F., Wu, M., Zhang, Y., Zheng, W., et al. (2020). Microbial bile acid metabolites modulate gut RORgamma(+) regulatory T cell homeostasis. *Nature*, *577*(7790), 410–415. https://doi.org/10.1038/s41586-019-1865-0.

Spear, T. T., Callender, G. G., Roszkowski, J. J., Moxley, K. M., Simms, P. E., Foley, K. C., et al. (2016). TCR gene-modified T cells can efficiently treat established hepatitis C-associated hepatocellular carcinoma tumors. *Cancer Immunology, Immunotherapy*, *65*(3), 293–304. https://doi.org/10.1007/s00262-016-1800-2.

Spiga, L., Winter, M. G., Furtado de Carvalho, T., Zhu, W., Hughes, E. R., Gillis, C. C., et al. (2017). An oxidative central metabolism enables salmonella to utilize microbiota-derived succinate. *Cell Host & Microbe*, *22*(3), 291–301, e296. https://doi.org/10.1016/j.chom.2017.07.018.
Stein, R. R., Tanoue, T., Szabady, R. L., Bhattarai, S. K., Olle, B., Norman, J. M., et al. (2018). Computer-guided design of optimal microbial consortia for immune system modulation. *eLife*, *7*, e30916. https://doi.org/10.7554/eLife.30916.
Suez, J., Zmora, N., Zilberman-Schapira, G., Mor, U., Dori-Bachash, M., Bashiardes, S., et al. (2018). Post-antibiotic gut mucosal microbiome reconstitution is impaired by probiotics and improved by autologous FMT. *Cell*, *174*(6), 1406–1423 e1416. https://doi.org/10.1016/j.cell.2018.08.047.
Sun, L., Guo, H., Jiang, R., Lu, L., Liu, T., & He, X. (2016). Engineered cytotoxic T lymphocytes with AFP-specific TCR gene for adoptive immunotherapy in hepatocellular carcinoma. *Tumour Biology*, *37*(1), 799–806. https://doi.org/10.1007/s13277-015-3845-9.
Sun, B., Yang, D., Dai, H., Liu, X., Jia, R., Cui, X., et al. (2019). Eradication of hepatocellular carcinoma by NKG2D-based CAR-T cells. *Cancer Immunology Research*, 7(11), 1813–1823. https://doi.org/10.1158/2326-6066.CIR-19-0026.
Sun, H. W., Yu, X. J., Wu, W. C., Chen, J., Shi, M., Zheng, L., et al. (2016). GLUT1 and ASCT2 as predictors for prognosis of hepatocellular carcinoma. *PLoS One*, *11*(12). https://doi.org/10.1371/journal.pone.0168907, e0168907.
Sung, H., Kim, S. W., Hong, M., & Suk, K. T. (2016). Microbiota-based treatments in alcoholic liver disease. *World Journal of Gastroenterology*, *22*(29), 6673–6682. https://doi.org/10.3748/wjg.v22.i29.6673.
Tajiri, K., Baba, H., Kawai, K., Minemura, M., Yasumura, S., Takahara, T., et al. (2016). Neutrophil-to-lymphocyte ratio predicts recurrence after radiofrequency ablation in hepatitis B virus infection. *Journal of Gastroenterology and Hepatology*, *31*(7), 1291–1299. https://doi.org/10.1111/jgh.13287.
Tang, Z., Mo, F., Liu, A., Duan, S., Yang, X., Liang, L., et al. (2019). A nanobody against cytotoxic T-lymphocyte associated antigen-4 increases the anti-tumor effects of specific CD8(+) T cells. *Journal of Biomedical Nanotechnology*, *15*(11), 2229–2239. https://doi.org/10.1166/jbn.2019.2859.
Tannahill, G. M., Curtis, A. M., Adamik, J., Palsson-McDermott, E. M., McGettrick, A. F., Goel, G., et al. (2013). Succinate is an inflammatory signal that induces IL-1beta through HIF-1alpha. *Nature*, *496*(7444), 238–242. https://doi.org/10.1038/nature11986.
Tanoue, T., Morita, S., Plichta, D. R., Skelly, A. N., Suda, W., Sugiura, Y., et al. (2019). A defined commensal consortium elicits CD8 T cells and anti-cancer immunity. *Nature*, *565*(7741), 600–605. https://doi.org/10.1038/s41586-019-0878-z.
Tao, R., de Zoeten, E. F., Ozkaynak, E., Chen, C., Wang, L., Porrett, P. M., et al. (2007). Deacetylase inhibition promotes the generation and function of regulatory T cells. *Nature Medicine*, *13*(11), 1299–1307. https://doi.org/10.1038/nm1652.
Tedesco, D., Thapa, M., Chin, C. Y., Ge, Y., Gong, M., Li, J., et al. (2018). Alterations in intestinal microbiota lead to production of interleukin 17 by intrahepatic gammadelta T-cell receptor-positive cells and pathogenesis of cholestatic liver disease. *Gastroenterology*, *154*(8), 2178–2193. https://doi.org/10.1053/j.gastro.2018.02.019.
Teng, D., Ding, L., Cai, B., Luo, Q., & Wang, H. (2019). Interleukin-7 enhances anti-tumor activity of CD8(+) T cells in patients with hepatocellular carcinoma. *Cytokine*, *118*, 115–123. https://doi.org/10.1016/j.cyto.2018.04.003.
The PLOS Genetics Staff. (2016). Correction: fMiRNA-192 and miRNA-204 directly suppress lncRNA HOTTIP and interrupt GLS1-mediated glutaminolysis in hepatocellular carcinoma. *PLoS Genetics*, *12*(1), e1005825. https://doi.org/10.1371/journal.pgen.1005825.

Tian, Z., Hou, X., Liu, W., Han, Z., & Wei, L. (2019). Macrophages and hepatocellular carcinoma. *Cell & Bioscience*, *9*, 79. https://doi.org/10.1186/s13578-019-0342-7.

Tian, W., Ma, J., Shi, R., Ren, C., He, J., & Zhao, H. (2018). gammadelta T cell-mediated individualized immunotherapy for hepatocellular carcinoma considering clinicopathological characteristics and immunosuppressive factors. *Oncology Letters*, *15*(4), 5433–5442. https://doi.org/10.3892/ol.2018.8026.

Timmerman, H. M., Rutten, N., Boekhorst, J., Saulnier, D. M., Kortman, G. A. M., Contractor, N., et al. (2017). Intestinal colonisation patterns in breastfed and formula-fed infants during the first 12 weeks of life reveal sequential microbiota signatures. *Scientific Reports*, 7(1), 8327. https://doi.org/10.1038/s41598-017-08268-4.

Tompkins, S. C., Sheldon, R. D., Rauckhorst, A. J., Noterman, M. F., Solst, S. R., Buchanan, J. L., et al. (2019). Disrupting mitochondrial pyruvate uptake directs glutamine into the TCA cycle away from glutathione synthesis and impairs hepatocellular tumorigenesis. *Cell Reports*, *28*(10), 2608–2619 e2606. https://doi.org/10.1016/j.celrep.2019.07.098.

Townsend, G. E., 2nd, Han, W., Schwalm, N. D., 3rd, Raghavan, V., Barry, N. A., Goodman, A. L., et al. (2019). Dietary sugar silences a colonization factor in a mammalian gut symbiont. *Proceedings of the National Academy of Sciences of the United States of America*, *116*(1), 233–238. https://doi.org/10.1073/pnas.1813780115.

Trefts, E., Gannon, M., & Wasserman, D. H. (2017). The liver. *Current Biology*, *27*(21), R1147–R1151. https://doi.org/10.1016/j.cub.2017.09.019.

Tripathi, A., Debelius, J., Brenner, D. A., Karin, M., Loomba, R., Schnabl, B., et al. (2018). The gut-liver axis and the intersection with the microbiome. *Nature Reviews. Gastroenterology & Hepatology*, *15*(7), 397–411. https://doi.org/10.1038/s41575-018-0011-z.

Tseng, P. L., Wu, W. H., Hu, T. H., Chen, C. W., Cheng, H. C., Li, C. F., et al. (2018). Decreased succinate dehydrogenase B in human hepatocellular carcinoma accelerates tumor malignancy by inducing the Warburg effect. *Scientific Reports*, *8*(1), 3081. https://doi.org/10.1038/s41598-018-21361-6.

Tsuda, Y., Fukui, H., Asai, A., Fukunishi, S., Miyaji, K., Fujiwara, S., et al. (2012). An immunosuppressive subtype of neutrophils identified in patients with hepatocellular carcinoma. *Journal of Clinical Biochemistry and Nutrition*, *51*(3), 204–212. https://doi.org/10.3164/jcbn.12-32.

Tu, J. F., Ding, Y. H., Ying, X. H., Wu, F. Z., Zhou, X. M., Zhang, D. K., et al. (2016). Regulatory T cells, especially ICOS(+) FOXP3(+) regulatory T cells, are increased in the hepatocellular carcinoma microenvironment and predict reduced survival. *Scientific Reports*, *6*. https://doi.org/10.1038/srep35056, 35056.

Tupin, E., Kinjo, Y., & Kronenberg, M. (2007). The unique role of natural killer T cells in the response to microorganisms. *Nature Reviews. Microbiology*, *5*(6), 405–417. https://doi.org/10.1038/nrmicro1657.

Turnbaugh, P. J., Hamady, M., Yatsunenko, T., Cantarel, B. L., Duncan, A., Ley, R. E., et al. (2009). A core gut microbiome in obese and lean twins. *Nature*, *457*(7228), 480–484. https://doi.org/10.1038/nature07540.

Tyler, C. J., McCarthy, N. E., Lindsay, J. O., Stagg, A. J., Moser, B., & Eberl, M. (2017). Antigen-presenting human gammadelta T cells promote intestinal CD4(+) T cell expression of IL-22 and mucosal release of calprotectin. *Journal of Immunology*, *198*(9), 3417–3425. https://doi.org/10.4049/jimmunol.1700003.

Uribe-Herranz, M., Bittinger, K., Rafail, S., Guedan, S., Pierini, S., Tanes, C., et al. (2018). Gut microbiota modulates adoptive cell therapy via CD8alpha dendritic cells and IL-12. *JCI Insight*, *3*(4). https://doi.org/10.1172/jci.insight.94952.

Valery, P. C., Laversanne, M., Clark, P. J., Petrick, J. L., McGlynn, K. A., & Bray, F. (2018). Projections of primary liver cancer to 2030 in 30 countries worldwide. *Hepatology*, *67*(2), 600–611. https://doi.org/10.1002/hep.29498.

van der Windt, D. J., Sud, V., Zhang, H., Varley, P. R., Goswami, J., Yazdani, H. O., et al. (2018). Neutrophil extracellular traps promote inflammation and development of hepatocellular carcinoma in nonalcoholic steatohepatitis. *Hepatology, 68*(4), 1347–1360. https://doi.org/10.1002/hep.29914.

Vandamme, T. F. (2014). Use of rodents as models of human diseases. *Journal of Pharmacy & Bioallied Sciences, 6*(1), 2–9. https://doi.org/10.4103/0975-7406.124301.

Vasilyeva, A., Durmus, S., Li, L., Wagenaar, E., Hu, S., Gibson, A. A., et al. (2015). Hepatocellular shuttling and recirculation of sorafenib-glucuronide is dependent on Abcc2, Abcc3, and Oatp1a/1b. *Cancer Research, 75*(13), 2729–2736. https://doi.org/10.1158/0008-5472.CAN-15-0280.

Vetizou, M., Pitt, J. M., Daillere, R., Lepage, P., Waldschmitt, N., Flament, C., et al. (2015). Anticancer immunotherapy by CTLA-4 blockade relies on the gut microbiota. *Science, 350*(6264), 1079–1084. https://doi.org/10.1126/science.aad1329.

Villanueva, A. (2019). Hepatocellular carcinoma. *The New England Journal of Medicine, 380*(15), 1450–1462. https://doi.org/10.1056/NEJMra1713263.

Vivier, E., Ugolini, S., Blaise, D., Chabannon, C., & Brossay, L. (2012). Targeting natural killer cells and natural killer T cells in cancer. *Nature Reviews. Immunology, 12*(4), 239–252. https://doi.org/10.1038/nri3174.

Wan, X., Cheng, C., Lin, Z., Jiang, R., Zhao, W., Yan, X., et al. (2015). The attenuated hepatocellular carcinoma-specific Listeria vaccine Lmdd-MPFG prevents tumor occurrence through immune regulation of dendritic cells. *Oncotarget, 6*(11), 8822–8838. https://doi.org/10.18632/oncotarget.3558.

Wan, M. L. Y., & El-Nezami, H. (2018). Targeting gut microbiota in hepatocellular carcinoma: Probiotics as a novel therapy. *HepatoBiliary Surgery and Nutrition*, 7(1), 11–20. https://doi.org/10.21037/hbsn.2017.12.07.

Wan, S., Wu, Y., Zhou, X., Chen, Y., An, J., Yu, X., et al. (2015). Polymorphisms in genes of tricarboxylic acid cycle key enzymes are associated with early recurrence of hepatocellular carcinoma. *PLoS One, 10*(4). https://doi.org/10.1371/journal.pone.0124471, e0124471.

Wan, S., Zhao, E., Kryczek, I., Vatan, L., Sadovskaya, A., Ludema, G., et al. (2014). Tumor-associated macrophages produce interleukin 6 and signal via STAT3 to promote expansion of human hepatocellular carcinoma stem cells. *Gastroenterology, 147*(6), 1393–1404. https://doi.org/10.1053/j.gastro.2014.08.039.

Wang, C., Chen, W. J., Wu, Y. F., You, P., Zheng, S. Y., Liu, C. C., et al. (2018). The extent of liver injury determines hepatocyte fate toward senescence or cancer. *Cell Death & Disease, 9*(5), 575. https://doi.org/10.1038/s41419-018-0622-x.

Wang, Q., Cheng, F., Ma, T. T., Xiong, H. Y., Li, Z. W., Xie, C. L., et al. (2016). Interleukin-12 inhibits the hepatocellular carcinoma growth by inducing macrophage polarization to the M1-like phenotype through downregulation of Stat-3. *Molecular and Cellular Biochemistry, 415*(1–2), 157–168. https://doi.org/10.1007/s11010-016-2687-0.

Wang, Z., & Dong, C. (2019). Gluconeogenesis in cancer: Function and regulation of PEPCK, FBPase, and G6Pase. *Trends Cancer, 5*(1), 30–45. https://doi.org/10.1016/j.trecan.2018.11.003.

Wang, X., He, Q., Shen, H., Lu, X. J., & Sun, B. (2019). Genetic and phenotypic difference in CD8(+) T cell exhaustion between chronic hepatitis B infection and hepatocellular carcinoma. *Journal of Medical Genetics, 56*(1), 18–21. https://doi.org/10.1136/jmedgenet-2018-105267.

Wang, X., He, Q., Shen, H., Xia, A., Tian, W., Yu, W., et al. (2019). TOX promotes the exhaustion of antitumor CD8(+) T cells by preventing PD1 degradation in hepatocellular carcinoma. *Journal of Hepatology, 71*(4), 731–741. https://doi.org/10.1016/j.jhep.2019.05.015.

Wang, S. T., Huang, S. W., Liu, K. T., Lee, T. Y., Shieh, J. J., & Wu, C. Y. (2020). Atorvastatin-induced senescence of hepatocellular carcinoma is mediated by downregulation of hTERT through the suppression of the IL-6/STAT3 pathway. *Cell Death Discovery*, *6*, 17. https://doi.org/10.1038/s41420-020-0252-9.

Wang, P., Qin, W., Liu, T., Jiang, D., Cui, L., Liu, X., et al. (2020). PiggyBac-engineered T cells expressing a glypican-3-specific chimeric antigen receptor show potent activities against hepatocellular carcinoma. *Immunobiology*, *225*(1). https://doi.org/10.1016/j.imbio.2019.09.009, 151850.

Wang, X., Shen, H., Zhangyuan, G., Huang, R., Zhang, W., He, Q., et al. (2018). 14-3-3zeta delivered by hepatocellular carcinoma-derived exosomes impaired anti-tumor function of tumor-infiltrating T lymphocytes. *Cell Death & Disease*, *9*(2), 159. https://doi.org/10.1038/s41419-017-0180-7.

Wang, C., Vegna, S., Jin, H., Benedict, B., Lieftink, C., Ramirez, C., et al. (2019). Inducing and exploiting vulnerabilities for the treatment of liver cancer. *Nature*, *574*(7777), 268–272. https://doi.org/10.1038/s41586-019-1607-3.

Wang, Y., Wang, Y., Mu, H., Liu, T., Chen, X., & Shen, Z. (2015). Enhanced specific antitumor immunity of dendritic cells transduced with the glypican 3 gene and co-cultured with cytokine-induced killer cells against hepatocellular carcinoma cells. *Molecular Medicine Reports*, *11*(5), 3361–3367. https://doi.org/10.3892/mmr.2015.3239.

Wang, S., Wang, Q., Wang, H., Qin, C., Cui, X., Li, L., et al. (2019). Induction of ROS and DNA damage-dependent senescence by icaritin contributes to its antitumor activity in hepatocellular carcinoma cells. *Pharmaceutical Biology*, *57*(1), 424–431. https://doi.org/10.1080/13880209.2019.1628073.

Wang, Y., Yao, R., Zhang, L., Xie, X., Chen, R., & Ren, Z. (2019). IDO and intra-tumoral neutrophils were independent prognostic factors for overall survival for hepatocellular carcinoma. *Journal of Clinical Laboratory Analysis*, *33*(5), e22872. https://doi.org/10.1002/jcla.22872.

Warburg, O. (1925). The metabolism of carcinoma cells. *The Journal of Cancer Research*, *9*(1), 148. https://doi.org/10.1158/jcr.1925.148.

Ward, P. S., & Thompson, C. B. (2012). Metabolic reprogramming: A cancer hallmark even warburg did not anticipate. *Cancer Cell*, *21*(3), 297–308. https://doi.org/10.1016/j.ccr.2012.02.014.

Wei, Y., Lao, X. M., Xiao, X., Wang, X. Y., Wu, Z. J., Zeng, Q. H., et al. (2019). Plasma cell polarization to the immunoglobulin g phenotype in hepatocellular carcinomas involves epigenetic alterations and promotes hepatoma progression in mice. *Gastroenterology*, *156*(6), 1890–1904, e1816. https://doi.org/10.1053/j.gastro.2019.01.250.

Wei, Y., Zeng, B., Chen, J., Cui, G., Lu, C., Wu, W., et al. (2016). Enterogenous bacterial glycolipids are required for the generation of natural killer T cells mediated liver injury. *Scientific Reports*, *6*. https://doi.org/10.1038/srep36365, 36365.

Wells, C. L., Jechorek, R. P., & Erlandsen, S. L. (1995). Inhibitory effect of bile on bacterial invasion of enterocytes: Possible mechanism for increased translocation associated with obstructive jaundice. *Critical Care Medicine*, *23*(2), 301–307. https://doi.org/10.1097/00003246-199502000-00016.

Weng, D. S., Zhou, J., Zhou, Q. M., Zhao, M., Wang, Q. J., Huang, L. X., et al. (2008). Minimally invasive treatment combined with cytokine-induced killer cells therapy lower the short-term recurrence rates of hepatocellular carcinomas. *Journal of Immunotherapy*, *31*(1), 63–71. https://doi.org/10.1097/CJI.0b013e31815a121b.

Wherry, E. J. (2011). T cell exhaustion. *Nature Immunology*, *12*(6), 492–499. https://doi.org/10.1038/ni.2035.

Wherry, E. J., & Kurachi, M. (2015). Molecular and cellular insights into T cell exhaustion. *Nature Reviews. Immunology*, *15*(8), 486–499. https://doi.org/10.1038/nri3862.

Wilson, C. L., Jurk, D., Fullard, N., Banks, P., Page, A., Luli, S., et al. (2015). NFkappaB1 is a suppressor of neutrophil-driven hepatocellular carcinoma. *Nature Communications*, *6*, 6818. https://doi.org/10.1038/ncomms7818.

Wingender, G., Hiss, M., Engel, I., Peukert, K., Ley, K., Haller, H., et al. (2012). Neutrophilic granulocytes modulate invariant NKT cell function in mice and humans. *Journal of Immunology*, *188*(7), 3000–3008. https://doi.org/10.4049/jimmunol.1101273.

Wolchok, J. D., Kluger, H., Callahan, M. K., Postow, M. A., Rizvi, N. A., Lesokhin, A. M., et al. (2013). Nivolumab plus ipilimumab in advanced melanoma. *The New England Journal of Medicine*, *369*(2), 122–133. https://doi.org/10.1056/NEJMoa1302369.

Wolf, M. J., Adili, A., Piotrowitz, K., Abdullah, Z., Boege, Y., Stemmer, K., et al. (2014). Metabolic activation of intrahepatic CD8+ T cells and NKT cells causes nonalcoholic steatohepatitis and liver cancer via cross-talk with hepatocytes. *Cancer Cell*, *26*(4), 549–564. https://doi.org/10.1016/j.ccell.2014.09.003.

Wong, S. H., Zhao, L., Zhang, X., Nakatsu, G., Han, J., Xu, W., et al. (2017). Gavage of fecal samples from patients with colorectal cancer promotes intestinal carcinogenesis in germ-free and conventional mice. *Gastroenterology*, *153*(6), 1621–1633, e1626. https://doi.org/10.1053/j.gastro.2017.08.022.

Wu, K., Kryczek, I., Chen, L., Zou, W., & Welling, T. H. (2009). Kupffer cell suppression of CD8+ T cells in human hepatocellular carcinoma is mediated by B7-H1/programmed death-1 interactions. *Cancer Research*, *69*(20), 8067–8075. https://doi.org/10.1158/0008-5472.CAN-09-0901.

Wu, T., Leboeuf, C., Durand, S., Su, B., Deschamps, M., Zhang, X., et al. (2016). Suicide gene-modified killer cells as an allogeneic alternative to autologous cytokine-induced killer cell immunotherapy of hepatocellular carcinoma. *Molecular Medicine Reports*, *13*(3), 2645–2654. https://doi.org/10.3892/mmr.2016.4811.

Wu, C., Lin, J., Weng, Y., Zeng, D. N., Xu, J., Luo, S., et al. (2020). Myeloid signature reveals immune contexture and predicts the prognosis of hepatocellular carcinoma. *The Journal of Clinical Investigation*, *130*(9), 4679–4693. https://doi.org/10.1172/JCI135048.

Wu, H., Singer, J., Kwan, T. K., Loh, Y. W., Wang, C., Tan, J., et al. (2020). Gut microbial metabolites induce donor-specific tolerance of kidney allografts through induction of T regulatory cells by short-chain fatty acids. *Journal of the American Society of Nephrology*, *31*(7), 1445–1461. https://doi.org/10.1681/ASN.2019080852.

Wu, G., Wang, N., Luo, Y., Zhang, Y., Wang, P., Zhu, Z., et al. (2017). Metabolic perturbation of epigenome by inhibiting S-adenosylhomocysteine hydrolase elicits senescence through DNA damage response in hepatoma cells. *Tumour Biology*, *39*(5), 1010428317699117. https://doi.org/10.1177/1010428317699117.

Wu, M., Xiao, H., Liu, G., Chen, S., Tan, B., Ren, W., et al. (2016). Glutamine promotes intestinal SIgA secretion through intestinal microbiota and IL-13. *Molecular Nutrition & Food Research*, *60*(7), 1637–1648. https://doi.org/10.1002/mnfr.201600026.

Wu, Y., Zhao, Q., Peng, C., Sun, L., Li, X. F., & Kuang, D. M. (2011). Neutrophils promote motility of cancer cells via a hyaluronan-mediated TLR4/PI3K activation loop. *The Journal of Pathology*, *225*(3), 438–447. https://doi.org/10.1002/path.2947.

Xiang, Y., Stine, Z. E., Xia, J., Lu, Y., O'Connor, R. S., Altman, B. J., et al. (2015). Targeted inhibition of tumor-specific glutaminase diminishes cell-autonomous tumorigenesis. *The Journal of Clinical Investigation*, *125*(6), 2293–2306. https://doi.org/10.1172/JCI75836.

Xiang, Q., Wu, X., Pan, Y., Wang, L., Cui, C., Guo, Y., et al. (2020). Early-life intervention using fecal microbiota combined with probiotics promotes gut microbiota maturation, regulates immune system development, and alleviates weaning stress in piglets. *International Journal of Molecular Sciences*, *21*(2), 503. https://doi.org/10.3390/ijms21020503.

Xiao, Y. S., Gao, Q., Xu, X. N., Li, Y. W., Ju, M. J., Cai, M. Y., et al. (2013). Combination of intratumoral invariant natural killer T cells and interferon-gamma is associated with prognosis of hepatocellular carcinoma after curative resection. *PLoS One*, *8*(8), e70345. https://doi.org/10.1371/journal.pone.0070345.

Xiao, X., Lao, X. M., Chen, M. M., Liu, R. X., Wei, Y., Ouyang, F. Z., et al. (2016). PD-1hi identifies a novel regulatory B-cell population in human hepatoma that promotes disease progression. *Cancer Discovery*, *6*(5), 546–559. https://doi.org/10.1158/2159-8290.CD-15-1408.

Xiao, P., Long, X., Zhang, L., Ye, Y., Guo, J., Liu, P., et al. (2018). Neurotensin/IL-8 pathway orchestrates local inflammatory response and tumor invasion by inducing M2 polarization of Tumor-Associated macrophages and epithelial-mesenchymal transition of hepatocellular carcinoma cells. *Oncoimmunology*, 7(7), e1440166. https://doi.org/10.1080/2162402X.2018.1440166.

Xiao, M., Yang, H., Xu, W., Ma, S., Lin, H., Zhu, H., et al. (2012). Inhibition of alpha-KG-dependent histone and DNA demethylases by fumarate and succinate that are accumulated in mutations of FH and SDH tumor suppressors. *Genes & Development*, *26*(12), 1326–1338. https://doi.org/10.1101/gad.191056.112.

Xie, G., Wang, X., Liu, P., Wei, R., Chen, W., Rajani, C., et al. (2016). Distinctly altered gut microbiota in the progression of liver disease. *Oncotarget*, 7(15), 19355–19366. https://doi.org/10.18632/oncotarget.8466.

Xie, J., Wu, H., Dai, C., Pan, Q., Ding, Z., Hu, D., et al. (2014). Beyond Warburg effect—Dual metabolic nature of cancer cells. *Scientific Reports*, *4*, 4927. https://doi.org/10.1038/srep04927.

Xu, P., Oosterveer, M. H., Stein, S., Demagny, H., Ryu, D., Moullan, N., et al. (2016). LRH-1-dependent programming of mitochondrial glutamine processing drives liver cancer. *Genes & Development*, *30*(11), 1255–1260. https://doi.org/10.1101/gad.277483.116.

Xu, M., Zhao, Z., Song, J., Lan, X., Lu, S., Chen, M., et al. (2017). Interactions between interleukin-6 and myeloid-derived suppressor cells drive the chemoresistant phenotype of hepatocellular cancer. *Experimental Cell Research*, *351*(2), 142–149. https://doi.org/10.1016/j.yexcr.2017.01.008.

Xu, Y., Zhao, W., Xu, J., Li, J., Hong, Z., Yin, Z., et al. (2016). Activated hepatic stellate cells promote liver cancer by induction of myeloid-derived suppressor cells through cyclooxygenase-2. *Oncotarget*, 7(8), 8866–8878. https://doi.org/10.18632/oncotarget.6839.

Xue, L., He, J., Gao, N., Lu, X., Li, M., Wu, X., et al. (2017). Probiotics may delay the progression of nonalcoholic fatty liver disease by restoring the gut microbiota structure and improving intestinal endotoxemia. *Scientific Reports*, *7, 45176*. https://doi.org/10.1038/srep45176.

Xue, H., Lin, F., Tan, H., Zhu, Z. Q., Zhang, Z. Y., & Zhao, L. (2016). Overrepresentation of IL-10-expressing B cells suppresses cytotoxic CD4+ T cell activity in HBV-induced hepatocellular carcinoma. *PLoS One*, *11*(5), e0154815. https://doi.org/10.1371/journal.pone.0154815.

Yamada, S., Kimura, M., Saito, Y., & Saito, H. (2018). Nrf2-mediated anti-oxidant effects contribute to suppression of non-alcoholic steatohepatitis-associated hepatocellular carcinoma in murine model. *Journal of Clinical Biochemistry and Nutrition*, *63*(2), 123–128. https://doi.org/10.3164/jcbn.17-125.

Yamada, S., Takashina, Y., Watanabe, M., Nagamine, R., Saito, Y., Kamada, N., et al. (2018). Bile acid metabolism regulated by the gut microbiota promotes non-alcoholic steatohepatitis-associated hepatocellular carcinoma in mice. *Oncotarget*, *9*(11), 9925–9939. https://doi.org/10.18632/oncotarget.24066.

Yan, J., Zhang, Y., Zhang, J. P., Liang, J., Li, L., & Zheng, L. (2013). Tim-3 expression defines regulatory T cells in human tumors. *PLoS One*, *8*(3), e58006. https://doi.org/10.1371/journal.pone.0058006.

Yang, T., Chakraborty, S., Saha, P., Mell, B., Cheng, X., Yeo, J. Y., et al. (2020). Gnotobiotic rats reveal that gut microbiota regulates colonic mRNA of Ace2, the receptor for SARS-CoV-2 infectivity. *Hypertension*, *76*(1), e1–e3. https://doi.org/10.1161/HYPERTENSIONAHA.120.15360.

Yang, H., & Duan, Z. (2016). Bile acids and the potential role in primary biliary cirrhosis. *Digestion*, *94*(3), 145–153. https://doi.org/10.1159/000452300.

Yang, J. D., Hainaut, P., Gores, G. J., Amadou, A., Plymoth, A., & Roberts, L. R. (2019). A global view of hepatocellular carcinoma: Trends, risk, prevention and management. *Nature Reviews. Gastroenterology & Hepatology*, *16*(10), 589–604. https://doi.org/10.1038/s41575-019-0186-y.

Yang, X., He, F., Zhang, Y., Xue, J., Li, K., Zhang, X., et al. (2019). Inulin ameliorates alcoholic liver disease via suppressing LPS-TLR4-Mpsi axis and modulating gut microbiota in mice. *Alcoholism, Clinical and Experimental Research*, *43*(3), 411–424. https://doi.org/10.1111/acer.13950.

Yang, J., Jin, X., Yan, Y., Shao, Y., Pan, Y., Roberts, L. R., et al. (2017). Inhibiting histone deacetylases suppresses glucose metabolism and hepatocellular carcinoma growth by restoring FBP1 expression. *Scientific Reports*, 7. https://doi.org/10.1038/srep43864, 43864.

Yang, L. Y., Luo, Q., Lu, L., Zhu, W. W., Sun, H. T., Wei, R., et al. (2020). Increased neutrophil extracellular traps promote metastasis potential of hepatocellular carcinoma via provoking tumorous inflammatory response. *Journal of Hematology & Oncology*, *13*(1), 3. https://doi.org/10.1186/s13045-019-0836-0.

Yang, L. N., Ning, Z. Y., Wang, L., Yan, X., & Meng, Z. Q. (2019). HSF2 regulates aerobic glycolysis by suppression of FBP1 in hepatocellular carcinoma. *American Journal of Cancer Research*, *9*(8), 1607–1621. Retrieved from https://www.ncbi.nlm.nih.gov/pubmed/31497345.

Yang, J., Wang, C., Zhao, F., Luo, X., Qin, M., Arunachalam, E., et al. (2017). Loss of FBP1 facilitates aggressive features of hepatocellular carcinoma cells through the Warburg effect. *Carcinogenesis*, *38*(2), 134–143. https://doi.org/10.1093/carcin/bgw109.

Yang, Z. Q., Yang, Z. Y., Zhang, L. D., Ping, B., Wang, S. G., Ma, K. S., et al. (2010). Increased liver-infiltrating CD8+FoxP3+ regulatory T cells are associated with tumor stage in hepatocellular carcinoma patients. *Human Immunology*, *71*(12), 1180–1186. https://doi.org/10.1016/j.humimm.2010.09.011.

Yang, Y., Ye, Y. C., Chen, Y., Zhao, J. L., Gao, C. C., Han, H., et al. (2018). Crosstalk between hepatic tumor cells and macrophages via Wnt/beta-catenin signaling promotes M2-like macrophage polarization and reinforces tumor malignant behaviors. *Cell Death & Disease*, *9*(8), 793. https://doi.org/10.1038/s41419-018-0818-0.

Yang, J., Zhang, J. X., Wang, H., Wang, G. L., Hu, Q. G., & Zheng, Q. C. (2012). Hepatocellular carcinoma and macrophage interaction induced tumor immunosuppression via Treg requires TLR4 signaling. *World Journal of Gastroenterology*, *18*(23), 2938–2947. https://doi.org/10.3748/wjg.v18.i23.2938.

Yang, T., Zhang, W., Wang, L., Xiao, C., Wang, L., Gong, Y., et al. (2018). Co-culture of dendritic cells and cytokine-induced killer cells effectively suppresses liver cancer stem cell growth by inhibiting pathways in the immune system. *BMC Cancer*, *18*(1), 984. https://doi.org/10.1186/s12885-018-4871-y.

Yang, J., Zhang, L., Yu, C., Yang, X. F., & Wang, H. (2014). Monocyte and macrophage differentiation: Circulation inflammatory monocyte as biomarker for inflammatory diseases. *Biomarker Research*, *2*(1), 1. https://doi.org/10.1186/2050-7771-2-1.

Yano, J. M., Yu, K., Donaldson, G. P., Shastri, G. G., Ann, P., Ma, L., et al. (2015). Indigenous bacteria from the gut microbiota regulate host serotonin biosynthesis. *Cell, 161*(2), 264–276. https://doi.org/10.1016/j.cell.2015.02.047.

Yao, W., Ba, Q., Li, X., Li, H., Zhang, S., Yuan, Y., et al. (2017). A natural CCR2 antagonist relieves tumor-associated macrophage-mediated immunosuppression to produce a therapeutic effect for liver cancer. *eBioMedicine, 22*, 58–67. https://doi.org/10.1016/j.ebiom.2017.07.014.

Yao, W., He, J. C., Yang, Y., Wang, J. M., Qian, Y. W., Yang, T., et al. (2017). The prognostic value of tumor-infiltrating lymphocytes in hepatocellular carcinoma: A systematic review and meta-analysis. *Scientific Reports*, 7(1), 7525. https://doi.org/10.1038/s41598-017-08128-1.

Yao, R. R., Li, J. H., Zhang, R., Chen, R. X., & Wang, Y. H. (2018). M2-polarized tumor-associated macrophages facilitated migration and epithelial-mesenchymal transition of HCC cells via the TLR4/STAT3 signaling pathway. *World Journal of Surgical Oncology, 16*(1), 9. https://doi.org/10.1186/s12957-018-1312-y.

Ye, L., Zhang, Q., Cheng, Y., Chen, X., Wang, G., Shi, M., et al. (2018). Tumor-derived exosomal HMGB1 fosters hepatocellular carcinoma immune evasion by promoting TIM-1(+) regulatory B cell expansion. *Journal for Immunotherapy of Cancer, 6*(1), 145. https://doi.org/10.1186/s40425-018-0451-6.

Ye, Y. C., Zhao, J. L., Lu, Y. T., Gao, C. C., Yang, Y., Liang, S. Q., et al. (2019). NOTCH signaling via WNT regulates the proliferation of alternative, CCR2-independent tumor-associated macrophages in hepatocellular carcinoma. *Cancer Research, 79*(16), 4160–4172. https://doi.org/10.1158/0008-5472.CAN-18-1691.

Yeung, O. W., Lo, C. M., Ling, C. C., Qi, X., Geng, W., Li, C. X., et al. (2015). Alternatively activated (M2) macrophages promote tumour growth and invasiveness in hepatocellular carcinoma. *Journal of Hepatology, 62*(3), 607–616. https://doi.org/10.1016/j.jhep.2014.10.029.

Yi, Y., He, H. W., Wang, J. X., Cai, X. Y., Li, Y. W., Zhou, J., et al. (2013). The functional impairment of HCC-infiltrating gammadelta T cells, partially mediated by regulatory T cells in a TGFbeta- and IL-10-dependent manner. *Journal of Hepatology, 58*(5), 977–983. https://doi.org/10.1016/j.jhep.2012.12.015.

Yildiz, G., Arslan-Ergul, A., Bagislar, S., Konu, O., Yuzugullu, H., Gursoy-Yuzugullu, O., et al. (2013). Genome-wide transcriptional reorganization associated with senescence-to-immortality switch during human hepatocellular carcinogenesis. *PLoS One, 8*(5), e64016. https://doi.org/10.1371/journal.pone.0064016.

Yin, Z., Ma, T., Lin, Y., Lu, X., Zhang, C., Chen, S., et al. (2018). IL-6/STAT3 pathway intermediates M1/M2 macrophage polarization during the development of hepatocellular carcinoma. *Journal of Cellular Biochemistry, 119*(11), 9419–9432. https://doi.org/10.1002/jcb.27259.

Yona, S., Kim, K. W., Wolf, Y., Mildner, A., Varol, D., Breker, M., et al. (2013). Fate mapping reveals origins and dynamics of monocytes and tissue macrophages under homeostasis. *Immunity, 38*(1), 79–91. https://doi.org/10.1016/j.immuni.2012.12.001.

Yoon, H. Y., Kim, H. N., Lee, S. H., Kim, S. J., Chang, Y., Ryu, S., et al. (2018). Association between neutrophil-to-lymphocyte ratio and gut microbiota in a large population: A retrospective cross-sectional study. *Scientific Reports, 8*(1). https://doi.org/10.1038/s41598-018-34398-4, 16031.

Yoon, J. S., Song, B. G., Lee, J. H., Lee, H. Y., Kim, S. W., Chang, Y., et al. (2019). Adjuvant cytokine-induced killer cell immunotherapy for hepatocellular carcinoma: A propensity score-matched analysis of real-world data. *BMC Cancer, 19*(1), 523. https://doi.org/10.1186/s12885-019-5740-z.

Yoshimoto, S., Loo, T. M., Atarashi, K., Kanda, H., Sato, S., Oyadomari, S., et al. (2013). Obesity-induced gut microbial metabolite promotes liver cancer through senescence secretome. *Nature, 499*(7456), 97–101. https://doi.org/10.1038/nature12347.

Yu, M., Luo, H., Fan, M., Wu, X., Shi, B., Di, S., et al. (2018). Development of GPC3-specific chimeric antigen receptor-engineered natural killer cells for the treatment of hepatocellular carcinoma. *Molecular Therapy*, *26*(2), 366–378. https://doi.org/10.1016/j.ymthe.2017.12.012.

Yu, S. J., Ma, C., Heinrich, B., Brown, Z. J., Sandhu, M., Zhang, Q., et al. (2019). Targeting the crosstalk between cytokine-induced killer cells and myeloid-derived suppressor cells in hepatocellular carcinoma. *Journal of Hepatology*, *70*(3), 449–457. https://doi.org/10.1016/j.jhep.2018.10.040.

Yu, A. I., Zhao, L., Eaton, K. A., Ho, S., Chen, J., Poe, S., et al. (2020). Gut microbiota modulate CD8 t cell responses to influence colitis-associated tumorigenesis. *Cell Reports*, *31*(1). https://doi.org/10.1016/j.celrep.2020.03.035, 107471.

Yu, X., Zhao, H., Liu, L., Cao, S., Ren, B., Zhang, N., et al. (2014). A randomized phase II study of autologous cytokine-induced killer cells in treatment of hepatocellular carcinoma. *Journal of Clinical Immunology*, *34*(2), 194–203. https://doi.org/10.1007/s10875-013-9976-0.

Yuan, Y. H., Zhou, C. F., Yuan, J., Liu, L., Guo, X. R., Wang, X. L., et al. (2016). NaHCO3 enhances the antitumor activities of cytokine-induced killer cells against hepatocellular carcinoma HepG2 cells. *Oncology Letters*, *12*(5), 3167–3174. https://doi.org/10.3892/ol.2016.5112.

Zawada, A. M., Rogacev, K. S., Schirmer, S. H., Sester, M., Bohm, M., Fliser, D., et al. (2012). Monocyte heterogeneity in human cardiovascular disease. *Immunobiology*, *217*(12), 1273–1284. https://doi.org/10.1016/j.imbio.2012.07.001.

Zhang, X., Fu, X., Li, T., & Yan, H. (2019). The prognostic value of myeloid derived suppressor cell level in hepatocellular carcinoma: A systematic review and meta-analysis. *PLoS One*, *14*(12), e0225327. https://doi.org/10.1371/journal.pone.0225327.

Zhang, H., He, G., Kong, Y., Chen, Y., Wang, B., Sun, X., et al. (2017). Tumour-activated liver stromal cells regulate myeloid-derived suppressor cells accumulation in the liver. *Clinical and Experimental Immunology*, *188*(1), 96–108. https://doi.org/10.1111/cei.12917.

Zhang, D., Li, Z., Li, T., Luo, D., Feng, X., Liu, Y., et al. (2018). miR-517a promotes Warburg effect in HCC by directly targeting FBP1. *Oncotargets and Therapy*, *11*, 8025–8032. https://doi.org/10.2147/OTT.S172084.

Zhang, Z., Ma, L., Goswami, S., Ma, J., Zheng, B., Duan, M., et al. (2019). Landscape of infiltrating B cells and their clinical significance in human hepatocellular carcinoma. *Oncoimmunology*, *8*(4), e1571388. https://doi.org/10.1080/2162402X.2019.1571388.

Zhang, Y., Pan, N., Sheng, Y., Zhou, M., Wen, Z., Chen, Y., et al. (2019). Hypoxia enhances IL-10-producing B cell generation through upregulating high-mobility group B1 on tumor cell-released autophagosomes. *Immunology Letters*, *216*, 36–42. https://doi.org/10.1016/j.imlet.2019.09.005.

Zhang, W., Song, Z., Xiao, J., Liu, X., Luo, Y., Yang, Z., et al. (2019). Blocking the PD-1/PD-L1 axis in dendritic cell-stimulated Cytokine-Induced Killer Cells with pembrolizumab enhances their therapeutic effects against hepatocellular carcinoma. *Journal of Cancer*, *10*(11), 2578–2587. https://doi.org/10.7150/jca.26961.

Zhang, Q., Wang, H., Mao, C., Sun, M., Dominah, G., Chen, L., et al. (2018). Fatty acid oxidation contributes to IL-1beta secretion in M2 macrophages and promotes macrophage-mediated tumor cell migration. *Molecular Immunology*, *94*, 27–35. https://doi.org/10.1016/j.molimm.2017.12.011.

Zhang, D., Wang, Y., Shi, Z., Liu, J., Sun, P., Hou, X., et al. (2015). Metabolic reprogramming of cancer-associated fibroblasts by IDH3alpha downregulation. *Cell Reports*, *10*(8), 1335–1348. https://doi.org/10.1016/j.celrep.2015.02.006.

Zhang, H. L., Yu, L. X., Yang, W., Tang, L., Lin, Y., Wu, H., et al. (2012). Profound impact of gut homeostasis on chemically-induced pro-tumorigenic inflammation and hepatocarcinogenesis in rats. *Journal of Hepatology*, *57*(4), 803–812. https://doi.org/10.1016/j.jhep.2012.06.011.

Zhang, Z., Zhang, J., He, P., Han, J., & Sun, C. (2020). Interleukin-37 suppresses hepatocellular carcinoma growth through inhibiting M2 polarization of tumor-associated macrophages. *Molecular Immunology*, *122*, 13–20. https://doi.org/10.1016/j.molimm.2020.03.012.

Zhang, J., Zhang, Q., Lou, Y., Fu, Q., Chen, Q., Wei, T., et al. (2018). Hypoxia-inducible factor-1alpha/interleukin-1beta signaling enhances hepatoma epithelial-mesenchymal transition through macrophages in a hypoxic-inflammatory microenvironment. *Hepatology*, *67*(5), 1872–1889. https://doi.org/10.1002/hep.29681.

Zhao, N., Dang, H., Ma, L., Martin, S. P., Forgues, M., Ylaya, K., et al. (2020). Intratumoral gammadelta T-cell infiltrates, CCL4/5 protein expression and survival in patients with hepatocellular carcinoma. *Hepatology*. https://doi.org/10.1002/hep.31412.

Zhao, S., Jang, C., Liu, J., Uehara, K., Gilbert, M., Izzo, L., et al. (2020). Dietary fructose feeds hepatic lipogenesis via microbiota-derived acetate. *Nature*, *579*(7800), 586–591. https://doi.org/10.1038/s41586-020-2101-7.

Zhao, W., Su, W., Kuang, P., Zhang, L., Liu, J., Yin, Z., et al. (2012). The role of hepatic stellate cells in the regulation of T-cell function and the promotion of hepatocellular carcinoma. *International Journal of Oncology*, *41*(2), 457–464. https://doi.org/10.3892/ijo.2012.1497.

Zhao, Q., Xiao, X., Wu, Y., Wei, Y., Zhu, L. Y., Zhou, J., et al. (2011). Interleukin-17-educated monocytes suppress cytotoxic T-cell function through B7-H1 in hepatocellular carcinoma patients. *European Journal of Immunology*, *41*(8), 2314–2322. https://doi.org/10.1002/eji.201041282.

Zhao, W., Xu, Y., Xu, J., Wu, D., Zhao, B., Yin, Z., et al. (2015). Subsets of myeloid-derived suppressor cells in hepatocellular carcinoma express chemokines and chemokine receptors differentially. *International Immunopharmacology*, *26*(2), 314–321. https://doi.org/10.1016/j.intimp.2015.04.010.

Zheng, D., Liwinski, T., & Elinav, E. (2020). Interaction between microbiota and immunity in health and disease. *Cell Research*, *30*(6), 492–506. https://doi.org/10.1038/s41422-020-0332-7.

Zheng, R., Wang, G., Pang, Z., Ran, N., Gu, Y., Guan, X., et al. (2020). Liver cirrhosis contributes to the disorder of gut microbiota in patients with hepatocellular carcinoma. *Cancer Medicine*, *9*(12), 4232–4250. https://doi.org/10.1002/cam4.3045.

Zheng, C., Zheng, L., Yoo, J. K., Guo, H., Zhang, Y., Guo, X., et al. (2017). Landscape of infiltrating T cells in liver cancer revealed by single-cell sequencing. *Cell*, *169*(7), 1342–1356. e1316 https://doi.org/10.1016/j.cell.2017.05.035.

Zhou, S., Chen, L., Qin, J., Li, R., Tao, H., Zhen, Z., et al. (2013). Depletion of CD4+ CD25+ regulatory T cells promotes CCL21-mediated antitumor immunity. *PLoS One*, *8*(9), e73952. https://doi.org/10.1371/journal.pone.0073952.

Zhou, S. L., Dai, Z., Zhou, Z. J., Wang, X. Y., Yang, G. H., Wang, Z., et al. (2012). Overexpression of CXCL5 mediates neutrophil infiltration and indicates poor prognosis for hepatocellular carcinoma. *Hepatology*, *56*(6), 2242–2254. https://doi.org/10.1002/hep.25907.

Zhou, X., Li, W., Wang, S., Zhang, P., Wang, Q., Xiao, J., et al. (2019). YAP aggravates inflammatory bowel disease by regulating M1/M2 macrophage polarization and gut microbial homeostasis. *Cell Reports*, *27*(4), 1176–1189. e1175 https://doi.org/10.1016/j.celrep.2019.03.028.

Zhou, J., Liu, M., Sun, H., Feng, Y., Xu, L., Chan, A. W. H., et al. (2018). Hepatoma-intrinsic CCRK inhibition diminishes myeloid-derived suppressor cell immunosuppression and enhances immune-checkpoint blockade efficacy. *Gut*, *67*(5), 931–944. https://doi.org/10.1136/gutjnl-2017-314032.

Zhou, J., Ma, P., Li, J., Cui, X., & Song, W. (2016). Improvement of the cytotoxic T lymphocyte response against hepatocellular carcinoma by transduction of cancer cells with an adeno-associated virus carrying the interferon-gamma gene. *Molecular Medicine Reports*, *13*(4), 3197–3205. https://doi.org/10.3892/mmr.2016.4884.

Zhou, D., Pan, Q., Shen, F., Cao, H. X., Ding, W. J., Chen, Y. W., et al. (2017). Total fecal microbiota transplantation alleviates high-fat diet-induced steatohepatitis in mice via beneficial regulation of gut microbiota. *Scientific Reports*, 7(1), 1529. https://doi.org/10.1038/s41598-017-01751-y.

Zhou, G., Sprengers, D., Boor, P. P. C., Doukas, M., Schutz, H., Mancham, S., et al. (2017). Antibodies against immune checkpoint molecules restore functions of tumor-infiltrating T cells in hepatocellular carcinomas. *Gastroenterology*, *153*(4), 1107–1119 e1110. https://doi.org/10.1053/j.gastro.2017.06.017.

Zhou, A., Tang, L., Zeng, S., Lei, Y., Yang, S., & Tang, B. (2020). Gut microbiota: A new piece in understanding hepatocarcinogenesis. *Cancer Letters*, *474*, 15–22. https://doi.org/10.1016/j.canlet.2020.01.002.

Zhou, Z. Q., Tong, D. N., Guan, J., Tan, H. W., Zhao, L. D., Zhu, Y., et al. (2016). Follicular helper T cell exhaustion induced by PD-L1 expression in hepatocellular carcinoma results in impaired cytokine expression and B cell help, and is associated with advanced tumor stages. *American Journal of Translational Research*, *8*(7), 2926–2936. Retrieved from https://www.ncbi.nlm.nih.gov/pubmed/27508013.

Zhou, M., Wen, Z., Cheng, F., Ma, J., Li, W., Ren, H., et al. (2016). Tumor-released autophagosomes induce IL-10-producing B cells with suppressive activity on T lymphocytes via TLR2-MyD88-NF-kappaB signal pathway. *Oncoimmunology*, *5*(7), e1180485. https://doi.org/10.1080/2162402X.2016.1180485.

Zhou, L., Zhang, M., Wang, Y., Dorfman, R. G., Liu, H., Yu, T., et al. (2018). Faecalibacterium prausnitzii produces butyrate to maintain Th17/Treg balance and to ameliorate colorectal colitis by inhibiting histone deacetylase 1. *Inflammatory Bowel Diseases*, *24*(9), 1926–1940. https://doi.org/10.1093/ibd/izy182.

Zhou, D., Zhang, Y., Xu, L., Zhou, Z., Huang, J., & Chen, M. (2015). A monocyte/granulocyte to lymphocyte ratio predicts survival in patients with hepatocellular carcinoma. *Scientific Reports*, *5*, *15263*. https://doi.org/10.1038/srep15263.

Zhou, S. L., Zhou, Z. J., Hu, Z. Q., Huang, X. W., Wang, Z., Chen, E. B., et al. (2016). Tumor-associated neutrophils recruit macrophages and T-regulatory cells to promote progression of hepatocellular carcinoma and resistance to sorafenib. *Gastroenterology*, *150*(7), 1646–1658 e1617. https://doi.org/10.1053/j.gastro.2016.02.040.

Zhou, T. Y., Zhou, Y. L., Qian, M. J., Fang, Y. Z., Ye, S., Xin, W. X., et al. (2018). Interleukin-6 induced by YAP in hepatocellular carcinoma cells recruits tumor-associated macrophages. *Journal of Pharmacological Sciences*, *138*(2), 89–95. https://doi.org/10.1016/j.jphs.2018.07.013.

Zhu, K., He, C., Liu, S. Q., Qu, M., Xie, T., Yang, X., et al. (2019). Lineage tracking the generation of T regulatory cells from microbial activated T effector cells in naive mice. *Frontiers in Immunology*, *10*, 3109. https://doi.org/10.3389/fimmu.2019.03109.

Zhu, W., Peng, Y., Wang, L., Hong, Y., Jiang, X., Li, Q., et al. (2018). Identification of alpha-fetoprotein-specific T-cell receptors for hepatocellular carcinoma immunotherapy. *Hepatology*, *68*(2), 574–589. https://doi.org/10.1002/hep.29844.

Zmora, N., Zilberman-Schapira, G., Suez, J., Mor, U., Dori-Bachash, M., Bashiardes, S., et al. (2018). Personalized gut mucosal colonization resistance to empiric probiotics is associated with unique host and microbiome features. *Cell*, *174*(6), 1388–1405 e1321. https://doi.org/10.1016/j.cell.2018.08.041.

Zong, L., Peng, H., Sun, C., Li, F., Zheng, M., Chen, Y., et al. (2019). Breakdown of adaptive immunotolerance induces hepatocellular carcinoma in HBsAg-tg mice. *Nature Communications*, *10*(1), 221. https://doi.org/10.1038/s41467-018-08096-8.

Zou, Y., Xue, W., Luo, G., Deng, Z., Qin, P., Guo, R., et al. (2019). 1,520 reference genomes from cultivated human gut bacteria enable functional microbiome analyses. *Nature Biotechnology*, *37*(2), 179–185. https://doi.org/10.1038/s41587-018-0008-8.

CHAPTER SIX

Systemic therapy of liver cancer

Tarik Demir, Sunyoung S. Lee, and Ahmed O. Kaseb*
Department of Gastrointestinal Medical Oncology, The University of Texas M.D. Anderson Cancer Center, Houston, TX, United States
*Corresponding author: e-mail address: akaseb@mdanderson.org

Contents

1. Introduction 258
2. Staging 259
3. Liver function 260
4. Current systemic treatment 262
   4.1 First-line treatment 262
   4.2 Second-line treatment 276
   4.3 Adjuvant and neoadjuvant approach 279
5. Histological variants 283
   5.1 Combined hepatocellular carcinoma and cholangiocarcinoma (cHCC-CC) 283
   5.2 Fibrolamellar hepatocellular carcinoma (FLHCC) 284
6. Conclusion 285
Acknowledgment 286
References 286

Abstract

Hepatocellular carcinoma (HCC) is a lethal malignancy with poor prognosis. More than 80% of patients are diagnosed at an advanced stage, and most patients with HCC also have liver cirrhosis that complicates cancer management. No targeted treatment options currently exist outside genomics-based clinical trials. Multiple tyrosine kinase inhibitors (mTKIs) such as sorafenib, lenvatinib, cabozantinib, and regorafenib have been used to treat advanced hepatocellular carcinoma (aHCC). Immune checkpoint inhibitors including nivolumab and pembrolizumab have shown survival benefit. More recently, atezolizumab in combination with bevacizumab resulted in improved overall survival and progression-free survival, compared with sorafenib in patients with aHCC in the first-line setting. The combination of nivolumab with ipilimumab as an alternative in the treatment of patients treated with sorafenib has inspired various combination studies of immune checkpoint inhibitors. Currently, ongoing studies of systemic therapy consist of various immune-based combination therapies. Finally, there is no established adjuvant and neoadjuvant therapy although a few early phase studies show promising results. In this chapter, we summarize current approaches of systemic treatment in patients with liver cancer.

Advances in Cancer Research, Volume 149
ISSN 0065-230X
https://doi.org/10.1016/bs.acr.2020.12.001

1. Introduction

Liver cancer is treated in a multidisciplinary fashion. Surgical resection, locoregional therapies (i.e., radiofrequency/microwave ablation, transarterial embolization, radiation therapy), systemic treatments (i.e., chemotherapy, multi-tyrosine kinase inhibitors (mTKIs), immunotherapy), and liver transplantation are available options (Couri & Pillai, 2019).

Liver function, tumor burden, tumor location, performance status, and portal hypertension are essential factors in the treatment decision (Villanueva, 2019). The Barcelona Clinic Liver Cancer (BCLC) stage is the staging system that involves these factors and is widely used for selecting treatment plans (Llovet, Bru, & Bruix, 1999). Transplant decisions frequently use criteria from the United Network for Organ Sharing (UNOS) and Model for End-Stage Liver Disease (MELD) scores (Kaseb, Abaza, & Roses, 2013; Martin, Bartels, Hauss, & Fangmann, 2007).

Hepatocellular carcinoma (HCC) is the most common liver cancer and is the fourth leading cause of cancer-related deaths worldwide (Siegel, Miller, & Jemal, 2020). The etiology of HCC consists of 71% preventable causes, such as obesity, alcohol abuse, and hepatitis B and C viruses (Islami et al., 2018). The 5-year overall survival (OS) is only 18%, making HCC the second-most lethal tumor after pancreatic cancer (Jemal et al., 2017). The poor prognosis is due to 70%–80% of patients diagnosed at an advanced stage (Montella, Palmieri, Addeo, & Del Prete, 2016), and HCC is usually accompanied by advanced liver cirrhosis (Pinato et al., 2017).

More recently, various systemic therapy options are applied in advanced HCC (aHCC) (Roderburg, Ozdirik, Wree, Demir, & Tacke, 2020). Systemic treatment options approved by the U.S. Food and Drug Administration (FDA) for aHCC include the following: sorafenib (both first and second-line, 2007), regorafenib (second-line, 2017), lenvatinib (first-line, 2018), nivolumab (both first and second-line, 2017), pembrolizumab (second-line, 2018), cabozantinib (second-line, 2019), ramucirumab (second-line, 2019, in patients with alpha-fetoprotein [AFP] $\geq$ 400 ng/mL) and nivolumab plus ipilimumab (second-line, 2020) (Jindal, Thadi, & Shailubhai, 2019). The IMbrave150 study comparing atezolizumab plus bevacizumab with sorafenib led to the approval of atezolizumab plus bevacizumab in the first-line treatment. This combination is the first systemic treatment that showed improved survival, compared with sorafenib, and was approved by the FDA in May 2020 (Finn et al., 2020; U. S. Food and Drug Administration/Center for Drug Evaluation and Research, 2020).

2. Staging

Staging is essential to determine the treatment plan and prognosis of HCC. However, which staging system best determines the prognosis of HCC is still controversial (Liu et al., 2016; Sherman, 2014).

The tumor, node, metastasis (TNM) staging system appears to be more useful than others in determining the prognosis of patients undergoing surgery (Shindoh et al., 2013). The BCLC system is the most widely accepted staging system (Maida, Orlando, Camma, & Cabibbo, 2014). Many studies have shown that the BCLC staging system is the most useful in predicting survival in HCC patients (Faria, Szklaruk, Kaseb, Hassabo, & Elsayes, 2014). Both the American Association for the Study of Liver Diseases (AASLD) and the European Association for the Study of the Liver (EASL) recommend the BCLC staging system (Torzilli et al., 2013). The Americas Hepato-Pancreato-Biliary Association recommends the TNM system to predict the prognosis of patients undergoing surgery and transplant and the BCLC system for patients who are not surgical candidates (Vauthey et al., 2010). The BCLC staging system incorporates liver function, performance status, and tumor burden (number and size of tumors, presence of macrovascular invasion, and extrahepatic spread) (Llovet et al., 1999). The Child-Turcotte-Pugh (CTP) score is widely used for the assessment of liver function in patients with liver cirrhosis (Table 1). The performance status is evaluated along with the Eastern Cooperative Oncology Group (ECOG) performance status (PS) (Table 2). Fig. 1 shows HCC treatment algorithm.

Table 1 Child-Turcotte-Pugh (CTP) score.

Criteria	1	2	3
Encephalopathy	None	Mild (Grade 1–2)	Severe (Grade 3–4)
Ascites	None	Mild (Diuretic response)	Severe (Refractory to diuretics)
Bilirubin	<2 mg/dL	2–3 mg/dL	>3 mg/dL
Albumin	>3.5 g/dL	2.8–3.5 g/dL	<2.8 g/dL
INR	<1.7	1.7–2.2	>2.2

CTP Class A: 5–6 points; *CTP Class B:* 7–9 points; *CTP Class C:* 10–15 points.

Table 2 Eastern Cooperative Oncology Group (ECOG) performance status (PS).

Score	
0	Fully active, able to carry on all pre-disease performance without restriction
1	Restricted in physically strenuous activities but ambulatory and able to carry out work of a light or sedentary nature
2	Ambulatory and capable of all self-care but unable to carry out any work activities; Up and about >50% of waking hours
3	Capable of only limited self-care, confined to bed or chair >50% of waking hours
4	Completely disabled; cannot carry on any self-care. Totally confined to bed or chair
5	Dead

3. Liver function

HCC most often develops with chronic liver damage. Deterioration of liver function is an important obstacle in the systemic treatment plan (Poh, Goh, Chang, & Tan, 2015).

A study of 3202 patients showed that the underlying liver function of aHCC patients receiving sorafenib impacted the survival and adverse events. Liver function was assessed using the CTP score and classified into three groups (A, B, and C). The median OS was 13.6 (12.8–14.7), 5.2 (4.6–6.3), and 2.6 (1.5–4.0) months, respectively, in CPT-A, B, and C groups. Patients with CTP-B had higher grade adverse events compared to those with CTP-A. Patients with CTP-B showed statistically different survival outcome in the subgroup analysis: the median OS of patients with CTP-B7, B8, and B9 was 6.2 (4.9–8.7), 4.8 (4.1–6.9), and 3.7 (3.0–5.1) months, respectively (Marrero et al., 2016).

Systemic treatment is recommended for aHCC patients with CTP-A, while best supportive care is suggested for those with CTP-C. The survival benefit from systemic therapy is controversial in patients with CTP-B (Bouattour, Mehta, He, Cohen, & Nault, 2019). Except for sorafenib and nivolumab, only CTP-A patients were included in HCC trials of systemic treatment options (Cheng et al., 2009; Kudo et al., 2019). The National Comprehensive Cancer Network (NCCN) guideline recommends sorafenib and nivolumab up to CTP-B (National Comprehensive Cancer Network).

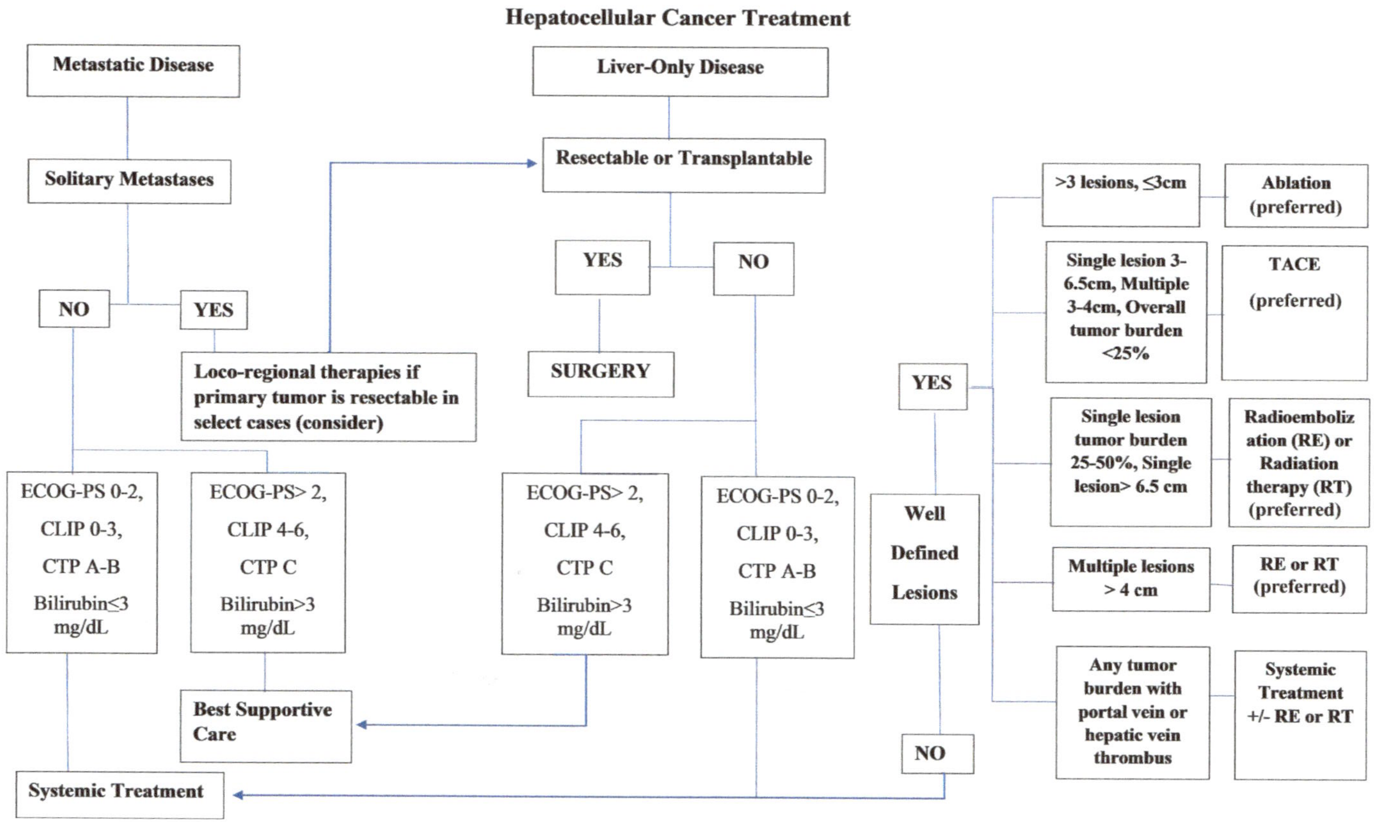

Fig. 1 Clinical algorithm for the management of hepatocellular carcinoma. CLIP, Cancer of the Liver Italian Program; CTP, Child-Turcotte-Pugh score; TACE, transarterial chemoembolization; ECOG-PS, Eastern Cooperative Oncology Group-Performance Status.

An important research topic is calculating liver function rationally and objectively to identify patients who will benefit from treatment. Many studies have shown that albumin-bilirubin (ALBI) grade can be an important alternative to the CTP score in liver function evaluation. The ALBI grade is calculated using the following formula: (log (bilirubin [mcmol/L]) $\times$ 0.66)—(albumin [g/L]) $\times$ 0.085. The resulting score is used to classify patients into three groups: A1 ≤ -2.6, A2 > -2.6 and ≤ -1.39, and A3 > -1.39 (Johnson et al., 2015). Kuo et al. estimated the median OS using the ALBI grade in HCC patients after sorafenib treatment: median OSs for A1, A2, and A3 were 14.8, 5.0, and 0.7 months, respectively. Patients whose ALBI grade improved after treatment had longer OS (6.4 vs 2.1 months, $P < 0.001$) (Kuo et al., 2017). Pinato et al. showed the importance of pre-treatment ALBI grade in patients who received immune checkpoint inhibitors: median OSs in A1, A2, and A3 were 22.5, 9.6, and 4.6 months (Pinato et al., 2020). Many studies show that the ALBI grade may better prognosticate aHCC than the CTP score (Edeline et al., 2016; Lee et al., 2018; Pinato et al., 2017).

The CTP score includes subjective criteria such as encephalopathy and ascites, while the ALBI grade evaluates liver function with bilirubin and albumin. The ALBI score may be a more objective method, although more extensive studies are needed.

4. Current systemic treatment

Systemic treatment is recommended with or without local treatment for BCLC-C and BCLC-B patients who are not appropriate for surgery (National Comprehensive Cancer Network). Table 3 shows the current systemic treatment options for HCC patients.

4.1 First-line treatment

The IMbrave150 study showed a survival advantage of atezolizumab plus bevacizumab over sorafenib, and it became the first-line therapy for aHCC.

Vogel et al. supported the superiority of atezolizumab plus bevacizumab over other options in first-line treatment of aHCC patients in a network meta-analysis. The records of 8783 patients from 55 studies were screened, and the data of 9 studies were found suitable. This analysis showed that atezolizumab plus bevacizumab improved survival over lenvatinib (OS, HR 0.63; progression-free survival [PFS], HR 0.91) and nivolumab (OS, HR 0.68; PFS, HR 0.63). This supports that atezolizumab plus bevacizumab

Table 3 Approved systemic therapies for advanced-stage HCC.

Drug	Target	Therapy-line	Approved year	N	PFS (median, months)	OS (median, months)
Sorafenib (SHARP)	VEGFR1–3, PDGFR-β, KIT kinases	1.line	2007	602	5.5 vs 2.8 mo	10.7 vs 7.9 mo HR: 0.69
(Asya-Pasific)		1.line		271	4.2	6.5 vs 4.2 mo HR: 0.68
Lenvatinib (REFLECT)	FGFR1–4, VEGFR1–3, PDGFRα, RET, KIT	1.line	2018	954	8.9 vs 3.7 mo	13.6 vs 12.3 mo HR: 0.92
Atezolizumab plus Bevacizumab (IMbrave 150)	PD-L1, VEGF	1.line	2020	501	6.8 vs 4.3 mo	N/A vs 13.2 mo HR: 0.58
Regorafenib (RESORCE)	Tie2, VEGFR, PDGFR, FGFR, KIT,RET, RAF-1, BRAF	2.line	2017	573	3.1 vs 1.5 mo	10.6 vs 7.8 mo HR: 0.63
Cabozantinib (CELESTIAL)	c-Met, VEGFR-2, AXL, RET	2.line	2019	707	5.2 vs 1.9 mo	10.2 vs 8.0 mo HR: 0.76
Nivolumab (CheckMate040)	PD-1	1./2.line	2017	214	4 mo	13.2 mo
Pembrolizumab (KEYNOTE224)	PD-1	2.line	2018	104	4.9 mo	12.9 mo
Ramucirumab (REACH-2)	VEGFR2	2.line	2019	292 AFP > 400 ng/mL	2.8 vs 1.6 mo	8.5 vs 7.3 mo HR: 0.710
Nivolumab plus Ipilimumab (CheckMate-040)	PD-1, CTLA-4	2.line	2020	148	N/A	25.5 mo

HR, hazard ratio; mo, month; N, number of patients; N/A, not available; OS, overall survival; PFS, progression-free survival.

improves OS and PFS over lenvatinib and nivolumab in the first-line treatment of aHCC patients (Vogel et al., 2020). Currently, multiple clinical trials of TKIs and immune checkpoint inhibitors are underway for the treatment of aHCC (Table 4).

4.1.1 Multi-tyrosine kinase inhibitor

There was no effective systemic treatment option until sorafenib was proven effective in aHCC (Llovet et al., 2008). Sorafenib is a multi-tyrosine kinase inhibitor (VEGFR1, -2, and 3; PDGFR-β; and KIT) (Ranieri et al., 2012). Two phase 3 trials (SHARP and Asia-Pacific) showed OS benefit of sorafenib over placebo. The SHARP study included 602 treatment-naïve patients with aHCC. Patients were divided into the sorafenib (400 mg twice daily) and placebo cohorts. Sorafenib showed better PFS and OS vs placebo (PFS 5.5 vs 2.8 months; OS 10.7 vs 7.9 months). Partial response (PR) was seen in 2% of patients in the sorafenib cohort. The most common treatment-related adverse events in patients treated with sorafenib included diarrhea, weight loss, palmar-plantar erythrodysesthesia, and hypophosphatemia. The Asia-Pacific study had 271 patients, and only Asian patients were included. This study showed that sorafenib was effective in the first-line setting (OS 6.5 vs 4.2 months, HR 0.68, $P=0.014$) (Llovet et al., 2008). In the subgroup analysis, sorafenib had better efficacy in patients with no extrahepatic disease, hepatitis C-associated disease, and a low neutrophil-lymphocyte ratio (Bruix et al., 2017). Other mTKIs (brivanib, sunitinib, erlotinib, and linifanib) have also been tested, and none of them showed superiority over sorafenib (Cainap et al., 2015; Cheng et al., 2013; Johnson et al., 2013; Zhu et al., 2015). Sorafenib was the only FDA-approved option for aHCC until April 2017, when regorafenib was approved for the second-line therapy, and the only approved first-line medication until lenvatinib was approved in August 2018.

Lenvatinib is an mTKI (VEGFR1, -2, and -3; FGFR1, -2, -3, and -4; PDGFR-α; RET; and KIT) (Kudo, 2017). Its use as a first-line treatment for aHCC was established by REFLECT, a noninferiority study comparing lenvatinib (12 mg/day for bodyweight ≥60 kg or 8 mg/day for bodyweight <60 kg) with sorafenib (400 mg twice daily). The study enrolled 954 treatment-naïve patients who were randomly assigned to receive lenvatinib ($n=478$) or sorafenib ($n=476$). The median OS was 13.6 months for patients treated with lenvatinib, which was non-inferior to that of patients treated with sorafenib (12.3 months). The median PFS for patients treated with lenvatinib was 7.4 months, the objective response rate (ORR) was

Table 4 Ongoing clinical trials for advanced HCC.

A: Ongoing clinical trials of immune checkpoint inhibitor monotherapy for advanced HCC

Trial identifier	Phase	Drug	Target	N	Patient group	ORR (%)	DCR (%)	PFS (median, months)	OS (median, months)
NCT01658878 (CheckMate040)	I/II	Nivolumab	PD-1	48[a] 214[b]	Naive/pre-treated	20%	64%	4	13.2
NCT02576509 (CheckMate459)	III	Nivolumab vs Sorafenib	PD-1	743	Naive	15% vs 7%	N/A	3.7 vs 3.8	16.4 vs 14.7
NCT02702414 (KEYNOTE224)	II	Pembrolizumab	PD-1	104	Pre-treated	17%	61%	4.9	12.9
NCT02702401 (KEYNOTE240)	III	Pembrolizumab vs placebo	PD-1	413	Pre-treated	18.3% vs 4.4%	62.2% vs 53.3%	3 vs 2.8	13.9 vs 10.6
NCT01693562	I/II	Durvalumab	PD-L1	39	Pre-treated	10.3%	33.3%	N/A	13.2
NCT01008358	II	Tremelimumab	CTLA-4	20	Pre-treated	17.6%	76.4%	6.48	8.2
NCT03412773 (Rationale301)	III	Tislelizumab vs sorafenib	PD-1	674	Naive	N/A	N/A	N/A	N/A
NCT03062358 (KEYNOTE394)	III	Pembrolizumab vs placebo	PD-1	N/A	Pre-treated	N/A	N/A	N/A	N/A
NCT02989922	II	Camrelizumab	PD-1	217	Pre-treated	14.7%	44.2%	N/A	N/A

Continued

Table 4 Ongoing clinical trials for advanced HCC.—cont'd

B: Ongoing clinical trials of immune checkpoint inhibitor combined therapy for advanced HCC

Trial identifier	Phase	Drug	Target	N	Patient group	ORR (%)	DCR (%)	PFS (median, mo)	OS (median, mo)
ICI plus ICI									
NCT01658878 (CheckMate-040)	I/II	Nivolumab plus Ipilimumab	PD-1 + CTLA-4	50	Pre-treated	32%	54%	N/A	22.8
NCT02519348 (TD300 + D)	I/II	Durvalumab ± Tremelimumab	PD-1 + CTLA-4	40	Naive/ pre-treated	22.7%	57.5%	N/A	18.7
NCT04039607 (CheckMate9DW)	III	Nivolumab plus Ipilimumab vs Sorafenib or Lenvatinib	PD-1 + CTLA-4	1084	Naive	N/A	N/A	N/A	N/A
NCT03298451 (HIMALAYA)	III	Durvalumab ± Tremelimumab vs Sorafenib or Lenvatinib	PD-1 + CTLA-4	1310	Naive	N/A	N/A	N/A	N/A
ICI plus anti-vascular endothelial growth factor (VEGF)									
NCT02715531 (GO30140)	Ib	Atezolizumab + Bevacizumab	PD-L1 + VEGF	60	Naive	27%	68%	5.6	N/R
NCT02572687	Ib	Ramucirumab +Durvalumab	PD-L1 + VEGF	28	Pre-treatment	11%	61%	4.4	10.7
NCT03434379 (IMbrave150)	III	Atezolizumab + Bevacizumab vs sorafenib	PD-L1 + VEGF	501	Naive	27.3% vs 11.9%	73.6% vs 55.3%	6.8 vs 4.3	N/R vs 13.2
NCT03794440 (ORIENT32)	III	Sintilimab + Bevacizumab biosimilar vs Sorafenib	PD-1 + VEGF	566	Naive	N/A	N/A	N/A	N/A

ICI plus multi-tyrosin kinase inhibitors (mTKI)									
NCT03006926 (KEYNOTE524)	Ib	Lenvatinib + Pembrolizumab	PD-1 + mTKI	104	Naive	36%	88%	8.6	22
NCT3463876 (RESCUE)	I/II	SHR-1210 (Camrelizumab) + Apatinib	PD-1 + mTKI	18	Naive	38.9%	83.3%	7.2	N/A
NCT01658878 (CheckMate040)	I/II	Cabozantinib + Nivolumab ± Ipilimumab	PD-1 + mTKI ± CTLA-4	71	Naive	17% vs 26%	81% vs 83%	5.5 vs 6.8	N/R
NCT03289533 (VEGFLiver100)	I/II	Avelumab + Axitinib	PD-L1 + mTKI	AFP > 400 ng/mL $n=22$	Naive	13.6%	68.2%	5.5	12.7
NCT03170960 (COSMIC021)	I/II	Cabozantinib + Atezolizumab	PD-L1 + mTKI	N/A	Naive	N/A	N/A	N/A	N/A
NCT03764293	III	SHR-1210 (Camrelizumab) + Apatinib vs Sorafenib	PD-1 + mTKI	510	Naive	N/A	N/A	N/A	N/A
NCT03755791 (COSMIC312)	III	Cabozantinib + Atezolizumab vsSorafenib vs Cabozantinib	PD-L1 + mTKI	740	Naive	N/A	N/A	N/A	N/A
NCT03713593 (LEAP002)	III	Lenvatinib + Pembrolizumab versus Lenvatinib	PD-1 + mTKI	750	Naive	N/A	N/A	N/A	N/A
NCT03347292	I/II	Regorafenib + Pembrolizumab	PD-1 + mTKI	57	Naive	N/A	N/A	N/A	N/A
NCT03539822 (CAMILLA)	I/II	Cabozantinib + Durvalumab	PD-L1 + mTKI	30	Pre-treated	N/A	N/A	N/A	N/A

Continued

Table 4 Ongoing clinical trials for advanced HCC.—cont'd

B: Ongoing clinical trials of immune checkpoint inhibitor combined therapy for advanced HCC

Trial identifier	Phase	Drug	Target	N	Patient group	ORR (%)	DCR (%)	PFS (median, mo)	OS (median, mo)
NCT03475953 (REGOMUNE)	II	Regorafenib + Avelumab	PD-L1 + mTKI	212	Pre-treatment	N/A	N/A	N/A	N/A
NCT03418922	I/II	Lenvatinib + Nivolumab	PD-1 + mTKI	30	Naive	N/A	N/A	N/A	N/A
ICI plus Chemotherapy (CT)									
NCT03092895	II	Camrelizumab + FOLFOX4 or GEMOX	PD-1 + CT	34	Naive	26.5%	79.4%	5.5	N/A
NCT00471965	III	FOLFOX4 vs Doxorubicin	CT vs CT	371	Naive	8.6% vs 1.4%	47.1% vs 26.6%	2.4 vs 1.7	5.9 vs 4.3
NCT04380545	I/II	Nivolumab + Fluorouracil + Interferon alfa 2b	PD-1 + CT + INF	15	Naive	N/A	N/A	N/A	N/A
NCT03605706	III	Camrelizumab + FOLFOX4 vs FOLFOX or Sorafenib	PD-1 + CT	448	Naive	N/A	N/A	N/A	N/A
ICI plus Yttrium-90 radioembolization (Y90-RE)									
NCT03033446	II	Nivolumab + Y90-RE	PD-1 + TARE	40	Naive	31%	58.3%	4.6	15.1
NCT04124991	I/II	Durvalumab + Y90-RE	PD-L1 + TARE	24	Pre-treatment	N/A	N/A	N/A	N/A
NCT03099564	I	Pembrolizumab + Y90-RE	PD-1 + TARE	30	Naive	N/A	N/A	N/A	N/A

Others ICI combinetherapy									
NCT01853618	I/II	Tremelimumab + ablation	CTLA-4	32	Pre-treatment	26%	85%	7.4	12.3
NCT03778957 (Emrald01)	III	Durvalumab + TACE vs Durvalumab + TACE + Bevacizumab vs TACE	PD-L1 + VEGF	600	Naive	N/A	N/A	N/A	N/A
NCT03316872	II	Pembrolizumab + SIRT	PD-1 + Radiation	30	Pre-treatment	N/A	N/A	N/A	N/A
NCT03695250	I/II	BMS986205 + Nivolumab	IDO-1 + PD-1	23	Naive/ pre-treatment	N/A	N/A	N/A	N/A

C: Ongoing clinical trials of non-ICI Immunotherapy for Advanced HCC

Trial identifier	Phase	Drug	Target	N	Patient group	ORR (%)	DCR (%)	PFS (median, mo)	OS (median, mo)
Tumor-Infiltrating Lymphocyte (TIL) Therapy									
NCT01174121	II	TILs + Aldesleukin + CT + Pembrolizumab	TILs + IL2 + PD-1		Pre-treatment	N/A	N/A	N/A	N/A
T Cell Receptor (TCR) Therapy									
NCT02719782	I	TCR-T, HBsAg +	TCR	10	Pre-treatment	N/A	N/A	N/A	N/A
NCT02686372	I	TCR-T, HBsAg +	TCR	10	Pre-treatment	N/A	N/A	N/A	N/A
NCT04368182	I	TCR-T, AFP +	TCR	3	Pre-treatment	N/A	N/A	N/A	N/A

Continued

Table 4 Ongoing clinical trials for advanced HCC.—cont'd

C: Ongoing clinical trials of non-ICI Immunotherapy for Advanced HCC

Trial identifier	Phase	Drug	Target	N	Patient group	ORR (%)	DCR (%)	PFS (median, mo)	OS (median, mo)
NCT03132792	I	TCR-T, AFP+	TCR	45	Pre-treatment	N/A	N/A	N/A	N/A
Chimeric Antigen Receptor (CAR)-T Cell Therapy									
NCT03993743	I	CD147-CART	CAR-T Cell	34	Pre-treatment	N/A	N/A	N/A	N/A
NCT04121273	I	GPC3-CART	CAR-T Cell	20	Pre-treatment	N/A	N/A	N/A	N/A
NCT02905188	I	GPC3-CART	CAR-T Cell	14	Pre-treatment	N/A	N/A	N/A	N/A
NCT03198546	I	GPC3/TGFβ-CART	CAR-T Cell	30	Pre-treatment	N/A	N/A	N/A	N/A
NCT03980288	I	GPC3-CART	CAR-T Cell	36	Pre-treatment	N/A	N/A	N/A	N/A
NCT03884751	I	GPC3-CART	CAR-T Cell	15	Pre-treatment	N/A	N/A	N/A	N/A
NCT03013712	I/II	EpCAM-CART	CAR-T Cell	60	Pre-treatment	N/A	N/A	N/A	N/A
NCT03130712	I/II	GPC3-CART	CAR-T Cell	10	Pre-treatment	N/A	N/A	N/A	N/A

NCT02715362	I/II	GPC3-CART	CAR-T Cell	30	Pre-treatment	N/A	N/A	N/A	N/A
Natural Killer (NK) Cell									
NCT03347123	I/II	Epacadostat + nivolumab + lirilumab	IDO1 + PD-1 + KIR	**11**	Pre-treatment	N/A	N/A	N/A	N/A
NCT03203876	I/II	İpilimumab + nivolumab + lirilumab	CTLA-4 + PD-1 + KIR	**21**	Pre-treatment	N/A	N/A	N/A	N/A
NCT04374877	I	SRF388	TIGIT	**92**	Pre-treatment	N/A	N/A	N/A	N/A
NCT02671435	I/II	Durvalimumab + monalizumab	PD-1 + NKG2A	**383**	Pre-treatment	N/A	N/A	N/A	N/A
Vaccines									
NCT03480152	I/II	mRNA vaccine	Vaccine	5	Pre-treatment	N/A	N/A	N/A	N/A
NCT04248569	I	DNAJB1-PRKACA peptide vaccine + nivolumab + lirilumab	CTLA-4 + PD-1 + vaccine	12	Pre-treatment	N/A	N/A	N/A	N/A
NCT02232490	III	Hepcortespenlisimut-L	Vaccine	120	Pre-treatment	N/A	N/A	N/A	N/A
NCT04147078	I	DC vaccine	Vaccine	80	Pre-treatment	N/A	N/A	N/A	N/A
NCT03311334	I/II	DSP-7888 + pembrolizumab or nivolumab	Vaccine + PD-1	84	Pre-treatment	N/A	N/A	N/A	N/A

Continued

Table 4 Ongoing clinical trials for advanced HCC.—cont'd

C: Ongoing clinical trials of non-ICI Immunotherapy for Advanced HCC

Trial identifier	Phase	Drug	Target	N	Patient group	ORR (%)	DCR (%)	PFS (median, mo)	OS (median, mo)
NCT03942328	I	ADC + EBRT + Pneumococcal 13-valent Conjugate Vaccine	Vaccine + radiotherapy	26	Pre-treatment	N/A	N/A	N/A	N/A
NCT01869088	III	RH Adenovirus T5 + TACE	Vaccine + CT	266	Pre-treatment	N/A	N/A	N/A	N/A
NCT01628640	I	R Vesicular Stomatitis Virus-expressing IF-b	Vaccine	17	Pre-treatment	N/A	N/A	N/A	N/A
T cell immunoglobulin and mucin domain 3 (TIM-3)									
NCT03680508	I/II	TSR-002 + TSR-042 (Dostarlimab)	TIM-3 + PD-1	42	naive	N/A	N/A	N/A	N/A
NCT03652077	I	INCAGN02390	TIM-3	41	Pre-treatment	N/A	N/A	N/A	N/A
Lymphocyte activation gene 3 (LAG-3)									
NCT03250832	I/II	TSR-033 + TSR-042	LAG-3 + PD-1	200	Pre-treatment	N/A	N/A	N/A	N/A
NCT04374877	I	SRF388	LAG-3	92	Pre-treatment	N/A	N/A	N/A	N/A
NCT03538028	I	INCAGN02385	LAG-3	40	Pre-treatment	N/A	N/A	N/A	N/A

NCT03849469	I	XmAb22841 + pembrolizumab	CTLA-4/LAG-3 + PD-1	242	Pre-treatment	N/A	N/A	N/A	N/A
Transforming growth factor-β (TGF-β)									
NCT02947165	I	NIS793 + PDR001	TGF-β + PD-1	220	Pre-treatment	N/A	N/A	N/A	N/A
NCT02178358	II	LY2157299 + Sorafenib vs sorafenib	TGF-β + mTKI	120	naive	N/A	N/A	N/A	N/A
NCT02423343	II	Galunisertib + Nivolumab	TGF-βR1 + PD-1	75	naive	N/A	N/A	N/A	N/A
NCT02240433	Ib	LY2157299 + Sorafenib	TGF-β + mTKI	9	naive	N/A	N/A	N/A	N/A
Invariant Natural Killer T (iNKT) Cell									
NCT03897543	I/II	ABX196 + Nivolumab	INKT + PD-1	48	Pre-treatment	N/A	N/A	N/A	N/A
NCT04011033	III	iNKT cells + TACE	INKT + CT	144	Pre-treatment	N/A	N/A	N/A	N/A

[a] Dose-escalation.
[b] Dose-expansion.
DCR, disease control rate; IC, immune checkpoint inhibitor; N, number of patients; N/A, not available; N/R, not reach; ORR, Objective response rate; OS, overall survival; PFS, progression-free survival.

24% (1% complete response [CR]), and the disease control rate (DCR) was 75%. A subgroup analysis showed that lenvatinib was more effective in Asian patients and those with hepatitis B and AFP $\geq$200 mg/dL. Common adverse events included hypertension (42%), diarrhea (39%), decreased appetite (34%), and reduced weight (31%) (Kudo et al., 2018).

A phase 3 study in 668 Chinese aHCC patients showed that those treated with donafenib had better OS outcomes than those treated with sorafenib (OS 12.1 vs 10.3 months, HR 0.831, $P=0.0363$). PFS, ORR, and DCR were not significantly different between the two groups (Bi et al., 2020). Donafenib is not available in the United States.

The current NCCN guideline recommends sorafenib or lenvatinib in the first-line treatment of aHCC patients who are not appropriate for atezolizumab plus bevacizumab (European Association for the Study of the Liver. Electronic address & European Association for the Study of the, 2018; National Comprehensive Cancer Network, n.d.).

4.1.2 Immunotherapy-based treatment

Atezolizumab is a fully humanized, monoclonal antibody of IgG1 isotype against the programmed cell death-ligand 1 (PD-L1). Bevacizumab is a humanized antibody binding to VEGF. The combination of atezolizumab and bevacizumab showed efficacy with ORR of 27% in the randomized portion of phase 1b GO30140 trial ($n=60$) (Lee et al., 2020). The combination was studied in a phase 3, randomized trial (IMbrave150), and it demonstrated the superiority of atezolizumab and bevacizumab over sorafenib in patients with treatment-naïve aHCC. Five-hundred one CTP-A patients were included in the study. Atezolizumab plus bevacizumab showed longer median OS (not reached vs 13.2 months, HR 0.58, $P=0.0006$) and PFS (6.8 vs 4.3 months), compared with sorafenib (Finn et al., 2020). The ORR was 33.3% (10.2% CR and 23.1% PR) vs 13.3% (1.9% CR and 11.4% PR), and the DCR was 72.3% and 55.1% for the combination and sorafenib, respectively. The most common grade 3–4 adverse events included hypertension (15.2%), increased aspartate aminotransferase (7%), thrombocytopenia (3.3%), and proteinuria (3%) in the combination arm. Based on this study, the FDA approved atezolizumab plus bevacizumab for the first-line treatment of aHCC patients in May 2020 (U. S. Food and Drug Administration/ Center for Drug Evaluation and Research, 2020).

The combination of lenvatinib plus pembrolizumab was safe and effective in a phase 1b trial (NCT03006926, $n=104$). Lenvatinib plus pembrolizumab showed an ORR of 36%, median OS of 22.0 months, and median

PFS of 8.6 months. Treatment-related adverse events occurred in 95% of patients (grade ≥ 3, 71%). The most common grade ≥ 3 adverse event was hypertension (17%). Three patients died from intestinal perforation, acute respiratory distress syndrome, and abnormal hepatic function (Zhu, Finn, et al., 2020; Zhu, Ikeda, et al., 2020).

A clinical trial of nivolumab plus cabozantinib with and without ipilimumab (NCT01658878) showed promising results in patients with aHCC. Seventy one patients were assigned to two groups: nivolumab (240 mg every 2 weeks) plus cabozantinib (40 mg daily) in group 1 ($n = 36$) and nivolumab (3 mg/kg every 2 weeks) plus ipilimumab (1 mg/kg every 6 weeks) plus cabozantinib (40 mg daily) in group 2 ($n = 35$). Compared with group 1, patients in group 2 had better investigator-assessed ORR (17% vs 26%), DCR (81% vs 83%), and PFS (5.5 vs 6.8 months). Median OS was not reached in both groups. Grade 3–4 adverse events and discontinuation of treatment were reported in group 1 (42% and 3%) and group 2 (71% and 20%) (Yau et al., 2020).

Avelumab is a fully human monoclonal antibody that binds to PD-L1, and axitinib is an mTKI selective for VEGFR1, 2, and 3. The VEGF Liver 100 (NCT03289533) was a phase 1b trial evaluating the efficacy and safety of avelumab (10 mg/kg every 2 weeks) plus axitinib (5 mg orally twice daily) in treatment-naïve patients with aHCC. The ORR was 31.8% (per Modified Response Evaluation Criteria in Solid Tumors [mRECIST]). Median PFS and OS were 5.5 and 12.7 months. The most common grade 3 adverse events were hypertension (50%) and palmar-plantar erythrodysesthesia (22.7%), and no grade 4/5 adverse events or grade ≥ 3 immune-related adverse events were reported (Kudo et al. (2019). Other clinical trials of combinations of immune checkpoint inhibitors and mTKIs are ongoing in patients with aHCC (Table 4).

4.1.3 Chemotherapy

HCC is a chemotherapy-refractory tumor (Marin et al., 2019). GEMOX (gemcitabine and oxaliplatin), FOLFOX (5-fluorouracil, leucovorin, and oxaliplatin), and single-agent therapy with capecitabine, cisplatin, and doxorubicin have been studied in many randomized clinical trials. Although these regimens showed only modest clinical benefit, chemotherapy is still an option for patients with treatment-refractory disease in many countries (Ikeda et al., 2018).

The EACH trial is a study comparing FOLFOX with doxorubicin in the first-line treatment of HCC patients: 371 treatment-naïve Chinese patients were randomized (1:1). FOLFOX showed better median OS (5.9 vs

4.3 months, HR 0.75, $P=0.03$) and disease-free survival (DFS, 2.4 vs 1.7 months, HR 0.55, $P=0.0002$) than doxorubicin. The ORR and DCR also were significantly higher in the FOLFOX group than in the doxorubicin group (ORR 8.6 vs 1.4%, $P=0.006$; DCR 47.1 vs 26.6%, $P=0.0004$) (Qin et al., 2014).

Camrelizumab plus oxaliplatin-based chemotherapy yielded a 26.5% ORR and 5.5-month median PFS, suggesting that immunotherapy plus chemotherapy combined treatment may be effective in the first-line treatment of aHCC (Qin et al., 2019). However, the combination of an mTKI and chemotherapy did not show survival benefit and increased toxicity (Abou-Alfa et al., 2019). Chemotherapy is not routinely recommended for the systemic treatment of aHCC.

4.2 Second-line treatment

4.2.1 Targeted therapy

Regorafenib and cabozantinib showed survival benefit over placebo as second-line treatments for aHCC (Abou-Alfa et al., 2018; Bruix et al., 2017). Other mTKIs (brivanib, everolimus, and axitinib) were not superior to placebo in this setting (Kang et al., 2015; Llovet et al., 2013; Zhu et al., 2014).

Regorafenib is an mTKI of VEGFR1, -2, and -3; PDGFR; FGFR; KIT; RET; RAF-1; and BRAF. The RESORCE study included 573 patients with sorafenib-refractory aHCC, who were assigned into regorafenib (160 mg once daily during weeks 1–3 of each 4-week cycle) and placebo cohorts. Regorafenib showed better median OS than placebo (10.6 vs 7.8 months, HR 0.63, $P<0.0001$). The most common grade 3–4 adverse events in patients treated with regorafenib included hypertension, palmar-plantar erythrodysesthesia, fatigue, and diarrhea (Bruix, Qin, et al., 2017). As a result, in 2017 the FDA approved regorafenib (160 mg once a day for 21 days on and 7 days off in each 28-day cycle) for patients with aHCC in the second-line setting.

Cabozantinib is an mTKI of VEGFR2, c-MET, TIE-2, and AXL. The CELESTIAL study showed OS benefit of cabozantinib vs placebo. The study included 707 patients with aHCC who had disease progression after at least one systemic treatment. Patients were assigned into cabozantinib (60 mg once daily) and placebo cohorts. Cabozantinib showed better median PFS (5.2 vs 1.9 months, HR 0.44, $P<0.001$) and OS (10.2 vs 8.0 months, HR 0.76, $P=0.005$) than did placebo. The ORR was 4%. The most common grade 3–4 adverse events were palmar-plantar erythrodysesthesia (17%), hypertension (16%), and increased aspartate aminotransferase level (12%). Based on this trial, cabozantinib was approved by the FDA in January 2019 in the second-line setting (Abou-Alfa et al., 2018).

A phase 3 study showed survival benefit of the anti-VEGFR2 TKI apatinib vs placebo in the second-line setting. The study included 393 Chinese patients, and apatinib showed better median OS (8.7 vs 6.8 months, HR 0.785, $P = 0.0476$) and DFS (4.5 vs 1.9 months, HR 0.471, $P < 0.001$) than did placebo (Li et al., 2020).

Ramucirumab is a recombinant monoclonal antibody inhibiting VEGFR2. In the REACH study, 565 patients with aHCC were randomly assigned to receive ramucirumab (8 mg/kg every 2 weeks; $n = 283$) or placebo ($n = 282$) as a second-line treatment. Ramucirumab showed better median PFS (2.8 vs 2.1 months, HR 0.63, $P < 0.0001$) and ORR (7% vs <1%, $P < 0.001$) than placebo. However, ramucirumab did not show a statistically significant OS benefit over placebo (9.2 vs 7.6 months, HR 0.87, $P = 0.14$). In a subgroup analysis, ramucirumab had better efficacy in patients with AFP ≥ 400 ng/mL (median OS 7.8 vs 4.2 months, HR 0.67, $P = 0.006$) (Zhu et al., 2015). As a result of this finding, the REACH-2 study was performed in patients with AFP ≥ 400 ng/mL. Ramucirumab showed better median OS (8.5 vs 7.3 months, HR 0.71, $P = 0.0199$) and PFS (2.8 vs 1.6 months, HR 0.45, $P < 0.0001$) than placebo in this group of patients (Zhu et al., 2019). Therefore, ramucirumab (8 mg/kg once every 2 weeks) was approved by the FDA in May 2019 for patients with aHCC and AFP of ≥400 ng/mL in the second-line setting. In a meta-analysis of HCC studies including REACH and REACH-2, ramucirumab provided survival benefit in patients who did or did not experience ascites in the second-line treatment of HCC. The OS was 6.7 vs 3.4 months in ramucirumab and placebo arms with HR of 0.3 (0.18, 0.49), and the PFS was 4.2 vs 2 months with HR 0.46 (0.29, 0.74) in patients with aHCC who experience ascites (Zhu, Finn, et al., 2020; Zhu, Ikeda, et al., 2020).

4.2.2 Immunotherapy-based treatment

Nivolumab is a fully human IgG4 antibody inhibiting programmed cell death-1 (PD-1) from binding to its ligands, PD-L1 and PD-L2. Checkmate-040 was a phase 1/2 trial of nivolumab as monotherapy for patients with aHCC. Two hundred sixty-two patients with CTP-A/B7 were included, 48 in the dose escalation phase and 214 in the dose expansion (3 mg/kg) phase. Fifty six patients were therapy naïve. The ORR was 15% (range 6%–28%) in the dose escalation phase and 20% (range 15%–26%) in the expansion phase.

Among all patients the DCR was 64%, and the median OS and PFS were 15.0 and 4.0 months, respectively. The most common grade 3–4 adverse events were elevated lipase, aspartate aminotransferase, and alanine aminotransferase levels as well as fatigue. This study demonstrated that nivolumab was effective and safe in the systemic treatment of aHCC patients (El-Khoueiry et al., 2017). However, a phase 3 trial, CheckMate-459 study ($n=743$), failed to show a survival benefit of nivolumab vs sorafenib in therapy-naïve patients (PFS 3.7 vs 3.8 months, OS 16.4 vs 14.7 months, HR 0.84, $P=0.0419$). The ORR was 15% (14 CR) and 7% (5 CR) in nivolumab and sorafenib cohorts. In patients with PD-L1 $\geq$ 1% and $<$ 1%, ORR was 28% and 12%, respectively (Yau et al., 2019). The FDA approved nivolumab (240 mg every 2 weeks) in the second line setting in September 2017. It is also recommended in the front-line setting if patients are considered ineligible or intolerant to other front-line medications.

Pembrolizumab, an anti-PD-1 antibody, was assessed in the single-arm Keynote-224 trial for efficacy and safety in the second-line setting. The study included 104 patients with aHCC. The median PFS and OS were 4.9 (range 3.4–7.2) and 12.9 (range 9.7–15.5) months, respectively. The ORR and DCR were 17% (range 11–26) and 67%. The duration of response ranged from 3.1 to 16.7 months. Grade 3 adverse events were reported in 25 (24%) patients: the most common were increased aspartate aminotransferase and alanine aminotransferase levels and fatigue. Hyperbilirubinemia was found in 1% of patients (Zhu et al., 2018). Pembrolizumab was approved by the FDA in the second line setting for patients with aHCC in November 2018.

In the combination arm of the CheckMate-040 study, nivolumab plus ipilimumab were administered in the second-line setting. The study included 148 patients who had previously been treated with sorafenib. Most patients had aHCC at baseline: 88% of patients had vascular invasion or extrahepatic spread, and 91% had BCLC-C disease. Patients were assigned into three arms: nivolumab (1 mg/kg) plus ipilimumab (3 mg/kg) every 3 weeks (4 doses), followed by nivolumab (240 mg) every 2 weeks; nivolumab (3 mg/kg) plus ipilimumab (1 mg/kg) every 3 weeks (4 doses), followed by nivolumab (240 mg) every 2 weeks; and nivolumab (3 mg/kg) every 2 weeks plus ipilimumab (1 mg/kg) every 6 weeks. A total of 50 and 49 patients, respectively, received nivolumab (1 mg/kg) plus ipilimumab (3 mg/kg) and nivolumab (3 mg/kg) plus ipilimumab (1 mg/kg) every 3 weeks followed by nivolumab (240 mg) every 2 weeks. Patients in these two arms, respectively, had an ORR of 32% and 31%, DCR of 54% and 43%, and median OS of 23 and 12 months. Response duration ranged from 4.6 to 30.5+

months, with 31% of responses lasting at least 24 months. Among patients in these two arms, 37% had grade 3–4 adverse events, most commonly pruritus and rash, as well as other known immune-related adverse events reported in clinical studies for other cancers (He et al., 2020). Based on these results, nivolumab (1 mg/kg) in combination with ipilimumab (3 mg/kg) every 3 weeks for four doses, followed by single-agent nivolumab (240 mg every 2 weeks or 480 mg every 4 weeks) was approved by the FDA in March 2020 for the treatment of aHCC in the second-line setting.

4.3 Adjuvant and neoadjuvant approach

There is no curative treatment option for HCC other than surgical resection and liver transplantation. Most patients are diagnosed at a locally advanced or metastatic stage, and curative treatments are applied to approximately 20% of patients (Kianmanesh, Regimbeau, & Belghiti, 2003). The 5-year relapse-free survival (RFS) rate of HCC patients who have undergone surgery or ablation is 30%–37% (Imamura et al., 2003; Lim et al., 2012; Zhong et al., 2015). Among patients whose disease recurs after surgery or ablation, more than 70% experience recurrence within 2 years, and patients have a poor prognosis (Jung et al., 2019; You et al., 2017). The recurrence rate in HCC patients undergoing liver transplantation is 10%–15%, these recurrences usually develop within the first year (Mehta et al., 2017). The presence of micrometastases not seen on conventional imaging or the dissemination of tumor cells during surgery are suggested etiologies of recurrence after resection. Patients initially considered surgical candidates may have significant disease progression before surgery, which excludes them from resection (Cheng et al., 2015). In this regard, the use of preoperative and postoperative therapy has been proposed and has theoretical benefit, including tumor downsizing and reduction in micrometastases (Yang et al., 2017). However, there is no established adjuvant or neoadjuvant treatment.

4.3.1 Adjuvant

Early clinical trials of adjuvant chemotherapy in HCC showed mixed results. A randomized study of adjuvant capecitabine showed improved time-to-recurrence and 5-year OS rate (Xia et al., 2010). Adjuvant 5-fluorouracil showed survival benefit in patients with stage 1 disease but not those with stage 2 disease (Yamamoto, Arii, Sugahara, & Tobe, 1996). Adjuvant studies of tegafur, epirubicin, and carmofur did not show survival benefit (Lai, Lo, Fan, Liu, & Wong, 1998; Ono, Yamanoi, Nazmy El Assal, Kohno, & Nagasue, 2001).

More recently, sorafenib was tested as adjuvant treatment in a phase 3, randomized clinical trial (STORM). The study included 1114 patients who underwent surgery or local ablation. Patients were randomly assigned to sorafenib ($n=556$) and placebo ($n=558$) groups. Sorafenib did not show a statistical difference in median RFS compared with placebo (33.3 vs 33.7 months, HR 0.94, 95% CI 0.78–1.13) (Bruix et al., 2015).

Adjuvant immunotherapy studies demonstrated mixed results. Hui et al. applied adjuvant cytokine-induced killer (CIK) cell immunotherapy to HCC patients. The study included 127 patients, who were grouped to three cohorts: patients in group 1 received three courses of CIK therapy ($n=41$); those in group 2, six courses of CIK therapy ($n=43$); and group 3 was the observation-only control group ($n=43$). The DFS was significantly higher in group 1 ($P=0.001$) and group 2 ($P=0.004$) than in the control group. However, no statistical significance was found in OS among the three groups (Hui, Qiang, Jian, Ti, & Da-Lu, 2009). An adjuvant phase 3 Korean study in HCC patients ($n=230$) showed that the treatment with activated CIK cells compared with observation only improved both median RFS (44 vs 30 months, HR 0.63, $P=0.01$) and OS ($P=0.02$) (Lee et al., 2015). An adjuvant, phase 2 glypican-3 (GPC3) vaccine study showed that one and 2-year recurrence rates were 24% and 48% in GPC3-positive HCC patients with surgery plus vaccination and 52.4% and 61.9% in patients with surgery alone. The GPC3 peptide vaccine improved the 1-year recurrence rate in patients with GPC3-positive HCC ($P=0.047$) (Sawada et al., 2016).

Currently, multiple clinical trials of TKIs and immune checkpoint inhibitors are underway, but no systemic therapy is considered standard of care in the adjuvant setting (Table 5).

4.3.2 Neoadjuvant

The BIOSPHERE study was an open-label, multicenter, phase 2 study that tested sorafenib in the neoadjuvant setting for patients with resectable HCC. The study included 25 patients who received sorafenib 400 mg for 4 weeks before surgery. The ORR was 32%. The R0 tumor resection rate and rate of pathologic response with necrosis ≥50% were achieved in 88% and 24% of patients, respectively (Bouattour et al., 2016). Transarterial chemoembolization (TACE) and TACE with sorafenib were compared in the neoadjuvant setting for HCC patients on the waiting list for liver transplantation.

Table 5 Ongoing adjuvant and neoadjuvant clinical trials in HCC patients.

Trial identifier	Phase	Drug	Target	N	Patient group
NCT03383458 (CheckMate 9DX)	III	Nivolumab vs Placebo	PD-1	530	Adjuvant
NCT03867084 (KEYNOTE-937)	III	Pembrolizumab vs Placebo	PD-1	950	Adjuvant
NCT04102098 (IMbrave050)	III	Atezolizumab + Bevacizumab vs Placebo	PD-L1 + VEGF	662	Adjuvant
NCT03847428 (EMERALD-2)	III	Durvalumab + Bevacizumab vs Bevacizumab	PD-L1 + VEGF	888	Adjuvant
NCT01681446	III	Interferon-alpha	IFN-α	296	Adjuvant
NCT02725996	II	NK Cells	NK	140	Adjuvant
NCT04233840	I/II	Ropeginterferon alfa-2b + Nivolumab	PD-1	72	Adjuvant
NCT03722875	Not Applicable	SHR-1210 + Apatinib	PD-1 + mTKI	45	Adjuvant
NCT03299946	I	Nivolumab + Cabozantinib	PD-1 + mTKI	15	Adjuvant
NCT03630640	II	Nivolumab Injection + Nivolumab IV	PD-1	50	Neoadjuvant/ Adjuvant
NCT03778957 (EMERALD-1)	III	Durvalumab + TACE or Durvalumab + Bevacizumab + TACE vs TACE alone	PD-L1 + VEGF	600	Locoregional (Naïve)

Continued

Table 5 Ongoing adjuvant and neoadjuvant clinical trials in HCC patients.—cont'd

Trial identifier	Phase	Drug	Target	N	Patient group
NCT03099564	I	Pembrolizumab + Y90 radioembolization I 30	PD-1 +radioembolization	30	Locoregional
NCT03337841	II	Pembrolizumab	PD-1	50	Neoadjuvant
NCT03916627	II	Cemiplimab	PD-1	94	Neoadjuvant
NCT03510871	II	Ipilimumab + Nivolumab	PD-1 + CTLA-4	40	Neoadjuvant
NCT04174781	II	Sintilimab Injection + TACE	PD-1 + CT	61	Neoadjuvant
NCT04123379	II	Nivolumab + BMS-813160 (CCR2/5-inhibitor) or BMS-986253 (Anti-IL-8)	PD-1 + CCR2/5 or IL-8	50	Neoadjuvant
NCT04297202	II	SHR-1210 Injection + Apatinib	PD-1 + mTKI	20	Neoadjuvant
NCT04443322	Not Applicable	Durvalumab Injection + Lenvatinib	PD-1 + mTKI	20	Neoadjuvant

N = Number of patients.

The median time to progression was 71 days in the sorafenib group and 85 days in the placebo group, with no statistical difference (HR 1.106) (Hoffmann et al., 2015).

A randomized phase 2 trial investigated neoadjuvant therapy with immune checkpoint inhibitors for patients with resectable HCC. The study had two cohorts: patients who received nivolumab ($n=13$) and those who received nivolumab plus ipilimumab ($n=14$). Of 13 patients who received nivolumab, three did not have surgery (two had disease progression; for one patient, surgery was aborted due to a small residual liver volume). Three patients had major pathologic responses (two pathologic CRs and one major tumor necrosis of 70%), accounting for a pathologic response rate of 23%. Of 14 patients who received both nivolumab and ipilimumab, three did not have surgery (two had disease progression; surgery was aborted for one patient due to a small residual liver volume). Three patients had major pathologic responses (three pathologic CRs), accounting for a pathologic response rate of 21% (Kaseb et al., 2019). Other ongoing trials with TKIs or immune checkpoint inhibitors will provide more information on the neoadjuvant treatment of HCC (Table 5).

5. Histological variants

5.1 Combined hepatocellular carcinoma and cholangiocarcinoma (cHCC-CC)

Combined hepatocellular carcinoma and cholangiocarcinoma (cHCC-CC) is a rare histological variant of primary liver cancer (Yeh, 2010). cHCC-CC comprises elements of both hepatocellular carcinoma (HCC) and cholangiocarcinoma (CC) and has a poorer prognosis than HCC (Gera, Ettel, Acosta-Gonzalez, & Xu, 2017). Liver resection and transplantation are the only known curative options in cHCC-CC (Connell, Harding, Shia, & Abou-Alfa, 2016). Three patients underwent liver transplantation as reported by Chan et al.: one patient died of distant metastases at 16.5 months after liver transplantation; no tumor recurrence was observed in the other two patients at 25 and 35 months after transplantation (Chan, Lo, Ng, & Fan, 2007). In a report by Panjala et al., the 5-year OS rate of 12 patients who underwent liver transplant was 16% (Panjala et al., 2010). Another study with 14 patients who received liver transplantation with a median follow-up period of 32 months showed a median DFS of 8 months and a 57% recurrence rate (Sapisochin, Fidelman, Roberts, & Yao, 2011). More data are needed to assess liver transplantation in patients with cHCC-CC.

In comparison to resection and liver transplantation, systemic therapy is the main backbone of treatment for unresectable cHCC-CC. However, there is no consensus about chemotherapy or molecular targeted therapy. In a retrospective study of seven patients, first-line systemic treatments showed poor survival responses. Regimens given included gemcitabine with or without bevacizumab, gemcitabine plus platinum-based agents (GP) with or without bevacizumab, and sorafenib. The DFS and median OS were 3.4 and 8.3 months, respectively. The DCR was 29% (in two patients treated with gemcitabine plus bevacizumab and GP) on the first radiographic scan. Disease control was not demonstrated in patients who received sorafenib as a first-line treatment (Rogers et al., 2017). Similarly, in a retrospective study including 36 patients with cHCC-CC, the survival outcomes of patients treated with chemotherapy were better than those of patients treated with sorafenib (HR 15.83, $P=0.006$). The median OS was 11.9, 10.2, and 3.5 months in three patients treated with gemcitabine plus cisplatin, fluorouracil plus cisplatin, and sorafenib, respectively (Kobayashi et al., 2018). Although sorafenib appears to be ineffective in cHCC-CC patients in these retrospective studies, remission with sorafenib treatment was reported in a hepatitis C-positive patient (Futsukaichi et al., 2019).

Currently, gemcitabine plus a platinum agent (cisplatin or oxaliplatin) with or without bevacizumab and sorafenib are the recommended options in the systemic treatment of cHCC-CC patients (Salimon et al., 2018).

5.2 Fibrolamellar hepatocellular carcinoma (FLHCC)

Fibrolamellar hepatocellular carcinoma (FLHCC) is a rare tumor, accounting for less than 1% of liver cancers (El-Serag & Davila, 2004). Clinical and pathological features of FLHCC significantly differ from HCC (Mayo et al., 2014).

Surgical resection is the standard of care in patients with resectable and locally advanced FLHCC (Mavros, Mayo, Hyder, & Pawlik, 2012). Also, similar to HCC patients, liver transplantation is used in FLHCC patients as a treatment option (El-Gazzaz et al., 2000). However, recurrence develops in all patients after resection or transplantation (Yamashita et al., 2016), and adjuvant and neoadjuvant therapies are not well defined for FLHCC patients (Ang et al., 2013).

Systemic therapy is the standard application of treatment for patients with unresectable FLHCC. There is no consensus opinion on systemic therapy as therapeutic options are not based on randomized studies.

FLHCC is known as a chemotherapy-refractory tumor (Lafaro & Pawlik, 2015), yet successful results have been reported using chemotherapy plus interferon-α2b (IFN-α2b) (Bower, Newlands, & Habib, 1996; Maniaci et al., 2009). The combination of 5-fluorouracil plus IFN-α2b was found to be safe and effective in a phase 2 trial ($n=8$ patients). In this trial, patients received 5-fluorouracil at a dose of 200 mg/m^2 infused continuously over 21 days with IFN-α2b at a dose of 4 million U/m^2 three times weekly during 5-fluorouracil infusion on a 28-day cycle. 5-fluorouracil plus IFN-α2b showed an ORR of 62.5% (one patient experienced CR, four patients experienced PR), and median OS of 23.1 months. The most common grade ≥ 3 adverse events were stomatitis (32.6%), fatigue (4.7%), and hematologic toxicity (9.3%) (Patt et al., 2003). Since that study's publication, the combination of 5-fluorouracil plus IFN-α2b has been established as the most frequently used systemic therapy. Another study tested 5-fluorouracil plus interferon-α2b and capecitabine plus interferon-α2b in 94 patients with FLHCC. The ORRs were 80% and 66% in 5-fluorouracil plus interferon-α2b and capecitabine plus interferon-α2b, respectively. In the same study, the patients who received resection plus neoadjuvant ($n=34$) or adjuvant therapy ($n=10$) had the longest median OS, 60, and 110.5 months, respectively, of the entire cohort (Kaseb et al., 2013).

The studies of immune checkpoint inhibitors in the treatment of FLHCC patients demonstrated mixed results (Bauer et al., 2019; De Toni & Roessler, 2020). However, the combination of immunotherapy and 5-fluorouracil plus IFN-α2b may contribute to survival in FLHCC patients (Bazhin, von Ahn, Fritz, Werner, & Karakhanova, 2018; Moschos et al., 2006). The ongoing phase 1/2 clinical trial (NCT04380545) of nivolumab, 5-fluorouracil, and IFN-α2b in patients with unresectable FLHCC will demonstrate the importance of immune checkpoint inhibitor treatments.

6. Conclusion

Sorafenib was the standard of care and the only systemic treatment for many years. After a decade of negative clinical trials, the IMbrave150 study showed survival benefit of atezolizumab plus bevacizumab over sorafenib, and it has become the standard first-line therapy for aHCC. All of the approved second-line treatments are based on studies conducted after sorafenib treatment against placebo, and no study has compared second-line options.

Most clinical studies included CTP-A patients only. The NCCN guideline recommends sorafenib and nivolumab up to CTP-B patients. The benefit of other agents in HCC patients with CTP-B and C remains unclear. Deterioration of liver function on systemic therapy is an essential obstacle in treatment planning, and liver function evaluation appears to be an area of research that needs more attention.

While neoadjuvant and adjuvant studies show promising results, standard treatment has not been established. The best treatment order for cHCC-CC and FLHCC still remains unclear.

Acknowledgment

Editorial support was provided by Bryan Tutt, Scientific Editor, Research Medical Library.

References

Abou-Alfa, G. K., Meyer, T., Cheng, A. L., El-Khoueiry, A. B., Rimassa, L., Ryoo, B. Y., et al. (2018). Cabozantinib in patients with advanced and progressing hepatocellular carcinoma. *The New England Journal of Medicine*, *379*(1), 54–63. https://doi.org/10.1056/NEJMoa1717002.

Abou-Alfa, G. K., Shi, Q., Knox, J. J., Kaubisch, A., Niedzwiecki, D., Posey, J., et al. (2019). Assessment of treatment with sorafenib plus doxorubicin vs sorafenib alone in patients with advanced hepatocellular carcinoma: Phase 3 CALGB 80802 randomized clinical trial. *JAMA Oncology*. https://doi.org/10.1001/jamaoncol.2019.2792.

Ang, C. S., Kelley, R. K., Choti, M. A., Cosgrove, D. P., Chou, J. F., Klimstra, D., et al. (2013). Clinicopathologic characteristics and survival outcomes of patients with fibrolamellar carcinoma: Data from the fibrolamellar carcinoma consortium. *Gastrointestinal Cancer Research*, *6*(1), 3–9.

Bauer, U., Mogler, C., Braren, R. F., Algul, H., Schmid, R. M., & Ehmer, U. (2019). Progression after immunotherapy for fibrolamellar carcinoma. *Visceral Medicine*, *35*(1), 39–42. https://doi.org/10.1159/000497464.

Bazhin, A. V., von Ahn, K., Fritz, J., Werner, J., & Karakhanova, S. (2018). Interferon-alpha up-regulates the expression of PD-L1 molecules on immune cells through STAT3 and p 38 signaling. *Frontiers in Immunology*, *9*, 2129. https://doi.org/10.3389/fimmu.2018.02129.

Bi, F., Qin, S., Gu, S., Bai, Y., Chen, Z., Wang, Z., et al. (2020). Donafenib versus sorafenib as first-line therapy in advanced hepatocellular carcinoma: An open-label, randomized, multicenter phase II/III trial. *Journal of Clinical Oncology*, *38*(15_Suppl), 4506. https://doi.org/10.1200/JCO.2020.38.15_suppl.4506.

Bouattour, M., Fartoux, L., Rosmorduc, O., Scatton, O., Vibert, E., Costentin, C., et al. (2016). BIOSHARE multicenter neoadjuvant phase 2 study: Results of pre-operative sorafenib in patients with resectable hepatocellular carcinoma (HCC)—From GERCOR IRC. *Journal of Clinical Oncology*, *34*(4_Suppl), 252. https://doi.org/10.1200/jco.2016.34.4_suppl.252.

Bouattour, M., Mehta, N., He, A. R., Cohen, E. I., & Nault, J. C. (2019). Systemic treatment for advanced hepatocellular carcinoma. *Liver Cancer*, *8*(5), 341–358. https://doi.org/10.1159/000496439.

Bower, M., Newlands, E. S., & Habib, N. (1996). Fibrolamellar hepatocellular carcinoma responsive to platinum-based combination chemotherapy. *Clinical Oncology*, *8*(5), 331–333. https://doi.org/10.1016/S0936-6555(05)80724-1.

Bruix, J., Takayama, T., Mazzaferro, V., Chau, G. Y., Yang, J., Kudo, M., et al. (2015). Adjuvant sorafenib for hepatocellular carcinoma after resection or ablation (STORM): A phase 3, randomised, double-blind, placebo-controlled trial. *The Lancet Oncology*, *16*(13), 1344–1354. https://doi.org/10.1016/s1470-2045(15)00198-9.

Bruix, J., Cheng, A. L., Meinhardt, G., Nakajima, K., De Sanctis, Y., & Llovet, J. (2017). Prognostic factors and predictors of sorafenib benefit in patients with hepatocellular carcinoma: Analysis of two phase III studies. *Journal of Hepatology*, *67*(5), 999–1008. https://doi.org/10.1016/j.jhep.2017.06.026.

Bruix, J., Qin, S., Merle, P., Granito, A., Huang, Y. H., Bodoky, G., et al. (2017). Regorafenib for patients with hepatocellular carcinoma who progressed on sorafenib treatment (RESORCE): A randomised, double-blind, placebo-controlled, phase 3 trial. *Lancet*, *389*(10064), 56–66. https://doi.org/10.1016/s0140-6736(16)32453-9.

Cainap, C., Qin, S., Huang, W. T., Chung, I. J., Pan, H., Cheng, Y., et al. (2015). Linifanib versus sorafenib in patients with advanced hepatocellular carcinoma: Results of a randomized phase III trial. *Journal of Clinical Oncology*, *33*(2), 172–179. https://doi.org/10.1200/JCO.2013.54.3298.

Chan, A. C.-Y., Lo, C. M., Ng, I. O.-L., & Fan, S. T. (2007). Liver transplantation for combined hepatocellular cholangiocarcinoma. *Asian Journal of Surgery*, *30*(2), 143–146. https://doi.org/10.1016/S1015-9584(09)60149-4.

Cheng, A. L., Kang, Y. K., Chen, Z., Tsao, C. J., Qin, S., Kim, J. S., et al. (2009). Efficacy and safety of sorafenib in patients in the Asia-Pacific region with advanced hepatocellular carcinoma: A phase III randomised, double-blind, placebo-controlled trial. *The Lancet Oncology*, *10*(1), 25–34. https://doi.org/10.1016/S1470-2045(08)70285-7.

Cheng, A. L., Kang, Y. K., Lin, D. Y., Park, J. W., Kudo, M., Qin, S., et al. (2013). Sunitinib versus sorafenib in advanced hepatocellular cancer: Results of a randomized phase III trial. *Journal of Clinical Oncology*, *31*(32), 4067–4075. https://doi.org/10.1200/JCO.2012.45.8372.

Cheng, Z., Yang, P., Qu, S., Zhou, J., Yang, J., Yang, X., et al. (2015). Risk factors and management for early and late intrahepatic recurrence of solitary hepatocellular carcinoma after curative resection. *HPB: The Official Journal of the International Hepato Pancreato Biliary Association*, *17*(5), 422–427. https://doi.org/10.1111/hpb.12367.

Connell, L. C., Harding, J. J., Shia, J., & Abou-Alfa, G. K. (2016). Combined intrahepatic cholangiocarcinoma and hepatocellular carcinoma. *Chinese Clinical Oncology*, *5*(5), 66. https://doi.org/10.21037/cco.2016.10.02.

Couri, T., & Pillai, A. (2019). Goals and targets for personalized therapy for HCC. *Hepatology International*, *13*(2), 125–137. https://doi.org/10.1007/s12072-018-9919-1.

De Toni, E. N., & Roessler, D. (2020). Using dual checkpoint blockade to treat fibrolamellar hepatocellular carcinoma. *Gut*, *69*(11), 2056–2058. https://doi.org/10.1136/gutjnl-2020-320604.

Edeline, J., Blanc, J. F., Johnson, P., Campillo-Gimenez, B., Ross, P., Ma, Y. T., et al. (2016). A multicentre comparison between child Pugh and albumin-bilirubin scores in patients treated with sorafenib for hepatocellular carcinoma. *Liver International*, *36*(12), 1821–1828. https://doi.org/10.1111/liv.13170.

El-Gazzaz, G., Wong, W., El-Hadary, M. K., Gunson, B. K., Mirza, D. F., Mayer, A. D., et al. (2000). Outcome of liver resection and transplantation for fibrolamellar hepatocellular carcinoma. *Transplant International*, *13*(Suppl. 1), S406–S409. https://doi.org/10.1007/s001470050372.

El-Khoueiry, A. B., Sangro, B., Yau, T., Crocenzi, T. S., Kudo, M., Hsu, C., et al. (2017). Nivolumab in patients with advanced hepatocellular carcinoma (CheckMate 040):

An open-label, non-comparative, phase 1/2 dose escalation and expansion trial. *Lancet*, *389*(10088), 2492–2502. https://doi.org/10.1016/s0140-6736(17)31046-2.

El-Serag, H. B., & Davila, J. A. (2004). Is fibrolamellar carcinoma different from hepatocellular carcinoma? A US population-based study. *Hepatology*, *39*(3), 798–803. https://doi.org/10.1002/hep.20096.

European Association for the Study of the Liver. Electronic address: easloffice@easloffice.eu; European Association for the Study of the Liver. (2018). EASL Clinical Practice Guidelines: Management of hepatocellular carcinoma. *Journal of Hepatology*, *69*(1), 182–236. https://doi.org/10.1016/j.jhep.2018.03.019.

Faria, S. C., Szklaruk, J., Kaseb, A. O., Hassabo, H. M., & Elsayes, K. M. (2014). TNM/Okuda/Barcelona/UNOS/CLIP international multidisciplinary classification of hepatocellular carcinoma: Concepts, perspectives, and radiologic implications. *Abdominal Imaging*, *39*(5), 1070–1087. https://doi.org/10.1007/s00261-014-0130-0.

Finn, R. S., Qin, S., Ikeda, M., Galle, P. R., Ducreux, M., Kim, T. Y., et al. (2020). Atezolizumab plus bevacizumab in unresectable hepatocellular carcinoma. *The New England Journal of Medicine*, *382*(20), 1894–1905. https://doi.org/10.1056/NEJMoa1915745.

Futsukaichi, Y., Tajiri, K., Kobayashi, S., Nagata, K., Yasumura, S., Takahara, T., et al. (2019). Combined hepatocellular-cholangiocarcinoma successfully treated with sorafenib: Case report and review of the literature. *Clinical Journal of Gastroenterology*, *12*(2), 128–134. https://doi.org/10.1007/s12328-018-0918-5.

Gera, S., Ettel, M., Acosta-Gonzalez, G., & Xu, R. (2017). Clinical features, histology, and histogenesis of combined hepatocellular-cholangiocarcinoma. *World Journal of Hepatology*, *9*(6), 300–309. https://doi.org/10.4254/wjh.v9.i6.300.

He, A. R., Yau, T., Hsu, C., Kang, Y.-K., Kim, T.-Y., Santoro, A., et al. (2020). Nivolumab (NIVO) + ipilimumab (IPI) combination therapy in patients (pts) with advanced hepatocellular carcinoma (aHCC): Subgroup analyses from CheckMate 040. *Journal of Clinical Oncology*, *38*(4_Suppl), 512. https://doi.org/10.1200/JCO.2020.38.4_suppl.512.

Hoffmann, K., Ganten, T., Gotthardtp, D., Radeleff, B., Settmacher, U., Kollmar, O., et al. (2015). Impact of neo-adjuvant Sorafenib treatment on liver transplantation in HCC patients—a prospective, randomized, double-blind, phase III trial. *BMC Cancer*, *15*, 392. https://doi.org/10.1186/s12885-015-1373-z.

Hui, D., Qiang, L., Jian, W., Ti, Z., & Da-Lu, K. (2009). A randomized, controlled trial of postoperative adjuvant cytokine-induced killer cells immunotherapy after radical resection of hepatocellular carcinoma. *Digestive and Liver Disease*, *41*(1), 36–41. https://doi.org/10.1016/j.dld.2008.04.007.

Ikeda, M., Morizane, C., Ueno, M., Okusaka, T., Ishii, H., & Furuse, J. (2018). Chemotherapy for hepatocellular carcinoma: Current status and future perspectives. *Japanese Journal of Clinical Oncology*, *48*(2), 103–114. https://doi.org/10.1093/jjco/hyx180.

Imamura, H., Matsuyama, Y., Tanaka, E., Ohkubo, T., Hasegawa, K., Miyagawa, S., et al. (2003). Risk factors contributing to early and late phase intrahepatic recurrence of hepatocellular carcinoma after hepatectomy. *Journal of Hepatology*, *38*(2), 200–207. https://doi.org/10.1016/s0168-8278(02)00360-4.

Islami, F., Goding Sauer, A., Miller, K. D., Siegel, R. L., Fedewa, S. A., Jacobs, E. J., et al. (2018). Proportion and number of cancer cases and deaths attributable to potentially modifiable risk factors in the United States. *CA: A Cancer Journal for Clinicians*, *68*(1), 31–54. https://doi.org/10.3322/caac.21440.

Jemal, A., Ward, E. M., Johnson, C. J., Cronin, K. A., Ma, J., Ryerson, B., et al. (2017). Annual report to the nation on the status of cancer, 1975-2014, featuring survival. *Journal of the National Cancer Institute*, *109*(9). https://doi.org/10.1093/jnci/djx030.

Jindal, A., Thadi, A., & Shailubhai, K. (2019). Hepatocellular carcinoma: Etiology and current and future drugs. *Journal of Clinical and Experimental Hepatology, 9*(2), 221–232. https://doi.org/10.1016/j.jceh.2019.01.004.

Johnson, P. J., Qin, S., Park, J. W., Poon, R. T., Raoul, J. L., Philip, P. A., et al. (2013). Brivanib versus sorafenib as first-line therapy in patients with unresectable, advanced hepatocellular carcinoma: Results from the randomized phase III BRISK-FL study. *Journal of Clinical Oncology, 31*(28), 3517–3524. https://doi.org/10.1200/JCO.2012.48.4410.

Johnson, P. J., Berhane, S., Kagebayashi, C., Satomura, S., Teng, M., Reeves, H. L., et al. (2015). Assessment of liver function in patients with hepatocellular carcinoma: A new evidence-based approach-the ALBI grade. *Journal of Clinical Oncology, 33*(6), 550–558. https://doi.org/10.1200/JCO.2014.57.9151.

Jung, S. M., Kim, J. M., Choi, G. S., Kwon, C. H. D., Yi, N. J., Lee, K. W., et al. (2019). Characteristics of early recurrence after curative liver resection for solitary hepatocellular carcinoma. *Journal of Gastrointestinal Surgery, 23*(2), 304–311. https://doi.org/10.1007/s11605-018-3927-2.

Kang, Y. K., Yau, T., Park, J. W., Lim, H. Y., Lee, T. Y., Obi, S., et al. (2015). Randomized phase II study of axitinib versus placebo plus best supportive care in second-line treatment of advanced hepatocellular carcinoma. *Annals of Oncology, 26*(12), 2457–2463. https://doi.org/10.1093/annonc/mdv388.

Kaseb, A. O., Abaza, Y. M., & Roses, R. E. (2013). Multidisciplinary management of hepatocellular carcinoma. *Recent Results in Cancer Research, 190*, 247–259. https://doi.org/10.1007/978-3-642-16037-0_16.

Kaseb, A. O., Shama, M., Sahin, I. H., Nooka, A., Hassabo, H. M., Vauthey, J. N., et al. (2013). Prognostic indicators and treatment outcome in 94 cases of fibrolamellar hepatocellular carcinoma. *Oncology, 85*(4), 197–203. https://doi.org/10.1159/000354698.

Kaseb, A. O., Pestana, R. C., Vence, L. M., Blando, J. M., Singh, S., Ikoma, N., et al. (2019). Randomized, open-label, perioperative phase II study evaluating nivolumab alone versus nivolumab plus ipilimumab in patients with resectable HCC. *Journal of Clinical Oncology, 37*(4_Suppl), 185. https://doi.org/10.1200/JCO.2019.37.4_suppl.185.

Kianmanesh, R., Regimbeau, J. M., & Belghiti, J. (2003). Selective approach to major hepatic resection for hepatocellular carcinoma in chronic liver disease. *Surgical Oncology Clinics of North America, 12*(1), 51–63. https://doi.org/10.1016/s1055-3207(02)00090-x.

Kobayashi, S., Terashima, T., Shiba, S., Yoshida, Y., Yamada, I., Iwadou, S., et al. (2018). Multicenter retrospective analysis of systemic chemotherapy for unresectable combined hepatocellular and cholangiocarcinoma. *Cancer Science, 109*(8), 2549–2557. https://doi.org/10.1111/cas.13656.

Kudo, M. (2017). Lenvatinib in advanced hepatocellular carcinoma. *Liver Cancer, 6*(4), 253–263. https://doi.org/10.1159/000479573.

Kudo, M., Finn, R. S., Qin, S., Han, K. H., Ikeda, K., Piscaglia, F., et al. (2018). Lenvatinib versus sorafenib in first-line treatment of patients with unresectable hepatocellular carcinoma: A randomised phase 3 non-inferiority trial. *Lancet, 391*(10126), 1163–1173. https://doi.org/10.1016/s0140-6736(18)30207-1.

Kudo, M., Matilla, A., Santoro, A., Melero, I., Gracian, A. C., Acosta-Rivera, M., et al. (2019). *Checkmate-040: Nivolumab (NIVO) in patients (pts) with advanced hepatocellular carcinoma (aHCC) and Child-Pugh B (CPB) status. Vol. 37.*

Kudo, M., Motomura, K., Wada, Y., Inaba, Y., Sakamoto, Y., Kurosaki, M., et al. (2019). First-line avelumab + axitinib in patients with advanced hepatocellular carcinoma: Results from a phase 1b trial (VEGF Liver 100). *Journal of Clinical Oncology, 37*(15_Suppl), 4072. https://doi.org/10.1200/JCO.2019.37.15_suppl.4072.

Kuo, Y. H., Wang, J. H., Hung, C. H., Rau, K. M., Wu, I. P., Chen, C. H., et al. (2017). Albumin-bilirubin grade predicts prognosis of HCC patients with sorafenib use. *Journal of Gastroenterology and Hepatology*, *32*(12), 1975–1981. https://doi.org/10.1111/jgh.13783.

Lafaro, K. J., & Pawlik, T. M. (2015). Fibrolamellar hepatocellular carcinoma: Current clinical perspectives. *Journal of Hepatocellular Carcinoma.*, *2*, 151–157. https://doi.org/10.2147/JHC.S75153.

Lai, E. C., Lo, C. M., Fan, S. T., Liu, C. L., & Wong, J. (1998). Postoperative adjuvant chemotherapy after curative resection of hepatocellular carcinoma: A randomized controlled trial. *Archives of Surgery*, *133*(2), 183–188. https://doi.org/10.1001/archsurg.133.2.183.

Lee, J. H., Lee, J. H., Lim, Y. S., Yeon, J. E., Song, T. J., Yu, S. J., et al. (2015). Adjuvant immunotherapy with autologous cytokine-induced killer cells for hepatocellular carcinoma. *Gastroenterology*, *148*(7), 1383–1391. e1386 https://doi.org/10.1053/j.gastro.2015.02.055.

Lee, P. C., Chen, Y. T., Chao, Y., Huo, T. I., Li, C. P., Su, C. W., et al. (2018). Validation of the albumin-bilirubin grade-based integrated model as a predictor for sorafenib-failed hepatocellular carcinoma. *Liver International*, *38*(2), 321–330. https://doi.org/10.1111/liv.13527.

Lee, M. S., Ryoo, B. Y., Hsu, C. H., Numata, K., Stein, S., Verret, W., et al. (2020). Atezolizumab with or without bevacizumab in unresectable hepatocellular carcinoma (GO30140): An open-label, multicentre, phase 1b study. *The Lancet Oncology*, *21*(6), 808–820. https://doi.org/10.1016/S1470-2045(20)30156-X.

Li, Q., Qin, S., Gu, S., Chen, X., Lin, L., Wang, Z., et al. (2020). Apatinib as second-line therapy in Chinese patients with advanced hepatocellular carcinoma: A randomized, placebo-controlled, double-blind, phase III study. *Journal of Clinical Oncology*, *38*(15_Suppl), 4507. https://doi.org/10.1200/JCO.2020.38.15_suppl.4507.

Lim, K. C., Chow, P. K., Allen, J. C., Siddiqui, F. J., Chan, E. S., & Tan, S. B. (2012). Systematic review of outcomes of liver resection for early hepatocellular carcinoma within the Milan criteria. *The British Journal of Surgery*, *99*(12), 1622–1629. https://doi.org/10.1002/bjs.8915.

Liu, P. H., Hsu, C. Y., Hsia, C. Y., Lee, Y. H., Su, C. W., Huang, Y. H., et al. (2016). Prognosis of hepatocellular carcinoma: Assessment of eleven staging systems. *Journal of Hepatology*, *64*(3), 601–608. https://doi.org/10.1016/j.jhep.2015.10.029.

Llovet, J. M., Bru, C., & Bruix, J. (1999). Prognosis of hepatocellular carcinoma: The BCLC staging classification. *Seminars in Liver Disease*, *19*(3), 329–338. https://doi.org/10.1055/s-2007-1007122.

Llovet, J. M., Ricci, S., Mazzaferro, V., Hilgard, P., Gane, E., Blanc, J. F., et al. (2008). Sorafenib in advanced hepatocellular carcinoma. *The New England Journal of Medicine*, *359*(4), 378–390. https://doi.org/10.1056/NEJMoa0708857.

Llovet, J. M., Decaens, T., Raoul, J. L., Boucher, E., Kudo, M., Chang, C., et al. (2013). Brivanib in patients with advanced hepatocellular carcinoma who were intolerant to sorafenib or for whom sorafenib failed: Results from the randomized phase III BRISK-PS study. *Journal of Clinical Oncology*, *31*(28), 3509–3516. https://doi.org/10.1200/JCO.2012.47.3009.

Maida, M., Orlando, E., Camma, C., & Cabibbo, G. (2014). Staging systems of hepatocellular carcinoma: A review of literature. *World Journal of Gastroenterology*, *20*(15), 4141–4150. https://doi.org/10.3748/wjg.v20.i15.4141.

Maniaci, V., Davidson, B. R., Rolles, K., Dhillon, A. P., Hackshaw, A., Begent, R. H., et al. (2009). Fibrolamellar hepatocellular carcinoma: Prolonged survival with multimodality therapy. *European Journal of Surgical Oncology*, *35*(6), 617–621. https://doi.org/10.1016/j.ejso.2008.12.009.

Marin, J. J. G., Cives-Losada, C., Asensio, M., Lozano, E., Briz, O., & Macias, R. I. R. (2019). Mechanisms of anticancer drug resistance in hepatoblastoma. *Cancers (Basel)*, *11*(3). https://doi.org/10.3390/cancers11030407.

Marrero, J. A., Kudo, M., Venook, A. P., Ye, S. L., Bronowicki, J. P., Chen, X. P., et al. (2016). Observational registry of sorafenib use in clinical practice across Child-Pugh subgroups: The GIDEON study. *Journal of Hepatology*, *65*(6), 1140–1147. https://doi.org/10.1016/j.jhep.2016.07.020.

Martin, A. P., Bartels, M., Hauss, J., & Fangmann, J. (2007). Overview of the MELD score and the UNOS adult liver allocation system. *Transplantation Proceedings*, *39*(10), 3169–3174. https://doi.org/10.1016/j.transproceed.2007.04.025.

Mavros, M. N., Mayo, S. C., Hyder, O., & Pawlik, T. M. (2012). A systematic review: Treatment and prognosis of patients with fibrolamellar hepatocellular carcinoma. *Journal of the American College of Surgeons*, *215*(6), 820–830. https://doi.org/10.1016/j.jamcollsurg.2012.08.001.

Mayo, S. C., Mavros, M. N., Nathan, H., Cosgrove, D., Herman, J. M., Kamel, I., et al. (2014). Treatment and prognosis of patients with fibrolamellar hepatocellular carcinoma: A national perspective. *Journal of the American College of Surgeons*, *218*(2), 196–205. https://doi.org/10.1016/j.jamcollsurg.2013.10.011.

Mehta, N., Heimbach, J., Harnois, D. M., Sapisochin, G., Dodge, J. L., Lee, D., et al. (2017). Validation of a risk estimation of tumor recurrence after transplant (RETREAT) score for hepatocellular carcinoma recurrence after liver transplant. *JAMA Oncology*, *3*(4), 493–500. https://doi.org/10.1001/jamaoncol.2016.5116.

Montella, L., Palmieri, G., Addeo, R., & Del Prete, S. (2016). Hepatocellular carcinoma: Will novel targeted drugs really impact the next future? *World Journal of Gastroenterology*, *22*(27), 6114–6126. https://doi.org/10.3748/wjg.v22.i27.6114.

Moschos, S. J., Edington, H. D., Land, S. R., Rao, U. N., Jukic, D., Shipe-Spotloe, J., et al. (2006). Neoadjuvant treatment of regional stage IIIB melanoma with high-dose interferon alfa-2b induces objective tumor regression in association with modulation of tumor infiltrating host cellular immune responses. *Journal of Clinical Oncology*, *24*(19), 3164–3171. https://doi.org/10.1200/JCO.2005.05.2498.

National Comprehensive Cancer Network. n.d. Hepatobiliary cancers. Version 5.2020. Retrieved from https://www.nccn.org/professionals/physician_gls/pdf/hepatobiliary.pdf.

Ono, T., Yamanoi, A., Nazmy El Assal, O., Kohno, H., & Nagasue, N. (2001). Adjuvant chemotherapy after resection of hepatocellular carcinoma causes deterioration of long-term prognosis in cirrhotic patients: Metaanalysis of three randomized controlled trials. *Cancer*, *91*(12), 2378–2385.

Panjala, C., Senecal, D. L., Bridges, M. D., Kim, G. P., Nakhleh, R. E., Nguyen, J. H. H., et al. (2010). The diagnostic conundrum and liver transplantation outcome for combined hepatocellular-cholangiocarcinoma. *American Journal of Transplantation*, *10*(5), 1263–1267. https://doi.org/10.1111/j.1600-6143.2010.03062.x.

Patt, Y. Z., Hassan, M. M., Lozano, R. D., Brown, T. D., Vauthey, J. N., Curley, S. A., et al. (2003). Phase II trial of systemic continuous fluorouracil and subcutaneous recombinant interferon Alfa-2b for treatment of hepatocellular carcinoma. *Journal of Clinical Oncology*, *21*(3), 421–427. https://doi.org/10.1200/JCO.2003.10.103.

Pinato, D. J., Sharma, R., Allara, E., Yen, C., Arizumi, T., Kubota, K., et al. (2017). The ALBI grade provides objective hepatic reserve estimation across each BCLC stage of hepatocellular carcinoma. *Journal of Hepatology*, *66*(2), 338–346. https://doi.org/10.1016/j.jhep.2016.09.008.

Pinato, D. J., Yen, C., Bettinger, D., Ramaswami, R., Arizumi, T., Ward, C., et al. (2017). The albumin-bilirubin grade improves hepatic reserve estimation post-sorafenib failure: Implications for drug development. *Alimentary Pharmacology & Therapeutics*, *45*(5), 714–722. https://doi.org/10.1111/apt.13904.

Pinato, D. J., Kaneko, T., Saeed, A., Pressiani, T., Kaseb, A., Wang, Y., et al. (2020). Immunotherapy in hepatocellular Cancer patients with mild to severe liver dysfunction: Adjunctive role of the ALBI grade. *Cancers (Basel)*, *12*(7). https://doi.org/10.3390/cancers12071862.

Poh, Z., Goh, B. B., Chang, P. E., & Tan, C. K. (2015). Rates of cirrhosis and hepatocellular carcinoma in chronic hepatitis B and the role of surveillance: A 10-year follow-up of 673 patients. *European Journal of Gastroenterology & Hepatology*, *27*(6), 638–643. https://doi.org/10.1097/MEG.0000000000000341.

Qin, S., Cheng, Y., Liang, J., Shen, L., Bai, Y., Li, J., et al. (2014). Efficacy and safety of the FOLFOX4 regimen versus doxorubicin in Chinese patients with advanced hepatocellular carcinoma: A subgroup analysis of the EACH study. *The Oncologist*, *19*(11), 1169–1178. https://doi.org/10.1634/theoncologist.2014-0190.

Qin, S., Chen, Z., Liu, Y., Xiong, J., Ren, Z., Meng, Z., et al. (2019). A phase II study of anti–PD-1 antibody camrelizumab plus FOLFOX4 or GEMOX systemic chemotherapy as first-line therapy for advanced hepatocellular carcinoma or biliary tract cancer. *Journal of Clinical Oncology*, *37*(15_Suppl), 4074. https://doi.org/10.1200/JCO.2019.37.15_suppl.4074.

Ranieri, G., Gadaleta-Caldarola, G., Goffredo, V., Patruno, R., Mangia, A., Rizzo, A., et al. (2012). Sorafenib (BAY 43-9006) in hepatocellular carcinoma patients: From discovery to clinical development. *Current Medicinal Chemistry*, *19*(7), 938–944. https://doi.org/10.2174/092986712799320736.

Roderburg, C., Ozdirik, B., Wree, A., Demir, M., & Tacke, F. (2020). Systemic treatment of hepatocellular carcinoma: From sorafenib to combination therapies. *Hepatology Oncology*, 7(2), HEP20. https://doi.org/10.2217/hep-2020-0004.

Rogers, J. E., Bolonesi, R. M., Rashid, A., Elsayes, K. M., Elbanan, M. G., Law, L., et al. (2017). Systemic therapy for unresectable, mixed hepatocellular-cholangiocarcinoma: Treatment of a rare malignancy. *The Journal of Gastrointestinal Oncology*, *8*(2), 347–351. https://doi.org/10.21037/jgo.2017.03.03.

Salimon, M., Prieux-Klotz, C., Tougeron, D., Hautefeuille, V., Caulet, M., Gournay, J., et al. (2018). Gemcitabine plus platinum-based chemotherapy for first-line treatment of hepatocholangiocarcinoma: An AGEO French multicentre retrospective study. *British Journal of Cancer*, *118*(3), 325–330. https://doi.org/10.1038/bjc.2017.413.

Sapisochin, G., Fidelman, N., Roberts, J. P., & Yao, F. Y. (2011). Mixed hepatocellular cholangiocarcinoma and intrahepatic cholangiocarcinoma in patients undergoing transplantation for hepatocellular carcinoma. *Liver Transplantation*, *17*(8), 934–942. https://doi.org/10.1002/lt.22307.

Sawada, Y., Yoshikawa, T., Ofuji, K., Yoshimura, M., Tsuchiya, N., Takahashi, M., et al. (2016). Phase II study of the GPC3-derived peptide vaccine as an adjuvant therapy for hepatocellular carcinoma patients. *Oncoimmunology*, *5*(5), e1129483. https://doi.org/10.1080/2162402X.2015.1129483.

Sherman, M. (2014). Staging for hepatocellular carcinoma: Complex and confusing. *Gastroenterology*, *146*(7), 1599–1602. https://doi.org/10.1053/j.gastro.2014.04.026.

Shindoh, J., Andreou, A., Aloia, T. A., Zimmitti, G., Lauwers, G. Y., Laurent, A., et al. (2013). Microvascular invasion does not predict long-term survival in hepatocellular carcinoma up to 2 cm: Reappraisal of the staging system for solitary tumors. *Annals of Surgical Oncology*, *20*(4), 1223–1229. https://doi.org/10.1245/s10434-012-2739-y.

Siegel, R. L., Miller, K. D., & Jemal, A. (2020). Cancer statistics, 2020. *CA: A Cancer Journal for Clinicians*, *70*(1), 7–30. https://doi.org/10.3322/caac.21590.

Torzilli, G., Belghiti, J., Kokudo, N., Takayama, T., Capussotti, L., Nuzzo, G., et al. (2013). A snapshot of the effective indications and results of surgery for hepatocellular carcinoma in tertiary referral centers: Is it adherent to the EASL/AASLD recommendations?: An observational study of the HCC East-West study group. *Annals of Surgery*, *257*(5), 929–937. https://doi.org/10.1097/SLA.0b013e31828329b8.

U. S. Food and Drug Administration/Center for Drug Evaluation and Research. (2020). Retrieved from https://www.fda.gov/drugs/drug-approvals-and-databases/fda-approves-atezolizumab-plus-bevacizumab-unresectable-hepatocellular-carcinoma.

Vauthey, J. N., Dixon, E., Abdalla, E. K., Helton, W. S., Pawlik, T. M., Taouli, B., et al. (2010). Pretreatment assessment of hepatocellular carcinoma: Expert consensus statement. *HPB: The Official Journal of the International Hepato Pancreato Biliary Association*, *12*(5), 289–299. https://doi.org/10.1111/j.1477-2574.2010.00181.x.

Villanueva, A. (2019). Hepatocellular Carcinoma. *The New England Journal of Medicine*, *380*(15), 1450–1462. https://doi.org/10.1056/NEJMra1713263.

Vogel, A., Rimassa, L., Sun, H.-C., Abou-Alfa, G. K., El-Khoueiry, A. B., Pinato, D. J., et al. (2020). Clinical value of atezolizumab + bevacizumab for first-line unresectable hepatocellular carcinoma (HCC): A network meta-analysis. *Journal of Clinical Oncology*, *38*(15_Suppl), 4585. https://doi.org/10.1200/JCO.2020.38.15_suppl.4585.

Xia, Y., Qiu, Y., Li, J., Shi, L., Wang, K., Xi, T., et al. (2010). Adjuvant therapy with capecitabine postpones recurrence of hepatocellular carcinoma after curative resection: A randomized controlled trial. *Annals of Surgical Oncology*, *17*(12), 3137–3144. https://doi.org/10.1245/s10434-010-1148-3.

Yamamoto, M., Arii, S., Sugahara, K., & Tobe, T. (1996). Adjuvant oral chemotherapy to prevent recurrence after curative resection for hepatocellular carcinoma. *The British Journal of Surgery*, *83*(3), 336–340. https://doi.org/10.1002/bjs.1800830313.

Yamashita, S., Vauthey, J. N., Kaseb, A. O., Aloia, T. A., Conrad, C., Hassan, M. M., et al. (2016). Prognosis of fibrolamellar carcinoma compared to non-cirrhotic conventional hepatocellular carcinoma. *Journal of Gastrointestinal Surgery*, *20*(10), 1725–1731. https://doi.org/10.1007/s11605-016-3216-x.

Yang, S. L., Luo, Y. Y., Chen, M., Zhou, Y. P., Lu, F. R., Deng, D. F., et al. (2017). A systematic review and meta-analysis comparing the prognosis of multicentric occurrence and vs. intrahepatic metastasis in patients with recurrent hepatocellular carcinoma after hepatectomy. *HPB: The Official Journal of the International Hepato Pancreato Biliary Association*, *19*(10), 835–842. https://doi.org/10.1016/j.hpb.2017.06.002.

Yau, T., Park, J. W., Finn, R. S., Cheng, A. L., Mathurin, P., Edeline, J., et al. (2019). LBA38_PR—CheckMate 459: A randomized, multi-center phase III study of nivolumab (NIVO) vs sorafenib (SOR) as first-line (1L) treatment in patients (pts) with advanced hepatocellular carcinoma (aHCC). *Annals of Oncology*, *30*, v874–v875. https://doi.org/10.1093/annonc/mdz394.029.

Yau, T., Zagonel, V., Santoro, A., Acosta-Rivera, M., Choo, S. P., Matilla, A., et al. (2020). Nivolumab (NIVO) + ipilimumab (IPI) + cabozantinib (CABO) combination therapy in patients (pts) with advanced hepatocellular carcinoma (aHCC): Results from CheckMate 040. *Journal of Clinical Oncology*, *38*(4_Suppl), 478. https://doi.org/10.1200/JCO.2020.38.4_suppl.478.

Yeh, M. M. (2010). Pathology of combined hepatocellular-cholangiocarcinoma. *Journal of Gastroenterology and Hepatology*, *25*(9), 1485–1492. https://doi.org/10.1111/j.1440-1746.2010.06430.x.

You, D. D., Kim, D. G., Seo, C. H., Choi, H. J., Yoo, Y. K., & Park, Y. G. (2017). Prognostic factors after curative resection hepatocellular carcinoma and the surgeon's role. *Annals of Surgical Treatment and Research*, *93*(5), 252–259. https://doi.org/10.4174/astr.2017.93.5.252.

Zhong, J. H., Rodriguez, A. C., Ke, Y., Wang, Y. Y., Wang, L., & Li, L. Q. (2015). Hepatic resection as a safe and effective treatment for hepatocellular carcinoma involving a single large tumor, multiple tumors, or macrovascular invasion. *Medicine (Baltimore)*, *94*(3), e396. https://doi.org/10.1097/MD.0000000000000396.

Zhu, A. X., Kudo, M., Assenat, E., Cattan, S., Kang, Y. K., Lim, H. Y., et al. (2014). Effect of everolimus on survival in advanced hepatocellular carcinoma after failure of sorafenib: The EVOLVE-1 randomized clinical trial. *JAMA*, *312*(1), 57–67. https://doi.org/10.1001/jama.2014.7189.

Zhu, A. X., Park, J. O., Ryoo, B. Y., Yen, C. J., Poon, R., Pastorelli, D., et al. (2015). Ramucirumab versus placebo as second-line treatment in patients with advanced hepatocellular carcinoma following first-line therapy with sorafenib (REACH): A randomised, double-blind, multicentre, phase 3 trial. *The Lancet Oncology*, *16*(7), 859–870. https://doi.org/10.1016/s1470-2045(15)00050-9.

Zhu, A. X., Rosmorduc, O., Evans, T. R., Ross, P. J., Santoro, A., Carrilho, F. J., et al. (2015). SEARCH: A phase III, randomized, double-blind, placebo-controlled trial of sorafenib plus erlotinib in patients with advanced hepatocellular carcinoma. *Journal of Clinical Oncology*, *33*(6), 559–566. https://doi.org/10.1200/JCO.2013.53.7746.

Zhu, A. X., Finn, R. S., Edeline, J., Cattan, S., Ogasawara, S., Palmer, D., et al. (2018). Pembrolizumab in patients with advanced hepatocellular carcinoma previously treated with sorafenib (KEYNOTE-224): A non-randomised, open-label phase 2 trial. *The Lancet Oncology*, *19*(7), 940–952. https://doi.org/10.1016/s1470-2045(18)30351-6.

Zhu, A. X., Kang, Y. K., Yen, C. J., Finn, R. S., Galle, P. R., Llovet, J. M., et al. (2019). Ramucirumab after sorafenib in patients with advanced hepatocellular carcinoma and increased α-fetoprotein concentrations (REACH-2): A randomised, double-blind, placebo-controlled, phase 3 trial. *The Lancet Oncology*, *20*(2), 282–296. https://doi.org/10.1016/s1470-2045(18)30937-9.

Zhu, A. X., Finn, R. F., Ikeda, M., Sung, M. W., Baron, A. D., Kudo, M., et al. (2020). A phase Ib study of lenvatinib (LEN) plus pembrolizumab (PEMBRO) in unresectable hepatocellular carcinoma (uHCC). *Journal of Clinical Oncology*, *38*(15_Suppl), 4519. https://doi.org/10.1200/JCO.2020.38.15_suppl.4519.

Zhu, A. X., Ikeda, M., Galle, P. R., Yamashita, T., Llovet, J., Liang, K., et al. (2020). Ramucirumab in patients with advanced HCC and elevated alpha-fetoprotein (AFP): Outcomes by treatment-emergent ascites. *Journal of Clinical Oncology*, *38*(15_Suppl), 4644. https://doi.org/10.1200/JCO.2020.38.15_suppl.4644.

CHAPTER SEVEN

Immunotherapy in hepatocellular cancer

Christos Fountzilas, Rachel Evans, Sabah Alaklabi, and Renuka Iyer*
Roswell Park Comprehensive Cancer Center, Buffalo, NY, United States
*Corresponding author: e-mail address: renuka.iyer@roswellpark.org

Contents

1. Introduction 295
2. Immune checkpoint inhibitors in hepatocellular carcinoma 298
   2.1 Monotherapy 298
   2.2 Immune checkpoint inhibitor combinations 302
   2.3 VEGF/VEGFR pathway and immune checkpoint inhibitor combinations 304
3. Future directions 307
4. Conclusion 312
References 312

Abstract

Hepatocellular carcinoma is a major global healthcare problem. It is closely related to chronic liver inflammation triggered by viral and non-viral insults, that can lead to exhaustion of effector T-cells. Furthermore, immune cells within the normal liver itself tend to be more immune tolerant in order to support the essential function of liver as the first processing station of molecules absorbed in the gastrointestinal tract. Dysregulation of the immune system is a hallmark of hepatocellular carcinoma. Immune checkpoint inhibitors targeting the programmed death-1 axis have shown promise as monotherapy in the management of advanced disease, but still most patients do not benefit from treatment. Most recently, combinatorial strategies with other immune checkpoint inhibitors or agents targeting the second hallmark of hepatocellular carcinoma, i.e., the activation of the vascular epithelial growth factor axis have been studied. In this paper, we review the current immunotherapy approaches for hepatocellular carcinoma and discuss novel immunotherapy approaches and optimal patient selection.

1. Introduction

Hepatocellular carcinoma (HCC) is the most common form of primary liver cancer (Villanueva, 2019). Worldwide, it is the 6th most common type of cancer (Villanueva, 2019). The incidence of HCC in

Advances in Cancer Research, Volume 149
ISSN 0065-230X
https://doi.org/10.1016/bs.acr.2020.12.002

the United States (US) is on the rise and, despite the overall decrease in cancer-related mortality over the past 2 decades, the incidence of death from HCC is increasing significantly (Jemal et al., 2017). From 2000 to 2016, the deaths from HCC increased by 43% and, with a 5-year overall survival (OS) of less than 20%, HCC is the 2nd leading cause of cancer-related death in the US (Jemal et al., 2017; Villanueva, 2019).

The role of vascular endothelial growth factor (VEGF) in HCC carcinogenesis has been well established and proven to be a valuable target for HCC therapy (Morse et al., 2019). HCC arises in an intensely inflammatory environment from chronic liver disease and cirrhosis (Fig. 1) (Villanueva, 2019). The tumor microenvironment (TME) in HCC is characterized by the presence of both effector cells, such as cytotoxic $CD8^+$ T-cells (CTLs) and natural killer (NK) cells and suppressive immune cells including tumor-associated macrophages (TAMs), regulatory T-cells (T_{regs}) and myeloid-derived suppressor cells (MDSCs) (Chew et al., 2017; Hoechst et al., 2008). Furthermore, fibrosis—a key characteristic of chronic liver disease—can negatively affect immunosurveillance and antigen recognition by sinusoidal defenestration and capillarization (Guidotti et al., 2015). Intratumoral CTL and T_{reg} infiltration in HCC has long been recognized as a significant prognostic feature (Gao et al., 2007; Unitt et al., 2006). Though present, CTLs in HCC are

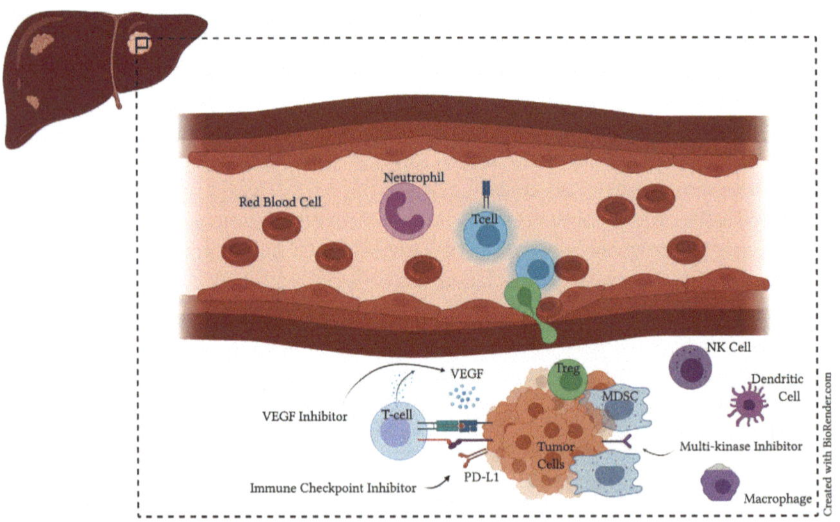

Fig. 1 Clinical Algorithm for the Management of Hepatocellular Carcinoma. CLIP: Cancer of the Liver Italian Program, CTP: Child-Turcotte-Pugh score, TACE: Transarterial chemoembolization ECOG-PS: Eastern Cooperative Oncology Group-Performance Status.

functionally impaired, with decreased interferon-gamma (INF-γ) production upon antigen stimulation; this is more pronounced in patients with advanced disease stage (Flecken et al., 2014). T_{regs} in patients with HCC suppress expression of cytolytic molecules such as perforin and granzyme A and B by CTLs (Fu et al., 2007). T_{reg} depletion can improve T-cell proliferation but cannot restore INF-γ production (Flecken et al., 2014). Further, $CD8^+$ T-cells within the TIME are characterized by increased expression of multiple T-cell exhaustion markers, such as programmed death-1 (PD1), lymphocyte activation gene-3 (LAG3), and T-cell immunoglobulin and mucin domain-containing protein-3 (TIM3) (Chew et al., 2017). Approximately 25% of HCC are characterized by increased expression of PD1/programmed death ligand-1 (PD-L1) and cytolytic molecules; these can be further subclassified to an increased adaptive T-cell response group and an increased T-cell exhaustion group (Sia et al., 2017). TAMs can secrete proinflammatory cytokines such as IL-1β, IL-6, IL-23, and tumor necrosis factor-alpha (TNF-α) leading to expansion of interleukin-17-producing, pro-tumor $CD4^+$ T-helper 17 cells (T_{h17}) that induce T-cell exhaustion (Kuang et al., 2010). Further, by secretion of IL-10, TAMs can lead to increase in intratumoral T_{regs} (Zhou et al., 2009). Hypoxia, a typical characteristic of TME in HCC, can trigger epithelial to mesenchymal transition. CCL20 secretion by HCC cells is increased during this process and can induce production of immunosuppressive indoleamine 2,3-dioxygenase (IDO) production from TAMs that suppresses CTL proliferation and promotes T_{reg} expansion (Ye et al., 2016). Hypoxia can also promote regulated on activation, normal T-cell expressed and secreted (RANTES)-dependent liver neutrophilic infiltration that promotes tumorigenesis (Koh et al., 2016). A subset of tumor-associated neutrophils (TANs) has a tumor-promoting phenotype and high neutrophil/lymphocyte ratio has been recognized as a negative predictive biomarker for benefit from immunotherapy targeting the PD1/PD-L1 axis in multiple tumor types (Ameratunga et al., 2018) including HCC (Dharmapuri et al., 2019). MDSCs are increased in patients with HCC (Nan et al., 2018), and contribute to decreased T-cell infiltration (Liu, Zhou, et al., 2020), proliferation, and interferon-gamma (IFN-γ) response (Nan et al., 2018) as well as induction of T_{reg} proliferation (Hoechst et al., 2008). In addition, MDSCs inhibit NK cell IFN-γ production (Zhang et al., 2017) and cytotoxicity (Hoechst et al., 2009). Similarly, TAMs in patients with advanced HCC interact with NK cells and lead to impaired IFN-γ production and NK cell exhaustion (Wu et al., 2013). Alpha-fetoprotein (AFP), frequently elevated in cirrhosis and HCC, can induce T_{reg} expansion (Alisa et al., 2008; Murgita, Goidl, Kontianen, & Wigzell, 1977), decrease the NK cell viability in a timed-dependent fashion (Vujanovic et al., 2017), and cytotoxic activity via inhibition of IL-12 production by dendritic cells (DCs) (Yamamoto et al., 2011) and increased DC apoptosis (Um et al., 2004).

Phenotyping of peripheral blood mononuclear cells (PBMCs) from patients with HCC by Kalathil et al. revealed increased numbers of T_{regs}, MDSCs, $PD1^+$ exhausted T-cells, and increased levels of immunosuppressive cytokines in patients with HCC compared to healthy normal controls; $CD4^+$ and $CD8^+$ T-cells have impaired proliferation, INF-γ and granzyme B secretion capacity (Kalathil, Lugade, Miller, Iyer, & Thanavala, 2013). This impaired T-cell functionally was only modestly improved with depletion of the inhibitory immune cell populations. T_{regs} from patients with HCC can inhibit secretion of IFN-γ and cytotoxicity of $CD8^+$ T-cells, a phenomenon that can be partially reversed by PD1/PDL1 axis inhibition (Langhans et al., 2019).

All these important studies to understand HCC carcinogenesis and immune dysfunction have helped develop trials and led to approvals of both VEGF-targeted agents and immune checkpoint inhibitors (ICIs). Up until 2017, the multi-kinase inhibitor (MKi) sorafenib was the only systemic therapy option available for patients with advanced HCC for over a decade (Llovet et al., 2008), with several trials of other potent VEGF-targeted MKis not proven superior to sorafenib (Cainap et al., 2015; Cheng et al., 2013; Johnson et al., 2013). In recent years, however, the systemic therapeutic armamentarium for HCC has expanded (Table 1) to include the MKis regorafenib (Bruix et al., 2017), cabozantinib (Abou-Alfa, Chan, et al., 2018; Abou-Alfa, Meyer, et al., 2018), and lenvatinib (Kudo et al., 2018) as well as ICIs targeting the PD1/PD-L1 axis (pembrolizumab (Zhu et al., 2018), nivolumab (El-Khoueiry et al., 2017), atezolizumab (Finn et al., 2020)) and cytotoxic T-cell lymphocyte associated protein-4 (CTLA4) pathways (ipilimumab (Yau et al., 2020)) and finally, monoclonal antibodies targeting the VEGF/VEGF receptor (VEGFR) pathway (ramucirumab (Zhu et al., 2019), bevacizumab (Finn, Qin, et al., 2020)).

2. Immune checkpoint inhibitors in hepatocellular carcinoma

2.1 Monotherapy

The first ICIs to be Food and Drug Administration (FDA)-approved for use in advanced HCC after sorafenib treatment were pembrolizumab and

Table 1 FDA-approved agents for the management of advanced hepatocellular carcinoma.

	Mechanism of action/drug target	Line of therapy	ORR (%)	OS	References
Sorafenib	Multikinase inhibitor	1st	2–12	10.7–12.3 months (median), 55% (12-month)	Llovet et al. (2008), Kudo et al. (2018, 2019), Finn, Ikeda, et al. (2020), Finn, Qin, et al. (2020), and Finn, Ryoo, et al. (2020)
Lenvatinib	Multikinase inhibitor	1st	24	13.6 months (median)	Kudo et al. (2018, 2019)
Atezolizumab/ Bevacizumab	Anti-PDL1 + Anti-VEGF-A	1st	27	72% (12-month)	Finn, Ikeda, et al. (2020), Finn, Qin, et al. (2020), and Finn, Ryoo, et al. (2020)
Regorafenib	Multikinase inhibitor	2nd	11	10.6 months (median)	Bruix et al. (2017)
Cabozantinib	Multikinase inhibitor	2nd or 3rd	4	10.2 months (median)	Abou-Alfa, Chan, et al. (2018) and Abou-Alfa, Meyer, et al. (2018)
Pembrolizumab	Anti-PD1	2nd	17–18	12.9–13.9 months (median)	Zhu et al. (2018, 2020, 2019), Finn, Ikeda, et al. (2020), Finn, Qin, et al. (2020), and Finn, Ryoo, et al. (2020)
Nivolumab	Anti-PD1	2nd	20	74% (9-month)	El-Khoueiry et al. (2017)
Nivolumab/ Ipilimumab	Anti-PD1 + Anti-CTLA4	2nd	32	22.8 months (median)	Yau et al. (2020, 2019)
Ramucirumab	Anti-VEGF	2nd	5	8.5 months (median)	Zhu et al. (2018, 2020, 2019)

nivolumab, both IgG4 monoclonal antibodies targeting PD1. Pembrolizumab was tested in the phase II, non-randomized Keynote-244 study (Zhu et al., 2018). In this study, 104 patients with Barcelona Clinic Liver Cancer stage (BCLC) C (or B not amenable to locoregional or curative therapy), Eastern Cooperative Oncology Group Performance Status (ECOG PS) 0–1, Child-Pugh score (CP) A and who had prior progressive disease on or were intolerable to sorafenib, were treated with pembrolizumab 200mg intravenous (IV) once every 3weeks. Seventeen percent of the patients had an objective response, 62% had disease control. Responses were attained within 8–10weeks for the majority of patients with response. The overall response rate (ORR) in PD-L1 positive patients (using both the combined positive score (CPS) and the tumor proportional score (TPS) with the 22C3 antibody) was 25%. The median OS was 12.9months and median progression-free survival (PFS) was 4.9months. Treatment-related adverse events (TRAEs) were common but mostly grade 1 or 2; the most common TRAEs of any grade were fatigue, diarrhea, rash, pruritus and increased transaminases. Liver toxicity was infrequent. Patients with viral hepatitis were eligible for enrollment (in case of hepatitis B (HBV) infection, patients should have a viral load of less than 100IU/mL and were treated with antiviral therapy); there was no flare of HBV or Hepatitis C (HCV). Immune related adverse events (IRAEs) were uncommon (14%), the most frequent was hypothyroidism in 8% of the patients. Overall, pembrolizumab was well tolerated with 25% of patients having dose interruptions for toxicity, mainly increased transaminase levels.

The CheckMate-040 phase I/II study, enrolled HCC patients with ECOG 0–1, CP A (and B7 only in the dose-escalation cohort), who had progressed in at least one prior line of systemic therapy (including sorafenib) or were intolerant to sorafenib (El-Khoueiry et al., 2017). In the dose-escalation phase, 48 patients were treated with nivolumab in doses ranging from 0.1 to 10mg/kg IV every 2weeks in 3 different cohorts (no viral hepatitis, HBV as in Keynote 224 study, and HCV) using a 3+3 design. The maximum tolerated dose (MTD) was not reached, and the 3mg/kg every 2weeks was selected for expansion. The dose-expansion phase enrolled 214 patients with advanced HCC. The ORR was 20%; 64% had disease control. Twenty-six percent of the patients with tumor PD-L1 expression at least 1% had an objective response vs. 19% with PD-L1 of less than 1%. The 9-month OS and PFS rate were 74% and 28%, respectively. Grade 3 or 4 TRAEs occurred in 19% of patients. Similar to pembrolizumab the most common TRAE of any grade were rash, diarrhea, fatigue, pruritus and transaminase elevation. Significant liver toxicity was uncommon and there were no cases of viral hepatitis reactivation.

In sum, both nivolumab and pembrolizumab have activity in advanced HCC after sorafenib treatment; their safety and activity profile appear similar and there is no biomarker to guide treatment selection. In the basis of Keynote-224 trial, the Keynote-240 phase III, randomized, placebo-controlled study enrolled patients with advanced HCC ($N=413$) (Finn, Ryoo, et al., 2020). The inclusion criteria were overall similar to Keynote-224; patients with recent hepatic encephalopathy, clinically-detected ascites and main portal or inferior vena cava invasion were excluded. Patients were randomized at a 2:1 ratio to pembrolizumab 200mg IV or matching placebo every 3weeks for up to 2years. The co-primary endpoints were OS and PFS. The study failed to reach the primary endpoint as the median OS was 13.9months and 10.6months in the pembrolizumab and placebo group, respectively ($P=0.02$), and the median PFS was 3months vs. 2.8months for pembrolizumab and placebo, respectively ($P=0.02$); both not meeting the prespecified P boundaries (0.0174 for OS and 0.002 for PFS). The ORR was 18.3% for pembrolizumab and the safety profile was as expected from prior studies.

The phase III CheckMate-459 study compared nivolumab and sorafenib as first-line therapy in patients with advanced HCC; 743 patients were randomized to nivolumab 240mg IV every 2weeks or sorafenib 400mg twice daily orally in 1:1 ratio (Yau et al., 2019). Similar to Keynote-240, CheckMate-459 did not meet its primary endpoint of OS with median OS 16.4months vs. 14.7months in the nivolumab and sorafenib groups respectively ($P=0.07$). The ORR for nivolumab was 15% and 7% for sorafenib. Grade 3 or 4 TRAEs were observed in 22% vs. 49% patients in the nivolumab and sorafenib arms, respectively, and led to discontinuation in 4% and 8% of patients, respectively. Forty-six percent of the patients in the sorafenib arm received subsequent systemic therapies vs. 38% in the nivolumab group.

Despite clear clinical activity, the negative results for both pembrolizumab and nivolumab in randomized controlled studies, beyond the design and statistical considerations and goals, have much to do with approval and availability of other life-prolonging therapies during the conduct of the studies that impacted OS results. Both remain appropriate standard of care therapies for advanced HCC after progression or intolerance to sorafenib. While currently published guidelines and clinical contraindications guide selection of ICI and MKi therapy, much work remains to be done in understanding biomarkers of response to ICIs. Both chronic HBV and HCV infection patients have been treated with and responded to ICIs but a significant percentage of patients with HCC overall has no benefit from ICI therapy. This can be related to both tumor genomic and TME factors. For example, HCC analysis from The Cancer Genome Atlas (TCGA) revealed three distinct immune HCC subgroups, cluster 1 (rich in high infiltration activated adaptive immune cells, IFN-γ and TNF-α production, TAMs, and T_{regs}), cluster 2 was (high infiltration of stromal cells, M2 macrophages, and low abundance of cytotoxic lymphocytes), and cluster 3 (characterized by inactivated immune cells and stromal cells) (Liu, Qin, et al., 2020). As such, cluster 1 is characterized by a both highly active antitumor and immunosuppressive immune microenvironment at the same time, while in cluster 2 TME is highly immunosuppressive and in cluster 3 resting. Interestingly, long-term outcomes were worse in cluster 1 and improved in cluster 3 (Liu, Qin, et al., 2020). In silico modeling of the two primary mechanisms of tumor immune evasion, i.e., the induction of T-cell dysfunction and prevention of T-cell infiltration (Tumor Immune Dysfunction and Exclusion (TIDE) tool) (Jiang et al., 2018) most responses to ICIs are expected in cluster 1 (Liu, Qin, et al., 2020). Furthermore, activating alterations in the WNT/β-catenin pathway in HCC are related to decreased activity of ICIs (Harding et al., 2019).

2.2 Immune checkpoint inhibitor combinations

Both $CD8^+$ effector T-cells and antigen presenting cells (APCs) in HCC express high levels of inhibitory immune checkpoint molecules (PD1/TIM3/LAG3/cytotoxic T-lymphocyte associated protein-4 [CTLA4] and their ligands PD-L1/GAL9/MHCII/CD86-CD80, respectively) (Zhou et al., 2017). The impaired effector T-cell functionality and proliferation can be reversed with inhibitory monoclonal antibodies targeting each pathway while combinations had a synergistic effect. Importantly, combinations of blocking antibodies toward different inhibitory checkpoint pathways have an additive effect in ex vivo responses of effector T-cells to tumor associated HCC antigens (Zhou et al., 2017).

Inhibition of PD1 and CTLA4 axis have distinct and non-overlapping immunological effects in the TME and lymphoid organs (Das et al., 2015; Melero et al., 2015). Building on the experience with dual anti-PD1/CTLA4 therapy in melanoma and other solid tumor types (Hodi et al., 2016; Motzer et al., 2018), in the CheckMate-040 (cohort 4) nivolumab was combined with the anti-CTLA4 IgG4 monoclonal antibody ipilimumab in patients with advanced HCC, not amenable to locoregional therapy, intolerance or progression on sorafenib, CP A and ECOG PS 0–1 (Yau et al., 2020). In this randomized study ($N = 148$), patients were randomized 1:1:1 to either nivolumab 1 mg/kg IV plus ipilimumab 3 mg/kg IV, administered every 3 weeks for 4 doses, followed by nivolumab 240 mg IV every 2 weeks (arm A); nivolumab 3 mg/kg IV plus ipilimumab 1 mg/kg IV, administered every 3 weeks for 4 doses, followed by nivolumab 240 mg IV every 2 weeks (arm B); or nivolumab 3 mg/kg IV every 2 weeks plus ipilimumab 1 mg/kg IV every 6 weeks (arm C). The ORR was 32%, 27%, and 29% in Arms A, B, and C, respectively; the median OS was 22.8, 12.5, and 12.7 months, respectively. More TRAEs of any grade were observed in Arm A vs. Arm B or C (94% vs. 71% vs. 79%) and more AEs led to treatment discontinuation in Arm A (18% vs. 6% vs. 2%). Similarly, the rate of IRAEs was higher in Arm A and led to more treatment discontinuations. Overall, the rate of AEs was higher compared with nivolumab monotherapy. Increases in HBV DNA or HCV RNA from baseline of at least 1-log was observed in 9% of HBV- and 10% of HCV-infected patients overall but changes in viral kinetics were not associated with hepatic TRAEs.

At another phase I/II study, the IgG1 anti-PD-L1 monoclonal antibody durvalumab was combined with the IgG4 anti-CTLA4 antibody tremelimumab in patients with unresectable HCC (with or without HBV or HCV infection) who had disease progression on or intolerance to or had refused treatment with sorafenib (Kelley et al., 2017). Treatment included durvalumab 20 mg/kg IV every 4 weeks and tremelimumab 1 mg/kg IV every 4 weeks for 4 doses followed by durvalumab alone at 20 mg/kg every 4 weeks. Forty patients were enrolled, 30% with no prior systemic therapy and almost all with CP A (93%). Sixty percent had at least one TRAE and 20% had at least one grade 3 or higher TRAE. The most common TRAEs were fatigue, pruritus and increased transaminases. Three patients discontinued therapy for TRAEs (one grade 3 pneumonitis, one grade 3 colitis, and one asymptomatic grade 4 transaminase elevation). The confirmed ORR was 15%. Taking into account the promising result of this safety run-in phase as well as data in other solid tumors treated with increasing doses of tremelimumab suggesting immune priming with a higher tremelimumab dose, the randomized expansion cohorts comprised 4 arms evaluating tremelimumab 750 mg IV every 4 weeks and durvalumab 1500 mg IV every 4 weeks as monotherapies and 2 combinations regimens (priming dose tremelimumab 300 mg IV plus durvalumab 1500 mg IV once followed by durvalumab every for weeks, or tremelimumab 75 mg IV plus durvalumab 1500 mg IV every 4 weeks for 4 doses followed by durvalumab every 4 weeks) (Kelley et al., 2020). The rate of grade 3 or 4 TRAEs was higher in the tremelimumab monotherapy arm (42%) vs. in priming combination (35%), low dose tremelimumab combination (24%) or durvalumab monotherapy (18%). The ORR was higher in the priming combination arm (23%) while it was less than 10% in the other arms. A unique proliferative T-cell profile, that patients with an objective response exhibited high CTL counts, was identified for patients in the priming combination arm, suggesting additive biologic activity. The encouraging clinical activity in the priming combination arm led to a randomized phase III study evaluating the combination priming strategy vs. durvalumab monotherapy vs. sorafenib in treatment-naïve patients with advanced HCC that is currently ongoing (Abou-Alfa, Chan, et al., 2018; Abou-Alfa, Meyer, et al., 2018).

In March 2020, the FDA approved the nivolumab/ipilimumab combination (dose as in Arm A) for patients with prior progression or intolerance to sorafenib. As this approach has not been validated in a randomized study and given the significant incidence of TRAEs (overall: 94%, grade 3 or 4: 53%), caution should be exercised in terms of patient selection. The priming anti-CTLA4 approach can potentially have a comparable antitumor effect with less toxicity.

2.3 VEGF/VEGFR pathway and immune checkpoint inhibitor combinations

Increased angiogenesis is a landmark characteristic of hypervascular tumors such as HCC (Morse et al., 2019). The effects of VEFR/VEGFR axis on tumor immunity has been extensively studied. Normalization of tumor neovasculature via inhibition of VEGF can increase tumor infiltration by CTLs (Shrimali et al., 2010). Furthermore, VEGF-A acting synergistically with IL-10 and prostaglandin E2 can promote apoptosis of CTLs via induction of Fas ligand (FasL) in tumor-related endothelium (Motz et al., 2014). Additionally, VEGF can inhibit dendritic cell (DC) maturation (Takahashi et al., 2004) and lead to increased expression of immunosuppressive enzymes by DCs (Marti et al., 2014). VEGF-A also has a direct effect on CTLs, increasing the expression of exhaustion markers, that can be reversed by disruption of the VEGF/VEGFR axis (Voron et al., 2015).

In the open label, multi-arm phase Ib GO30140 study, 104 patients with systemic therapy naïve, advanced HCC were enrolled in group A and treated with the IgG1 anti-PDL1 atezolizumab at 1200 mg IV and the anti-VEGF-A monoclonal antibody bevacizumab at 15 mg/kg IV, both every 3 weeks (Lee et al., 2020). In group F of this study, 119 patients were randomized at a 1:1 ratio to atezolizumab/bevacizumab vs. atezolizumab monotherapy at the same doses as in group A (Lee et al., 2020). The ORR in group A was 36%. The most common grade 3 or 4 TRAEs were hypertension (13%) and proteinuria (7%); serious TRAEs occurred in 24% of patients. In group F, the combination therapy led to improved PFS compared to atezolizumab monotherapy (median PFS 5.6 vs. 3.4 months with monotherapy, $P=0.011$). As expected, cases of significant hypertension and proteinuria were higher in the combination arm (both less than 10%) and serious TRAEs occurred in 2% patients in the atezolizumab/bevacizumab group vs. 3% in the atezolizumab monotherapy group. The promising results of this study created the premise for a randomized, open-label phase III study in patients with treatment-naïve, advanced HCC comparing the atezolizumab/bevacizumab combination, dosed as in group A of GO30140 study, to sorafenib 400 mg orally twice daily in a 2:1 randomization ($N=501$) (Finn, Qin, et al., 2020). Patients had histologically or cytologically confirmed HCC, CP A, and not amenable to locoregional or curative therapy and ECOG PS 0–1; patients with HBV and HCV coinfection and untreated esophageal varices were excluded. The study was positive for the coprimary endpoints of OS and PFS, with a 12-month OS of 67% in the atezolizumab/bevacizumab arm vs. 55% in the sorafenib arm ($P<0.001$) and median PFS 6.8 vs. 4.3 months respectively ($P<0.001$). The ORR also was in favor of the combination arm (27% vs. 12%). Serious AEs were more frequent in the combination arm (38% vs. 31%), 15% of patients in the combination arm had to discontinue therapy for AEs vs. 10% in the sorafenib arm. Based on the results of this study, the FDA approved the atezolizumab/bevacizumab combination as first-line therapy in patients with advanced HCC in May 2020.

Translational work from our group has clearly defined the ability of MKis to reverse immunosuppression in HCC by their inhibitory effect on targets other than VEGFR2. We have shown that treatment with sorafenib in patients with advanced HCC can significantly decrease MDSCs and T_{reg} populations and improve the $CD4^+CD127^+PD\text{-}1^-$ effector T-cell/$CD4^+Foxp3^+PD\text{-}1^+$ T_{reg} and $CD8^+CD127^+$ effector T-cell/T_{reg} ratios in peripheral blood, with a more pronounced benefit in terms of OS for patients with the deepest T_{reg} responses (Kalathil, Hutson, Barbi, Iyer, & Thanavala, 2019; Kalathil, Lugade, Miller, Iyer, & Thanavala, 2016).

In addition, CTLA-4$^+$ T_{regs} decreased after sorafenib therapy (Kalathil et al., 2019). We noted sustained decrease in downstream ERK phosphorylation in FLT3$^+$ MDSCs and T_{regs} that is pointing toward a direct effect of sorafenib on immunosuppressive cells via FLT3 inhibition (Kalathil et al., 2019). Beyond favorable effects in reducing pro-tumor immune cell populations, sorafenib can lead to improvements in proliferation, INF-γ producing capacity of CD8$^+$ and CD4$^+$ T-cells that is long-lasting as well as production of cytolytic molecules (Kalathil et al., 2019). Tivozanib, a novel MKi currently in development for HCC, has shown promising anti-neoplastic activity and a favorable AE profile (Fountzilas et al., 2020). Similar to sorafenib, tivozanib can decrease the number of T_{regs} and MDSCs; ERK phosphorylation was decreased in c-kit$^+$ T_{regs} and MDSCs (Kalathil, Wang, Hutson, Iyer, & Thanavala, 2020). The expression of inhibitory immune checkpoints also decreased in CD8$^+$ and CD4$^+$ T-cells. When compared to sorafenib, tivozanib led to more pronounced decrease in the T_{regs}, MDSCs and exhausted T-cells (Kalathil et al., 2020). Furthermore, greater increase in CD4$^+$ T-cell/Treg ratio with tivozanib was associated with significant improvement in OS compared to sorafenib treatment (Kalathil et al., 2020). This can be related to dose- and time-dependent immunomodulatory effects of sorafenib, with low doses of sorafenib having an antitumor immune effect while high doses are associated with more immunosuppression (Iyer et al., 2019).

Kimura and colleagues evaluated the immunomodulator effects of lenvatinib, a MKI currently approved as first-line therapy in patients with advanced HCC (Kudo et al., 2018), in an in vivo preclinical HCC model (Kimura et al., 2018). Similar to published work by Kalathil et al. regarding immunomodulation of sorafenib, treatment with lenvatinib decreased TAMs and increased CTLs. Lenvatinib had superior antitumor efficacy compared to sorafenib and this effect was dependent on the presence of an intact and functional immune system. Finally, the antitumor effect of lenvatinib was potentiated when combined with anti-PD1 inhibition.

Taken together, there is strong preclinical and translational rationale for combinations of ICIs with anti VEGF/VEGFR and MKis in HCC. In the single-arm, phase Ib KEYNOTE-524/Study 116 trial, lenvatinib 12 mg (weight $\geq$ 60 kg) or 8 mg (weight $<$ 60 kg) was combined with pembrolizumab in patients with previously untreated, BCLC stage B (not amenable to locoregional therapy) or C HCC, CP A (score, 5–6); and ECOG PS 0–1 (Finn et al., 2020). No dose-limiting toxicities (DLTs) were observed in the 6 patients enrolled in the DLT phase; 98 more patients were enrolled in the dose-expansion phase. The ORR was 36%; the median PFS was 8.6 months and the median OS was 22 months. The most common TRAEs (any grade) were hypertension (36%), diarrhea (35%), fatigue (30%), decreased appetite (28%), and hypothyroidism (25%); TRAEs grade 3 or higher occurred in 67% of patients, the most common being hypertension (17%). Similarly, the combination of the MKi axitinib plus the anti-PD-L1 monoclonal antibody avelumab has promising preliminary results with tumor shrinkage being observed in the majority of patients and objective responses attained in 14%–32% of patients (depending on the response evaluation criteria); there were no new safety signals (Kudo et al., 2019). The multicenter, randomized, double-blind phase 3 LEAP-002 trial is currently exploring the efficacy of the lenvatinib/pembrolizumab combination vs. lenvatinib monotherapy as first-line treatment in the same patient population (NCT03713593). Ongoing phase I/II studies are also evaluating sorafenib plus pembrolizumab (NCT03211416) and tivozanib plus durvalumab (NCT03970616) in patients with advanced HCC.

3. Future directions

Despite the large success of ICI combinations, advanced HCC remains an incurable disease with almost half of the patients progressing within 6 months and 30%–50% of patients dying within a year from treatment initiation (Finn, Qin, et al., 2020; Yau et al., 2020). Identification and targeting resistance mechanisms is of the utmost importance.

In addition to over- or under-expression of cell surface checkpoint molecules, immune dysregulation in cancer involves multiple pathways, both cellular and humoral. The indoleamine 2,3-dioxygenase (IDO) is a key enzyme in the tryptophane degradation pathway that attenuates immune responses (Munn & Mellor, 2016); IDO activation leads to tryptophane depletion, decrease in T-cell proliferation, induction of T-cell apoptosis, and activation of T_{regs} (Lee et al., 2002; Mellor & Munn, 2003; Munn & Mellor, 2016). IDO is expressed in HCC and correlates with poor survival (Pan et al., 2008). Anti-CTLA4 and anti-PD1 therapy can induce IDO via INF-γ in ICI resistant HCC in vivo and combinations of ICI with IDO inhibitors resulted in significant antitumor synergism in both heterotopic and orthotopic HCC animal models (Brown et al., 2018). A study evaluating the IDO inhibitor BMS-986205 plus nivolumab is ongoing (NCT03695250). Further, combination strategies of anti-PD1 or anti-CTLA4 therapy with other surface immune checkpoint targeting agents can potentially be of benefit. For example, HCC cells overexpress polyoma virus receptor (PVR) ligand-1 (PVRL1) that stabilizes PVR in the surface of cancer cells (Chiu et al., 2020). PVR is the main ligand of T-cell immunoreceptor with Ig and ITIM domains (TIGIT), an important inhibitory immune checkpoint for both CTLs and NK cells. Treatment with anti-PD1 and anti-TIGIT has a synergistic antitumor effect in vivo, reducing tumor growth, improving survival, and increasing the CTL/T_{reg} ratio in tumors (Chiu et al., 2020). Studies of anti-PD1/PD-L1 combinations with anti-LAG3, anti-TIM3 or anti-TIGIT agents in advanced solid tumors including HCC are ongoing (NCT01968109, NCT03680508, NCT04047862). As discussed in an earlier section of this review, MKis are potent immunomodulatory drugs, able to decrease systemic immunosuppression (Kalathil et al., 2019, 2016, 2020), and combinations of MKis and anti-PD1 agents are an active area of clinical investigation with promising results from early clinical trials (Finn, Ikeda, et al., 2020). Interestingly, increased intratumoral hypoxia after MKi treatment can promote in situ accumulation of immunosuppressive T_{regs} and M2 macrophages in preclinical in vivo HCC models that can offset the favorable systemic immunomodulatory properties of MKis (Chen et al., 2015). This effect is partly mediated by CXCR4, with CXCR4 inhibition improving the antitumor efficacy of MKi and anti-PD1 combination therapy (Chen et al., 2015).

Activation of the WNT/β-catenin pathway is associated with non-T-cell inflamed phenotype and immune exclusion in multiple tumor types (Li et al., 2019; Luke, Bao, Sweis, Spranger, & Gajewski, 2019). In HCC, presence of activating alterations in WNT/β-catenin pathway are similarly associated with worse disease control and overall survival with use of ICI (Harding et al., 2019). Both preclinical in vivo and human data suggest a significant defect in dendritic cell (DC) recruitment in WNT/β-catenin activated HCC which impairs antitumor immune responses by immune exclusion (Ruiz de Galarreta et al., 2019). As suggested by retrospective human studies, in in vivo preclinical models these tumors are resistant to anti-PD1 therapy (Ruiz de Galarreta et al., 2019). OMP-54F28, a recombinant fusion protein that acts as a scavenger receptor for WNT preventing binding to Frizzled receptors and thus inhibiting WNT signaling, has preclinical activity in HCC alone and in combination with sorafenib (Yeung et al., 2014), a clinical trial evaluating the combination in patients with HCC is complete and results are eagerly awaited (NCT02069145). A clinical trial evaluating the WNT inhibitor DKN-01 alone or in combination with sorafenib in patients with HCC harboring WNT alterations is also ongoing (NCT03645980). Combinations of WNT/β-catenin pathway inhibitors with MKis and ICIs is the logical next step. The observed immune exclusion in β-catenin–driven HCC can be related to CCL5 down-regulation, with restoration of CCL5 expression leading to the restoration of immune surveillance (Ruiz de Galarreta et al., 2019) and pointing toward a potential role for modulation of the intratumoral cytokine milieu to improve ICI therapy in immunotherapy refractory tumors. For example, though not specifically tested in WNT/β-catenin activated tumors, inhibition of IL-6 produced by cancer-associated fibroblasts in HCC can reverse local immunosuppression and improve the efficacy of ICIs (Liu, Shen, & Lu, 2017). Furthermore, locoregional therapies can induce favorable immunomodulatory changes in HCC secondary to release of highly immunogenic intracellular proteins, increase in cytokine levels driving effector T-cell and NK cell accumulation and function, and intratumoral infiltration of antigen-specific $CD8^+$ T-cells (reviewed by Lee, Loecher, & Iyer, 2018).

Similarly, favorable immunomodulation can be achieved with systemic or localized use of oncolytic viruses, tumor vaccines, and adoptive T-cell immunotherapy. Oncolytic viruses are naturally occurring or genetically engineered viral strains that can selectively infect and kill cancer cells (direct cytotoxicity), induce anti-tumor immune responses and target tumor neovasculature (indirect cytotoxicity, reviewed in Fountzilas, Patel, & Mahalingam, 2017). Pexa-Vec, a thymidine kinase-deactivated oncolytic vaccinia virus expressing granulocyte/monocyte colony stimulating factor (GM-CSF), was evaluated as a single IV infusion in patients with metastatic tumor to the liver prior to surgical resection (Anthoney et al., 2018). Beyond evidence of targeted oncolytic action, Pexa-Vec led to robust activation of NK and professional APC cells as well as $CD4^+$ and $CD8^+$ T-cells. Pexa-Vec is evaluated in patients with HCC in combination with nivolumab (NCT03071094). Peptide-based vaccines have been evaluated in HCC. GV1001, a telomerase peptide vaccine, did not lead to any objective responses or long-term immunity in HCC though it did decrease T_{regs} in patients with advanced HCC (Greten et al., 2010). DSP-7888, a WT1-specific polypeptide vaccine that can induce T-cell responses against WT1-expressing tumors, is also under investigation in combination with ICIs in multiple tumor types including HCC (NCT03311334). Ex vivo stimulation of DCs (DC-based vaccines) with HCC cell lysate or AFP-derived peptides can induce more robust T-cell responses against tumor antigens and may be a better platform for ICI combinations in HCC (Butterfield et al., 2006; Lee et al., 2005; Palmer et al., 2009).

Early experience in the post-operative setting reveals the potential for active autologous T-cell immunotherapy in HCC with autologous, ex vivo, IL-2 and anti-CD3 stimulated T-cells and NK-cells with or without DC vaccines to decrease recurrence rates and improve disease-specific survival and or OS after surgical resection (Lee et al., 2015; Shimizu et al., 2014; Takayama et al., 2000). Adaptive T-cell immunotherapy using engineered T-cell receptors targeting a specific tumor associated antigen (chimeric T-cell receptor T-cells, CAR-T) can be potentially more efficacious. Glypican-3 (GPC3) is a member of the heparan sulfate proteoglycan family that is associated with worse prognosis in HCC, can differentiate malignant vs. dysplastic cells and thus be a useful target for CAR-T (Kaseb et al., 2016). Studies in preclinical in vivo HCC models reveal that GPC3-targeting CAR-T can successfully eradicate GPC3-high HCC (Jiang et al., 2016). A first-in-human trial of GPC3-targeting CAR-T was recently reported (Shi et al., 2020). In this study, 13 patients with advanced HCC were treated with split dose GPC3-targeting CAR-T consisting of a humanized anti-GPC3 single-chain variable fragment, CD8α hinge domain, CD8α transmembrane domain, CD28 intracellular domain, and CD3ζ intracellular signaling domain that were cloned into a lentiviral backbone. Partial response was attained in 2 patients; 9 patients had cytokine release syndrome, there was no significant neurotoxicity or infusion-related reactions. Novel CAR-T constructs, incorporating soluble PD1, can prevent T-cell exhaustion and improves antitumor efficacy in vivo (Pan et al., 2018). CAR-T plus ICI/anti-VEGF therapy may be beneficial for patients with immune excluded tumors (cluster 2) after careful patient selection (i.e., excluding patients with rapidly progressive disease, very high tumor burden, and early signs of hepatic decompensation) (Liu, Qin, et al., 2020). Clinical trials using CAR-T targeting GPC3 and AFP are ongoing (NCT02905188, NCT03132792, NCT03998033).

In addition to the development of novel combination strategies for ICI-resistant HCC, identification of biomarkers of response to single agent ICI therapy is of utmost importance as the current cost of care for advanced HCC is indeed substantial (Sherrow et al., 2020). So far, we lack a reliable biomarker for patient selection. PDL1 detection by immunohistochemistry, though a useful biomarker in other disease types, has not been proven informative in advanced HCC (El-Khoueiry et al., 2017; Zhu et al., 2018). In a correlative tissue analysis of the CheckMate 040 study, patients with objective response had a high baseline percentage of $PD1^+$ cells in the tumor and a 4-gene inflammatory signature (PDL1, CD8A, STAT1, LAG3) was also associated with improved objective response and OS (Sangro et al., 2020). A similar analysis from the phase 1 GO30140 study revealed that there is an association between baseline immune status and clinical outcomes with atezolizumab/bevacizumab in advanced HCC, with high effector T-cell 3-gene signature (GZMB, PRF1, CXCL9) and PDL1 expression correlating with improved ORR and PFS while Notch pathway activation predicted lack of response (Zhu et al., 2020). Further, high VEGFR2 expression and T_{reg} and myeloid inflammation signatures were associated with longer PFS in patients treated with the combination compared to single agent atezolizumab. The tumor mutation burden (TMB) was not associated with outcomes. As discussed above, detection of activating mutations in WNT/β-catenin pathway, present in up to a third of patients with HCC (Torbenson, Ng, Park, Roncalli, & Sakamoto, 2019), is a promising biomarker of response to ICI (Harding et al., 2019) but may not be sufficient to identify all patients with primary resistance (von Felden et al., 2021).

The success of ICI in advanced disease has prompted studies to evaluate efficacy in earlier disease, when locoregional therapies are typically used and results are awaited. In one such study, a phase II study combining tremelimumab and locoregional therapy, significant increases in peripheral blood $CD4^+$ and $CD8^+$ T-cells, both activated $ICOS^+$ and exhausted $PD1^+$, were noted after one priming dose of tremelimumab (Agdashian et al., 2019). T-cell receptor sequencing revealed that one priming dose of tremelimumab can lead to a significant and sustained decrease peripheral clonality with repeated dosing, indicating broadening of the T-cell repertoire. Patients with response tended to have both high tumor clonality and T-cell fraction compared to non-responders. Tremelimumab increased intratumoral T-cell infiltration.

Beyond the sparsity of available tumor tissue for gene expression profiling and next-generation sequencing (NGS) in HCC, partly related to very commonly establishing diagnosis using radiographic criteria only (Marrero et al., 2018), temporospatial tumor heterogeneity can be a further obstacle in development and interpretation of predictive biomarkers (Dominguez & Wang, 2020). Novel, non-invasive sophisticated radiomic biomarkers can be of potential benefit as these are already evaluated in melanoma and non-small cell lung cancer (He et al., 2020; Trebeschi et al., 2019) especially when combined with blood-based biomarkers (Basler et al., 2020). In patients with resectable HCC, a radiomic model incorporating AFP and AST was able to predict long-term survival (Wang et al., 2020). Importantly, using baseline CT texture analysis, portal venous phase-derived tumor entropy (a radiomic marker that reflects irregularity and inhomogeneity of the gray-level intensity values within a predefined region of interest) was shown to be a prognostic biomarker for OS in patients with advanced HCC treated with sorafenib (Mule et al., 2018). In addition, MRI texture analysis can be as useful as non-invasive methods to assess the TME constitution and expression of immunotherapy targets at the protein and mRNA level (Hectors et al., 2020); use of Gd-EOB-DTPA (a liver-specific MRI contrast agent that accumulates in tumor cells through WNT/β-catenin–induced OATP1B3) can be used to guide immunotherapy decision-making in HCC (reviewed in Kudo, 2020). Radiomic signatures can be supplemented with blood-based biomarkers. In a phase 2 study of pembrolizumab in patients with advanced HCC, high baseline soluble TGF-β predicted poor outcomes (Feun et al., 2019). Non-responders had a higher TGF-β concentration compared to responders; patients with TGF-β concentration $\geq$ 200 pg/mL had lower median OS and PFS compared to patients with $<$200 pg/mL (7 and 2 months vs. $>$25 months, respectively).

4. Conclusion

Immunotherapy with ICI has provided new, active therapeutic options for patients with advanced HCC. Identification of patients who will benefit from single agent or combination (with anti-CTLA4- or anti-VEGF agents) treatment and incorporation of novel combination strategies that can overcome resistance mechanisms, immune exhaustion and immune exclusion if of utmost importance.

References

Abou-Alfa, G. K., Chan, S. L., Furuse, J., Galle, P. R., Kelley, R. K., Qin, S., et al. (2018). A randomized, multicenter phase 3 study of durvalumab (D) and tremelimumab (T) as first-line treatment in patients with unresectable hepatocellular carcinoma (HCC): HIMALAYA study. *Journal of Clinical Oncology*, *36*(15_suppl), TPS4144-TPS4144. https://doi.org/10.1200/JCO.2018.36.15_suppl.TPS4144.

Abou-Alfa, G. K., Meyer, T., Cheng, A. L., El-Khoueiry, A. B., Rimassa, L., Ryoo, B. Y., et al. (2018). Cabozantinib in patients with advanced and progressing hepatocellular carcinoma. *The New England Journal of Medicine*, *379*(1), 54–63. https://doi.org/10.1056/NEJMoa1717002.

Agdashian, D., ElGindi, M., Xie, C., Sandhu, M., Pratt, D., Kleiner, D. E., et al. (2019). The effect of anti-CTLA4 treatment on peripheral and intra-tumoral T cells in patients with hepatocellular carcinoma. *Cancer Immunology, Immunotherapy*, *68*(4), 599–608. https://doi.org/10.1007/s00262-019-02299-8.

Alisa, A., Boswell, S., Pathan, A. A., Ayaru, L., Williams, R., & Behboudi, S. (2008). Human CD4(+) T cells recognize an epitope within alpha-fetoprotein sequence and develop into TGF-beta-producing CD4(+) T cells. *Journal of Immunology*, *180*(7), 5109–5117. https://doi.org/10.4049/jimmunol.180.7.5109.

Ameratunga, M., Chenard-Poirier, M., Moreno Candilejo, I., Pedregal, M., Lui, A., Dolling, D., et al. (2018). Neutrophil-lymphocyte ratio kinetics in patients with advanced solid tumours on phase I trials of PD-1/PD-L1 inhibitors. *European Journal of Cancer*, *89*, 56–63. https://doi.org/10.1016/j.ejca.2017.11.012.

Anthoney, A., Samson, A., West, E., Turnbull, S. J., Scott, K., Tidswell, E., et al. (2018). Single intravenous preoperative administration of the oncolytic virus Pexa-Vec to prime anti-tumor immunity. *Journal of Clinical Oncology*, *36*(15_suppl), 3092. https://doi.org/10.1200/JCO.2018.36.15_suppl.3092.

Basler, L., Gabrys, H. S., Hogan, S. A., Pavic, M., Bogowicz, M., Vuong, D., et al. (2020). Radiomics, Tumor Volume, and Blood Biomarkers for Early Prediction of Pseudoprogression in Patients with Metastatic Melanoma Treated with Immune Checkpoint Inhibition. *Clinical Cancer Research*, *26*(16), 4414–4425. https://doi.org/10.1158/1078-0432.CCR-20-0020.

Brown, Z. J., Yu, S. J., Heinrich, B., Ma, C., Fu, Q., Sandhu, M., et al. (2018). Indoleamine 2,3-dioxygenase provides adaptive resistance to immune checkpoint inhibitors in hepatocellular carcinoma. *Cancer Immunology, Immunotherapy*, *67*(8), 1305–1315. https://doi.org/10.1007/s00262-018-2190-4.

Bruix, J., Qin, S., Merle, P., Granito, A., Huang, Y. H., Bodoky, G., et al. (2017). Regorafenib for patients with hepatocellular carcinoma who progressed on sorafenib treatment (RESORCE): a randomised, double-blind, placebo-controlled, phase 3 trial. *Lancet*, *389*(10064), 56–66. https://doi.org/10.1016/S0140-6736(16)32453-9.

Butterfield, L. H., Ribas, A., Dissette, V. B., Lee, Y., Yang, J. Q., De la Rocha, P., et al. (2006). A phase I/II trial testing immunization of hepatocellular carcinoma patients with dendritic cells pulsed with four alpha-fetoprotein peptides. *Clinical Cancer Research*, *12*(9), 2817–2825. https://doi.org/10.1158/1078-0432.CCR-05-2856.

Cainap, C., Qin, S., Huang, W. T., Chung, I. J., Pan, H., Cheng, Y., et al. (2015). Linifanib versus Sorafenib in patients with advanced hepatocellular carcinoma: results of a randomized phase III trial. *Journal of Clinical Oncology*, *33*(2), 172–179. https://doi.org/10.1200/JCO.2013.54.3298.

Chen, Y., Ramjiawan, R. R., Reiberger, T., Ng, M. R., Hato, T., Huang, Y., et al. (2015). CXCR4 inhibition in tumor microenvironment facilitates anti-programmed death receptor-1 immunotherapy in sorafenib-treated hepatocellular carcinoma in mice. *Hepatology*, *61*(5), 1591–1602. https://doi.org/10.1002/hep.27665.

Cheng, A. L., Kang, Y. K., Lin, D. Y., Park, J. W., Kudo, M., Qin, S., et al. (2013). Sunitinib versus sorafenib in advanced hepatocellular cancer: results of a randomized phase III trial. *Journal of Clinical Oncology*, *31*(32), 4067–4075. https://doi.org/10.1200/JCO.2012.45.8372.

Chew, V., Lai, L., Pan, L., Lim, C. J., Li, J., Ong, R., et al. (2017). Delineation of an immunosuppressive gradient in hepatocellular carcinoma using high-dimensional proteomic and transcriptomic analyses. *Proceedings of the National Academy of Sciences of the United States of America*, *114*(29), E5900–E5909. https://doi.org/10.1073/pnas.1706559114.

Chiu, D. K., Yuen, V. W., Cheu, J. W., Wei, L. L., Ting, V., Fehlings, M., et al. (2020). Hepatocellular carcinoma cells up-regulate PVRL1, stabilizing PVR and inhibiting the cytotoxic T-cell response via TIGIT to mediate tumor resistance to PD1 inhibitors in mice. *Gastroenterology*, *159*(2), 609–623. https://doi.org/10.1053/j.gastro.2020.03.074.

Das, R., Verma, R., Sznol, M., Boddupalli, C. S., Gettinger, S. N., Kluger, H., et al. (2015). Combination therapy with anti-CTLA-4 and anti-PD-1 leads to distinct immunologic changes in vivo. *Journal of Immunology*, *194*(3), 950–959. https://doi.org/10.4049/jimmunol.1401686.

Dharmapuri, S., Özbek, U., Lin, J. Y., Schwartz, M., Branch, A., & Ang, C. (2019). Predictive value of neutrophil-lymphocyte ratio (NLR) and platelet-lymphocyte ratio (PLR) In hepatocellular carcinoma (HCC) patients treated with nivolumab (N). *Annals of Oncology*, *30*, v285–v286. https://doi.org/10.1093/annonc/mdz247.071.

Dominguez, D. A., & Wang, X. W. (2020). Impact of next-generation sequencing on outcomes in hepatocellular carcinoma: How precise are we really? *Journal of Hepatocellular Carcinoma*, 7, 33–37. https://doi.org/10.2147/JHC.S217948.

El-Khoueiry, A. B., Sangro, B., Yau, T., Crocenzi, T. S., Kudo, M., Hsu, C., et al. (2017). Nivolumab in patients with advanced hepatocellular carcinoma (CheckMate 040): an open-label, non-comparative, phase 1/2 dose escalation and expansion trial. *Lancet*, *389*(10088), 2492–2502. https://doi.org/10.1016/S0140-6736(17)31046-2.

Feun, L. G., Li, Y. Y., Wu, C., Wangpaichitr, M., Jones, P. D., Richman, S. P., et al. (2019). Phase 2 study of pembrolizumab and circulating biomarkers to predict anticancer response in advanced, unresectable hepatocellular carcinoma. *Cancer*, *125*(20), 3603–3614. https://doi.org/10.1002/cncr.32339.

Finn, R. S., Ikeda, M., Zhu, A. X., Sung, M. W., Baron, A. D., Kudo, M., et al. (2020). Phase Ib study of lenvatinib plus pembrolizumab in patients with unresectable hepatocellular carcinoma. *Journal of Clinical Oncology*, *38*(26), 2960–2970. https://doi.org/10.1200/jco.20.00808.

Finn, R. S., Qin, S., Ikeda, M., Galle, P. R., Ducreux, M., Kim, T. Y., et al. (2020). Atezolizumab plus Bevacizumab in Unresectable Hepatocellular Carcinoma. *The New England Journal of Medicine*, *382*(20), 1894–1905. https://doi.org/10.1056/NEJMoa1915745.

Finn, R. S., Ryoo, B. Y., Merle, P., Kudo, M., Bouattour, M., Lim, H. Y., et al. (2020). Pembrolizumab As second-line therapy in patients with advanced hepatocellular carcinoma in KEYNOTE-240: A randomized, double-blind, phase III trial. *Journal of Clinical Oncology*, *38*(3), 193–202. https://doi.org/10.1200/JCO.19.01307.

Flecken, T., Schmidt, N., Hild, S., Gostick, E., Drognitz, O., Zeiser, R., et al. (2014). Immunodominance and functional alterations of tumor-associated antigen-specific CD8+ T-cell responses in hepatocellular carcinoma. *Hepatology*, *59*(4), 1415–1426. https://doi.org/10.1002/hep.26731.

Fountzilas, C., Gupta, M., Lee, S., Krishnamurthi, S., Estfan, B., Wang, K., et al. (2020). A multicentre phase 1b/2 study of tivozanib in patients with advanced inoperable hepatocellular carcinoma. *British Journal of Cancer*, *122*(7), 963–970. https://doi.org/10.1038/s41416-020-0737-6.

Fountzilas, C., Patel, S., & Mahalingam, D. (2017). Review: Oncolytic virotherapy, updates and future directions. *Oncotarget*, *8*(60), 102617–102639. https://doi.org/10.18632/oncotarget.18309.

Fu, J., Xu, D., Liu, Z., Shi, M., Zhao, P., Fu, B., et al. (2007). Increased regulatory T cells correlate with CD8 T-cell impairment and poor survival in hepatocellular carcinoma patients. *Gastroenterology*, *132*(7), 2328–2339. https://doi.org/10.1053/j.gastro.2007.03.102.

Gao, Q., Qiu, S. J., Fan, J., Zhou, J., Wang, X. Y., Xiao, Y. S., et al. (2007). Intratumoral balance of regulatory and cytotoxic T cells is associated with prognosis of hepatocellular carcinoma after resection. *Journal of Clinical Oncology*, *25*(18), 2586–2593. https://doi.org/10.1200/JCO.2006.09.4565.

Greten, T. F., Forner, A., Korangy, F., N'Kontchou, G., Barget, N., Ayuso, C., et al. (2010). A phase II open label trial evaluating safety and efficacy of a telomerase peptide vaccination in patients with advanced hepatocellular carcinoma. *BMC Cancer*, *10*, 209. https://doi.org/10.1186/1471-2407-10-209.

Guidotti, L. G., Inverso, D., Sironi, L., Di Lucia, P., Fioravanti, J., Ganzer, L., et al. (2015). Immunosurveillance of the liver by intravascular effector CD8(+) T cells. *Cell*, *161*(3), 486–500. https://doi.org/10.1016/j.cell.2015.03.005.

Harding, J. J., Nandakumar, S., Armenia, J., Khalil, D. N., Albano, M., Ly, M., et al. (2019). Prospective genotyping of hepatocellular carcinoma: clinical implications of next-generation sequencing for matching patients to targeted and immune therapies. *Clinical Cancer Research*, *25*(7), 2116–2126. https://doi.org/10.1158/1078-0432.CCR-18-2293.

He, B., Dong, D., She, Y., Zhou, C., Fang, M., Zhu, Y., et al. (2020). Predicting response to immunotherapy in advanced non-small-cell lung cancer using tumor mutational burden radiomic biomarker. *Journal for Immunotherapy of Cancer, 8*(2). https://doi.org/10.1136/jitc-2020-000550.

Hectors, S. J., Lewis, S., Besa, C., King, M. J., Said, D., Putra, J., et al. (2020). MRI radiomics features predict immuno-oncological characteristics of hepatocellular carcinoma. *European Radiology, 30*(7), 3759–3769. https://doi.org/10.1007/s00330-020-06675-2.

Hodi, F. S., Chesney, J., Pavlick, A. C., Robert, C., Grossmann, K. F., McDermott, D. F., et al. (2016). Combined nivolumab and ipilimumab versus ipilimumab alone in patients with advanced melanoma: 2-year overall survival outcomes in a multicentre, randomised, controlled, phase 2 trial. *The Lancet Oncology, 17*(11), 1558–1568. https://doi.org/10.1016/S1470-2045(16)30366-7.

Hoechst, B., Ormandy, L. A., Ballmaier, M., Lehner, F., Kruger, C., Manns, M. P., et al. (2008). A new population of myeloid-derived suppressor cells in hepatocellular carcinoma patients induces CD4(+)CD25(+)Foxp3(+) T cells. *Gastroenterology, 135*(1), 234–243. https://doi.org/10.1053/j.gastro.2008.03.020.

Hoechst, B., Voigtlaender, T., Ormandy, L., Gamrekelashvili, J., Zhao, F., Wedemeyer, H., et al. (2009). Myeloid derived suppressor cells inhibit natural killer cells in patients with hepatocellular carcinoma via the NKp30 receptor. *Hepatology, 50*(3), 799–807. https://doi.org/10.1002/hep.23054.

Iyer, R. V., Maguire, O., Kim, M., Curtin, L. I., Sexton, S., Fisher, D. T., et al. (2019). Dose-dependent Sorafenib-induced immunosuppression is associated with aberrant NFAT activation and expression of PD-1 in T cells. *Cancers (Basel), 11*(5). https://doi.org/10.3390/cancers11050681.

Jemal, A., Ward, E. M., Johnson, C. J., Cronin, K. A., Ma, J., Ryerson, B., et al. (2017). Annual report to the nation on the status of cancer, 1975-2014, featuring survival. *Journal of the National Cancer Institute, 109*(9). https://doi.org/10.1093/jnci/djx030.

Jiang, P., Gu, S., Pan, D., Fu, J., Sahu, A., Hu, X., et al. (2018). Signatures of T cell dysfunction and exclusion predict cancer immunotherapy response. *Nature Medicine, 24*(10), 1550–1558. https://doi.org/10.1038/s41591-018-0136-1.

Jiang, Z., Jiang, X., Chen, S., Lai, Y., Wei, X., Li, B., et al. (2016). Anti-GPC3-CAR T cells suppress the growth of tumor cells in patient-derived xenografts of hepatocellular carcinoma. *Frontiers in Immunology, 7*, 690. https://doi.org/10.3389/fimmu.2016.00690.

Johnson, P. J., Qin, S., Park, J. W., Poon, R. T., Raoul, J. L., Philip, P. A., et al. (2013). Brivanib versus sorafenib as first-line therapy in patients with unresectable, advanced hepatocellular carcinoma: results from the randomized phase III BRISK-FL study. *Journal of Clinical Oncology, 31*(28), 3517–3524. https://doi.org/10.1200/JCO.2012.48.4410.

Kalathil, S. G., Hutson, A., Barbi, J., Iyer, R., & Thanavala, Y. (2019). Augmentation of IFN-gamma+ CD8+ T cell responses correlates with survival of HCC patients on sorafenib therapy. *JCI Insight, 4*(15). https://doi.org/10.1172/jci.insight.130116.

Kalathil, S., Lugade, A. A., Miller, A., Iyer, R., & Thanavala, Y. (2013). Higher frequencies of GARP(+)CTLA-4(+)Foxp3(+) T regulatory cells and myeloid-derived suppressor cells in hepatocellular carcinoma patients are associated with impaired T-cell functionality. *Cancer Research, 73*(8), 2435–2444. https://doi.org/10.1158/0008-5472.CAN-12-3381.

Kalathil, S. G., Lugade, A. A., Miller, A., Iyer, R., & Thanavala, Y. (2016). PD-1(+) and Foxp3(+) T cell reduction correlates with survival of HCC patients after sorafenib therapy. *JCI Insight, 1*(11). https://doi.org/10.1172/jci.insight.86182.

Kalathil, S. G., Wang, K., Hutson, A., Iyer, R., & Thanavala, Y. (2020). Tivozanib mediated inhibition of c-Kit/SCF signaling on Tregs and MDSCs and reversal of tumor induced immune suppression correlates with survival of HCC patients. *OncoImmunology, 9*(1), 1824863. https://doi.org/10.1080/2162402X.2020.1824863.

Kaseb, A. O., Hassan, M., Lacin, S., Abdel-Wahab, R., Amin, H. M., Shalaby, A., et al. (2016). Evaluating clinical and prognostic implications of Glypican-3 in hepatocellular carcinoma. *Oncotarget*, 7(43), 69916–69926. https://doi.org/10.18632/oncotarget.12066.

Kelley, R. K., Abou-Alfa, G. K., Bendell, J. C., Kim, T.-Y., Borad, M. J., Yong, W.-P., et al. (2017). Phase I/II study of durvalumab and tremelimumab in patients with unresectable hepatocellular carcinoma (HCC): Phase I safety and efficacy analyses. *Journal of Clinical Oncology*, *35*(15_suppl), 4073. https://doi.org/10.1200/JCO.2017.35.15_suppl.4073.

Kelley, R. K., Sangro, B., Harris, W. P., Ikeda, M., Okusaka, T., Kang, Y.-K., et al. (2020). Efficacy, tolerability, and biologic activity of a novel regimen of tremelimumab (T) in combination with durvalumab (D) for patients (pts) with advanced hepatocellular carcinoma (aHCC). *Journal of Clinical Oncology*, *38*(15_suppl), 4508. https://doi.org/10.1200/JCO.2020.38.15_suppl.4508.

Kimura, T., Kato, Y., Ozawa, Y., Kodama, K., Ito, J., Ichikawa, K., et al. (2018). Immunomodulatory activity of lenvatinib contributes to antitumor activity in the Hepa1-6 hepatocellular carcinoma model. *Cancer Science*, *109*(12), 3993–4002. https://doi.org/10.1111/cas.13806.

Koh, M. Y., Gagea, M., Sargis, T., Lemos, R., Jr., Grandjean, G., Charbono, A., et al. (2016). A new HIF-1alpha/RANTES-driven pathway to hepatocellular carcinoma mediated by germline haploinsufficiency of SART1/HAF in mice. *Hepatology*, *63*(5), 1576–1591. https://doi.org/10.1002/hep.28468.

Kuang, D. M., Peng, C., Zhao, Q., Wu, Y., Chen, M. S., & Zheng, L. (2010). Activated monocytes in peritumoral stroma of hepatocellular carcinoma promote expansion of memory T helper 17 cells. *Hepatology*, *51*(1), 154–164. https://doi.org/10.1002/hep.23291.

Kudo, M. (2020). Gd-EOB-DTPA-MRI Could Predict WNT/beta-Catenin Mutation and Resistance to Immune Checkpoint Inhibitor Therapy in Hepatocellular Carcinoma. *Liver Cancer*, *9*(5), 479–490. https://doi.org/10.1159/000509554.

Kudo, M., Finn, R. S., Qin, S., Han, K. H., Ikeda, K., Piscaglia, F., et al. (2018). Lenvatinib versus sorafenib in first-line treatment of patients with unresectable hepatocellular carcinoma: a randomised phase 3 non-inferiority trial. *Lancet*, *391*(10126), 1163–1173. https://doi.org/10.1016/S0140-6736(18)30207-1.

Kudo, M., Motomura, K., Wada, Y., Inaba, Y., Sakamoto, Y., Kurosaki, M., et al. (2019). First-line avelumab + axitinib in patients with advanced hepatocellular carcinoma: Results from a phase 1b trial (VEGF Liver 100). *Journal of Clinical Oncology*, *37*(15_suppl), 4072. https://doi.org/10.1200/JCO.2019.37.15_suppl.4072.

Langhans, B., Nischalke, H. D., Kramer, B., Dold, L., Lutz, P., Mohr, R., et al. (2019). Role of regulatory T cells and checkpoint inhibition in hepatocellular carcinoma. *Cancer Immunology, Immunotherapy*, *68*(12), 2055–2066. https://doi.org/10.1007/s00262-019-02427-4.

Lee, J. H., Lee, J. H., Lim, Y. S., Yeon, J. E., Song, T. J., Yu, S. J., et al. (2015). Adjuvant immunotherapy with autologous cytokine-induced killer cells for hepatocellular carcinoma. *Gastroenterology*, *148*(7), 1383–1391. e1386. https://doi.org/10.1053/j.gastro.2015.02.055.

Lee, S., Loecher, M., & Iyer, R. (2018). Immunomodulation in hepatocellular cancer. *Journal of Gastrointestinal Oncology*, *9*(1), 208–219. https://doi.org/10.21037/jgo.2017.06.08.

Lee, G. K., Park, H. J., Macleod, M., Chandler, P., Munn, D. H., & Mellor, A. L. (2002). Tryptophan deprivation sensitizes activated T cells to apoptosis prior to cell division. *Immunology*, *107*(4), 452–460. https://doi.org/10.1046/j.1365-2567.2002.01526.x.

Lee, M. S., Ryoo, B. Y., Hsu, C. H., Numata, K., Stein, S., Verret, W., et al. (2020). Atezolizumab with or without bevacizumab in unresectable hepatocellular carcinoma (GO30140): An open-label, multicentre, phase 1b study. *The Lancet Oncology*, *21*(6), 808–820. https://doi.org/10.1016/S1470-2045(20)30156-X.

Lee, W. C., Wang, H. C., Hung, C. F., Huang, P. F., Lia, C. R., & Chen, M. F. (2005). Vaccination of advanced hepatocellular carcinoma patients with tumor lysate-pulsed dendritic cells: a clinical trial. *Journal of immunotherapy (Hagerstown, Md.: 1997), 28*(5), 496–504. https://doi.org/10.1097/01.cji.0000171291.72039.e2.

Li, X., Xiang, Y., Li, F., Yin, C., Li, B., & Ke, X. (2019). WNT/beta-Catenin Signaling Pathway Regulating T Cell-Inflammation in the Tumor Microenvironment. *Frontiers in Immunology, 10*, 2293. https://doi.org/10.3389/fimmu.2019.02293.

Liu, F., Qin, L., Liao, Z., Song, J., Yuan, C., Liu, Y., et al. (2020). Microenvironment characterization and multi-omics signatures related to prognosis and immunotherapy response of hepatocellular carcinoma. *Experimental Hematology & Oncology, 9*, 10. https://doi.org/10.1186/s40164-020-00165-3.

Liu, H., Shen, J., & Lu, K. (2017). IL-6 and PD-L1 blockade combination inhibits hepatocellular carcinoma cancer development in mouse model. *Biochemical and Biophysical Research Communications, 486*(2), 239–244. https://doi.org/10.1016/j.bbrc.2017.02.128.

Liu, M., Zhou, J., Liu, X., Feng, Y., Yang, W., Wu, F., et al. (2020). Targeting monocyte-intrinsic enhancer reprogramming improves immunotherapy efficacy in hepatocellular carcinoma. *Gut, 69*(2), 365–379. https://doi.org/10.1136/gutjnl-2018-317257.

Llovet, J. M., Ricci, S., Mazzaferro, V., Hilgard, P., Gane, E., Blanc, J. F., et al. (2008). Sorafenib in advanced hepatocellular carcinoma. *The New England Journal of Medicine, 359*(4), 378–390. https://doi.org/10.1056/NEJMoa0708857.

Luke, J. J., Bao, R., Sweis, R. F., Spranger, S., & Gajewski, T. F. (2019). WNT/beta-catenin Pathway Activation Correlates with Immune Exclusion across Human Cancers. *Clinical Cancer Research, 25*(10), 3074–3083. https://doi.org/10.1158/1078-0432.CCR-18-1942.

Marrero, J. A., Kulik, L. M., Sirlin, C. B., Zhu, A. X., Finn, R. S., Abecassis, M. M., et al. (2018). Diagnosis, staging, and management of hepatocellular carcinoma: 2018 Practice guidance by the american association for the study of liver diseases. *Hepatology, 68*(2), 723–750. https://doi.org/10.1002/hep.29913.

Marti, L. C., Pavon, L., Severino, P., Sibov, T., Guilhen, D., & Moreira-Filho, C. A. (2014). Vascular endothelial growth factor-A enhances indoleamine 2,3-dioxygenase expression by dendritic cells and subsequently impacts lymphocyte proliferation. *Memórias do Instituto Oswaldo Cruz, 109*(1), 70–79. https://doi.org/10.1590/0074-0276130252.

Melero, I., Berman, D. M., Aznar, M. A., Korman, A. J., Perez Gracia, J. L., & Haanen, J. (2015). Evolving synergistic combinations of targeted immunotherapies to combat cancer. *Nature Reviews. Cancer, 15*(8), 457–472. https://doi.org/10.1038/nrc3973.

Mellor, A. L., & Munn, D. H. (2003). Tryptophan catabolism and regulation of adaptive immunity. *Journal of Immunology, 170*(12), 5809–5813. https://doi.org/10.4049/jimmunol.170.12.5809.

Morse, M. A., Sun, W., Kim, R., He, A. R., Abada, P. B., Mynderse, M., et al. (2019). The Role of Angiogenesis in Hepatocellular Carcinoma. *Clinical Cancer Research, 25*(3), 912–920. https://doi.org/10.1158/1078-0432.Ccr-18-1254.

Motz, G. T., Santoro, S. P., Wang, L. P., Garrabrant, T., Lastra, R. R., Hagemann, I. S., et al. (2014). Tumor endothelium FasL establishes a selective immune barrier promoting tolerance in tumors. *Nature Medicine, 20*(6), 607–615. https://doi.org/10.1038/nm.3541.

Motzer, R. J., Tannir, N. M., McDermott, D. F., Aren Frontera, O., Melichar, B., Choueiri, T. K., et al. (2018). Nivolumab plus Ipilimumab versus Sunitinib in Advanced Renal-Cell Carcinoma. *The New England Journal of Medicine, 378*(14), 1277–1290. https://doi.org/10.1056/NEJMoa1712126.

Mule, S., Thiefin, G., Costentin, C., Durot, C., Rahmouni, A., Luciani, A., et al. (2018). Advanced hepatocellular carcinoma: Pretreatment contrast-enhanced CT texture parameters as predictive biomarkers of survival in patients treated with Sorafenib. *Radiology, 288*(2), 445–455. https://doi.org/10.1148/radiol.2018171320.

Munn, D. H., & Mellor, A. L. (2016). IDO in the tumor microenvironment: Inflammation, counter-regulation, and tolerance. *Trends in Immunology*, *37*(3), 193–207. https://doi.org/10.1016/j.it.2016.01.002.
Murgita, R. A., Goidl, E. A., Kontianen, S., & Wigzell, H. (1977). alpha-Fetoprotein induces suppressor T cells in vitro. *Nature*, *267*(5608), 257–259. https://doi.org/10.1038/267257a0.
Nan, J., Xing, Y. F., Hu, B., Tang, J. X., Dong, H. M., He, Y. M., et al. (2018). Endoplasmic reticulum stress induced LOX-1(+) CD15(+) polymorphonuclear myeloid-derived suppressor cells in hepatocellular carcinoma. *Immunology*, *154*(1), 144–155. https://doi.org/10.1111/imm.12876.
Palmer, D. H., Midgley, R. S., Mirza, N., Torr, E. E., Ahmed, F., Steele, J. C., et al. (2009). A phase II study of adoptive immunotherapy using dendritic cells pulsed with tumor lysate in patients with hepatocellular carcinoma. *Hepatology*, *49*(1), 124–132. https://doi.org/10.1002/hep.22626.
Pan, Z., Di, S., Shi, B., Jiang, H., Shi, Z., Liu, Y., et al. (2018). Increased antitumor activities of glypican-3-specific chimeric antigen receptor-modified T cells by coexpression of a soluble PD1-CH3 fusion protein. *Cancer Immunology, Immunotherapy*, *67*(10), 1621–1634. https://doi.org/10.1007/s00262-018-2221-1.
Pan, K., Wang, H., Chen, M. S., Zhang, H. K., Weng, D. S., Zhou, J., et al. (2008). Expression and prognosis role of indoleamine 2,3-dioxygenase in hepatocellular carcinoma. *Journal of Cancer Research and Clinical Oncology*, *134*(11), 1247–1253. https://doi.org/10.1007/s00432-008-0395-1.
Ruiz de Galarreta, M., Bresnahan, E., Molina-Sanchez, P., Lindblad, K. E., Maier, B., Sia, D., et al. (2019). Beta-catenin activation promotes immune escape and resistance to anti-PD-1 therapy in hepatocellular carcinoma. *Cancer Discovery*, *9*(8), 1124–1141. https://doi.org/10.1158/2159-8290.CD-19-0074.
Sangro, B., Melero, I., Wadhawan, S., Finn, R. S., Abou-Alfa, G. K., Cheng, A. L., et al. (2020). Association of inflammatory biomarkers with clinical outcomes in nivolumab-treated patients with advanced hepatocellular carcinoma. *Journal of Hepatology*, *73*(6), 1460–1469. https://doi.org/10.1016/j.jhep.2020.07.026.
Sherrow, C., Attwood, K., Zhou, K., Mukherjee, S., Iyer, R., & Fountzilas, C. (2020). Sequencing systemic therapy pathways for advanced hepatocellular carcinoma: A cost effectiveness analysis. *Liver Cancer*, *9*(5), 549–562. https://doi.org/10.1159/000508485.
Shi, D., Shi, Y., Kaseb, A. O., Qi, X., Zhang, Y., Chi, J., et al. (2020). Chimeric antigen receptor-glypican-3 T-cell therapy for advanced hepatocellular carcinoma: Results of phase I Trials. *Clinical Cancer Research*, *26*(15), 3979–3989. https://doi.org/10.1158/1078-0432.CCR-19-3259.
Shimizu, K., Kotera, Y., Aruga, A., Takeshita, N., Katagiri, S., Ariizumi, S., et al. (2014). Postoperative dendritic cell vaccine plus activated T-cell transfer improves the survival of patients with invasive hepatocellular carcinoma. *Human Vaccines & Immunotherapeutics*, *10*(4), 970–976. https://doi.org/10.4161/hv.27678.
Shrimali, R. K., Yu, Z., Theoret, M. R., Chinnasamy, D., Restifo, N. P., & Rosenberg, S. A. (2010). Antiangiogenic agents can increase lymphocyte infiltration into tumor and enhance the effectiveness of adoptive immunotherapy of cancer. *Cancer Research*, *70*(15), 6171–6180. https://doi.org/10.1158/0008-5472.CAN-10-0153.
Sia, D., Jiao, Y., Martinez-Quetglas, I., Kuchuk, O., Villacorta-Martin, C., Castro de Moura, M., et al. (2017). Identification of an immune-specific class of hepatocellular carcinoma, based on molecular features. *Gastroenterology*, *153*(3), 812–826. https://doi.org/10.1053/j.gastro.2017.06.007.
Takahashi, A., Kono, K., Ichihara, F., Sugai, H., Fujii, H., & Matsumoto, Y. (2004). Vascular endothelial growth factor inhibits maturation of dendritic cells induced by lipopolysaccharide, but not by proinflammatory cytokines. *Cancer Immunology, Immunotherapy*, *53*(6), 543–550. https://doi.org/10.1007/s00262-003-0466-8.

Takayama, T., Sekine, T., Makuuchi, M., Yamasaki, S., Kosuge, T., Yamamoto, J., et al. (2000). Adoptive immunotherapy to lower postsurgical recurrence rates of hepatocellular carcinoma: a randomised trial. *Lancet, 356*(9232), 802–807. https://doi.org/10.1016/S0140-6736(00)02654-4.

Torbenson, M., Ng, I., Park, Y., Roncalli, M., & Sakamoto, M. (2019). Hepatocellular carcinoma. In WHO, Classification, of, Tumours, Editorial, & Board (Ed.), *Digestive system tumours: WHO classification of tumours* (5th ed., pp. 229–239). WHO Press.

Trebeschi, S., Drago, S. G., Birkbak, N. J., Kurilova, I., Calin, A. M., Delli Pizzi, A., et al. (2019). Predicting response to cancer immunotherapy using noninvasive radiomic biomarkers. *Annals of Oncology, 30*(6), 998–1004. https://doi.org/10.1093/annonc/mdz108.

Um, S. H., Mulhall, C., Alisa, A., Ives, A. R., Karani, J., Williams, R., et al. (2004). Alpha-fetoprotein impairs APC function and induces their apoptosis. *Journal of Immunology, 173*(3), 1772–1778. https://doi.org/10.4049/jimmunol.173.3.1772.

Unitt, E., Marshall, A., Gelson, W., Rushbrook, S. M., Davies, S., Vowler, S. L., et al. (2006). Tumour lymphocytic infiltrate and recurrence of hepatocellular carcinoma following liver transplantation. *Journal of Hepatology, 45*(2), 246–253. https://doi.org/10.1016/j.jhep.2005.12.027.

Villanueva, A. (2019). Hepatocellular Carcinoma. *The New England Journal of Medicine, 380*(15), 1450–1462. https://doi.org/10.1056/NEJMra1713263.

von Felden, J., Craig, A. J., Garcia-Lezana, T., Labgaa, I., Haber, P. K., D'Avola, D., et al. (2021). Mutations in circulating tumor DNA predict primary resistance to systemic therapies in advanced hepatocellular carcinoma. *Oncogene, 40*(1), 140–151. https://doi.org/10.1038/s41388-020-01519-1.

Voron, T., Colussi, O., Marcheteau, E., Pernot, S., Nizard, M., Pointet, A. L., et al. (2015). VEGF-A modulates expression of inhibitory checkpoints on CD8+ T cells in tumors. *The Journal of Experimental Medicine, 212*(2), 139–148. https://doi.org/10.1084/jem.20140559.

Vujanovic, L., Stahl, E. C., Pardee, A. D., Geller, D. A., Tsung, A., Watkins, S. C., et al. (2017). Tumor-Derived alpha-Fetoprotein Directly Drives Human Natural Killer-Cell Activation and Subsequent Cell Death. *Cancer Immunology Research, 5*(6), 493–502. https://doi.org/10.1158/2326-6066.CIR-16-0216.

Wang, X. H., Long, L. H., Cui, Y., Jia, A. Y., Zhu, X. G., Wang, H. Z., et al. (2020). MRI-based radiomics model for preoperative prediction of 5-year survival in patients with hepatocellular carcinoma. *British Journal of Cancer, 122*(7), 978–985. https://doi.org/10.1038/s41416-019-0706-0.

Wu, Y., Kuang, D. M., Pan, W. D., Wan, Y. L., Lao, X. M., Wang, D., et al. (2013). Monocyte/macrophage-elicited natural killer cell dysfunction in hepatocellular carcinoma is mediated by CD48/2B4 interactions. *Hepatology, 57*(3), 1107–1116. https://doi.org/10.1002/hep.26192.

Yamamoto, M., Tatsumi, T., Miyagi, T., Tsunematsu, H., Aketa, H., Hosui, A., et al. (2011). alpha-Fetoprotein impairs activation of natural killer cells by inhibiting the function of dendritic cells. *Clinical and Experimental Immunology, 165*(2), 211–219. https://doi.org/10.1111/j.1365-2249.2011.04421.x.

Yau, T., Kang, Y. K., Kim, T. Y., El-Khoueiry, A. B., Santoro, A., Sangro, B., et al. (2020). Efficacy and safety of nivolumab plus ipilimumab in patients with advanced hepatocellular carcinoma previously treated with Sorafenib: The CheckMate 040 randomized clinical trial. *JAMA Oncology, 6*(11), e204564. https://doi.org/10.1001/jamaoncol.2020.4564.

Yau, T., Park, J. W., Finn, R. S., Cheng, A. L., Mathurin, P., Edeline, J., et al. (2019). CheckMate 459: A randomized, multi-center phase III study of nivolumab (NIVO) vs sorafenib (SOR) as first-line (1L) treatment in patients (pts) with advanced hepatocellular carcinoma (aHCC). *Annals of Oncology, 30*, v874–v875. https://doi.org/10.1093/annonc/mdz394.029.

Ye, L. Y., Chen, W., Bai, X. L., Xu, X. Y., Zhang, Q., Xia, X. F., et al. (2016). Hypoxia-induced epithelial-to-mesenchymal transition in hepatocellular carcinoma induces an immunosuppressive tumor microenvironment to promote metastasis. *Cancer Research*, *76*(4), 818–830. https://doi.org/10.1158/0008-5472.CAN-15-0977.

Yeung, P., Beviglia, L., Cancilla, B., Dee-Hoskins, C., Evans, J. W., Fischer, M. M., et al. (2014). Abstract 1907: Wnt pathway antagonist OMP-54F28 (FZD8-Fc) inhibits tumor growth and reduces tumor-initiating cell frequency in patient-derived hepatocellular carcinoma and ovarian cancer xenograft models. *Cancer Research*, *74*(19 Supplement), 1907. https://doi.org/10.1158/1538-7445.Am2014-1907.

Zhang, H., Li, Z., Wang, L., Tian, G., Tian, J., Yang, Z., et al. (2017). Critical role of myeloid-derived suppressor cells in tumor-induced liver immune suppression through inhibition of NKT cell function. *Frontiers in Immunology*, *8*, 129. https://doi.org/10.3389/fimmu.2017.00129.

Zhou, J., Ding, T., Pan, W., Zhu, L. Y., Li, L., & Zheng, L. (2009). Increased intratumoral regulatory T cells are related to intratumoral macrophages and poor prognosis in hepatocellular carcinoma patients. *International Journal of Cancer*, *125*(7), 1640–1648. https://doi.org/10.1002/ijc.24556.

Zhou, G., Sprengers, D., Boor, P. P. C., Doukas, M., Schutz, H., Mancham, S., et al. (2017). Antibodies against immune checkpoint molecules restore functions of tumor-infiltrating T cells in hepatocellular carcinomas. *Gastroenterology*, *153*(4), 1107–1119. e1110. https://doi.org/10.1053/j.gastro.2017.06.017.

Zhu, A. X., Finn, R. S., Edeline, J., Cattan, S., Ogasawara, S., Palmer, D., et al. (2018). Pembrolizumab in patients with advanced hepatocellular carcinoma previously treated with sorafenib (KEYNOTE-224): A non-randomised, open-label phase 2 trial. *The Lancet Oncology*, *19*(7), 940–952. https://doi.org/10.1016/S1470-2045(18)30351-6.

Zhu, A. X., Guan, Y., Abbas, A. R., Koeppen, H., Lu, S., Hsu, C.-H., et al. (2020). Abstract CT044: Genomic correlates of clinical benefits from atezolizumab combined with bevacizumab vs. atezolizumab alone in patients with advanced hepatocellular carcinoma (HCC). *Cancer Research*, *80*(16 Supplement), CT044. https://doi.org/10.1158/1538-7445.Am2020-ct044.

Zhu, A. X., Kang, Y. K., Yen, C. J., Finn, R. S., Galle, P. R., Llovet, J. M., et al. (2019). Ramucirumab after sorafenib in patients with advanced hepatocellular carcinoma and increased alpha-fetoprotein concentrations (REACH-2): A randomised, double-blind, placebo-controlled, phase 3 trial. *The Lancet Oncology*, *20*(2), 282–296. https://doi.org/10.1016/S1470-2045(18)30937-9.

CHAPTER EIGHT

Intrahepatic cholangiocarcinoma: Morpho-molecular pathology, tumor reactive microenvironment, and malignant progression

Alphonse E. Sirica[a,*,†], Mario Strazzabosco[b,†], and Massimiliano Cadamuro[c,†]

[a]Department of Pathology, Virginia Commonwealth University School of Medicine, Richmond, VA, United States

[b]Liver Center and Section of Digestive Diseases, Department of Internal Medicine, Section of Digestive Diseases, Yale University School of Medicine, New Haven, CT, United States

[c]Department of Molecular Medicine, University of Padova, Padova, Italy

[*]Corresponding author: e-mail address: alphonse.sirica@vcuhealth.org

Contents

1. Introduction 322
2. iCCA classification, cell origins, and molecular subtypes 324
   2.1 Macroscopic and microscopic classification 324
   2.2 Precursor lesions of iCCA subtypes and cell origins 328
   2.3 Molecular classification of iCCA subtypes 337
3. Fibroinflammatory cholangiopathies, mechanisms, and cholangiocarcinogenesis 339
   3.1 Primary sclerosing cholangitis 340
   3.2 Fluke infestations 342
4. The desmoplastic reaction and iCCA progression 344
   4.1 CAF origins and heterogeneity in iCCA 347
   4.2 PDGF-D and TGF-β and the desmoplastic reaction in iCCA 349
   4.3 The matrisome, tumor stiffness, and hypovascularity in iCCA 354
5. Immune milieu in iCCA 355
   5.1 The innate immune system in iCCA 355
   5.2 Adaptive immune system and tumor-infiltrating lymphocytes in iCCA 357
6. Challenges and clinical implications 359
Grant Support 367
Conflict of interest statement 367
References 367

† Contributed equally.

Advances in Cancer Research, Volume 149
ISSN 0065-230X
https://doi.org/10.1016/bs.acr.2020.10.005

Abstract

Intrahepatic cholangiocarcinoma (iCCA) is a relatively rare, but highly lethal and biologically complex primary biliary epithelial cancer arising within liver. After hepatocellular carcinoma, iCCA is the second most common primary liver cancer, accounting for approximately 10–20% of all primary hepatic malignancies. Over the last 10–20 years, iCCA has become the focus of increasing concern largely due to its rising incidence and high mortality rates in various parts of the world, including the United States. The challenges posed by iCCA are daunting and despite recent progress in the standard of care and management options for iCCA, the prognosis for this cancer continues to be dismal. In an effort to provide a framework for advancing our understanding of iCCA malignant aggressiveness and therapy resistance, this review will highlight key etiological, biological, molecular, and microenvironmental factors hindering more effective management of this hepatobiliary cancer. Particular focus will be on critically reviewing the cell origins and morpho-molecular heterogeneity of iCCAs, providing mechanistic insights into high risk fibroinflammatory cholangiopathies associated with iCCA development, and notably discussing the deleterious role played by the tumor reactive desmoplastic stroma in regulating iCCA malignant progression, lymphangiogenesis, and tumor immunobiology.

1. Introduction

Intrahepatic cholangiocarcinoma (iCCA) denotes a rare heterogeneous class of epithelial cancers arising within liver, which exhibits characteristics of cholangiocyte differentiation (Sirica et al., 2009). After hepatocellular carcinoma (HCC), iCCA is the second most common primary liver cancer, accounting for approximately 3.0% all gastrointestinal malignancies (Banales et al., 2020; Khan, Tavolari, & Brandi, 2019) and comprising around 10–20% hepatobiliary cancers (Banales et al., 2020; Blechacz, 2017; Gupta & Dixon, 2017; Massarweh & El-Serag, 2017) and an estimated 8–20% of total biliary tract cholangiocarcinomas (Gupta & Dixon, 2017; Khan et al., 2019; Rizvi, Khan, Hallemeier, Kelley, & Gores, 2018; Vijgen, Terris, & Rubbia-Brandt, 2017).

Over the past few decades, iCCA has become the focus of heightened interest and concern, largely due to its increasing incidence in various countries around the world (Banales et al., 2016; Cardinale et al., 2018; Saha, Zhu, Fuchs, & Brooks, 2016), together with its globally high mortality rates also rising in several areas of the world (Bertuccio et al., 2019), including the Unites States (Beal et al., 2018; Yao, Jabbour, Parekh, Lin, & Moss, 2016). Specific high risk factors for CCA include liver fluke infection with *Opisthorchis viverrini* (Thailand) or *Clonorchis sinensis* (Korea), primary sclerosing cholangitis (PSC), hepatolithiasis, and fibropolycystic liver diseases, such as Caroli disease (Khan et al., 2019; Massarweh & El-Serag, 2017). The highest rates of iCCA are reported in Southeast and East Asia, particularly where parasite infection with liver flukes is endemic (Sripa, Tangkawattana, & Brindley, 2018), whereas PSC with or without ulcerative colitis has been established as having the strongest association with CCA development in Western populations (Khan et al., 2019; Kirstein & Vogel, 2016). Hepatitis B (HBV), hepatitis C, and cirrhosis have also been identified as strong risk factors for iCCA (Clements, Eliahoo, Kim, Taylor-Robinson, & Khan, 2020; Gupta & Dixon, 2017; Kirstein & Vogel, 2016) and may account in part for the rising incidence of iCCA observed in different countries of the world.

In Western countries, most iCCAs (i.e., 50–70%) arise sporadically without any identifiable risk factors except for advancing age (Khan et al., 2019; Kirstein & Vogel, 2016). Furthermore, in contrast to HCC, iCCAs usually develop in non-cirrhotic liver (Lee & Lee, 2017), but it is likely that the presence of liver cirrhosis in iCCA may have been underestimated (Jesper et al., 2018). It is also becoming increasingly evident that existing global trends in iCCA incidence rates associated with geographically variable risk factors need to be interpreted with caution based on recently reported concerns about historical under-reporting, country-based variations in data recording, and notably, iCCA misclassification in cancer registries (Cardinale, 2019; Khan et al., 2019; Labib, Goodchild, & Pereira, 2019).

The vast majority of iCCAs are diagnosed at an advanced incurable stage [multicentric disease within the liver, lymph node and/or peritoneal metastasis] (Sirica et al., 2019), accounting for the dismally high mortality rates for this hepatobiliary cancer. Moreover, the indolence of the early disease, the malignant aggressiveness and therapeutic refractoriness of the advanced disease, and the high reported rates (>60%) of iCCA recurrence following curative-intent surgical resection (Chan et al., 2018; Hu et al., 2019; Wang, Pang, et al., 2019; Zhang et al., 2018) all underscore the importance of advancing our current understanding of interactive cellular and molecular mechanisms that drive the development and progression of iCCA, with the ultimate aim of devising more effective biomarker-driven targeted approaches for managing and/or preventing this challenging and most often fatal hepatobiliary cancer.

In sharp contrast to conventional hepatocellular carcinomas (HCC), iCCAs are typically characterized by a prominent desmoplastic and usually hypovascularized tumor stroma (Bösmüller et al., 2018; Sirica & Gores, 2014), which in many cases represents the dominant histological feature of the tumor. This complex tumor reactive stroma is largely composed of a dense collagen-fiber enriched extracellular matrix (ECM) together with matricellular proteins and proteinases, and containing an abundance of activated cancer-associated fibroblasts (CAFs), and to a lesser extent tumor associated macrophages (TAMs) and varying numbers of vascular and innate immune cells (Affo, Yu, & Schwabe, 2017; Brivio, Cadamuro, Strazzabosco, & Fabris, 2017; Sirica & Gores, 2014). The cellular and ECM components of the desmoplastic stroma of iCCA are not structurally nor functionally static, but should be viewed as providing an evolving, interactive and mutable microenvironment that plays a preeminent role in promoting iCCA progression and invasive growth, cancer cell survival, and resistance to chemotherapeutic drugs, targeted agents, and immunotherapies (Cadamuro et al., 2018; Fabris, Perugorria, et al., 2019; Gentilini, Pastore, Marra, & Raggi, 2018; Loeuillard, Conboy, Gores, & Rizvi, 2019; Sirica et al., 2019).

This review will highlight timely research findings aimed at providing mechanistic insights into the biological and translational significance of the desmoplastic microenvironment in iCCA. Our focus will be on (1) iCCA pathological and molecular classification and cell origins, (2) fibroinflammatory cholangiopathies and mechanisms, (3) the desmoplastic reaction and iCCA progression, (4) the immune milieu of iCCA and immune resistance, and (5) clinical implications and challenges.

2. iCCA classification, cell origins, and molecular subtypes

2.1 Macroscopic and microscopic classification

iCCAs are macroscopically and microscopically diverse. Classically, iCCA can arise from both small intrahepatic bile ducts (septal & interlobular BD) and large intrahepatic bile ducts (segmental and area BD) within liver (Kendall et al., 2019; Nakanuma et al., 2010; Vijgen et al., 2017). Peribiliary glands associated with large intrahepatic BD may also give rise to iCCA (Nakagawa, Hayata, Yamada, Kawamura, Suzuki, & Koike, 2018; Nakanuma et al., 2010).

Based on macroscopic growth patterns, iCCAs have been grossly classified as mass-forming (MF), periductular infiltrating (PI), intraductal growing (IG) and MF + PI types (Guglielmi et al., 2009; Kendall et al., 2019; Nakanuma et al., 2010; Shimada et al., 2007; Vijgen et al., 2017). The MF growth pattern is the most common form of the macroscopic types, accounting for about 65% of all iCCAs (Guglielmi et al., 2009; Kendall et al., 2019; Vijgen et al., 2017). The PI and IG types are much less common, comprising 6.0%–14%% and 4.0%, respectively, of all iCCAs (Guglielmi et al., 2009; Kendall et al., 2019; Tsukahara et al., 2016; Vijgen et al., 2017). Mixed MF + PI iCCAs, on the other hand, account for about 25% of CCA in liver (Guglielmi et al., 2009; Kendall et al., 2019; Vijgen et al., 2017). The MF + PI type has the worst prognosis with higher rates of early recurrence following curative-intent surgical resection when compared with other types of iCCA (Bagante et al., 2018; Blechaz, Komuta, Roskams, & Gores, 2012; Guglielmi et al., 2009; Shimada et al., 2007; Vijgen et al., 2017).

The MF type of iCCA is generally thought to preferentially arise from small intrahepatic bile ducts (Cardinale et al., 2018; Kendall et al., 2019). This type commonly presents as a singular solid nodular mass that is gray to gray-white in color, firm and solid, non-encapsulated, and polylobulated with no macroscopically discernable connection with a bile duct (Nakanuma et al., 2010; Vijgen et al., 2017). MF-iCCAs can be quite large, often ranging between 5 and 10cm in diameter at the time of diagnosis (Buettner, van Vugt, IJzermans, & Koerkamp, 2017). Advanced MF type of iCCA may also exhibit satellite or multifocal tumor growth within liver, consisting of various sized nodules that may coalesce. In comparison, iCCAs arising within the large intrahepatic bile ducts are usually of the PI, IG, or the more frequent MF + PI type (Vijgen et al., 2017). The PI type of iCCA does not form a nodular mass, but rather grows longitudinally along the wall of the large intrahepatic bile ducts and spreads along the portal tracts, resulting in strictures of the affected bile ducts and dilatation of the smaller proximal bile ducts (Kendall et al., 2019; Nakanuma et al., 2010; Vijgen et al., 2017). The IG type of iCCA typically presents as a slow growing polypoid or papillary tumor growing within a dilated bile duct lumen (Kendall et al., 2019; Nakanuma et al., 2010; Vijgen et al., 2017). IG-iCCA can be divided into two groups, those with tumors showing a predominantly papillary growth pattern and those whose tumors exhibited a predominantly tubular growth pattern (Chung et al., 2009; Tsukahara et al., 2016). Both the papillary and tubular subtypes of IG-iCCA have been found to have similar favorable survival outcomes, although tumors with the tubular growth pattern showed increased incidences of lymphatic and venous invasion and a higher rate of liver metastasis than those with the papillary growth pattern. Bagante et al. (2018) have also reported that while patients who had undergone curative-intent surgery for either MF or MF + PI types of iCCA were more likely to experience early recurrence, IG-iCCA patients were more at risk for late recurrence after 5-years from surgery.

Obstructive jaundice may be manifested in patients with the PI and IG types (Vijgen et al., 2017), but is significantly less common in MF type of iCCA originating at the periphery of the biliary tree (Akita et al., 2018; Liau et al., 2014). Jaundice is significantly associated with the more advanced MF + PI type (Shimada et al., 2007). The MF type of iCCA tends to invade the hepatic parenchyma via the portal vein system and at advanced stages via the lymphatics, whereas the PI type has a tendency to spread along the Glisson sheath via lymphatics (Blechaz et al., 2012; Meng, Pan, Hong, Chen, & Chen, 2017; Yang & Yan, 2008).

Histologically, 90–95% of iCCAs are classified as adenocarcinomas (typically well-to-moderately differentiated) displaying variable architectural patterns [i.e., tubular, acinar, papillary, tubular-papillary, anastomosing glandular] and characterized by varying degrees (usually prominent) of desmoplasia, which is a hallmark feature of iCCA (Akita et al., 2017; Blechacz, 2017; Kendall et al., 2019; Nakanuma et al., 2010; Sirica, 2012; Sirica et al., 2009). In a classification scheme proposed by Nakanuma et al. (2010) classic or conventional iCCA is subclassified according to histological features and anatomic location into two main types: small bile duct type and large bile duct type. Grossly, the small bile duct type of iCCA is almost exclusively of the MF type and is often associated with chronic non-bile duct liver diseases, such as viral hepatitis (B & C) and cirrhosis (Akita et al., 2019; Cardinale et al., 2018; Kendall et al., 2019), whereas the large bile duct type typically characterizes the PI and PI + MF types of iCCAs, such as those associated with predisposing high risk fibroinflammatory cholangiopathies like primary sclerosing cholangitis (PSC), hepatolithiasis, liver-fluke infestation, and Caroli's disease. Small bile duct type iCCAs, also described as "mixed type" (Cardinale et al., 2018) or "cholangiolar type" (Liau et al., 2014), display no or low mucin production, while large duct type iCCAs and those arising from associated peribiliary glands are mucin-producing adenocarcinomas (Cardinale et al., 2012; Kendall et al., 2019; Nakagawa et al., 2018; Sigel et al., 2018; Vijgen et al., 2017).

Cholangiolocellular carcinoma (CLC) is a rare malignancy first described by Steiner & Higginson (Steiner & Higginson, 1959), which accounts for 0.6–1.0% of primary liver cancers (Yamamoto, Oshita, Nishisaka, Nakahara, & Itamoto, 2018). CLC is considered to originate from cells within the ductules (cholangioles)/canals of Hering of liver (Brunt et al., 2018; Komuta et al., 2008). This tumor consists of thin, malignant ductular-like structures characterized by low grade cellular atypia that may appear to radiate from or surround a portal tract in a tubular, cord-like, anastomosing pattern within a dense hyalinized (fibrous) stroma, and which may show trabecular and replacing growth at its interface with the surrounding nontumorous liver (Brunt et al., 2018). CLC often exhibits a similar MF type at the periphery of the liver as small duct type of iCCA, and like small duct type MF-iCCA, is mucin negative (Brunt et al., 2018) and associated with cirrhosis or chronic viral hepatitis (Ariizumi & Yamamoto, 2015; Yamamoto et al., 2018). Patients with CLC who have undergone curative-intent hepatectomy have been reported to have a significantly better survival rates than those with classic iCCA (Ariizumi et al., 2014; Kusano et al., 2020). Integrative genomic analysis revealed CLC to be a distinct biliary-derived entity with chromosomal stability and active transforming growth factor-β (TGF-β) signaling (Moeini et al., 2017). More recent immunochemical features and mutational profiling of CLC (see Section 2.3) have further supported classifying CLC as a histological subtype of well differentiated iCCA (Balitzer et al., 2019).

Classic combined hepatocellular-cholangiocarcinoma (cHCC-CCA) is a rare primary MF-type liver cancer, which presents with a more aggressive malignant behavior and a grim prognosis with worse survival outcomes than for HCC and similar to or worse than for iCCA (Gera, Ettel, Acosta-Gonzales, & Xu, 2017; Stavraka, Rush, & Ross, 2019). Previous reported incidence has ranged between 0.4% and 14.2% (Stavraka et al., 2019; Wang, Li, et al., 2019) and 1.0–4.7% (Gera et al., 2017) of primary liver tumors, but the true incidence of cHCC-CCA is unclear due to misdiagnosis and a previous lack of consensus on the nomenclature Stavraka et al., 2019; Wang, Li, et al., 2019). Similar to iCCA, the incidence and mortality trends overall for cHCC-CCA have also been reported to have increased in the United States from 2000 to 2014 (Wang, Li, et al., 2019).

As the term implies, classic cHCC-CCA contain areas of both unequivocally differentiated HCC and typical iCCA (Brunt et al., 2018; Gera et al., 2017). The HCC component usually is in the form of thickened trabeculae composed of polygonal cells with abundant granular eosinophilic cytoplasm and scant stroma, whereas the iCCA component distinctly presents as an adenocarcinoma with malignant glands comprised of small bile duct-like cells supported by a conspicuous desmoplastic stroma (Brunt et al., 2018; Gera et al., 2017). The HCC and iCCA components may be intermixed or occupy separate regions of a tumor, but focal areas of intermingling are often observed (Brunt et al., 2018). Mucin production is most often negative. Both MF-iCCA and the cholangiocarcinoma component of cHCC-iCCA in cirrhotic liver were further shown to exhibit higher density of arteries and microvessels than classical iCCA in non-cirrhotic liver, which usually presents with a hypovascular stroma (Xu et al., 2012).

cHCC-iCCA with "stem/progenitor" cell features/phenotypes and intermediate cell carcinoma comprised of tumor cells intermediate between hepatocytes and cholangiocytes are considered non-classical subtypes of HCC or iCCA (Brunt et al., 2018; Gera et al., 2017). CLC, which may be entirely distinct as a primary liver cancer can also be observed variably as a component in either HCC, iCCA, or cHCC-iCCA (Brunt et al., 2018). Other unconventional subtypes of iCCA include rare variants such as adenosquamous type, squamous type, intestinal type, signet ring type, clear cell type, lymphoepithelial type, and others (Kendall et al., 2019; Vijgen et al., 2017). The intestinal-type of iCCA is a particularly interesting variant exemplifying the cellular plasticity of biliary cells in cholangiocarcinogenesis. Reported cases of this rare subtype of large duct iCCA showed absorptive columnar cells, Paneth cell and goblet cell metaplasia, and occasional neuroendocrine cells within carcinomatous epithelium of the tumor (Bae et al., 2002; Kozuka, Kurashina, Tsubone, Hachisuka, & Yasui, 1984). A similar pattern of intestinal-type differentiation has also been demonstrated in mucin producing iCCA induced by furan in rat livers (Elmore & Sirica, 1991, 1993; Ren, Silberg, & Sirica, 2000).

Lymphoepithelioma-like iCCA is also a distinctive rare sub-type. This primary hepatic tumor is composed of undifferentiated epithelial cells with a dense lymphoplasmacytic infiltration and is characterized by lower rates of recurrence and better overall survival after surgical resection than classic iCCA (Aosasa et al., 2015; Kendall et al., 2019). A recently identified "inflamed" iCCA subtype that presented with a massive T-lymphocyte infiltration and activation of inflammatory and immune checkpoint pathways was also shown to be associated with enhanced patient survival (Job et al., 2020).

2.2 Precursor lesions of iCCA subtypes and cell origins

2.2.1 Precursor lesions

Presently, no precursor lesions of the MF type of iCCA are known (Kendall et al., 2019), although bile duct adenoma-like and atypical bile duct lesions involving the small bile ducts have been suggested as possibly related to the development of the MF type of iCCA (Nakanuma et al., 2014). However, three preinvasive lesions: (1) biliary epithelial neoplasia (BillN),

(2) intraductal papillary neoplasms of the bile duct (IPNBs), and (3) intraductal tubulopapillary neoplasms of the bile duct (ITPNs), which are most frequently observed in high risk conditions affecting the large intrahepatic bile ducts, are now considered as morphologically established precursor lesions for iCCA (Kendall et al., 2019; Nakanuma, Kakuda, & Uesaka, 2019; Vijgen et al., 2017; Zaccari et al., 2019). Pancreatic counterparts to these premalignant intraepithelial neoplasms of the biliary tract are also well recognized (Nakanuma et al., 2019; Zaccari et al., 2019).

BillN is a microscopically identifiable dysplastic lesion exhibiting a flat, micropapillary, or pseudopapillary pattern of growth (Nakanuma et al., 2019; Zaccari et al., 2019), which is graded as BillN1 (low grade dysplasia), Billn2 (intermediate dysplasia), and BillN3 (carcinoma in situ). BillNs are frequently observed in areas of mucosa proximal to classic nodular/sclerosing iCCA and in the context of preneoplastic cholangiopathies, such as PSC, hepatolithiasis, liver fluke infection, and Caroli's disease (Kendall et al., 2019; Zaccari et al., 2019). BillN has also been documented in patients with non-biliary chronic diseases, such as HCV (Kendall et al., 2019; Vijgen et al., 2017), and may also be seen in peribiliary glands (Vijgen et al., 2017; Zaccari et al., 2019).

IPNB can be considered as the biliary counterpart to pancreatic intraductal papillary mucinous neoplasms, while ITNB appears to be the biliary counterpart to intraductal tubular neoplasm of the pancreas (Kendall et al., 2019; Nakanuma et al., 2019; Vijgen et al., 2017). IPNBs precede the growth of IG type and papillary type iCCAs (Nakanuma, Miyata, & Uchida, 2016) and are more commonly encountered in the Far East in patients with hepatolithiasis and liver fluke infection, and less frequently in the West (Kendall et al., 2019; Wan et al., 2013). Macroscopically, IPNBs grow as single or multiple yellow, friable papillary mass within large size intrahepatic and extrahepatic bile ducts, which may show unilocular or multilocular cystic dilation (Hucl, 2019; Kendall et al., 2019; Nakanuma et al., 2019). Histologically, IPNBs are classified into four subtypes based on the lining epithelial cells and architecture, including fine fibrovascular stalks: pancreaticobiliary, intestinal, gastric, and oncocytic. The pancreaticobiliary and intestinal subtypes are the most common, with IPNBs often graded as high grade (Kendall et al., 2019; Nakanuma et al., 2019). Invasive iCCA can be found in association with each of the IPNB subtypes, although it is more often associated with the pancreaticobiliary type (Kendall et al., 2019). The invasive parts of IPNBs usually show tubular adenocarcinoma with desmoplastic reaction and only occasionally show oncocytic or colloid (mucinous) carcinoma (Nakanuma et al., 2019). However, invasive iCCAs associated with the intestinal subtype of IPNB are often mucinous adenocarcinomas (Kendall et al., 2019). With respect to mucin types, MUC1, which is associated with iCCA aggressiveness and poor prognosis, is commonly expressed in the pancreaticobiliary subtype, whereas MUC2 is a goblet cell mucin characteristic of the intestinal phenotype of IPNB and of mucinous adenocarcinoma. The gastric and oncocytic subtypes are positive for MUC6, while MUC5AC can be detected in either of the four subtypes (Ishikawa et al., 2004; Kasprzak & Adamek, 2019; Rocha et al., 2012).

ITPN is also a polypoid lesion with a predominant tubular growth pattern. Key histological features of ITPNs are tubular glands densely packed back to back with abortive papillary elements, uniform high grade dysplasia throughout the tumor, no cytoplasmic mucin, and focal necrosis (Kendall et al., 2019; Nakanuma et al., 2019; Vijgen et al., 2017) Although ITPNs have a high risk of malignancy, they generally have a more favorable prognosis when compared with IPNBs (Kendall et al., 2019).

2.2.2 Cell origins

Histopathological observations of precancerous and borderline lesions that precede the development of iCCA livers of patients with fibroinflammatory cholangiopathies (Cannito et al., 2018) have led to the recognition that iCCA develops through a multistage carcinogenic process, which like other epithelial cancers, may proceed through a hyperplasia-dysplasia-carcinoma in situ sequence (Kendall et al., 2019; Shimonishi, Sasaki, & Nakanuma, 2000). Hyperplasia and dysplasia of the lining biliary epithelial cells and peribiliary glands are characteristically observed in patients with hepatolithiasis and those with liver fluke infection, and also in biliary lining epithelium of patients with PSC and Caroli's disease (Carpino, Cardinale, et al., 2019; Jang, Lee, & Kim, 2014; Ludwig, Wahlstrom, Batts, & Wiesner, 1992; Nakagawa et al., 2018; Shimonishi et al., 2000; Zimmermann, 2017). It has also become apparent that the epidemiological, morphological and molecular heterogeneity of iCCA subtypes is linked to potentially diverse cellular origins (Bragazzi et al., 2018; Carpino, Cardinale, Reid, Alvaro, & Gaudio, 2012; Overi et al., 2018; Sia, Villanueva, Friedman, & Llovet, 2017).

Much of our current understanding of the cell origins of iCCA comes from studies performed using genetic mouse models (Erice et al., 2019; Vicent et al., 2019; Zhu & Kwong, 2020); cell lineage tracing (Fan et al., 2012; Guest et al., 2013; Wang, Dong, et al., 2018), and immunophenotyping of "stem/progenitor" cells (Nakagawa et al., 2018;

Overi et al., 2018; Sia et al., 2017; Vicent et al., 2019). As exemplified in Table 1, and as proposed by Cardinale, Carpino, Reid, Gaudio, and Alvaro (2012), iCCA can potentially develop from multiple cells of origin, including immature or more mature cholangiocytes, hepatic stem/progenitor cells localized to the canals of Hering and bile ductules, biliary tree stem/progenitor cells (BTSCs) found in the peribiliary glands, and from transdifferentiation of mature hepatocytes.

While the experimental findings favor a diverse cellular origin of iCCA, it remains unclear as to the extent to which human iCCAs are developed from cholangiocytes versus those being derived from hepatic stem/progenitor cells or transdifferentiated hepatocytes. As previously noted, the majority of human iCCAs develop sporadically in livers in the absence of discernable chronic inflammatory liver injury. Also, as has been mentioned, a multistep carcinogenic process indicative of a hyperplasia-dysplasia-cancer in situ sequence is well recognized in the pathogenesis of iCCA arising from cholangiocytes within different levels of the hepatic biliary tree. However, the concept of a hepatic stem/progenitor cell origin is increasingly recognized as being relevant to the development of human MF-iCCA arising in a background of chronic viral hepatitis and cirrhosis, and is particularly apparent in the development of cHCC-CCA with stem cell features (Brunt et al., 2018; Cai, Li, & Kaplan, 2020; Stavraka et al., 2019; Terada, 2013). In a limited number of molecular studies involving tissue microarray, gene set enrichment analysis, and integrative genomics, expression of hepatic stem/progenitor traits have been demonstrated in cohorts of surgically resected cHCC-CCA (Coulouarn et al., 2012; Woo et al., 2010). Coulouarn et al. (2012) have further shown that cHCC-CCA exhibit downregulation of the hepatocyte differentiation program and a commitment to the biliary lineage, as well as an activation of the Wnt/β-catenin and TGF-β signaling pathways. Both of these pathways are involved in biliary differentiation. More recently, Xue et al. (2019) performing a large-scale integrative analysis of 133 cases of human cHCC-CCA, which included comprehensive genomic and transcriptomic profiling, showed that tumors with clearly defined areas of HCC and iCCA (denoted as combined-type of cHCC-CCA) and those with intimately mixed components of HCC and iCCA in the same tumor without clear boundaries (defined as mixed-type HCC-CCA) both showed stem-like features, were monoclonal in origin, and exhibited a high level of expression of Nestin, a liver stem cell/progenitor cell protein (Gleiberman et al., 2005). Interestingly, the HCC components of the combined-type cHCC-CCA

Table 1 Representative genetic mouse models supporting different cells of origin of intrahepatic cholangiocarcinoma.

Model	Method	Tumor	Suggested Cell Origin[a]	Reference
AhCreER^{T};*Kras*$^{V12/+}$;*Pten*$^{f/f}$ (*PTEN* loss; *KRAS* activation)	GEMM[b]	Non-invasive papillary neoplasms of intrahepatic biliary tract, including major interlobular bile ducts and small bile duct radicles	BEC/HPC	Marsh, Davies, Williams, and Clarke (2013)
Alb-Cre^{+};*LSL-Kras*$^{G12D/+}$;*Pten*$^{f/f}$	GEMM	Well differentiated, desmoplastic intrahepatic cholangiocarcinoma	BEC/HPC	Ikenoue et al. (2016); Lin et al. (2018)
Sox9-Cre^{ERT2};*LSL-Kras*$^{G12D/+}$;*Pten*$^{f/f}$	GEMM	Intrahepatic and extrahepatic cholangiocarcinomas[c]	BEC	Lin et al. (2018)
Alb-Cre,Smad4$^{f/f}$; *Pten*$^{f/f}$ (*Smad4* and *PTEN* ablation)	GEMM	Bile duct dysplasia, carcinoma *in situ*, well differentiated, desmoplastic intrahepatic cholangiocarcinoma	BEC/HPC	Xu et al. (2006)
Ck19-CreER^{T}*eYFP*^{R26}*p53*$^{f/f}$ + thioacetamide (*p53* deletion + chronic liver injury)	GEMM	Intrahepatic cholangiocarcinoma	BEC[d]	Guest et al. (2013)
Sox9-Cre^{ERT2};*Kras*$^{LSL\text{-}G12D}$;*Tp53*$^{f/f}$ (*KRAS* activation; *Tp53 deletion*)	GEMM	BilIN lesions[e]; desmoplastic intrahepatic cholangiocarcinoma	BEC	Hill et al. (2018)
AAV8-TBG-Cre;Kras$^{LSL\text{-}G12D}$;*Tp53*$^{f/f}$ plus 2 week dietary treatment with DCC[f] (KRAS activation; Tp53 deletion; dietary-induced liver injury)	GEMM	Intrahepatic cholangiocarcinoma; hepatocellular carcinoma; mixed hepatocellular carcinoma-cholangiocarcinoma	HEP[d]	Hill et al. (2018)

Alb-Cre;LSL-IDH2R172K; Kras^{G12D} (mutant *IDH* expression; *KRAS* activation)	GEMM	Intrahepatic cholangiocarcinoma	HPC	Saha et al. (2014)
Alb-Cre;NotchIC (Notch overexpression)	GEMM	Desmoplastic intrahepatic cholangiocarcinoma	HPC	Zender et al. (2013)
AAV8-TBG-Cre;Pten$^{f/f}$*;TBR2*$^{f/f}$ (*PTEN* and *TGFβR2* ablation)	GEMM	Desmoplastic intrahepatic cholangiocarcinoma	HEP[d]	Mu et al. (2016)
Prom1-CreERT2;Pten$^{f/f}$*;TBR2*$^{f/f}$ or *K19-CreERT; Pten*$^{f/f}$*;TBR2*$^{f/f}$	GEMM	Desmoplastic intrahepatic cholangiocarcinoma	BEC[d]	Mu et al. (2016)
Alb-Cre;Nf2$^{lox/lox}$ N*eurofibromatosis type 2 gene deletion*)	GEMM	Oval cell proliferation; hepatocellular carcinoma; intrahepatic cholangiocarcinoma	HPC	Benhamouche et al. (2010)
K19-Cre^{ERT}*;LSL-Kras*^{GD12D}*; Tgfbr2*$^{f/f}$ IL-33 (*KRAS* activation, *TGFβR2* loss, IL-33 mediated biliary epithelial injury response	GEMM	Extrahepatic cholangiocarcinoma	PBG[d]	Nakagawa et al. (2017)
AAV8-Tte-Cre; Hydrodynamic tail vein injection NICD/AKT plasmids (activated Notch and AKT)	SB[g]transposon-transfection	Intrahepatic cholangiocarcinoma lacking desmoplastic stroma	HEP[d]	Fan et al. (2012)

Continued

Table 1 Representative genetic mouse models supporting different cells of origin of intrahepatic cholangiocarcinoma.—cont'd

Model	Method	Tumor	Suggested Cell Origin[a]	Reference
Hydrodynamic tail vein injection of HA-tagged pT3-EF5α-PIK3CAH1047R and Flag-tagged pT3-EF5αYapS127A (Co-activation of PI3CA and Yap)	SB transposon transfection	Hepatocellular carcinoma; intrahepatic cholangiocarcinoma; mixed hepatocellular carcinoma-intrahepatic cholangiocarcinoma	HPC/HEP	Li et al. (2015)
Intrabiliary injection of murine myr-AKT and human YAP127A combined with systemic IL-33 (or IL-6) injection and lobar bile duct ligation (Transduction of constituitively activated AKT and Yap facilitated by a IL-6-mediated process and lobar bile duct obstruction)	SB transposon transfection	Desmoplastic intrahepatic cholangiocarcinoma	BEC	Yamada et al. (2015)

[a]BEC, biliary epithelial cell; HPC, hepatic progenitor cell; HEP, hepatocyte; PBG, peribiliary gland cell
[b]GEMM, genetically engineered mouse model
[c]Pancreatic ductal adenocarcinomas also developed in this model
[d]Lineage tracing employed
[e]BilIN, biliary intraepithelial neoplasia, which only preceded development of intrahepatic cholangiocarcinoma from BEC and not hepatocytes in this model
[f]DDC, 3,5-diethoxycarbonyl-1,4 dihydrocollidine
[g]SB, Sleeping Beauty

were not clustered into classic HCC subgroups nor were the iCCA components clustered into classic iCCA subgroups, but rather clustered together, suggesting that the combined-type of cHCC-CCA is not identical to traditional HCC and iCCA, but derived from a common progenitor cell. In this context, is of interest that loss of *p53* has been shown facilitate dedifferentiation of mature hepatocytes into nestin-positive liver progenitor-like cells, which then may be poised to potentially develop into HCC, iCCA, or cHCC-CCA (Tschaharganeh et al., 2014).

Experimental studies have also supported the possibility of adult hepatocytes transdifferentiating into cholangiocytes under conditions of severe chronic hepatic injury and cholestasis (Ko, Russell, Molina, & Monga, 2020; Michalopoulos, Barua, & Bowen, 2005; Michalopoulos & Khan, 2015; Sadri, Jeschke, & Amini-Nik, 2016). Morphological and phenotypic evidence for putative transdifferentiation of rare ductular hepatocytes from bile duct epithelium within interlobular bile ducts has also been described in human liver in various severe hepatic disease states, including chronic viral hepatitis and end stage cirrhosis (Nomoto, Uchikosi, Kajikazawa, Tanaka, & Asakura, 1992), as well as in hyperplastic bile ductular structures formed in rat liver in associated with extreme hepatotoxic injury induced by furan (Sirica, Gainey, & Mumaw, 1994). Of particular interest are the findings of Seehawer et al. (2018) who showed that the hepatic microenvironment epigenetically shapes lineage commitment in mosaic mouse models of liver tumorigenesis. Specifically, these investigators observed that a necroptosis-associated hepatic cytokine microenvironment promotes oncogenetically-transformed hepatocytes to give rise to iCCAs. In contrast, when hepatocytes containing the same oncogenic drivers were surrounded by an apoptotic microenvironment, they formed HCCs. Notably, the transcription factors *Prdm5* and *Tbx3* were singled out as being major microenvironment-dependent and epigenetically regulated lineage commitment factors for ICCA and HCC, respectively (i.e., *Prdm5* overexpression and *Thx3* knockdown resulted in HNF4α-negative and cytokeratin 19 (CK-19)-positive iCCA, whereas *Tbx3* overexpression and *Prdm5* knockdown yielded HNF4α-positive, CK19-negative HCC). Hepatocyte associated transcription factors (HNF4α and HNF6) have been demonstrated in cholangiocytes in chronic HCV infected liver with end stage cirrhosis (Limaye et al., 2008). However, while the apparent transdifferentiation of hepatocytes to cholangiocytes or of rare cholangiocytes to hepatocytes have been observed to occur in various relevant human liver diseases, it remains to be determined if this facultative progenitor cell mechanism plays a plausible role in human iCCA or cHCC-CCA development, since it is more likely to be predominantly manifested in end stage liver diseases where the normal regenerative capacity of liver is severely impaired.

Peribiliary glands are tubuloaveolar glands with mucinous and serous acini that are located in the wall of the large intrahepatic bile ducts and extrahepatic bile ducts, being largely distributed at the branching points of the biliary tree. They are connected to the bile ducts through small canals/tubes and are considered to modulate bile composition by secreting serous and mucinous components, although their pathophysiological role has not been definitively established (Cardinale, Carpino, et al., 2012; Nakagawa et al., 2018). Peribiliary glands have been implicated in the origin of human large duct mucin-producing CCA and IPNB (Cardinale, Wang, et al., 2012; Carpino, Cardinale, et al., 2019; Nakanuma & Sato, 2012; Sato, Harada, Sasaki, & Nakanuma, 2014).

BTSCs have been localized to the bottom of human peribiliary glands near the fibromuscular layer and have been demonstrated to express specific stem/progenitor cell markers, including Sox9, Sox17, Pdx1, Foxa2, CD133, CD44, CXCR4, EpCAM, NCAM, Sall4, and Lgr5 (Cardinale et al., 2011; Nakagawa et al., 2018; Overi et al., 2018), but lacked markers of mature cells, such as secretin receptor-SR, albumin, and insulin (Overi et al., 2018). A subpopulation of BTSCs were also found to express markers used to characterize pluripotent stem cells (i.e., Oct4, Nanog, Sox2) (Overi et al., 2018). That the peribiliary gland is a niche for multipotent BTSMs is further highlighted by the demonstration that BTSCs present in peribiliary glands of human extrahepatic biliary tree can give rise to hepatocytes, cholangiocytes, and beta-islet cells in culture and in vivo (Cardinale et al., 2011). $Sox17^{+}$ and/or $Pdx1^{+}$ cells of the peribiliary glands and peribiliary network of mice were also shown to proliferate in response to viral-induced biliary injury and to biliary obstructive cholestasis caused by bile duct ligation (DiPaola et al., 2013). A recently established novel mouse model of biliary injury-related extrahepatic CCA from peribiliary glands developed by ductal cell-specific activation of KRAS, and deletion of TGF-β receptor 2 and of E-cadherin, together with IL-33 treatment further adds to the likelihood of large duct CCAs arising in humans with underlying fibroinflammatory risk conditions as originating from peribiliary glands harboring BTSCs (Nakagawa et al., 2017, 2018; Nakagawa, Suzuki, & Koike, 2019). Additional support for this possibility also comes from the recent findings of Carpino, Cardinale, et al. (2019), who demonstrated marked proliferation of BTSCs, expansion of peribiliary glands and dysplasia, and high expression of BTSC markers in mucin-producing CCAs, which emerged in the PSC patients. However, while these findings are compelling, definitive evidence establishing a direct cell lineage relationship between BTSCs and cholangiocarcinoma cells arising within large duct CCAs has yet not been conclusively demonstrated.

2.3 Molecular classification of iCCA subtypes

Molecular profiling of human iCCA subtypes has revealed a complex mutational landscape with broad inter- and intra-tumor heterogeneity that are attributed in part to diverse multifactorial etiologies, histological tumor differences and cell origins, and evolving malignant progression (Sirica et al., 2019). However, small duct type and large duct type of iCCAs can be grouped into two distinctive morpho-molecular groups based on unique differences in their gene mutation profiles (Kendall et al., 2019). Isocitrate dehydrogenase 1/2 (*IDH1/2*) mutations and Fibroblast Growth Factor Receptor 2 (*FGFR2*) fusions are almost exclusively detected in small duct type of iCCAs when compared with large type iCCAs (Akita et al., 2018; Goeppert et al., 2019; Kendall et al., 2019; Liau et al., 2014; Ma, Meng, et al., 2020; Wang, Drill, et al., 2019) In contrast to small duct type iCCAs, large duct type iCCAs, as well as extrahepatic CCAs (perihilar and distal) typically show higher mutation frequencies for oncogenes (i.e., *KRAS*) and tumor suppressor genes (i.e., *TP53*) (Jusakul et al., 2017; Kendall et al., 2019; Liau et al., 2014). Loss of *BAP1* was also reported to be restricted to the small bile duct type of iCCA, whereas in the same study, loss of *SMAD4*, as well as *MDM2* amplification were more frequently detected in the large bile duct type (Akita et al., 2019). Genomic profiling of iCCA in cirrhosis ($n = 10$) demonstrated alterations similar to those of iCCA in non-cirrhotic liver, including *IDH1/2* mutations (30%), *FGFR2* fusions (20%), as well as *BAP1* (10%) and *ARID1A* (10%) mutations (Joseph et al., 2019). The genomic profile (*IDH1/2* mutations, *FGFR2* fusions, chromatin-remodeling gene mutations, such as *ARID1A* and *PBRM1* and copy number alterations) were also found to be similar in a small cohort of analyzed cases of human CLC ($n = 5$), iCCA ($n = 7$), and mixed CLC-iCCA ($n = 5$), with mutations typical of small duct type of iCCA, including *IDH1/2* mutations and *FGFR2* fusions shown to be present in 90% of the cases with a CLC component (Balitzer et al., 2019). Unlike small duct type iCCAs, a limited number of studies have revealed human cHCC-CCA to harbor highly recurrent *TP53* and *TERT* promoter mutations (Joseph et al., 2019; Liu et al., 2018; Sasaki, Sato, & Nakanuma, 2017), with *TP53* mutations frequently present in both the HCC and CCA components of the tumor (Joseph et al., 2019). None of the cases of cHCC-CCA ($n = 20$) analyzed by Joseph et al. (2019) demonstrated *IDH1/2*, *FGFR2*, or *BAP1* mutations. On the other hand, the mutational landscape of cHCC-CCA appears to be more similar to that of HCC and distinctly different from that of iCCA, even in cirrhosis (Joseph et al., 2019; Liu et al., 2018). Mutations in *TERT* promoter, *ARID1A*, *KRAS* and *TP53* were further shown to correlate with different clinical phenotypes of cHCC-iCCA. *TERT* promoter mutations correlated with HBV, an intermediate subtype-predominant histology, higher clinical stage, and higher N-factor (lymph node involvement). *ARID1A* mutations correlated with alcoholic liver disease, smaller tumor size, a lower grade of coexistent HCC, and α-fetoprotein positivity, and were associated with a CLC subtype predominance. *KRAS* mutations correlated with greater histological diversity and a higher M-factor (distant metastasis), while *TP53* mutations correlated with α-fetoprotein positivity (Sasaki et al., 2017).

Integrative molecular analysis was also used to identify two distinct molecular subclasses of human iCCA, respectively, designated as an inflammation subclass and a proliferation subclass (Sia et al., 2013, 2017). The inflammation subclass iCCAs was characterized by activation of inflammatory signaling pathways, overexpression of cytokines, including interleukin (IL)-6, IL-10, and IL-17, and STAT3 constitutive activation, whereas the proliferation subclass was based on activation of oncogenic signaling pathways (i.e., KRAS/RAF/ERK, IGF1R, EGFR, MET, NOTCH, ErbB2, PI3K/AKT/mTOR), oncogenic mutations (i.e., *KRAS, BRAF, EGFR, TP53*), DNA hypermethylation, and focal aberrations, including DNA amplifications at Chr 11q13.2 and deletions at Chr 14q22. The proliferation class also included a subtype of iCCA with stem cell-like features, chromosome instability and IDH1/2 mutations. This study also revealed a resemblance between the proliferation class of iCCA and HCC subtypes with poor prognosis and stem cell features (Sia et al., 2017).

The molecular classification profiles described above have significant relevance for the prognosis and treatment of patients with iCCA subtypes. For example, the proliferation subclass of iCCA was demonstrated to exhibit a more aggressive clinical behavior, where as the inflammation subclass defined a class of iCCA with a more favorable prognosis (Sia et al., 2013, 2017). In a separate study, iCCA with CLC differentiation whose transcriptomic profile resembled that of an inflammation-related subtype exhibited less aggressive histopathological features and more favorable survival outcomes and time to recurrence than classic iCCA resembling the proliferation subtype (Rhee et al., 2018). Integrative clustering of genetic and epigenetic data from an analysis of 40 small duct type iCCAs and 12 large duct type iCCAs identified four subgroups with prognostic relevance. These were designated as IDH, high (H), medium (M), and low (L) genetic and epigenetic alteration groups. The IDH group consisted of all samples with *IDH1* or *IDH2* mutations and showed together with the H group a highly disruptive genome. DNA methylation was highest in both the IDH and H subgroup, intermediate in the M subgroup, and low in the L subgroup, which also had few mutations and a lack of copy number alterations. Patient 3-year survival rates were 91% for the L subgroup, 65% for the IDH group, 50% for the H group, and 36% for the M group (Goeppert et al., 2019). An impressive integrated whole-genome analysis and epigenetic analysis of 489 CCAs from 10 countries carried out by the International Cancer Genome Consortium further demonstrated that an anatomic subset of tumors composed of almost entirely intrahepatic iCCAs, which were mostly liver-fluke negative and characterized by *IDH1/2* and *BAP1* mutations and *FGFR* alterations, had significantly better overall survival relative to other subsets comprised mostly of liver fluke positive tumors (large duct type) and which were enriched in *TP53* mutations and *ERBB2* amplifications (Jusakul et al., 2017). The clinical significance of morpho-molecular subclassification of iCCAs is further highlighted by encouraging phase III clinical trial data supporting the use of IDH1 inhibitors after progression to chemotherapy for the treatment of iCCA with *IDH1* mutations and phase II trial data demonstrating FGFR inhibitors to show therapeutic activity against iCCAs harboring *FGFR2* fusions (Lamarca, Barriuso, McNamara, & Valle, 2020).

3. Fibroinflammatory cholangiopathies, mechanisms, and cholangiocarcinogenesis

Cholangiopathies are a class of liver diseases of various etiologies that specifically affect the biliary tree (Cannito et al., 2018; Fabris, Fiorotto, et al., 2019). Chronic biliary inflammation, bile duct hyperplasia, fibrosis, and cholestasis are common characteristics of cholangiopathies that have been identified as high-risk conditions for cholangiocarcinoma (Cannito et al., 2018; Guicciardi, Trussoni, LaRusso, & Gores, 2020; Roy, Glaser, & Chakraborty, 2019). As already noted, PSC and liver fluke infestations are two of the most important examples of definitive high risk cholangiopathies for which progressive biliary inflammation and fibrosis have been linked to cholangiocarcinoma development. The main features of PSC and liver fluke infestations are described below in the context of their relationship to cholangiocarcinogenesis.

3.1 Primary sclerosing cholangitis

PSC is a rare chronic biliary disease typically diagnosed in the third and fourth decade of life (Lazaridis & LaRusso, 2016; Lindor et al., 2015). PSC is characterized by strong peribiliary fibrosis and inflammation, which can affect both the small and large bile ducts, leading to large bile duct biliary strictures, impaired bile flow, and ultimately to cirrhosis complicated by portal hypertension. Histopathological features of PSC include fibroinflammatory destruction of the interlobular bile ducts with thick concentric fibrosis (onion skinning) associated with progressive bile ductopenia, bile duct hyperplasia, and varying degrees of cholestasis (Cannito et al., 2018; Guicciardi et al., 2020). Based on experimental data and anatomic location, it has recently been proposed that in PSC, senescent cholangiocytes and portal fibroblasts participate mainly in the development of persistent peribiliary fibrosis, whereas ductular reactive cholangiocytes and hepatic stellate cells would be involved in the formation of liver parenchymal fibrosis (Guicciardi et al., 2020).

Patients with PSC have been reported to have a 400- to 1500-fold increased lifetime risk of developing cholangiocarcinoma than those without PSC (Fung, Lindor, & Tabibian, 2019; Roy et al., 2019), with 15–25% of PSC cases resulting in cholangiocarcinoma (Aron & Bowlus, 2009; Buckholz & Brown, 2020; Fung et al., 2019; Karlsen & Boberg, 2013; Karlsen, Schrumpf, & Boberg, 2010; Williamson & Chapman, 2015). The etiology and pathogenesis of PSC remains unclear (Karlsen, Folseraas, Thorburn, & Vesterhus, 2017; Vesterhus & Karlsen, 2020), but several studies have implicated chronic biliary inflammation as a critical component sustaining the pathogenesis of this disease (Yan et al., 2020). The known association of PSC with inflammatory bowel diseases, the changes in the bile and gut microbiotas, together with genome wide association study data demonstrating the presence of genetic variants related to immune pathways (i.e., IL2, IL2RA, CARD9, and REL) all suggest the involvement of innate immune mechanisms in PSC's pathogenesis (Janse et al., 2011; Liu et al., 2013; Mells, Kaser, & Karlsen, 2013; Melum et al., 2011).

The role of changes in the microbiota is still debated (Hov & Karlsen, 2017). Bacteria and bacterial components, also known as pathogen associated molecular patterns (PAMPs), could be transported to the liver through the enterohepatic circulation, stimulating sustained innate immune responses targeting cholangiocytes (Karlsen et al., 2010; Strazzabosco et al., 2018; Trauner et al., 2014). Katt et al. (2013) have further shown that IL-17a is upregulated in livers from patients with PSC and that Th17 cells are increased in peripheral blood from patients with PSC following challenges with candida extracts or toll-like receptor 5 agonists (i.e., flagellin). More recently Tedesco et al. (2018) showed in the *Mdr2*$^{-/-}$ mouse model of PSC that gut dysbiosis leads to the translocation of *Lactobacillus gasseri* to the liver and to the stimulation of IL-17 secretion by γδ TCR^+ cells. Notably, this response was blocked by inhibition of the γδ TCR^+ cell receptor, resulting in attenuation of peribiliary fibrosis. In addition, Nakamoto et al. (2019) reported that the liver of PSC patients is colonized *by Klebsiella pneumonia, Proteus mirabilis, and Enterococcus gallinarum*, stimulating Th17 immune responses and suggesting a possible role of IL-17 and the biliary epithelial cell response to this cytokine in the pathogenesis of PSC.

The persistence of the inflammatory insult in PSC lead to the secretion of numerous fibroinflammatory mediators, in particular of IL-6, interferon (IFN)γ, and tumor necrosis factor (TNF)-α, which, in turn, are able to stimulate inducible nitric oxide synthase (iNOS) expression and activity (Jaiswal, LaRusso, Burgart, & Gores, 2000; Jaiswal, LaRusso, Shapiro, Billiar, & Gores, 2001; Spirlì et al., 2003). Increased nitric oxide (NO) synthesized by iNOS stimulates inflammatory cells present in the portal areas to secrete a wide range of cyto- and chemokines, including chemokine (C-C motif) ligand (CCL) 1, 2, and 3, chemokine (C-X-C motif) ligand 2 (CXCL2), IFNβ, IL-6, IL-8, TGFβ, and TNF-α (Cadamuro, Girardi, Gores, Strazzabosco, & Fabris, 2020). Proliferating cholangiocytes also display increased secretions of proinflammatory cytokines, such as TNF-α, and IL-6, as well as chemokines (monocyte chemoattractant protein 1) and growth factors [i.e., platelet derived growth factor (PDGF), vascular endothelial growth factor (VEGF] (Guicciardi et al., 2020).The local accumulation of NO_2^- generate reactive oxidative species (ROS), such as peroxynitrites ($ONOO^-$), by interacting with superoxidedismutase-derived O_2^- (Fang, 2004; Jaiswal et al., 2000). ROS and NO are involved in the neoplastic transformation inducing lipid peroxidation of the cell membrane, nitrosylation of several proteins, single or double-strand breaks in DNA, and inactivating enzymes involved in

DNA proofreading and repair, such as 8-oxo-deoxyguanine DNA glycosylase 1 (Jaiswal et al., 2000). These mechanisms may favor the accumulation of genetic mutations in biliary epithelial cells, prodromal to neoplastic transformation. Moreover, NO and ROS could generate two mutagenic compounds, 8-oxo-7,8-dihydro-20-deoxyguanosine (8-oxodG) and 8-nitroguanine (Correia da Costa et al., 2014; Jaiswal et al., 2000; Jaiswal et al., 2001). The nitrosylation of cysteine residues of the active site of caspase 3, 8, and 9 could inhibit their activity, making tumor cells resistant to apoptosis. NO accumulation also activates the p38 MAPK/JNK pathway, stimulating cyclooxygenase (COX) 2 to generate prostaglandin E_2, which can directly stimulate the proliferation of cholangiocarcinoma cells through interaction with its receptor EP1, or by transactivating the epidermal growth factor receptor (EGFR) through Src protein (Zhang et al., 2007). The resultant activation of the PI3K/AKT encourages increased cell proliferation and chemoresistance. Carpino, Cardinale, et al. (2019) have further shown that neoplastic ducts of PSC patients are surrounded by an expanded peribiliary plexus and secrete higher amounts of IL-6, IL-8, TGFβ, and VEGF compared to non-neoplastic ducts.

Cholangiocarcinoma developed on a background of PSC is associated with *KRAS* and *TP53* mutations (Ahrendt et al., 2000; Boberg et al., 2000), mutations of the $p16^{INK4a}$ promoter (Taniai, Higuchi, Burgart, & Gores, 2002), and *FGFR2* fusion IDH 1 and 2 mutations (Wang et al., 2013). Another important mechanism in cholangiocarcinogenesis is the deregulation of specific miRNAs involved in cell proliferation, stemness, and migration. IL-6 is, in fact, able to depress the expression of the tumor suppressor NF2, by upregulating miR-let-7a, and of $p16^{INK4a}$ by increasing miR-148a and miR-152 (Braconi, Huang, & Patel, 2010), while stimulating p38 MAPK phosphorylation by reducing miR-370 (Meng, Wehbe-Janek, Henson, Smith, & Patel, 2008).

3.2 Fluke infestations

Liver fluke infestations of *Opisthorchis viverrini* and *Clonochis sinensis* are endemic in several Asian countries (Thailand, Cambodia, Vietnam, Korea, and China), whereas *Opisthorchis felineus* is prevalent in Eastern Europe [Russia, Siberia, Ukraine, Belarus, and Kazakhstan] (Fedorova et al., 2017; Qian, Utzinger, Keiser, & Zhou, 2016; Sripa et al., 2018). The infestation of the biliary network of human patients with *O. viverrini* is due to the ingestion of fluke metacercariae in undercooked or raw fish. Parasites proliferate in bile ducts, releasing eggs that are expelled with stools that contaminate water ways. Here, the eggs are eaten by water snails, the main food source for the fish (Qian et al., 2016; Sripa et al., 2007).

The association between fluke infestation and cholangiocarcinoma is well documented and the International Agency for Research on Cancer (IARC) has included *O. viverrini* and *C. sinensis* as Group I agents/biological carcinogens (van Tong, Brindley, Meyer, & Velavan, 2017). Within the biliary tree, liver flukes generate a strong immune response and induce an exacerbated chronic inflammatory response that is determined by duration of the infection, by number of infesting flukes, and probably by their modulation of the gut and liver microbiome (Plieskatt et al., 2013; Saltykova, Petrov, & Brindley, 2018). This chronic fibroinflammatory cholangiopathy, as exemplified by *O. viverrini* infestation, is characterized by, cholangitis, eosinophilia, and progressive peribiliary fibrosis (Sripa et al., 2018). Reactive biliary epithelial changes in response to *O. viverrini* include cholangiocyte hyperplasia, goblet cell metaplasia, adenomatous hyperplasia, and dysplasia. Extracellular vesicles from *O. viverrini* have been also shown to be actively internalized by human cholangiocytes in vitro, where they promote cell proliferation and stimulate secretion of the proinflammatory cytokine IL-6 (Chaiyadet, Smout, et al., 2015; Chaiyadet, Sotillo, et al., 2015).

The *O. viverrini* excretory/secretory products (OvESP), similarly to PAMPs, stimulate the signal cascade initiated by TLR4 activation that induce the NF-kB-mediated hypersecretion of several cytokines, including IL-6 and IL-8. This mechanism was developed as a host defense to microbial or parasitic infection, because it stimulates the action of COX2 and iNOS, leading to the local generation free oxygen and nitric radicals with antimicrobial activity, but which could also promote the malignant transformation of the cholangiocytes (Prueksapanich et al., 2018). *C. sinensis* infestation could also stimulate liver fibrogenesis by activating both TLR2 and 4, leading to the secretion of a wide range of ILs (1β, 4, 6, and 10), TNF- α, and IFNγ. These fibroinflammatory mediators contribute to the recruitment of inflammatory cells and to the proliferation and activation of resident fibroblasts (Prueksapanich et al., 2018; Sripa et al., 2007) leading to chronic inflammation and CCA development. However, it should also be noted that *O. viverrini* infection alone rarely shows cholangiocarcinoma development (Sripa et al., 2018). In this context, fluke-induced carcinogenesis has some peculiar characteristics. Specifically, opisthorchiasis is characterized by the accumulation of hormone-like molecules, such as oxysterol, a compound derived from the oxidation of conjugated bile acids by cytochrome P450.

Its metabolites mediate pro-tumorigenic responses in bile ducts generating DNA adducts, increasing nitrosylative and oxidative damages due the hyper-expression and activation of iNOS and COX2, and deregulating the proliferation/death ratio of cholangiocytes (Vale, Gouveia, Gärtner, & Brindley, 2020). Similar effects were also induced by catechol estrogen quinone-like hormones of fluke origin, which were able to inhibit the DNA editing action of 8-oxodG and 8-nitroguanine (Correia da Costa et al., 2014). In addition, opisthorchiasis may enhance colonization of the biliary tree by species of *Helicobacter*, including *H. pylori*, a known GI cancer risk factor (Sripa et al., 2018).

Lastly, using a whole genome sequencing approach, Chan-on et al. (2013) depicted the mutational patterns of iCCA on a background of *O. viverrini* infection relative to iCCA without fluke infestation. In a cohort of 108 patients, these authors reported that non-*O. viverrini* related iCCAs showed a higher incidence of mutation in *BAP1*, *IDH1* and *IDH2* genes, while *O. viverrini*-related iCCAs were characterized by a low frequency of *IDH1/2* mutations and a significantly higher frequency of *TP53* and *SMAD4* mutations. Consistent with the findings of Chan-on et al., and as previously noted, Jusakul et al. (2017) had also identified a cluster of liver fluke-associated iCCAs uniquely enriched in *TP53* mutations (and ERBB2 amplifications) together with a low frequency of expression of *BAP1* and *IDH1/2* mutations. Conversely, these investigators also found fluke-negative cholangiocarcinomas to be characterized by increased *BAP1* and *IDH1/2* mutations, together with *FGFR2* gene rearrangements. Collectively, these molecular genomic findings indicate that non-*O. viverrini* cholangiocarcinomas and *O. viverrini*-positive tumors have distinct mutational profiles, strongly implying that different etiologies may induce distinct mutational landscapes even within the same tumor type.

4. The desmoplastic reaction and iCCA progression

Irrespective of etiology (sporadic versus known risk factors), iCCAs are most often characterized by a prominent desmoplastic reaction typified by the formation of a hard, stiff, and hypovascularized scaffold comprised of a dense fibro-collagenous enriched matrix along with other ECM proteins, and containing an abundance of activated CAFs positive for α-SMA (a biomarker of myofibroblast differentiation), together with variable numbers inflammatory and vascular cell types (Fig. 1, Banales et al., 2020; Brivio et al., 2017; Cadamuro et al., 2018; Sirica & Gores, 2014). In recent years,

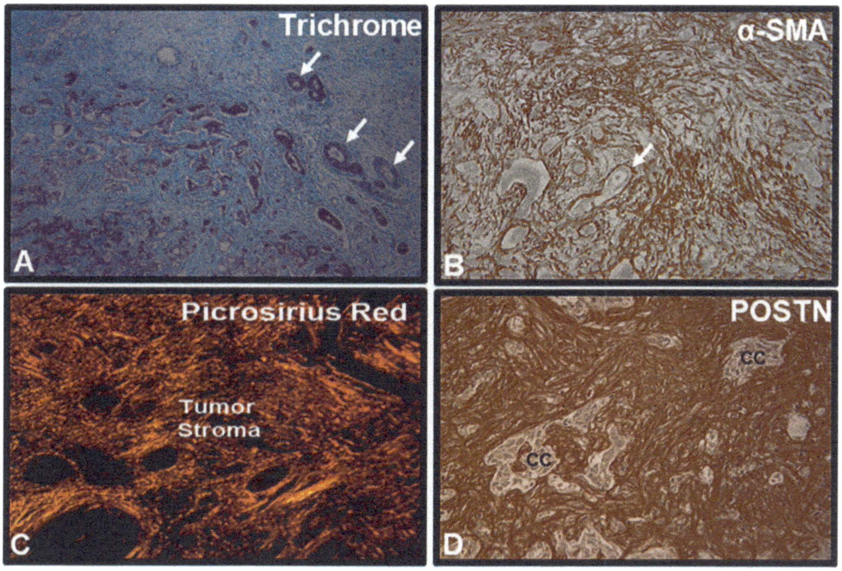

Fig. 1 Representative histological images illustrating characteristic features of the desmoplastic microenvironment in human iCCA. (A) Masson trichrome staining of a moderately to poorly-differentiated mass-forming iCCA demonstrating the tumor to be largely comprised of a prominent desmoplastic stroma strongly stained for collagen (blue staining). Arrows point to representative small clusters of cholangiocarcinoma. (B) CAFs comprising the vast majority of cell types populating the desmoplastic stroma of iCCA are seen to be strongly immunoreactive for α-smooth muscle actin (α-SMA), a biomarker of myofibroblast differentiation, whereas cholangiocarcinoma cells (arrow) are negatively stained for α-SMA. (C) Picrosirius red staining for collagen (orange-staining under polarized light) typically reveals the extracellular matrix of desmoplastic CCA to be comprised of thick collagen fiber bundles that are largely comprised of collagen type I. (D) Immunostaining for matricellular periostin (Postn), produced by α-SMA +CAFs and which has a binding site for collagen, is exclusively localized to the desmoplastic stroma of iCCA. **cc**, cholangiocarcinoma. Increased numbers of α-SMA +CAFs within the iCCA microenvironment together with corresponding strong stromal immunoreactivity for Postn have been shown to be predictors of poor survival outcomes for iCCA patients following curative-intent surgical resection.

the desmoplastic stroma of iCCA, also known as tumor reactive stroma or TRS (Brivio et al., 2017), has become the focus of a growing research interest due to its prognostic potential and possibilities for therapeutic targeting. Moreover, there is now a growing body of evidence to convincingly implicate the tumor reactive microenvironment in iCCA as playing an active and crucial role in promoting and or modulating the major hallmarks of cancer (Hanahan & Weinberg, 2011). The stromal cellular and ECM components of iCCA, which frequently comprises the bulk of the tumor, provide critical molecular and biomechanical signals for perpetuating cholangiocarcinoma cell proliferation, promoting iCCA progression, facilitating iCCA invasion and metastasis, resisting apoptotic cell death, regulating iCCA angiogenesis and lymphangiogenesis, mediating epigenome and metabolic reprogramming, enabling resistance to chemo- and targeted agent therapies, and avoiding immune destruction (Affo et al., 2017; Banales et al., 2020; Brivio et al., 2017; Cadamuro et al., 2019, 2018; Fabris, Perugorria, et al., 2019; Mancinelli et al., 2019; O'Rourke, Lafuente-Barquero, & Andersen, 2019; Pant, Richard, Peixoto, & Gradilone, 2020; Sirica, 2012; Sirica & Gores, 2014).

A critical aspect of the relationship between the desmoplastic reaction in iCCA and increased malignant behavior is the deleterious interplay between cholangiocarcinoma cells and α-SMA + CAFs, which are considered the most important host-derived stromal cells within the tumor microenvironment. Activated CAFs in iCCA are a major source of secreted growth factors [e.g., hepatocyte growth factor (HGF), PDGFs (-BB & -D), heparin-binding epidermal growth factor (HB-EGF)], chemokines/cytokines [e.g., IL-6, TGF-β, stromal derived factor-1 (SDF-1), also known as CXCL12,] ECM proteins [e.g., fibrillar collagens, fibronectin], matricellular proteins [e.g., periostin (Postn), tenascin-C, and osteopontin], proteinases [e.g., metalloproteinases (MMPs)-2 and -9], and angiogenic modifiers [e.g., VEGFs A & C, thrombospondin-1], which collectively function to promote or modulate cholangiocarcinoma cell behavior and progression (Affo et al., 2017; Banales et al., 2020; Brivio et al., 2017; Cadamuro et al., 2019, 2013, 2018; Clapéron et al., 2013; Fingas et al., 2011; Gentilini et al., 2018; Labib et al., 2019; Nissen, Karsdal, & Willumsen, 2019; Sirica, 2012; Sirica et al., 2009; Sirica & Gores, 2014; Vaquero, Aoudjehane, & Fouassier, 2020). Exosome crosstalk (Goulet et al., 2018; Yang, Li, Zou, & Zhu, 2019), exchange of metabolites, such as pyruvate and lactate (Fiori et al., 2019; Yoshida, Azuma, Miura, & Orimo, 2019), activation of mechanosensitive signaling cascades associated with activation of Yes-associated protein (YAP)/transcriptional coactivator with PDZ-binding motif (TAZ), desmoplastic ECM stiffness, and hypoxia (Banales et al., 2020; Sahai et al., 2020), and aberrant integrin, sonic hedgehog, and Wnt/β-catenin signaling (Banales et al., 2020; Gascard & Tlsty, 2016; Labib et al., 2019; Utispan et al., 2012) are also important to the interactive relationships that exist between cancer cells and CAFs, as well as other cell types of the iCCA tumor reactive microenvironment.

4.1 CAF origins and heterogeneity in iCCA

CAFs in iCCA are phenotypically heterogenous and are likely derived from various cell lineages (Affo et al., 2017; Sirica, Campbell, & Dumur, 2011; Sirica & Gores, 2014; Vaquero et al., 2020), although the actual cellular origins of these tumor stromal cells remains unclear due largely to a lack of in vivo cell fate tracing data. Moreover, CAFs are not normal fibroblasts, and their differentiation and phenotypic biomarker profiles can be influenced by the tumor microenvironment and molecular cross talk with cancer cells. However, based on what is known of fibrogenic cell responses in cirrhosis and fibroinflammatory/cholestatic liver diseases like PSC, it is generally believed that the most likely major source of CAFs in iCCA are likely originated from resident hepatic nonparenchymal cells, notably hepatic stellate cells and activated portal fibroblasts (Affo et al., 2017; Manzanares et al., 2017; Okabe et al., 2016; Vaquero et al., 2020).

Taking into account the limitations of using potential phenotypic biomarkers to identify the cell origins of CAFs in iCCA, such analyses can still be of value in suggesting possible cell lineage relationships, as well as potential prognostic relevance. Okabe et al. (2009) first reported that hepatic stellate cell markers (i.e., desmin, glial fibrillary acidic protein) were expressed in α-SMA + CAFs in human iCCA, suggesting a hepatic stellate cell origin. In contrast, Itou et al. (2019) more recently showed that α-SMA + CAFs in the stroma of primary human iCCAs expressed portal fibroblast markers [i.e., fibulin-2, thymus cell antigen-1 (Thy-1], and except for a small subset, were largely negative for fascin, a hepatic stellate cell biomarker. It was also reported by these investigators that α-SMA + CAFs in metastatic lymph nodes from iCCA patients, while positive for Thy-1 and negative for fascin, were also negative for fibulin-2. On the other hand, hepatic stellate cells in human cirrhotic liver were found to be negative for both Thy-1 and fibulin-2. In addition, Itou et al. also detected a small number of α-SMA + CAFs in the primary human iCCAs that were immunoreactive for bone-marrow derived fibrocyte biomarkers (CD34, CD45), suggesting that bone marrow derived cells may also contribute to some population of iCCA α-SMA + CAFs. Overall the findings by Itoh et al. imply that CAFs at the primary site of iCCA are similar to activated portal fibroblasts, but different from hepatic stellate cells and from CAFs in metastatic lymph nodes. Further, small subsets of CAFs may have also been derived from bone marrow derived fibrocytes, and in the case of the metastases, from resident cells of lymph nodes.

Manzanares et al. (2017) further showed by transcriptomic analysis and by immunophenotyping that α-SMA + CAFs purified from a rat orthotopic desmoplastic iCCA prominently overexpressed portal fibroblast biomarkers (e.g., gremlin 1, fibulin-2, Thy-1, cofilin-1, NTPDase2, elastin), but were negative for desmin, and lacked biomarkers for bone marrow derived fibrocytes or hematopoietic stem cells. Moreover, in this study, only a small number of desmin-positive fibroblastic cells were detected in the desmoplastic stroma of iCCA from which the α-SMA + CAFs were isolated from, as well as in human iCCA stroma. Biomarker and karyotypic analyses further supported a resident liver fibroblastic origin of the orthotopic iCCA-derived α-SMA + CAFs and argued against epithelial-mesenchymal transitional (EMT) as a source of these myofibroblastic cells. Lineage cell tracing has also presented evidence against the possibility of epithelial mesenchymal transition of neoplastic cholangiocytes as being a likely contributing source for α-SMA + CAFs accumulating within desmoplastic stroma of iCCA (Cadamuro et al., 2013).

It is also not known to what extent other cells, such as endothelial cells, may give rise to iCCAs. Recently, Zhang et al. (2020) using a droplet-based single-cell RNA sequencing platform to profile single cells from human iCCA identified six distinct fibroblast clusters, five of which were mainly enriched in the iCCA tissues and one primarily present in the adjacent tissues. All six subclusters expressed high levels of *ACTA2* that encodes for α-SMA, in addition to other canonical CAF markers. Interestingly, one of these subclusters (vCAFs), which accounted for 57.6% of the iCCA fibroblast population, was characterized by microvascular signature genes, including CD146 (an endothelial cell marker), as well as high levels of IL-6 expression. CD146 expression has also been demonstrated in human liver endothelial cells and pericytes surrounding the portal vasculature (Strauss, Phillips, Ruggiero, Bartlett, & Dunbar, 2017), as well as in mouse liver Thy-1-positive periportal fibroblasts and hepatic stellate cells, although at more variable levels in the hepatic stellate cells (Katsumata, Miyajima, & Itoh, 2017). Without cell fate tracing, however, it is not possible to conclude that vCAFs were derived from resident portal fibroblasts, hepatic stellate cells or subsets of hepatic endothelial cells and vascular associated pericytes.

In addition to α-SMA, other biomarkers used to delineate CAFs in iCCA include fibroblast specific protein 1 [FSP-1, also known as S100A4], fibroblast activating factor [FAP] and platelet-derived growth factor receptor-β [PDGFR-β] (Itou et al., 2019; Sha et al., 2018; Vaquero et al., 2020; Zhang et al., 2017). In a number of independent studies, iCCA patients whose tumor stroma was enriched in α-SMA + CAFs were demonstrated to have significantly shorter overall survival and worse recurrence-free survival rates following surgical resection than those whose iCCAs expressed low or negative levels of stromal α-SMA (Chuaysri et al., 2009; Okabe et al., 2009; Sha et al., 2018). High expression of FAP in iCCA has also been reported to be predictive of poor prognosis in iCCA patients (Yang et al., 2016). Zhang et al. (2017) have further shown histological categorization of CAFs in human iCCA to be a potentially useful predictor for prognosis, with surgically resected iCCA patients whose tumors showed a more normal fibroblastic-like morphology (termed mature CAFs) having a significantly better overall survival than those that were characterized by a myofibroblast morphology (termed immature CAFs). In this study, high stromal α-SMA expression was found to be associated with poor iCCA differentiation, but surprisingly, not with improved overall survival. This discrepancy may be due to a relatively small sample size and use of a different immunohistochemistry scoring system than those used by others.

Survival rates of iCCA resected patients with stromal CAFs immunoreactive for the transmembrane glycoprotein podoplanin (D2-40-positive myofibroblasts) were found to be significantly lower than those of patients without podoplanin expressing CAFs (Aishima et al., 2008). Podoplanin has also been shown to be a potentially useful stromal prognostic marker for perihilar cholangiocarcinoma (Obulkasim et al., 2018). Furthermore, CD10 (neprilysine), a membrane bound metalloprotease expressed in CAFs, may be more involved with the progression of perihilar and extrahepatic CCAs than of iCCAs (Nishihara et al., 2009). High tissue and serum Postn, which in both human and rat is produced solely by α-SMA + CAFs in iCCA tissue (Dumur, Campbell, DeWitt, Oyesanya, & Sirica, 2010; Utispan et al., 2010) have also been demonstrated to be a particularly promising predictor of poor prognosis in human iCCA patients (Utispan et al., 2010; Thuwajit et al., 2017) and increased malignant tumor grade and progression in a well established syngeneic rat model of desmoplastic iCCA (Manzanares, Campbell, Maldonado, & Sirica, 2018). In this same rat model, high accumulation of α-SMA + CAFs in the tumor stroma was also found to be associated with increased iCCA tumor grade and enhanced malignant aggressiveness (Sirica et al., 2011).

4.2 PDGF-D and TGF-β and the desmoplastic reaction in iCCA

PDGF-D and TGF-β play key roles in driving the recruitment and expansion of CAFs in iCCA. Cadamuro et al. (2013) have presented compelling data supporting a model of CAF recruitment into iCCA based on a PDGF-D-mediated paracrine fibroblastic cell recruitment by cholangiocarcinoma cells. In this model, PDGF-D, which is up-regulated in cholangiocarcinoma cells by a hypoxia-mediated mechanism, is secreted into the tumor cell microenvironment, where it in its active ligand form binds to and activates its cognate receptor PDGFRβ expressed by resident liver fibroblastic cells/myofibroblasts and CAFs. This interaction, in turn, promotes proliferation and elicits a strong migratory response in these cells via activation of Rho GTPases (i.e., Rac1, Cdc42) and JNK, thereby providing a mechanism for CAF recruitment in iCCA as illustrated in Sirica et al. (2019).

It has also been shown that PDGF-D, and also PDGFBB, each of which are abundantly produced by cholangiocarcinoma cells, can sensitize activated CAFs to apoptotic cell death triggered by BH3 mimetics, such as navitoclax and ABT-199 (Rizvi et al., 2014). Similar apoptotic sensitization was demonstrated in co-cultures of myofibroblasts and cholangiocarcinoma cells. Furthermore, this study also showed that PDGF-linked apoptotic priming of CAFs occurs via Puma-mediated Bak activation, which can then be converted to full-blown apoptosis induced by navitoclax or ABT-199. Notably, navitoclax or ABT199 have been demonstrated to significantly reduce iCCA tumor burden in mouse (Rizvi et al., 2014) and rat cholangiocarcinoma models (Cadamuro et al., 2019; Mertens et al., 2013) by depleting CAFs from the tumor stroma. Navitoclax treatment was also demonstrated to be associated with decreased lymphatic vascularization and lymph node metastasis in the rat cholangiocarcinoma model (Cadamuro et al., 2019).

TGF-β plays a key role in the acquisition of the CAF phenotype and is well known inducer of myofibroblast differentiation (i.e., α-SMA induction) and fibrosis (Caja et al., 2018; Gascard & Tlsty, 2016). Exosomal TGF-β has also been identified as a molecular mechanism involved in CAF activation (Goulet et al., 2018). In relation to iCCA, TGF-β, expressed in both cholangiocarcinoma cells and in α-SMA + CAFs, was demonstrated to be essential for provoking a dramatic desmoplastic-like reaction in vitro (prominent overproduction of dense fibrocollagenous matrix concomitant with significantly enhanced accumulation and proliferation of α-SMA + CAFs) within the matrix of a three-dimensional organotypic cholangiocarcinoma culture model that closely resembled the desmoplastic features of the in situ tumor. This model was established by co-culturing of cholangiocarcinoma cells derived from a syngeneic rat iCCA with α-SMA + CAFs exhibiting portal fibroblast biomarkers, obtained from the same tumor type, within a dilute collagen type I hydrogel (Manzanares et al., 2017). Three-dimensional co-culturing of the activated CAFs with the cholangiocarcinoma cells was further found to markedly increase mature TGF-β production within the gel cultures over that elaborated by CAFs alone, to cause marked gel contraction (shrinkage), an indicator of increased substratum stiffness, and to promote significantly increased cholangiocarcinoma cell growth and elevated expression of phenotypic markers of malignant progression (i.e., Muc1). Cholangiocarcinoma spheroid ductal-like structures which formed in the gel cultures were further observed to become increasingly more anaplastic, as well as invasive when co-cultured in the presence of α-SMA+CAFs (Campbell, Dumur, Lamour, DeWitt, & Sirica, 2012; Manzanares et al., 2018, 2017; Sirica et al., 2019). TGF-α, an epidermal growth factor receptor (EGFR) ligand expressed in iCCA, was further determined to be a key factor for further enhancing cholangiocarcinoma anaplasia, hyperproliferation, and higher malignant grading in this 3-D culture model (Manzanares et al., 2017).

The interplay between activated CAFs and cholangiocarcinoma cells through TGF-β expression and EGFR activation in relation to cholangiocarcinoma progression is further emphasized by the interesting findings of Clapéron et al. (2013), who showed that cholangiocarcinoma cells produced TGF-β1, which, in turn, induced HB-EGF expression in CAFs leading to paracrine activation of EGFR on the cancer cells, promoting cholangiocarcinoma cell invasion. Of further interest, TGF-β1 expression in cholangiocarcinoma cells was also found to be enhanced by HB-EGF stimulation. These results support a model as first proposed by Clapéron et al. depicting a reciprocal paracrine loop between cholangiocarcinoma cells and CAFs through the HD-EGF/EGFR axis that contributes to cholangiocarcinoma progression and also triggers TGF-β1 production in cholangiocarcinoma cells, which stimulates CAF activation and functions. TGF-β1 has also been demonstrated to promote iCCA tumor growth and metastasis in a rat cholangiocarcinoma model (Huang et al., 2016).

Mesothelin (Msln) is a glycosylphosphatidylinositol-anchored cell membrane glycoprotein that is expressed in liver portal myofibroblasts (Fausther, Lavoie, & Dranoff, 2017; Koyama et al., 2017), as well as shown to be highly overexpressed in human and rat cholangiocarcinoma cells, but less prominently expressed or not detected in CAFs in desmoplastic human or rat iCCA (Manzanares et al., 2018, 2017; Tang, Qian, & Ho, 2013; Yu et al., 2010). Msln, however, has been detected by Western blotting in α-SMA+CAFs expressing portal myofibroblast biomarkers, including Thy-1, which were derived from a syngeneic rat iCCA, but expressed at a lower level than Msln protein amounts detected in cell lysates from intrahepatic and extrahepatic metastatic cholangiocarcinoma cells obtained from the same iCCA model (Manzanares et al., 2018).

Msln overexpression has been suggested to be a prognostic factor for patients with intrahepatic cholangiocarcinoma (Nomura et al., 2013) and more recently demonstrated to predict malignant progression in the rat iCCA model described above (Manzanares et al., 2018). In relation to cholestatic liver fibrosis associated with early activation of portal fibroblasts, Koyama et al. (2017) have shown Msln to regulate TGF-β1-inducible activation of portal fibroblasts by disrupting the formation of an inhibitory Thy-1-TGFβR1 complex, as well as to facilitate fibroblast growth factor-mediated proliferation of these cells. However, while it may be assumed, it remains to be determined if Msln effects Thy-1 and TGFβR1 in portal fibroblast-derived CAFs in cholangiocarcinoma in a similar manner. Of added interest is the identification of two distinct molecular weight forms of Msln being expressed in rat cholangiocarcinoma cells in vivo and in 3-dimensional culture, notably a 40 kDa form (mature form), which was associated with a cytoplasmic and diffuse cell membrane immunostaining pattern, and whose increased expression predicted a more aggressive iCCA phenotype compared with a more heavily glycosylated 50 kDa form that was expressed predominantly at the apical luminal surface of well differentiated cholangiocarcinoma ducts and which predicted a less aggressive iCCA phenotype. Complementary to these findings, it could be further shown in 3-dimensional culture that co-culturing α-SMA + CAFs with cholangiocarcinoma cells contributed to a significant increase in the production of the 40 kDa form of Msln, which correlated with an increase in cholangiocarcinoma cell anaplasia and disruption of cholangiocarcinoma "ductal-like" polarity (Manzanares et al., 2018). In comparison, highly tumorigenic rat cholangiocytes harboring mutationally activated rat *neu* oncogene (homolog of *erbB2*) were found to primarily express the 40 kDa form of Msln, which was associated with enhanced malignant cell features (e.g., higher malignant grade, disruption of polarized morphogenesis and underexpression of cell polarity regulating proteins, oncogenic *neu* activation with corresponding overexpression of phosphorylated Erk1/2 and AKT, and invasiveness) even when cultured in the absence of CAFs (Manzanares et al., 2018; Yao, Qian, Campbell, Wang, & Sirica, 2018). The relationship between Msln isoforms, CAFs, and cholangiocarcinoma progression is illustrated in Fig. 2. Further studies are needed now to extend these preclinical findings to human iCCA.

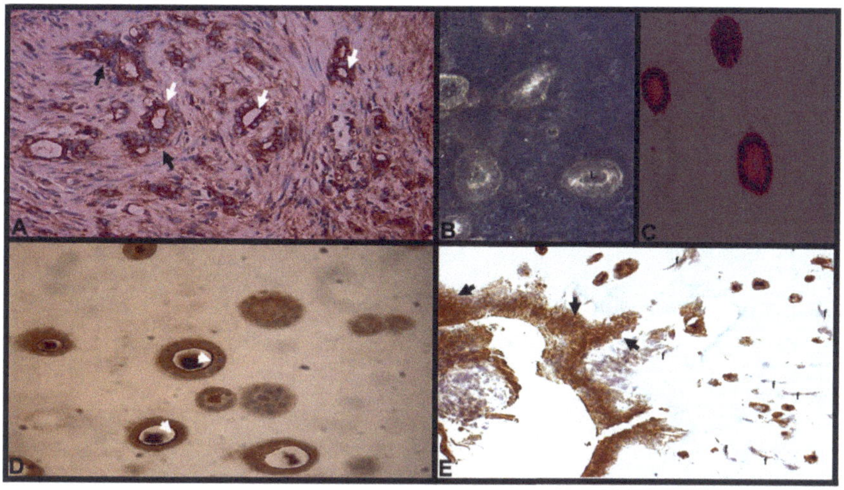

Fig. 2 Relationship between mesothelin (Msln) immunostaining pattern, disruption of polarized morphogenesis, and invasiveness in a syngeneic orthotopic rat tumor model and corresponding 3-dimensional (3-D) organotypic culture model of cholangiocarcinoma progression. (A) Immunohistochemical demonstration of luminal Msln immunoreactivity expressed in differentiated cholangiocarcinoma ductal structures (white arrows) contrasted with diffuse cell membrane/cytoplasmic Msln immunoreactivity (black arrows) expressed in invasive cholangiocarcinoma emanating from the ductal structures in an orthotopic rat iCCA (TDE_{CC} iCCA). (B) Phase contrast image of viable spheroids and well-differentiated ductal-like structures with distinct lumens (L) formed in 3-D organotypic culture of a cholangiocarcinoma cell strain (TDE_{CC} cells) derived from rat TDE_{CC} iCCA. (C) Hematoxylin & eosin stained section demonstrating polarized ductal-like structure formed from rat TDE_{CC} cholangiocarcinoma cells in 3-D culture. (D) In the absence of CAFs, polarized ductal-like structures formed from TDE_{CC} cholangiocarcinoma cells in 3-D culture exhibit strong luminal surface immunoreactivity for Msln (white arrow heads). (E) In sharp contrast, polarized morphogenesis is dramatically disrupted when TDE_{CC} cholangiocarcinoma cells are maintained in 3-D co-cultured with α- SMA+CAFs, also derived from a rat TDE_{CC} iCCA. Under these co-culture conditions, the cholangiocarcinoma cells can organized into hyperproliferative, invasive structures exhibiting diffuse cell membrane/cytoplasmic immunoreactivity for Msln (black arrows). In both the orthotopic tumor and organotypic culture models, a more heavily glycosylated 50-kDa form of Msln was demonstrated to be associated with the apical surface pattern of expression of Msln, whereas a 40-kDA Msln form was shown to be associated with the diffuse cell membrane/ cytoplasmic immunostaining pattern. Data based on results presented in Manzanares et al., 2018. Images in 2A and 2E are slightly modified versions of those published as Fig. 2B and Fig. 7B, respectively, in Manzanares et al., 2018.

TGF-β is also a known inducer of α-SMA + CAF-derived Postn, which is highly germane to supporting and sustaining iCCA's desmoplastic microenvironment and promoting ICCA invasiveness (Sirica, Almenara, & Li, 2014; Utispan et al., 2012) Moreover, Postn along with other matricellular proteins, such as tenascin-C and Msln, may be playing a pivotal role in conditioning metastatic niches and the formation of metastases (Coelho et al., 2020; González-Gonzáles & Alonzo, 2018; Lowy & Oskarsson, 2015; Ma, Tang, Esser, Pastan, & Xia, 2012). Postn, like TGF-β, has also been shown to stimulate liver fibrogenesis by activating lysyl oxidase in hepatic stellate cells, inducing the expression of collagen type I and fibronectin, and stimulating the phosphorylation of SMAD2/3, as well as of focal adhesion kinase and AKT in hepatic stellate cells (Kumar et al., 2018). A novel TGF-β induced long noncoding RNA has further been demonstrated to promote an inflammatory microenvironment in human intrahepatic cholangiocarcinoma (Merdrignac et al., 2018). In addition, preclinical targeting of TGF-β in rat cholangiocarcinoma models has been demonstrated attenuate the desmoplastic reaction generated by co-culturing α-SMA + CAFs with cholangiocarcinoma cells in 3-dimensional culture (Manzanares et al., 2017) and to reverse pre-existing fibrosis induced by thioacetamide in vivo (Ling et al., 2013), leading in both cases to a significant reduction in cholangiocarcinoma.

4.3 The matrisome, tumor stiffness, and hypovascularity in iCCA

In a recent (albeit limited) analysis by Carpino, Overi, et al. (2019) to characterize the matrisome or ECM proteome of human iCCA, the proteomic profile revealed a unique signature of ECM proteins in iCCA when compared with that of paired samples of non-cancerous liver tissue (NCT). As expected, the iCCA ECM exhibited high levels of fibrillar collagen type I components (COL1A1 and COL1A2) and of the matricellular protein Postn, which is known to interact with and regulate collagen fibrillogenesis (Norris et al., 2012; Sirica et al., 2014), as well as bind to other ECM proteins (i.e., tenascin-C) promoting iCCA malignant behavior (Sirica et al., 2014). Col3A1, a component of fibrillar collagen type III was also observed to be overexpressed in 6 of 9 analyzed samples of iCCA ECM when compared to NCT and further shown to be a component of tumor-associated aligned collagen fibers. Also, in contrast to the NCT, the iCCA ECM was characterized by an almost total absence of elastic fibers and a low expression of proteoglycans and downregulation of laminins, type IV collagen, and basement membrane-specific heparan sulphate proteoglycan. The angiogenesis factors angiopoietin-related protein 6 and sushi repeat protein X-linked 2 were also seen to be dramatically downregulated in the iCCA samples. Taken together, these findings are compatible with an increased tumor stiffness, a breakdown of basement membrane leading to disruption of cholangiocarcinoma cell polarity, and a hypovascularized iCCA stroma.

Increased tumor stiffness, as exemplified by desmoplastic epithelial cancers like iCCA not only affects malignant cell progression by triggering cell survival, proliferation and motility signaling cascades within the cancer cells (Kalli & Stylianopoulos, 2018; Noguchi, Saito, & Nagase, 2018; Sahai et al., 2020), but also impacts on surrounding stromal cells, whereby tumor stiffness mechanoactivates YAP/TAZ in fibroblastic cells, leading them to acquire a CAF phenotype (Noguchi et al., 2018; Yoshida et al., 2019). Moreover, YAP activation is enhanced by ECM matrix stiffness, thus establishing a feed-forward self-reinforcing loop to maintain the CAF phenotype. Increased mechanical stress associated with tumor stiffness produced by desmoplasia can also lead to collapse of micro-blood vessels within the tumor, leading to hypovascularization, hypoxia, and reduced bioavailability for therapeutic drugs (Sahai et al., 2020). Enhanced expression of thrombospondin-1, immunohistochemically localized to both CAFs cholangiocarcinoma cells in iCCA, has also been reported to be related to diminished microvessels in human iCCA (Aishima et al., 2002; Kawahara et al., 1998; Tang et al., 2006). It is also relevant here that thrombospondin-1 is an activator of latent TGF-β (Murphy-Ullrich & Suto, 2018).

5. Immune milieu in iCCA

5.1 The innate immune system in iCCA

The iCCA tumor microenvironment is populated to various degrees by cells of the innate immune response, including macrophages (in particular M2 or alternatively activated macrophages), neutrophils, and natural killer (NK) cells. In the TRS of iCCA, two populations of macrophages are present, those derived from Kupffer cells localized to the perisinusoidal space of Disse, and TAMs.

TAMs are $CD86^+/CD206^+$ M2 macrophages, mainly derived from $CD14^+/CD16^+$ circulating monocytes, which play a major role in suppressing T lymphocyte activity and proliferation, thereby having a pronounced anti-inflammatory action (Rőszer, 2015). M2 TAMs promote angiogenesis, induce tissue remodeling, and stimulate the apoptosis of M1 macrophages (Duluc et al., 2007; Rőszer, 2015). Monocytes accumulated within TRS are promoted by tumor cells through the local secretion of a variety of cytokines, including IL-1β, IL-4, IL-8, IL-10, IL-13, colony-stimulating factor 1, and growth factors (i.e., VEGF-A) and through the secretion of CCL2 by Kupffer cells (Li et al., 2018).

In iCCA, TAMs mainly localize to the tumor invasive front, with high counts of M2-TAMs in iCCA patients shown to correlate with poor disease-free survival (Hasita et al., 2010) and poor overall survival in surgically resected iCCA patients (Sun et al., 2020). M2 TAMs could, in fact, sustain tumor invasive ICCA growth and promote metastasis by various mechanisms, including (1) via the release of proangiogenic growth factors (e.g., angiopoietin 1, IL-8, VEGF-A) (Corliss, Azimi, Munson, Peirce, & Murfee, 2016; Leyva-Illades, McMillin, Quinn, & Demorrow, 2012), (2) secretion of soluble factors that stimulate the migration of neoplastic cells (e.g., VEGF, fibroblast growth factor (FGF)-1/-2, PDGF, and TGF-β) (Roy et al., 2019), and (3) through the action of MMP-9, a key metalloproteinase involved in tumor spread (Raggi et al., 2017). Furthermore, TAMs promote cholangiocarcinoma cell proliferation by secreting IL-6 and VEGF into the tumor microenvironment, which, in turn, can stimulate the activation of the p42/p44 MAPK and AKT/mTOR signaling pathways, (Banales et al., 2016; Roy et al., 2019).TAMs may drive iCCA progression by their increased expression of WNT ligands, such as WNT3a and WNT7b (Boulter et al., 2015; Loilome et al., 2014; Saito et al., 2018) and enzymes, such as COX2 and iNOS (Chariyalertsak et al., 2001; Wójcik et al., 2012). ECM modification in iCCA may also modulate the accumulation of M2 TAMs within the tumor microenvironment. Notably, Postn has been shown to stimulate the migration of TAMs in vitro and to spatially correlate with M2 TAM density in primary human iCCA tissue. (Zeng et al., 2018).

Tumor associated neutrophils (TANs) represent another cell population belonging to the tumor reactive microenvironment that is of increasing importance in cancer (Jaillon et al., 2020). $CD66b^{+}$ neutrophils are recruited through the local secretion of CXCL5 (Zhou et al., 2014), but very little is known about the pathogenic mechanisms of these cells in iCCA. Current data show that accumulation of TANs within the TRS of iCCA is predictor of poor prognosis (Mao, Zhu, Xiong, Ren, & Bai, 2015), and that a high neutrophil-to-lymphocyte ratio is an independent predictor of poor survival outcome and insensitivity to chemotherapy and radioembolization (Filippi et al., 2020; Lin et al., 2016; Tan et al., 2016; Wang et al., 2018). One may speculate that accumulation of TANs contributes to the generation of a microenvironment prone to immunosuppression that favors neoplastic cell growth through the local release of CCL2 and CCL17 and recruitment of M2 macrophages.

NK cells, immunophenotypically recognized as $CD3^{-}CD56^{+}$ cells, represent between 30% and 40% of the total normal human liver lymphocyte population (Björkström, Ljunggren, & Michaëlsson, 2016; Martín-Sierra et al., 2019). Their biological function is to protect against bacterial invasion via the release of cytotoxic granules, which also produce lytic death of tumor cells. In iCCA, NK are recruited in the tumor area in response to the secretion by cholangiocarcinoma cells of CXCL9, a INFγ-dependent chemokine. Notably, the ablation of CXCL9 in a mouse model of iCCA was observed to lead to a reduced accumulation of NK cells in peritumoral area, resulting in an increased volume of the tumor (Fukuda et al., 2020). In a xenotransplant model of human cholangiocarcinoma, the infusion of mice with NK cells was also shown to significantly reduce tumor growth as a result of the lytic and cytotoxic actions of the infused cells (Jung et al., 2018). The mechanisms regulating the antitumor effects of NK cells in iCCA are not well understood, but may involve the secretion of several immunomodulatory mediators, such as adenosine, PGE_2, and TGF-β (Melaiu, Lucarini, Cifaldi, & Fruci, 2020).

Dendritic cells (DCs) represent a bridge between innate and adaptive immune response due to their ability to function as antigen presenting cells to prime $CD4^+$ and $CD8^+$ T-lymphocytes. $CD83^+$ DCs accumulated at the invasive front of iCCA have been shown to significantly correlate with the number of $CD4^+$ and $CD8^+$ T effector cells in the cancerous region, leading to a significantly lower incidence of lymph node metastasis and a better survival outcome for the $CD83^+$ patients than that of CD83-negative patients (Takagi et al., 2004). Suppression of the expression of receptors for TGF-β and IL10 (immunosuppressive cytokines) on DCs was further demonstrated to result in an increased activation of the adaptive immune response, consequently leading to an enhanced secretion of IFNγ and increased cytolytic action of T-lymphocytes against cholangiocarcinoma cells (Thepmalee et al., 2020; Thepmalee, Panya, Junking, Chieochansin, & Yenchitsomanus, 2018). Furthermore, a limited number of studies on small cohorts of cholangiocarcinoma patients have yielded promising results when coupling chemotherapy (i.e., with gemcitabine) with immunotherapy using protein or RNA-pulsed DCs (Higuchi et al., 2006; Junking, Grainok, Thepmalee, Wongkham, & Yenchitsomanus, 2017; Kobayashi et al., 2013; Sawasdee et al., 2020).

5.2 Adaptive immune system and tumor-infiltrating lymphocytes in iCCA

The adaptive immune response, which involves tumor infiltrating lymphocytes (TILs) and their modulating effects on iCCA growth and spread, is becoming more apparent. This is particularly highlighted by the fact that reduced recruitment of TILs or their inactivation has been shown to correlate with worse patient survival outcomes and increased tumor growth and malignant aggressiveness (Zheng et al., 2020). In an analysis of human iCCA biopsies, TILs accounted for about 8.0% of the inflammatory cells present in the tumor sample (Martín-Sierra et al., 2019). $CD4^+$ T cells have been shown to be mainly localized at the tumor invasive front of iCCA, whilst $CD8^+$ T cells were usually, but not exclusively, recruited into the tumor mass (Kasper, Drebber, Stippel, Dienes, & Gillessen, 2009). Patients with iCCA whose tumors exhibited an increased presence of $CD3^+$ and $CD8^+$ infiltrating T cells were also found to exhibit higher overall and disease free survival rates, as well as a lower tumor recurrence risk after curative-intent tumor resection (Tian et al., 2019; Tian et al., 2020; Vigano et al., 2019). In contrast, patients whose iCCAs were characterized by increased infiltration of $Foxp3^+$ regulatory T cells (Tregs) exhibited worse overall survival rates (Vigano et al., 2019), likely due to the ability of Tregs to secrete IL-10 and TGF-β, cytokines able repress the cytotoxic activity of $CD8^+$ T and of NK cells in the tumor microenvironment (Tu, Chia, & Chen, 2014). $CD20^+$ B lymphocytes are unfrequently detected in iCCA (Kasper et al., 2009), but their presence also seems to be predictive of a better overall survival when compared with iCCA patients in which B cells were absent (Goeppert et al., 2013).

One of the salient mechanisms used by the cancer cells to survive is to escape immune surveillance of immune cells by modulating the expression of several immune checkpoints, such as programmed death-1 (PD-1) and its ligand PD-L1, or cytotoxic T-lymphocyte antigen-4 (CTLA-4). PD-L1 is variably expressed by cholangiocarcinoma cells, (Kitano et al., 2020; Sabbatino et al., 2016). The rational for the use of a PD-1 based immunotherapy in iCCA is that cholangiocarcinoma cells could promote their survival by inducing depletion of PD-1-expressiong T cells through the PD-L1/PD-1 route. Moreover, CTLA-4 is expressed by Tregs that are known to depress the activity of DCs by binding to its specific surface receptor CD80, thus blocking the cytotoxic action of DCs (Goeppert et al., 2013). In addition, it was shown that targeting of PD-1 and CTLA-4 expressed by iCCA-derived TILs could induce, ex vivo, an increased activation of tumor-infiltrating T cells (Zhou et al., 2019). Notwithstanding this, data regarding the importance of PD-L1 expression in cholangiocarcinoma cells are conflicting. In conflict with studies implicating the expression of this ligand as an independent prognostic factor for predicting poor overall survival of iCCA patients (Kitano et al., 2020; Sabbatino et al., 2016), a recent met-analysis of 1066 cholangiocarcinoma patients subdivided in 11 studies showed that the association among PD-L1 expression and overall or disease free survival was not significant (Xu et al., 2019).

To date, various clinical trials for iCCA based on PD-1/PD-L1 modulation are now in the recruiting phase (e.g., NCT04157985, NCT04440943, NCT02834013, NCT04295317, NCT03111732), but no data are yet available. As alluded to in a previous section of this review, Job et al. (2020) reclassified ICCA on the basis of the different inflammatory infiltrates into 4 subtypes that included immune desert (non-inflammatory), lymphoid, myeloid, and mesenchymal types. The lymphoid (immunogenic) subtype, which represented 11% of the total iCCAs analyzed, was characterized by a massive T-lymphocyte infiltration, an activation of inflammatory and immune checkpoint pathways, and was associated with the best patient survival outcome when compared with the other subtypes. Job et al. (2020) further suggested that this distinct inflamed iCCA subtype could be amenable for treatment with checkpoint blocking immunotherapy due to the unique composition of the inflammatory infiltrate.

6. Challenges and clinical implications

Despite improvements in the standard of care and multidisciplinary management of iCCA, the overall global 5-year survival of patients with iCCA has not changed appreciably over the past 20–30 years, currently reported as $\leq$10% (Dhanasekaran et al., 2013; Mody et al., 2018; Simile et al., 2019). As already noted in the Introduction, the majority of patients with iCCA (70–80%) present with advanced disease where therapeutic options are limited to palliative treatments. Gemcitabine plus cisplatin remains the standard of care as a first line systemic treatment for non-resectable cholangiocarcinoma. This palliative treatment, however, provides only a modest survival benefit (Buettner et al., 2017; Plentz & Malek, 2016; Vienot & Neuzillet, 2019). Furthermore, iCCAs, which most often develop sporadically in the absence of known risk factors, are asymptomatic in their early stage. Even in patients with known risk factors, early diagnosis is problematic, and further, most patients that are diagnosed with late disease (locally advanced or metastatic iCCA) usually present with ambiguous or nonspecific symptoms.

Currently there are no curative medical treatments for iCCA and the only hope for a potentially curative therapy is complete surgical resection. However, only a minority of iCCA patients (i.e., 10% to ~25%) are eligible for curative-intent surgical resection. (Banales et al., 2020; Buettner et al.,

2017; Ma, Hwang, et al., 2020). Moreover, even with curative-intent resection, an estimated median disease free survival in various patient series has been reported to range from 12 to 36 months (Rizvi et al., 2018), with overall 5-year survival rates after resection ranging between 20% and 35% in some series (Chun & Javle, 2017; Ma, Hwang, et al., 2020; Mavros, Economopoulos, Alexiou, & Pawlik, 2014; Waisberg et al., 2018).

The BILCAP phase 3 randomized trial (NCT00363584) aimed at evaluating capecitabine as an adjuvant treatment for biliary tract cancers of all types (including iCCA) following curative-intent surgical resection did not meet its end point of improving overall survival, but demonstrated a median overall survival of 51.1 months in the capecitabine group versus 36.4 months in the observational group not receiving capecitabine (Primrose et al., 2019). The PRODIGE12-ACCORD 18-UNICANCER GI phase 3 trial (NCT01313377), which evaluated adjuvant gemcitabine and oxiplatin in resected biliary tract cancers of all types, failed to show any demonstrable benefit (Edeline et al., 2019).

Molecular profiling of iCCA has revealed a complex and evolving genomic and epigenomic landscape characterized by extensive inter-tumor and intra-tumor heterogeneity that has been attributed in part to diverse multifactorial etiologies, histological tumor differences, adverse stromal microenvironments, and evolving tumor progression (Banales et al., 2020; Goeppert et al., 2019; Jusakul et al., 2017; Ma et al., 2019; Pellino et al., 2018; Rhee et al., 2018; Sirica et al., 2019). This inter-tumor and intra-tumor heterogeneity serves as a major obstacle to the development of molecular targeted and immune-based therapies for iCCA. It has also hindered the development of non-invasive serological tests for iCCA early diagnosis, as well as for monitoring iCCA treatment responses and prognosis. Research aimed at developing noninvasive serological tests based on multiplexing of discreet cholangiocarcinoma cell and tumor stromal markers that may include tumor exosomes, mRNAs, and distinct protein biomarkers linked to malignant progression, such as Muc1, Postn and Msln, could be highly useful towards advancing this clinical need.

The results from earlier clinical trials aimed molecular therapeutic targeting of altered signaling pathways in iCCA, including EGFR, ERBB2, MET, PDGFR, VEGF, and VEGFR have, to date, been disappointing (Labib et al., 2019; Mertens, Rizvi, & Gores, 2018; Pellino et al., 2018; Simile et al., 2019). Direct therapeutic inhibition of activated *KRAS* has also proved elusive (Mertens et al., 2018). However, with the advent of next-generation sequencing technology to identify oncogenic driver mutations, there is now a powerful tool to identify novel molecular targets and to allow personalized targeted approaches to iCCA therapy (Dabney, Khalife, Shahid, & Phan, 2019; Mertens et al., 2018; Pellino et al., 2018).

The value of the personalized approach to iCCA targeted therapy is exemplified by the rapid translation of FGFR2 fusion/translocation and IDH1/2 mutations, respectively, into promising therapeutic targets for particular subsets of iCCA patients (Dabney et al., 2019; Krook et al., 2020; Pellino et al., 2018). Mutated FGFR2 has been described in approximately 10–23% of patients with iCCAs (Dabney et al., 2019; Krook et al., 2020; Nakamura et al., 2015; Pellino et al., 2018; Simile et al., 2019). Sia et al. (2015), using RNA- and exome-sequencing analysis, reported a higher rate of FGFR2 fusions (45%) in a cohort of 107 resected iCCA patients, which represented the most recurrent targetable alteration. IDH1/2 mutations also commonly occur in patients with iCCA at reported rates of 13% up to 30% (Abou-Alfa et al., 2020; Goeppert et al., 2019; Lowery et al., 2019; Mertens et al., 2018; Pellino et al., 2018).

The efficacy of pemigatinib, a selective oral FGFR 1,2, and 3 inhibitor, was recently investigated in FIGHT-202 (NCT02924376), a multicenter open-label single arm phase II study that included a cohort of 107 cholangiocarcinoma patients with locally advanced or metastatic cholangiocarcinoma harboring FGFR2 gene rearrangements/fusions and who had received at least one systemic cancer therapy that did not include selective FGFR inhibitors. Among these 107 patients, permigatinib elicited an overall response rate of ~36% (38 of 107 patients), with 3 patients having a complete response. Kaplan-Meier estimates of overall survival further demonstrated a 12-month survival rate of 68% in the permigatinib-treated cholangiocarcinoma patients with FGFR2 fusions or rearrangements compared to 23% for treated cholangiocarcinoma patients with other FGF/FGFR alterations and 13% for those with no FGF/FGFR alterations (Abou-Alfa et al., 2020). Based on these results, the FDA fast tracked the approval of pemigatinib for the treatment of adult patients with previously treated, unresectable, locally advanced or metastatic cholangiocarcinoma with FGFR2 fusions or rearrangements, making it the first targeted agent FDA approved for cholangiocarcinoma therapy. Of further note, in a limited study involving six patients with advanced FGFR2 fusion-positive iCCA, treatment with TAS-120 (an irreversible FGFR inhibitor) was shown to overcome acquired resistance that had developed in these patients to previous treatments with the ATP-competitive FGFR inhibitors BGJ39 or Debio (Goyal et al., 2019).

ClarIDHy, an international multicenter, randomized, double-blind, placebo-controlled, phase III study, evaluated the therapeutic efficacy and safety of ivosidenib (AG-120), a small molecule targeted inhibitor of mutated IDH1, when administered to patients with chemotherapy-refractory advanced IDH1 mutant iCCA (Abou-Alfa, Macarulla, et al., 2020). This clinical trial demonstrated encouraging therapeutic benefits for the IDH-1 mutant iCCA patients over those treated with matched placebo. Specifically, in the ivosidenib-treated group, progression free survival was reported to be 32% at 6months and 22% at 12months, versus no patients in the placebo group being free from progression at 6months or more.

The complex microenvironment of desmoplastic iCCA and the deleterious interplay between activated CAFs, inflammatory and immune cells, ECM, and vascular components clearly provides a rationale for devising novel therapeutic strategies aimed at targeting stromal components of iCCA. Activated CAFs, by virtue of being the most prominent cell component and driver of the desmoplastic reaction, are emerging as an appealing target for iCCA therapy. As previously cited, preclinical results with TGF-β inhibitors and with BH3 mimetics offer proof of concept that depleting CAFs and ablating the desmoplastic stroma can potentially have significant therapeutic efficacy for iCCA. In further support of this concept, photothermal depletion of CAFs preferentially loaded with hybrid iron oxide-gold nanoparticles was recently shown to normalize tumor stiffness and produce tumor regression in desmoplastic human EGI-1 cholangiocarcinoma formed in a mouse xenograft model (Nicholás-Boluda et al., 2020). The FDA-approved antifibrotic drug nintedanib has also been recently demonstrated to reduce xenografted iCCA growth and activated CAFs expressing α-SMA, and further, combinational treatment with nintedanib and gemcitabine against CAFs and iCCA cells was shown to exhibit the strongest inhibition of tumor growth compared with control and single-treatment groups (Yamanaka et al., 2020).

Another potential strategy suggested for targeting CAFs for clinical benefit relates to developing treatments that would cause activated CAFs to revert to a phenotype comparable to that of normal fibroblasts, as exemplified by the targeting of the vitamin D receptor in pancreatic cancer (Sahai et al., 2020), or by the normalizing effect of dasatinib, a FDA approved inhibitor of PDGFR, on CAFs from primary lung cancer (Haubeiss et al., 2010). Targeting specific tumor promoting factors produced by CAFs, such as Postn, SDF-1, or FAP may also provide a potentially useful strategy for iCCA therapy (Liepelt & Tacke, 2016; Lin et al., 2019; Sirica et al., 2014).

Activated CAFs have also emerged as central players in immune regulation that shapes the immune microenvironment, including that of iCCA

(Fabris, Perugorria, et al., 2019; Monteran & Erez, 2019). FAP^+ CAFs have been shown to be a major source of CCL2, which promotes recruitment of myeloid-derived suppressor cells (MDSC) and immunosuppression via a STAT3-CCL2 signaling. FAP, phospho-STAT3 and CCL2 are expressed in the desmoplastic stroma of iCCA (Yang et al., 2016). Furthermore, TGF-β and VEGF, each of which are hypersecreted into the iCCA microenvironment, are both potent suppressors of the tumor immune response (Batlle & Massagué, 2019; Fukumura, Kloepper, Amoozgar, Duda, & Jain, 2018; Ma et al., 2019; Thepmalee et al., 2018).

As alluded to earlier in this Review, a majority of desmoplastic iCCAs have been found to exhibit a sub-optimal response rate of less than 10% to immune check point blockade monotherapy (Loeuillard et al., 2020). This is consistent with data already cited that most iCCAs are characterized by an immune desert phenotype (Job et al., 2020), with a non-T cell infiltrated tumor immune microenvironment. Loeuillard et al. (2020) have recently shown that TAMs are the major source of PD-L1 in murine and human cholangiocarcinomas and dual inhibition of TAMs and granulocytic-MDSC potentiated PD-1 blockade in a murine cholangiocarcinoma model.

With respect to other gastrointestinal cancer types, but relevant to iCCA, combined targeting of TGF-β and PD-L1 receptors has been shown to promote T-cell mediated tumor regression and improved survival in a genetic mouse model (KPC) of advanced pancreatic ductal adenocarcinoma (Principe et al., 2019). When tested against the highly immunogenic mouse MC38 colon adenocarcinoma model, combinational inhibition of TGF-β signaling and PD-L1 blockade was also found to be significantly more effective than corresponding single agent treatments in improving survival, which was associated with an influx of $CD8^+$ T-cells in the tumor microenvironment (Sow, Ren, Camps, Ossendorp, & ten Dijke, 2019). In addition, anti-PD-L1 therapy was demonstrated to sensitize RT2-PNET murine pancreatic neuroendocrine tumors to antiangiogenic therapy, and conversely to improve anti-PD-L1 therapy specifically though the generation of high endothelial venules in the tumor, which in turn, promoted cytotoxic T-cell infiltration, activity, and tumor destruction (Allen et al., 2017). Of added interest, inhibition of the ROS-producing enzyme NOX4, which is upregulated in activated CAFs, has most recently been shown in murine colorectal (as well as lung and breast tumor models), to normalize CAF myofibroblastic cells to a quiescent state. This effect, in turn, was found to re-sensitize the CAF-rich tumors to PD-1 check point inhibition by overcoming activated CAF-mediated $CD8^+$ T-cell exclusion (Ford et al., 2020).

As implied by the findings described above, therapeutic strategies that incorporate targeting of the tumor stroma and novel immunotherapies in conjunction with standardized chemotherapy or targeted agents against cholangiocarcinoma cells offer the possibility of increased survival outcomes and possibly even cures, particularly when considered as a second line treatment for distinct subsets of iCCA patients with recurrent disease. Mou et al. (2018) ostensibly was the first to describe in a case report the complete remission of a postoperatively recurred metastatic PD-L1-positive iCCA with a high tumor mutational burden following combination treatment with the anti-PD-1 antibody pembrolizumab together with chemotherapy with oxiplatin, and tegafur. A significant response to anti-PD-1 based immunotherapy with nivolumab administered in combination with the multi-kinase tyrosine kinase inhibitor lenvatinib was also demonstrated against recurrent iCCA with bone metastasis (Chen et al., 2019). The patient's initial primary iCCA, which had a high tumor mutation burden and actionable mutations in *KIT*, *NRAS*, *TP53*, *MET*, and *PDGFR*, recurred at 5 months after surgery and adjuvant chemotherapy. Following combinational treatment with nivolumab plus lenvatinib, the recurred liver tumor completely resolved and the bone metastases were stable after 9 months of therapy.

Bintrafusp alpha is a bifunctional fusion protein composed of a monoclonal antibody against PD-L1 fused to a TGF-β "trap" (Lind et al., 2020). In various preclinical mouse carcinoma models, bintrafusp alpha, previously designated M7824, demonstrated superior anti-tumor activity over that produced by treatments with anti-PD-L1 antibody alone or TGF-β trap alone by a mechanism involving simultaneous blockade of immunosuppressive PD-L1, as well as of TGF-β in the tumor microenvironment (Lan et al., 2018). In this study, treatment with bintrafusp alpha was also shown to reduce α-SMA expression and stromal collagen fiber density in syngeneic EMT-6 mouse breast tumors. In addition, combinations of bintrafusp alpha with chemotherapy (oxaliplatin/5-fluorouracil) or radiotherapy was demonstrated to be effective at improving antitumor efficacy and enhancing $CD8^+$ T-cell activation in various preclinical mouse carcinoma models (Lan et al., 2018; Lind et al., 2020).

An open-label Phase I clinical trial of bintrafusp (MSB0011359C) has been completed, showing early signs of efficacy in patients with heavily pretreated advanced solid cancers without prior immune checkpoint inhibitor treatment (Strauss et al., 2018). More recently, Yoo et al. (*2020*) described results from an expansion cohort of a Phase I open-label trial of bintrafusp in 30 Asian patients with metastatic or locally advanced biliary tract cancers. Included in this trial were 10 patients with iCCA. The objective response rate for the iCCA patients treated with bintrafusp was reported to be 30%, with one patient showing a complete response. While these findings are encouraging, they are limited by the fact that this study lacked of a comparator group, had a small patient enrollment, and did not include non-Asian patients as part of the cohort. Multiple clinical trials are currently ongoing involving the use of bistrafusp alpha as a monotherapy or in combination with other immunotherapeutics, chemotherapy, or radiation therapy in patients with a variety of solid cancer types, including breast, prostate, biliary, and pancreas (Lind et al., 2020). The results of these trials should hopefully provide new insights concerning the effectiveness of targeting the tumor microenvironment as a strategy for enhancing immunotherapies and other treatment modalities against desmoplastic cancers, such as iCCA.

A variety of combination regimens with PD-1/PD-L1 immune checkpoint inhibitors (ICI) are also now being tested in ongoing key clinical trials against various advanced gastrointestinal malignancies, including biliary cancers (Wang, Lin, et al., 2019). In this context, there is increasing preclinical and clinical trial evidence to support combined antiangiogenic therapy and immunotherapy with ICIs as a promising strategy for advanced solid cancer treatment (Fukamura, Kloepper, Amoozgar, Duda, & Jain, 2018; Song et al., 2020). VEGF, which is abundantly expressed in iCCA, impairs dendritic cell maturation, resulting in reduced T-cell priming, as well as stimulates expansion of Tregs, M2-TAMs and MDSCs. These actions, in turn, lead to an immunosuppressive tumor microenvironment and weakened anti-tumor immune response, together with dysregulating tumor neovascularization associated with impaired T cell infiltration and cell death (Fukumura et al., 2018; Kudo, 2020). Anti-VEGF agents have been shown to reprogram the tumor milieu from an immunosuppressive to an immune permissive microenvironment by normalizing the tumor vasculature, decreasing the activity of MDSCs, Tregs, and TAMs, and by suppressing dendritic cell maturation (Fukamura, Kloepper, Amoozgar, Duda, & Jain, 2018; Kudo, 2020; Song et al., 2020). Immune checkpoint blockade has also been shown to facilitate anti-angiogenesis by downregulating VEGF and alleviating hypoxia (Song et al., 2020). These findings, in turn, have served as the basis for multiple clinical trials, many of which have demonstrated improved anti-cancer efficacy and prolonged survival when anti-angiogenic agents were combined with ICIs in the treatment of variety of advanced carcinoma types (Fukumura et al., 2018; Song et al., 2020).

The therapeutic benefit of combining anti-angiogenic therapy with ICIs is exemplified by the results of the phase III IMbrave 150 study-NCT03434379 (Cheng et al., 2019; Kudo, 2020; Nakano et al., 2020). This study evaluated combinational therapy with the anti-PD-L1 antibody atezolizumab + the anti-VEGF antibody bevacizumab for unresectable HCC without prior systemic therapy versus sorafenib monotherapy (the standard of care for advanced HCC). The results revealed that atezolizumab + bevacizumab-treated group had a significantly improved overall response rate and progression free survival compared to the sorafenib-treated group. Moreover, 6% of the HCC patients treated with the atezolizumab + bevacizumab combination were found to be in complete remission. The patients in the combination group also reported a considerably better quality of life than those in the sorafenib group.

In July 2019, the FDA granted a Breakthrough Therapy designation for lenvatinib, a multikinase inhibitor targeting VEGFR 1–3, FGFR 1–4, PDGFR-α, RET, and KIT) in combination with the PD-1inhibitor pembrolizumab for the first-line treatment of patients with unresectable HCC, based on the results from the phase 1b KEYNOTE-524/Study 116 trial-NCT03006926 (Finn et al., 2020; Nakano et al., 2020). An ongoing double-blind randomized controlled phase III study of lenvatinib plus pembrolizumab as a first-line treatment for HCC versus lenvatinib plus placebo (NCT03713593) to further assess the efficacy and safety of this combination in patients with HCC is now ongoing (Finn et al., 2020), the results of which are also likely to have potential implications for cHCC-CCA and iCCA, notably those formed in association with viral hepatitis and cirrhosis.

Other suggested considerations for targeting of tumor stroma and immune checkpoints in iCCA include the use of BH3 mimetics such as navitoclax to deplete activated CAFs, together with PD-L1 inhibition to reconstruct T-cell response and anti-tumor immunity (Mertens et al., 2018). A number of phase I/II clinical trials (e.g., NCT02341625, NCT03644550, NCT03371381) assessing targeted therapies for Msln expressing cancers based on antibody-based drugs (8BMS-98,614; LMB-100) or vaccines (live attenuated *Lysteria monocytogenes*) in combination with immune checkpoint inhibitors (nivolumab; pembrolizumab) are currently recruiting or ongoing. In addition, a limited number of clinical trials (e.g., NCT03615313; NCT03030001) for targeted therapy of Msln positive advanced solid tumors with PD-1 antibody expressing mesothelin specific chimeric antigen receptor T (CAR-T) cells have also been reported to be in development (Lv & Li, 2019). In support of this strategy, Lal et al. (2019) have recently demonstrated that CAR-T cells targeting Msln and secreting PD-L1 antibodies enhanced antitumor efficacy when compared with Msln-targeted CAR-T cells in xenograft mouse models of mesothelioma, pancreatic, and ovarian cancers. The results of these trials are awaited with anticipation, since they may have important implications for the treatment Msln-expressing iCCA.

Last, a better understanding of cellular heterogeneity and functions of the desmoplastic reaction in iCCA is needed inorder to successfully exploit its potential as a therapeutic target while avoiding any possibility of accelerating malignant aggressiveness. Multiple CAF subtypes demonstrating inter- and intra-tumoral heterogeneity have been demonstrated in human desmoplastic pancreatic desmoplastic adenocarcinoma, which may account for inconsistencies in preclinical results and the failure of some stromal targeted agents (Neuzillet et al., 2019). Two separate classes of CAFs, one expressing myofibroblastic features (α-SMA) and the other expressing inflammatory fibroblast features (i.e., IL-6 secretion) have also been shown to coexist in the desmoplastic stroma of pancreatic cancer and to show distinct transcriptional profiles (Öhlund et al., 2017). Furthermore, as previously alluded to, the reactive microenvironment in primary iCCA is distinct from that of metastatic iCCA. In this context, careful attention needs to be given to advancing the development of in vivo models closely reproducing the morpho-molecular pathology and TRS of early, advanced, and metastatic human iCCA in an effort to precisely identify and test primary and adjuvant combinational therapeutic strategies against advanced disease that can predictively translate into more novel systemic targeted agent and immunotherapeutic treatments for iCCA patients, as well as for those with other more common desmoplastic cancers.

Grant Support

NIH grant R01DK101528 to M.S., and support from the Clinical and Translational Core of the Yale Liver Center- grant DK034989, Silvio Conte Digestive Disease Centers.

Conflict of interest statement

The authors (A.E.S., M.S., and M.C.) have no financial or personal disclosures relevant to this manuscript.

References

Abou-Alfa, G. K., Macarulla, T., Javle, M. M., Kelley, R. K., Lubner, S. J., Adeva, J., et al. (2020). Ivosidenib in IDH1-mutant, chemotherapy-refractory cholangiocarcinoma (ClarIDHy): A multicentre, randomized, double-blind, placebo-controlled, phase 3 study. *Lancet Oncology*, *21*, 796–807.

Abou-Alfa, G. K., Sahai, V., Hollebecque, A., Vaccaro, G., Melisi, D., Al-Rajabi, R., et al. (2020). Pemigatinib for previously treated, locally advanced or metastatic cholangiocarcinoma: A multicentre, open-label, phase 2 study. *Lancet Oncology*, *21*, 671–684.

Affo, S., Yu, L.-Y., & Schwabe, R. F. (2017). The role of cancer-associated fibroblasts and fibrosis in liver cancer. *Annual Review of Pathology: Mechanisms of Disease*, *12*, 153–186.

Ahrendt, S. A., Rashid, A., Chow, J. T., Eisenberger, C. F., Pitt, H. A., & Sidransky, D. (2000). p53 overexpression and K-ras gene mutations in primary sclerosing cholangitis-associated biliary tract cancer. *Journal of Hepato-Biliary-Pancreatic Surgery*, *7*, 426–431.

Aishima, S., Nishihara, Y., Iguchi, T., Taguchi, K., Taketomi, A., Maehara, Y., et al. (2008). Lymphatic spread is related to VEGF-C expression and D2-40-positive myofibroblasts in intrahepatic cholangiocarcinoma. *Modern Pathology*, *21*, 256–264. https://doi.org/10.1038/modpathol.3800985.

Aishima, S., Taguchi, K., Sugimachi, K., Asayama, Y., Nishi, H., Shimada, M., et al. (2002). The role of thymidine phosphorylase and thrombospondin-1 in angiogenesis and progression of intrahepatic cholangiocarcinoma. *International Journal of Surgical Pathology*, *10*, 47–56.

Akita, M., Fujikura, K., Ajiki, T., Fukumoto, T., Otani, K., Azuma, T., et al. (2017). Dichotomy in intrahepatic cholangiocarcinomas based on histological similarities to hilar cholangiocarcinomas. *Modern Pathology*, *30*, 986–997.

Akita, M., Sofue, K., Fujikura, K., Otani, K., Itoh, T., Ajiki, T., et al. (2018). Histological and molecular characterization of intrahepatic bile duct cancers suggest an expanded definition of perihilar cholangiocarcinoma. *HPB*, *21*, 226–234.

Allen, E., Jabouille, A., Rivera, L. B., Lodewijckx, I., Missiaen, R., Steri, V., et al. (2017). Combined antiangiogenic and anti-PD-L1 therapy stimulates tumor immunity through HEV formation. *Science Translational Medicine*, *9*(385), eaak9679.

Aosasa, S., Maejima, T., Kimura, A., Nishiyama, K., Edo, H., Shinmoto, H., et al. (2015). Intrahepatic cholangiocarcinoma with lymphoepithelioma-like carcinoma components not associated with Epstein-Barr virus: Report of a case. *International Surgery*, *100*, 689–695.

Ariizumi, S., Kotera, Y., Katagiri, S., Nakano, M., Nakanuma, Y., Saito, A., et al. (2014). Long-term survival of patients with cholangiolocellular carcinoma after curative surgery. *Annals of Surgical Oncology*, *21*, 451–458.

Ariizumi, S., & Yamamoto, M. (2015). Intrahepatic cholangiocarcinoma and cholangiolocellular carcinoma in cirrhosis and chronic viral hepatitis. *Surgery Today*, *45*, 682–687.

Aron, J. H., & Bowlus, C. L. (2009). The immunobiology of primary sclerosing cholangitis. *Seminars in Immunopathology*, *31*, 383–397.

Bae, J. Y., Park, Y. N., Nakanuma, Y., Lee, W. J., Kim, Y., Park, C., et al. (2002). Intestinal type cholangiocarcinoma of intrahepatic large bile duct associated with hepatolithiasis—A new histologic subtype for further investigation. *Hepato-Gastroenterology*, *49*, 628–630.

Bagante, F., Weiss, M., Alexandrescu, S., Marques, H. P., Aldrighetti, L., Maithel, S. K., et al. (2018). Long-term outcomes of patients with intraductal growth sub-type of intrahepatic cholangiocarcinoma. *HPB*, *20*, 1189–1197.

Balitzer, D., Joseph, N. M., Ferrell, L., Shafizadeh, N., Jain, D., Zhang, X., et al. (2019). Immunohistochemical and molecular features of cholangiolocellular carcinoma are similar to well-differentiated intrahepatic cholangiocarcinoma. *Modern Pathology*, *32*, 1486–1494.

Banales, J. M., Cardinale, V., Carpino, G., Marzioni, M., Andersen, J. B., Invernizzi, P., et al. (2016). Expert consensus document: Cholangiocarcinoma: Current knowledge and future perspectives consensus statement from the European Network for the Study of Cholangiocarcinoma (ENS-CCA). *Nature Reviews. Gastroenterology & Hepatology*, *13*, 261–280.

Banales, J. M., Marin, J. J., Lamarca, A., Rodrigues, P. M., Khan, S. A., Roberts, L. R., et al. (2020). Cholangiocarcinoma 2020: The next horizon in mechanisms and management. *Nature Reviews Gastroenterology & Hepatology*, *17*, 557–588.

Batlle, E., & Massagué. (2019). Transforming growth factor-β signaling in immunity and cancer. *Immunity*, *50*, 924–940.

Beal, E. W., Tumin, D., Moris, D., Zhang, X.-F., Chakedis, J., Dilhoff, M., et al. (2018). Cohort contributions to trends in the incidence and mortality of intrahepatic cholangiocarcinoma. *HepatoBiliary Surgery and Nutrition*, 7, 270–276.

Benhamouche, S., Curto, M., Saotome, I., Gladden, A. B., Liu, C.-H., Giovannini, M., et al. (2010). *Nf2*/Merlin controls progenitor homeostasis and tumorigenesis in the liver. *Genes & Development*, *24*, 1718–1730.

Bertuccio, P., Malvezzi, M., Carioli, G., Hashim, D., Boffetta, P., El-Serag, H. B., et al. (2019). Global trends in mortality from intrahepatic and extrahepatic cholangiocarcinoma. *Journal of Hepatology*, *71*, 104–114.

Björkström, N. K., Ljunggren, H. G., & Michaëlsson, J. (2016). Emerging insights into natural killer cells in human peripheral tissues. *Nature Reviews Immunology*, *16*, 310–320.

Blechacz, B. (2017). Cholangiocarcinoma: Current knowledge and new developments. *Gut and Liver*, *11*, 13–26.

Blechacz, B., Komuta, M., Roskams, T., & Gores, G. J. (2012). Clinical diagnosis and staging of cholangiocarcinoma. *Nature Reviews Gastroenterology & Hepatology*, *8*, 512–522.

Boberg, K. M., Schrumpf, E., Bergquist, A., Broomé, U., Pares, A., Remotti, H., et al. (2000). Cholangiocarcinoma in primary sclerosing cholangitis: K-ras mutations and Tp53 dysfunction are implicated in the neoplastic development. *Journal of Hepatology*, *32*, 374–380.

Bösmüller, H., Pfefferle, V., Bittar, Z., Scheble, V., Horger, M., Sipos, B., et al. (2018). Microvessel density and angiogenesisin primary hepatic malignancies: Differential expression of CD31 and VEGFR-2 in hepatocellular carcinoma and intrahepatic cholangiocarcinoma. *Pathology Research and Practice*, *214*, 1136–1141.

Boulter, L., Guest, R. V., Kendall, T. J., Wilson, D. H., Wojtacha, D., Robson, A. J., et al. (2015). WNT signaling drives cholangiocarcinoma growth and can be pharmacologically inhibited. *The Journal of Clinical Investigation*, *125*, 1269–1285.

Braconi, C., Huang, N., & Patel, T. (2010). MicroRNA-dependent regulation of DNA methyltransferase-1 and tumor suppressor gene expression by interleukin-6 in human malignant cholangiocytes. *Hepatology*, *51*, 881–890.

Bragazzi, M. C., Ridola, L., Safarikia, S., Di Matteo, S., Costantini, D., Nevi, L., et al. (2018). New insights into cholangiocarcinoma: Multiple stems and related cell lineages of origin. *Annals of Gastroenterology*, *31*, 42–55.

Brivio, S., Cadamuro, M., Strazzabosco, M., & Fabris, L. (2017). Tumor reactive stroma in cholangiocarcinoma: The fuel behind cancer aggressiveness. *World Journal of Hepatology*, *9*, 455–468.

Brunt, E., Aishima, S., Clavien, P.-A., Fowler, K., Goodman, Z., Gores, G., et al. (2018). cHCC-CCA: Consensus terminology for primary liver carcinomas with both hepatocytic and cholangiocytic differentiation. *Hepatology*, *68*, 113–126.

Buckholz, A. P., & Brown, R. S., Jr. (2020). Cholangiocarcinoma: Diagnosis and management. *Clinics in Liver Disease*, *24*, 421–436.

Buettner, S., van Vugt, J. L. A., IJzermans, J. N. M., & Koerkamp, B. G. (2017). Intrahepatic cholangiocarcinoma: Current perspectives. *OncoTargets and Therapy*, *10*, 1131–1142.

Cadamuro, M., Brivio, S., Mertens, J., Vismara, M., Moncsek, A., Milani, C., et al. (2019). Platelet-derived growth factor-D enables liver myofibroblasts to promote tumor lymphangiogenesis in cholangiocarcinoma. *Journal of Hepatology*, *70*, 700–709.

Cadamuro, M., Girardi, N., Gores, G. J., Strazzabosco, M., & Fabris, L. (2020). The emerging role of macrophages in chronic cholangiopathies featuring biliary fibrosis: An attractive therapeutic target for orphan diseases. *Frontiers in Medicine*, 7, 115.

Cadamuro, M., Nardo, G., Indraccolo, S., Dall'Olmo, L., Sambado, L., Moserle, L., et al. (2013). Platelet-derived growth factor-D and Rho GTPases regulate recruitment of cancer-associated fibroblasts in cholangiocarcinoma. *Hepatology*, *58*, 1042–1053.

Cadamuro, M., Stecca, T., Brivio, S., Mariotti, V., Fiorotto, R., Spirli, C., et al. (2018). The deleterious interplay between tumor epithelia and stroma in cholangiocarcinoma. *Biochimica et Biophysica Acta-Molecular Basis of Disease*, *1864*(4 Pt. B), 1435–1443.

Cai, X., Li, H., & Kaplan, D. E. (2020). Murine hepatoblast-derived liver tumors resembling human combined hepatocellular-cholangiocarcinoma with stem cell features. *Cell & Bioscience*, *10*, 38. https://doi.org/10.1186/s13578-020-00395-2.

Caja, L., Dituri, F., Mancarella, S., Caballero-Diaz, D., Moustakas, A., Giannelli, G., et al. (2018). TGF-β and the tissue microenvironment: Relevance in fibrosis and cancer. *International Journal of Cancer*, *19*, 1294.

Campbell, D. J. W., Dumur, C. I., Lamour, N. F., DeWitt, J. L., & Sirica, A. E. (2012). Novel organotypic culture model of cholangiocarcinoma progression. *Hepatology Research*, *42*, 1119–1130.

Cannito, S., Milani, C., Cappon, A., Parola, M., Strazzabosco, M., & Cadamuro, M. (2018). Fibroinflammatory liver injuries as preneoplastic conditions in cholangiopathies. *International Journal of Molecular Sciences*, *19*, 3875.

Cardinale, V. (2019). Classification and misclassification in cholangiocarcinoma. *Liver International*, *39*, 260–262.

Cardinale, V., Bragazzi, M. C., Carpino, G., Di Matteo, S., Overi, D., Nevi, L., et al. (2018). Intrahepatic cholangiocarcinoma: Review and update. *Hepatoma Research*, *4*, 20.

Cardinale, V., Carpino, G., Reid, L., Gaudio, E., & Alvaro, D. (2012). Multiple cells of origin in cholangiocarcinoma underlie biological, epidemiological, and clinical heterogeneity. *World Journal of Gastrointestinal Oncology*, *4*, 94–102.

Cardinale, V., Wang, Y., Carpino, G., Cui, C.-B., Gatto, M., Rossi, M., et al. (2011). Multipotent stem/progenitor cells in human biliary tree give rise to hepatocytes, cholangiocytes, and pancreatic islets. *Hepatology*, *54*, 2159–2172.

Cardinale, V., Wang, Y., Carpino, G., Reid, L. M., Gaudio, E., & Alvaro, D. (2012). Mucin-producing cholangiocarcinoma might derive from biliary tree stem/progenitor cells located in peribiliary glands. *Hepatology*, *55*, 2041–2042.

Carpino, G., Cardinale, V., Folseraas, T., Overi, D., Grzyb, K., Costantini, D., et al. (2019). Neoplastic transformation of the peribiliary stem cell niche in cholangiocarcinoma arisen in primary sclerosing cholangitis. *Hepatology*, *69*, 622–638.

Carpino, G., Cardinale, V., Reid, L., Alvaro, D., & Gaudio, E. (2012). Cells of origin and cancer stem cells in cholangiocarcinoma. *Translational Gastrointestinal Cancer*, *1*, 33–43.

Carpino, G., Overi, D., Melandro, F., Grimaldi, A., Cardinale, V., Matteo, S., et al. (2019). Matrisome analysis of intrahepatic cholangiocarcinoma unveils a peculiar cancer-associated extracellular matrix structure. *Clinical Proteomics*, *16*, 37.

Chaiyadet, S., Smout, M., Johnson, M., Whitchurch, C., Turnbull, L., Kaewkes, S., et al. (2015). Excretory/secretory products of the carcinogenic liver fluke are endocytosed by human cholangiocytes and drive cell proliferation and IL6 production. *International Journal for Parasitology*, *45*, 773–781.

Chaiyadet, S., Sotillo, J., Smout, M., Cantacessi, C., Jones, M. K., Johnson, M. S., et al. (2015). Carcinogenic liver fluke secretes extracellular vesicles that promote cholangiocytes to adopt a tumorigenic phenotype. *The Journal of Infectious Diseases*, *212*, 1636–1645.

Chan, K.-M., Tsai, C.-Y., Yeh, C.-N., Yeh, T.-S., Lee, W.-C., Jan, Y.-Y., et al. (2018). Characterization of intrahepatic cholangiocarcinoma after curative resection: Outcome, prognostic factor, and recurrence. *BMC Gastroenterology*, *18*, 180.

Chan-on, W., Nairismägi, M.-L., Ong, C. K., Lim, W. K., Dima, S., Pairojkul, C., et al. (2013). Exome sequencing identifies distinct mutational patterns in liver fluke-related and non-infection-related bile duct cancers. *Nature Genetics*, *45*, 1474–1478.

Chariyalertsak, S., Sirikulchayanonta, V., Mayer, D., Kopp-Schneider, A., Fürstenberger, G., Marks, F., et al. (2001). Aberrant cyclooxygenase isozyme expression in human intrahepatic cholangiocarcinoma. *Gut*, *48*, 80–86.

Chen, W., Li, G., Hu, Z., Zhu, P., Zhang, B., & Ding, Z. (2019). Significant response to anti-PD-1 based immunotherapy plus lenvatinib for recurrent intrahepatic cholangiocarcinoma with bone metastasis—A case report and literature review. *Medicine*, *98*(45), e17832.

Cheng, A.-L., Qin, S., Ikeda, M., Galle, P., Ducreux, M., Zhu, A., et al. (2019). LBA3-IMbrave150: Efficacy and safety results from a ph III study evaluating atezolizumab (atezo) + bevacizumab (bev) vs sorafenib (Sor) as first treatment (tx) for patients (pts) with unresectable hepatocellular carcinoma. *Annals of Oncology*, *30*(Suppl. 9), ix186–ix187.

Chuaysri, C., Thuwajit, P., Paupairoj, A., Chau-In, S., Suthiphongchai, T., & Thuwajit, C. (2009). Alpha-smooth muscle actin positive fibroblasts promote biliary cell proliferation and correlate with poor survival in cholangiocarcinoma. *Oncology Reports*, *21*, 957–969.

Chun, Y. S., & Javle, M. (2017). Systemic and adjuvant therapies for intrahepatic cholangiocarcinoma. *Cancer Control*, *24*, 1–7.

Chung, Y. E., Kim, M.-J., Park, Y. N., Choi, J.-Y., Pyo, J. Y., Kim, Y. C., et al. (2009). Varying appearances of cholangiocarcinoma: Radiologic-pathologic correlation. *RadioGraphics*, *29*, 683–700.

Claperon, A., Mergey, M., Aoudjehane, L., Ho-Bouldoires, T. H. N., Wendum, D., Prignon, A., et al. (2013). Hepatic myofibroblasts promote the progression of human cholangiocarcinoma through activation of epidermal growth factor receptor. *Hepatology*, *58*, 2001–2011.

Clements, O., Eliahoo, J., Kim, J. U., Taylor-Robinson, S. D., & Khan, S. A. (2020). Risk factors for intrahepatic and extrahepatic cholangiocarcinoma: A systematic review and meta-analysis. *Journal of Hepatology*, *72*, 95–103.

Coelho, R., Ricardo, S., Amaral, A. L., Huang, Y.-L., Nunes, M., Neves, J. P., et al. (2020). Regulation of invasion and peritoneal dissemination of ovarian cancer by mesothelin manipulation. *Oncogenesis*, *9*, 61.

Corliss, B. A., Azimi, M. S., Munson, J. M., Peirce, S. M., & Murfee, W. L. (2016). Macrophages: An inflammatory link between angiogenesis and lymphangiogenesis. *Microcirculation*, *23*, 95–121.

Correia da Costa, J. M., Vale, N., Gouveia, M. J., Botelho, M. C., Sripa, B., Santos, L. L., et al. (2014). Schistosome and liver fluke derived catechol-estrogens and helminth associated cancers. *Frontiers in Genetics*, *5*, 444.

Coulouarn, C., Cavard, C., Rubbia-Brandt, L., Audebourg, A., Dumont, F., Jacques, S., et al. (2012). Combined hepatocellular-cholangiocarcinomas exhibit progenitor features and activation of Wnt and TGFβ signaling pathways. *Carcinogenesis*, *33*, 1791–1796.

Dabney, R. S., Khalife, M., Shahid, K., & Phan, A. T. (2019). Molecular pathways and targeted therapy in cholangiocarcinoma. *Clinical Advances in Hematology & Oncology*, *17*, 630–637.

Dhanasekaran, R., Hemming, A. W., Zendejas, I., George, T., Nelson, D. R., Soldevila-Pico, C., et al. (2013). Treatment outcomes and prognostic factors of intrahepatic cholangiocarcinoma. *Oncology Reports*, *29*, 1259–1267.

DiPaola, F., Shivakumar, P., Pfister, J., Walters, S., Sabla, G., & Bezerra, J. A. (2013). Identification of intramural epithelial networks linked to peribiliary glands that express progenitor cell markers and proliferate after injury. *Hepatology*, *58*, 1486–1496.

Duluc, D., Delneste, Y., Tan, F., Moles, M.-P., Grimaud, L., Lenoir, J., et al. (2007). Tumor-associated leukemia inhibitory factor and IL-6 skew monocyte differentiation into tumor-associated macrophage-like cells. *Blood*, *110*, 4319–4330.

Dumur, C. I., Campbell, D. J. W., DeWitt, J. L., Oyesanya, R. A., & Sirica, A. E. (2010). Differential gene expression profiling of cultured *neu*-transformed *versus* spontaneously-transformed rat cholangiocytes and of corresponding cholangiocarcinomas. *Experimental and Molecular Pathology, 89*, 227–235.

Edeline, J., Benabdelghani, M., Bertaut, A., Watelet, J., Hammel, P., Joly, J.-P., et al. (2019). Gemcitabine and oxaliplatin chemotherapy or surveillance in resected biliary tract cancer (PRODGE 12-ACCORD 18-UNICANCER GI): A randomized phase III study. *Journal of Clinical Oncology, 37*, 658–667.

Elmore, L. W., & Sirica, A. E. (1991). Phenotypic characterization of metaplastic intestinal glands and ductular hepatocytes in cholangiofibrotic lesions rapidly induced in the caudate liver lobe of rats treated with furan. *Cancer Research, 51*, 5752–5759.

Elmore, L. W., & Sirica, A. E. (1993). "Intestinal-type" of adenocarcinoma preferentially induced in right/caudate liver lobes of rats treated with furan. *Cancer Research, 53*, 254–259.

Erice, O., Vallejo, A., Ponz-Sarvise, M., Saborowski, M., Vogel, A., Calvisi, D. F., et al. (2019). Genetic mouse models as in vivo tools for cholangiocarcinoma research. *Cancers, 11*, 1868. https://doi.org/10.3390/cancers11121868.

Fabris, L., Fiorotto, R., Spirli, C., Cadamuro, M., Mariotti, V., Perugorria, M. J., et al. (2019). Pathobiology of inherited biliary diseases: A roadmap to understand acquired liver diseases. *Nature Reviews Gastroenterology & Hepatology, 16*, 497–511.

Fabris, L., Perugorria, M. J., Mertens, J., Björkström, N. K., Cramer, T., Lleo, A., et al. (2019). The tumour microenvironment and immune milieu of cholangiocarcinoma. *Liver International, 39*, 63–78.

Fan, B., Malato, Y., Calvisi, D. F., Nagqi, S., Razumilava, N., Ribback, S., et al. (2012). Cholangiocarcinoma can originate from hepatocytes in mice. *Journal of Clinical Investigation, 122*, 2911–2915.

Fang, F. C. (2004). Antimicrobial reactive oxygen and nitrogen species: Concepts and controversies. *Nature Reviews Microbiology, 2*, 820–832.

Fausther, M., Lavoie, E. G., & Dranoff, J. A. (2017). Liver myofibroblasts of murine origins express mesothelin: Identification of novel rat mesothelin splice variants. *PLoS One, 12*, e0184499.

Fedorova, O. S., Kovshirina, Y. V., Kovshirina, A. E., Fedotova, M. M., Deev, I. A., Petrovskiy, F. I., et al. (2017). Opisthorchis felineus infection and cholangiocarcinoma in the Russian Federation: A review of medical statistics. *Parasitology International, 66*, 365–371.

Filippi, L., Di Costanzo, G. G., Tortora, R., Pelle, G., Saltarelli, A., Marino Marsilia, G., et al. (2020). Prognostic value of neutrophil-to-lymphocyte ratio and its correlation with fluorine-18-fluorodeoxyglucose metabolic parameters in intrahepatic cholangiocarcinoma submitted to 90Y-radioembolization. *Nuclear Medicine Communications, 41*, 78–86.

Fingas, C. D., Bronk, S. F., Werneburg, N. W., Mott, J. L., Guicciardi, M. E., Cazanave, S. C., et al. (2011). Myofibroblast-derived PDGF-BB promotes hedgehog survival signaling in cholangiocarcinoma cells. *Hepatology, 54*, 2076–2088.

Finn, R. S., Ikeda, M., Zhu, A. X., Sung, M. W., Baron, A. D., Kudo, M., et al. (2020). Phase Ib study of lenvatinib plus pembrolizumab in patients with unresectable hepatocellular carcinoma. *Journal of Clinical Oncology, 38*, 2960–2970.

Fiori, M. E., Di Franco, S., Villanova, L., Bianci, P., Stassi, G., & De Maria, R. (2019). Cancer-associated fibroblasts as abettors of tumor progression at the crossroads of EMT and therapy resistance. *Molecular Cancer, 18*, 70.

Ford, K., Hanley, C. J., Mellone, M., Szyndralewiez, C., Heitz, F., Wiesel, P., et al. (2020). NOX4 inhibition potentiates immunotherapy by overcoming cancer-associated fibroblast-mediated CD8 T-cell exclusion from tumors. *Cancer Research, 80*, 1846–1860.

Fukamura, D., Kloepper, J., Amoozgar, Z., Duda, D. G., & Jain, R. K. (2018). Enhancing cancer immunotherapy using antiangiogenics: Opportunities and challenges. *Nature Reviews Clinical Oncology*, *15*, 325–340. https://doi.org/10.1038/nrclinonc.2018.29.

Fukuda, Y., Asaoka, T., Eguchi, H., Yokota, Y., Kubo, M., Kinoshita, M., et al. (2020). Endogenous CXCL9 affects prognosis by regulating tumor-infiltrating natural killer cells in intrahepatic cholangiocarcinoma. *Cancer Science*, *111*, 323–333.

Fung, B. M., Lindor, K. D., & Tabibian, J. H. (2019). Cancer risk in primary sclerosing cholangitis: Epidemiology, prevention, and surveillance strategies. *World Journal of Gastroenterology*, *25*, 659–671.

Gascard, P., & Tlsty, T. D. (2016). Carcinoma-associated fibroblasts: Orchestrating the composition of malignancy. *Genes & Development*, *30*, 1002–1019.

Gentilini, A., Pastore, M., Marra, F., & Raggi, C. (2018). The role of stroma in cholangiocarcinoma: The intriguing interplay between fibroblastic component, immune cell subsets, and tumor epithelium. *International Journal of Molecular Sciences*, *19*, 2885.

Gera, S., Ettel, M., Acosta-Gonzales, G., & Xu, R. (2017). Clinical features, histology, and histogenesis of combined hepatocellular-cholangiocarcinoma. *World Journal of Hepatology*, *9*, 300–309.

Gleiberman, A. S., Encinas, J. M., Mignone, J. L., Michurina, T., Rosenfeld, M. G., & Enikolopov, G. (2005). Expression of nestin-GFP transgene marks oval cells in the adult liver. *Developmental Dynamics*, *234*, 413–421.

Goeppert, B., Frauenschuh, L., Zucknick, M., Stenzinger, A., Andrulis, M., Klauschen, F., et al. (2013). Prognostic impact of tumour-infiltrating immune cells on biliary tract cancer. *British Journal of Cancer*, *109*, 2665–2674.

Goeppert, B., Toth, R., Singer, S., Albrecht, T., Lipka, D. B., Lutsik, P., et al. (2019). Integrative analysis defines distinct prognostic subgroups of intrahepatic cholangiocarcinoma. *Hepatology*, *69*, 2091–2106.

González-Gonzáles, L., & Alonzo, J. (2018). Periostin: A matricellular protein with multiple functions in cancer development and progression. *Frontiers in Oncology*, *8*, 225.

Goulet, C. R., Bernard, G., Tremblay, S., Chabaud, S., Bolduc, S., & Pouliot, F. (2018). Exosomes induce fibroblast differentiation into cancer-associated fibroblasts through TGFβ signaling. *Molecular Cancer Research*, *16*, 1196–1204.

Goyal, L., Shi, L., Liu, L. Y., Fece de la Cruz, F., Lennerz, J. K., Raghavan, S., et al. (2019). TAS-120 overcomes resistance to ATP-competitive FGFR inhibitors in patients with FGFR2 fusion-positive intrahepatic cholangiocarcinoma. *Cancer Discovery*, *9*, 1064–1079.

Guest, R. V., Boulter, L., Kendall, T. J., Minnis-Lyons, S. E., Walker, R., Wigmore, S. J., et al. (2013). Cell lineage tracing reveals a biliary origin of intrahepatic cholangiocarcinoma. *Cancer Research*, *74*, 1005–1010.

Guglielmi, A., Ruzzenente, A., Campagnaro, T., Pachera, S., Valdegamberi, A., Nicoli, P., et al. (2009). Intrahepatic cholangiocarcinoma: Prognostic factors after surgical resection. *World Journal of Surgery*, *33*, 1247–1254.

Guicciardi, M. E., Trussoni, C. E., LaRusso, N. F., & Gores, G. J. (2020). The spectrum of reactive cholangiocytes in primary sclerosing cholangitis. *Hepatology*, *71*, 741–748.

Gupta, A., & Dixon, E. (2017). Epidemiology and risk factors: Intrahepatic cholangiocarcinoma. *HepatoBiliary Surgery and Nutrition*, *6*, 101–104.

Hanahan, D., & Weinberg, R. A. (2011). Hallmarks of cancer: The next generation. *Cell*, *144*, 646–674.

Hasita, H., Komohara, Y., Okabe, H., Masuda, T., Ohnishi, K., Lei, X. F., et al. (2010). Significance of alternatively activated macrophages in patients with intrahepatic cholangiocarcinoma. *Cancer Science*, *101*, 1913–1919.

Haubeiss, S., Schmid, J. O., Mürdter, T. E., Sonnenberg, M., Friedel, G., van der Kuip, H., et al. (2010). Dasatinib reverses cancer-associated fibroblasts (CAFs) from primary lung carcinomas to a phenotype comparable to that of normal fibroblasts. *Molecular Cancer*, *9*, 168. https://doi.org/10.1186/1476-4598-9-168.

Higuchi, R., Yamamoto, M., Hatori, T., Shimizu, K., Imai, K., & Takasaki, K. (2006). Intrahepatic cholangiocarcinoma with lymph node metastasis successfully treated by immunotherapy with CD3-activated T cells and dendritic cells after surgery: Report of a case. *Surgery Today*, *36*, 559–562.
Hill, M. A., Alexander, W. B., Guo, B., Kato, Y., Patra, K., O'Dell, M. R., et al. (2018). *Kras* and *Tp53* mutations cause cholangiocyte- and hepatocyte-derived cholangiocarcinoma. *Cancer Research*, *78*, 4445–4451.
Hov, J. R., & Karlsen, T. H. (2017). The microbiome in primary sclerosing cholangitis: Current evidence and potential concepts. *Seminars in Liver Disease*, *37*, 314–331.
Hu, L.-S., Zhang, X.-F., Weiss, M., Popescu, I., Marques, H. P., Aldrighetti, L., et al. (2019). Recurrence patterns and timing courses following curative-intent resection for intrahepatic cholangiocarcinoma. *Annuals of Surgical Oncology*, *26*, 2549–2557.
Huang, C.-K., Aihara, A., Iwagami, Y., Yu, T., Carlson, R., Koga, H., et al. (2016). Expression of transforming growth factor β1 promotes cholangiocarcinoma development and progression. *Cancer Letters*, *380*, 153–162.
Hucl, T. (2019). Precursors to cholangiocarcinoma. *Gastroenterology Research and Practice*, *2019*. https://doi.org/10.1155/2019/1389289, 1389289.
Ikenoue, T., Terakado, Y., Nakagawa, H., Hikiba, Y., Fujii, T., Matsubara, D., et al. (2016). A novel mouse model of intrahepatic cholangiocarcinoma induced by liver-specific *Kras* activation and *Pten* deletion. *Scientific Reports*, *6*, 23899.
Ishikawa, A., Sasaki, M., Ohira, S., Ohta, T., Oda, K., Nimura, Y., et al. (2004). Aberrant expression of CDX2 is closely related to the intestinal metaplasia and MUC2 expression in intraductal papillary neoplasms of the liver in hepatolithiasis. *Laboratory Investigation*, *84*, 629–638.
Itou, R. A., Uyama, N., Hirota, S., Kawada, N., Wu, S., Miyashita, S., et al. (2019). Immunohistochemical characterization of cancer-associated fibroblasts at the primary sites and in the metastatic lymph nodes of human intrahepatic cholangiocarcinoma. *Human Pathology*, *83*, 77–89.
Jaillon, S., Ponzetta, A., Di Mitri, D., Santoni, A., Bonecchi, R., & Mantovani, A. (2020). Neutrophil diversity and plasticity in tumour progression and therapy. *Nature Reviews Cancer*, *20*, 485–503.
Jaiswal, M., LaRusso, N. F., Burgart, L. J., & Gores, G. J. (2000). Inflammatory cytokines induce DNA damage and inhibit DNA repair in cholangiocarcinoma cells by a nitric oxide-dependent mechanism. *Cancer Research*, *60*, 184–190.
Jaiswal, M., LaRusso, N. F., Shapiro, R. A., Billiar, T. R., & Gores, G. J. (2001). Nitric oxide-mediated inhibition of DNA repair potentiates oxidative DNA damage in cholangiocytes. *Gastroenterology*, *120*, 190–199.
Jang, M. H., Lee, Y. J., & Kim, H. (2014). Intrahepatic cholangiocarcinoma arising in Caroli's disease. *Clinical and Molecular Hepatology*, *20*, 402–405.
Janse, M., Lamberts, L. E., Franke, L., Raychaudhuri, S., Ellinghaus, E., Muri Boberg, K., et al. (2011). Three ulcerative colitis susceptibility loci are associated with primary sclerosing cholangitis and indicate a role for IL2, REL, and CARD9. *Hepatology*, *53*, 1977–1985.
Jesper, D., Heyn, S. G., Schellhass, B., Pfeifer, L., Goertz, R., Ruediger, S., et al. (2018). Effects of liver cirrhosis and patient condition on clinical outcomes in intrahepatic cholangiocarcinoma: A retrospective analysis of 156 cases in a single center. *European Journal of Gastroenterology & Hepatology*, *30*, 552–556.
Job, S., Rapoud, D., Dos Santos, A., Gonzalez, P., Desterke, C., Pascal, G., et al. (2020). Identification of four immune subtypes characterized by distinct composition and functions of tumor microenvironment in intrahepatic cholangiocarcinoma. *Hepatology*, *72*, 965–981. https://doi.org/10.1002/hep.31092.

Joseph, N. M., Tsokos, C. G., Umetsu, S. E., Shain, A. H., Kelley, R. K., Onodera, C., et al. (2019). Genomic profiling of combined hepatocellular-cholangiocarcinoma reveals similar genetics to hepatocellular carcinoma. *Journal of Pathology*, *248*, 164–178.

Jung, I. H., Kim, D. H., Yoo, D. K., Baek, S. Y., Jeong, S. H., Jung, D. E., et al. (2018). In vivo study of natural killer (NK) cell cytotoxicity against cholangiocarcinoma in a nude mouse model. *In Vivo*, *32*, 771–781.

Junking, M., Grainok, J., Thepmalee, C., Wongkham, S., & Yenchitsomanus, P. T. (2017). Enhanced cytotoxic activity of effector T-cells against cholangiocarcinoma by dendritic cells pulsed with pooled mRNA. *Tumour Biology*, *39*, 1010428317733367.

Jusakul, A., Cutcutache, I., Yong, C. H., Lim, J. Q., Huang, M. N., Padmanabhan, N., et al. (2017). Whole-genome and epigenomic landscapes of etiologically distinct subtypes of cholangiocarcinoma. *Cancer Discovery*, *7*, 1116–1135.

Kalli, M., & Stylianopoulos, T. (2018). Defining the role of solid stress and matrix stiffness in cancer cell proliferation and metastasis. *Frontiers in Oncology*, *8*, 55.

Karlsen, T. H., & Boberg, K. M. (2013). Update on primary sclerosing cholangitis. *Journal of Hepatology*, *59*, 571–582.

Karlsen, T. H., Folseraas, T., Thorburn, D., & Vesterhus, M. (2017). Primary sclerosing cholangitis—A comprehensive review. *Journal of Hepatology*, *67*, 1298–1323.

Karlsen, T. H., Schrumpf, E., & Boberg, K. M. (2010). Update on primary sclerosing cholangitis. *Digestive and Liver Disease*, *42*, 390–400.

Kasper, H. U., Drebber, U., Stippel, D. L., Dienes, H.-P., & Gillessen, A. (2009). Liver tumor infiltrating lymphocytes: Comparison of hepatocellular and cholangiolar carcinoma. *World Journal of Gastroenterology*, *15*, 5053–5057.

Kasprzak, A., & Adamek, A. (2019). Mucins: The old, the new and the promising factors in hepatobiliary carcinogenesis. *International Journal of Molecular Sciences*, *20*, 1288.

Katsumata, L. W., Miyajima, A., & Itoh, T. (2017). Portal fibroblasts marked by the surface antigen Thy1 contribute to fibrosis in mouse models of cholestatic liver injury. *Hepatology Communications*, *1*, 198–214.

Katt, J., Schwinge, D., Schoknecht, T., Quaas, A., Sobottka, I., Burandt, E., et al. (2013). Increased T helper type 17 response to pathogen stimulation in patients with primary sclerosing cholangitis. *Hepatology*, *58*, 1084–1093.

Kawahara, N., Ono, M., Taguchi, K., Okamoto, M., Shimada, M., Takenaka, K., et al. (1998). Enhanced expression of thrombospondin-1 and hypovascularity in human cholangiocarcinoma. *Hepatology*, *28*, 1512–1517. https://doi.org/10.1002/hep.510280610.

Kendall, T., Verheij, J., Gaudio, E., Evert, M., Guido, M., Goeppert, B., et al. (2019). Anatomical, histomorphological and molecular classification of cholangiocarcinoma. *Liver International*, *39*, 7–18.

Khan, S. A., Tavolari, S., & Brandi, G. (2019). Cholangiocarcinoma: Epidemiology and risk factors. *Liver International*, *39*, 19–31.

Kirstein, M. M., & Vogel, A. (2016). Epidemiology and risk factors of cholangiocarcinoma. *Visceral Medicine*, *32*, 395–400.

Kitano, Y., Yamashita, Y.-I., Nakao, Y., Itoyama, R., Yusa, T., Umezaki, N., et al. (2020). Clinical significance of PD-L1 expression in both cancer and stroma cells of cholangiocarcinoma patients. *Annals of Surgical Oncology*, *27*, 599–607.

Ko, S., Russell, J. O., Molina, L. M., & Monga, S. P. (2020). Liver progenitors and adult cell plasticity in hepatic injury and repair: Knowns and unknowns. *Annual Review of Pathology: Mechanisms of Disease*, *15*, 23–50.

Kobayashi, M., Sakabe, T., Abe, H., Tanii, M., Takahashi, H., Chiba, A., et al. (2013). Dendritic cell-based immunotherapy targeting synthesized peptides for advanced biliary tract cancer. *Journal of Gastrointestinal Surgery*, *17*, 1609–1617.

Komuta, M., Spee, B., Borght, S. V., De Vos, R., Verslype, C., Aerts, R., et al. (2008). Clinicopathological study on cholangiolocellular carcinoma suggesting hepatic progenitor cell origin. *Hepatology*, *47*, 1544–1556.

Koyama, Y., Wang, P., Liang, S., Iwaisako, K., Liu, X., Xu, J., et al. (2017). Mesothelin/mucin16 signaling in activated portal fibroblasts regulates cholestatic liver fibrosis. *The Journal of Clinical Investigation*, *127*, 1254–1270.

Kozuka, S., Kurashina, M., Tsubone, M., Hachisuka, K., & Yasui, A. (1984). Significance of intestinal metaplasia for the evolution of cancer in the biliary tract. *Cancer*, *54*, 2277–2285.

Krook, M. A., Lenyo, A., Wilberding, M., Barker, H., Dantuono, M., Bailey, K. M., et al. (2020). Efficacy of FGFR inhibitors and combination therapies for acquired resistance in FGFR2-fusion cholangiocarcinoma. *Molecular Cancer Therapeutics*, *19*, 847–857.

Kudo, M. (2020). A new era in systemic therapy for hepatocellular carcinoma: Atezolizumab plus bevacizumab combination therapy. *Liver Cancer*, *9*, 119–137.

Kumar, P., Smith, T., Raeman, R., Chopyk, D. M., Brink, H., Liu, Y., et al. (2018). Periostin promotes liver fibrogenesis by activating lysyl oxidase in hepatic stellate cells. *Journal of Biological Chemistry*, *293*, 12781–12792.

Kusano, H., Naito, Y., Mihara, Y., Kondo, R., Ogasawara, S., Akiba, J., et al. (2020). Distinctive clinicopathological features and KRAS and IDH1/2 mutation status of cholangiolocellular carcinoma. *Hepatology Research*, *50*, 84–91.

Labib, P. L., Goodchild, G., & Pereira, S. P. (2019). Molecular pathogenesis of cholangiocarcinoma. *BMC Cancer*, *19*, 185.

Lal, S., Wang, G., Chen, Z., Xiang, X., Liang, C., Huang, X. F., et al. (2019). Co-expression of PD-L1 antibodies enhances the anti-tumor efficacy of chimeric antigen receptor T cells. *Journal of Cancer Research and Immuno-oncology*, *5*, 119.

Lamarca, A., Barriuso, J., McNamara, M. G., & Valle, J. W. (2020). Molecular targeted therapies: Ready for "prime time" in biliary tract cancer. *Journal of Hepatology*, *73*, 170–185.

Lan, Y., Zhang, D., Xu, C., Hance, K. W., Marelli, B., Qi, J., et al. (2018). Enhanced preclinical antitumor activity of M7824, a bifunctional fusion protein simultaneously targeting PD-L1 and TGF-β. *Science Translational Medicine*, *10*, eaan5488.

Lazaridis, K. N., & LaRusso, N. F. (2016). Primary sclerosing cholangitis. *The New England Journal of Medicine*, *375*, 2501–2502.

Lee, D. H., & Lee, J. M. (2017). Primary malignant tumors in non-cirrhotic liver. *European Journal of Radiology*, *95*, 349–361.

Leyva-Illades, D., McMillin, M., Quinn, M., & Demorrow, S. (2012). Cholangiocarcinoma pathogenesis: Role of the tumor microenvironment. *Translational Gastrointestinal Cancer*, *1*, 71–80.

Li, X., Tao, J., Cigliano, A., Sini, M., Calderaro, J., Azoulay, D., et al. (2015). Co-activation of PIK3CA and Yap promotes development of hepatocellular and cholangiocellular tumors in mouse and human liver. *Oncotarget*, *6*, 10102–10115.

Li, L., Wei, W., Li, Z., Chen, H., Li, Y., Jiang, W., et al. (2018). The spleen promotes the secretion of CCL2 and supports an M1 dominant phenotype in hepatic macrophages during liver fibrosis. *Cellular Physiology and Biochemistry*, *51*, 557–574.

Liau, J.-Y., Tsai, J.-H., Yuan, R.-H., Chang, C.-N., Lee, H.-J., & Jeng, Y.-M. (2014). Morphological subclassification of intrahepatic cholangiocarcinoma: Etiological, clinicopathological, and molecular features. *Modern Pathology*, *27*, 1163–1173.

Liepelt, A., & Tacke, F. (2016). Stromal cell-derived factor-1 (SDF-1) as a target in liver diseases. *American Journal of Physiology—Gastrointestinal and Liver Physiology*, *311*, G203–G209.

Limaye, P. B., Alarcón, G., Walls, A. L., Nalesnik, M. A., Michalopoulos, G. K., Demetris, A. J., et al. (2008). Expression of specific hepatocyte and cholangiocyte transcription factors in human liver disease and embryonic development. *Laboratory Investigation*, *88*, 865–872.

Lin, Y.-K., Fang, Z., Jiang, T.-Y., Wan, Z.-H., Pan, Y.-F., Ma, Y.-H., et al. (2018). Combination of Kras activation and PTEN deletion contributes to murine hepatopancreatic ductal malignancy. *Cancer Letters*, *421*, 161–169.

Lin, Y., Li, B., Yang, X., Cai, Q., Liu, W., Tian, M., et al. (2019). Fibroblastic FAP promotes intrahepatic cholangiocarcinoma growth via MDSCs recruitment. *Neoplasia, 21*, 1133–1142. https://doi.org/10.1016/j.neo.2019.10.005.

Lin, G., Liu, Y., Li, S., Mao, Y., Wang, J., Shuang, Z., et al. (2016). Elevated neutrophil-to-lymphocyte ratio is an independent poor prognostic factor in patients with intrahepatic cholangiocarcinoma. *Oncotarget*, 7, 50963–50971.

Lind, H., Gameiro, S. R., Jochems, C., Donahue, R. N., Strauss, J., Gulley, J. L., et al. (2020). Dual targeting of TGF-β and PD-L1 via a bifunctional anti-PD-L1/TGF-BRII agent: Status of preclinical and clinical advances. *Journal for Immunotherapy of Cancer, 8*, e000433, e1426519.

Lindor, K. D., Kowdley, K. V., Harrison, M. E., & American College of Gastroenterology. (2015). ACG clinical guideline: Primary sclerosing cholangitis. *The American Journal of Gastroenterology, 110*, 646–660.

Ling, H., Roux, E., Hempel, D., Tao, J., Smith, M., Lonning, S., et al. (2013). Transforming growth factor β neutralization ameliorates pre-existing hepatic fibrosis and reduces cholangiocarcinoma in thioacetamide-treated rats. *PLoS One, 8*, e54499.

Liu, J. Z., Hov, J. R., Folseraas, T., Ellinghaus, E., Rushbrook, S. M., Doncheva, N. T., et al. (2013). Dense genotyping of immune-related disease regions identifies nine new risk loci for primary sclerosing cholangitis. *Nature Genetics, 45*, 670–675.

Liu, Z.-H., Lian, B.-F., Dong, Q.-Z., Sun, H., Wei, J.-W., Sheng, Y.-Y., et al. (2018). Whole-exome mutational and transcriptional landscapes of combined hepatocellular cholangiocarcinoma and intrahepatic cholangiocarcinoma reveal molecular diversity. *Biochimica et Biophysica Acta-Molecular Basis of Disease, 1864*, 2360–2368.

Loeuillard, E., Conboy, C. B., Gores, G. J., & Rizvi, S. (2019). Immunobiology of cholangiocarcinoma. *JHEP Reports, 1*, 297–311.

Loeuillard, E., Yang, J., Buckarma, E., Wang, J., Liu, Y., Conboy, C., et al. (2020). Targeting tumor-associated macrophages and granulocytic-myeloid-derived suppressor cells augments PD-1 blockade in cholangiocarcinoma. *The Journal of Clinical Investigation, 130*, 5380–5396.

Loilome, W., Bungkanjana, P., Techasen, A., Namwat, N., Yongvanit, P., Puapairoj, A., et al. (2014). Activated macrophages promote Wnt/β-catenin signaling in cholangiocarcinoma cells. *Tumour Biology, 35*, 5357–5367.

Lowery, M. A., Burris, H. A., III, Janku, F., Shroff, R. T., Cleary, J. M., Azad, N. S., et al. (2019). Safety and activity of ivosidenib in patients with IDH1-mutant advanced cholangiocarcinoma: A phase 1 study. *The Lancet Gastroenterology & Hepatology, 4*, 711–720.

Lowy, C. M., & Oskarsson, T. (2015). Tenascin C in metastasis: A view from the invasive front. *Cell Adhesion & Migration, 9*, 112–124.

Ludwig, J., Wahlstrom, H. E., Batts, K. P., & Wiesner, R. H. (1992). Papillary bile duct dysplasia in primary sclerosing cholangitis. *Gastroenterology, 102*, 2134–2138.

Lv, J., & Li, P. (2019). Mesothelin as a biomarker for targeted therapy. *Biomarker Research*, 7, 18.

Ma, L., Hernandes, M. O., Zhao, Y., Mehta, M., Tran, B., Kelly, M., et al. (2019). Tumor cell biodiversity drives microenvironmental reprogramming in liver cancer. *Cancer Cell, 36*, 418–430.

Ma, C. H., Hwang, D. W., Song, K. B., Kim, S. C., Shin, S. H., Lee, J. H., et al. (2020). Prognostic factors predicting survival rate over 10 years of patients with intrahepatic cholangiocarcinoma after hepatic resection. *Annals of Surgical Treatment and Research, 98*, 116–123.

Ma, B., Meng, H., Tian, Y., Wang, Y., Song, T., Zhang, T., et al. (2020). Distinct clinical and prognostic implications of IDH1/2 mutation and other most frequent mutations in large duct and small duct subtypes of intrahepatic cholangiocarcinoma. *BMC Cancer, 20*, 318.

Ma, J., Tang, W. K., Esser, L., Pastan, I., & Xia, D. (2012). Recognition of mesothelin by the therapeutic antibody MORAb-009: Structural and mechanistic insights. *Journal of Biological Chemistry*, *287*, 33123–33131.

Mancinelli, R., Mammola, C. L., Sferra, R., Pompili, S., Vetuschi, A., & Pannarale, L. (2019). Role of the angiogenic factors in cholangiocarcinoma. *Applied Sciences*, *9*, 1393.

Manzanares, M.Á., Campbell, D. J. W., Maldonado, G. T., & Sirica, A. E. (2018). Overexpression of periostin and distinct mesothelin forms predict malignant progression in a rat cholangiocarcinoma model. *Hepatology Communications*, *2*, 155–172.

Manzanares, M.Á., Usui, A., Campbell, D. J., Dumur, C. I., Maldonado, G. T., Fausther, M., et al. (2017). Transforming growth factors α and β are essential for modeling cholangiocarcinoma desmoplasia and progression in a three-dimensional organotypic culture model. *American Journal of Pathology*, *187*, 1068–1092.

Mao, Z. Y., Zhu, G. Q., Xiong, M., Ren, L., & Bai, L. (2015). Prognostic value of neutrophil distribution in cholangiocarcinoma. *World Journal of Gastroenterology*, *21*, 4961–4968.

Marsh, V., Davies, E. J., Williams, G. T., & Clarke, A. R. (2013). PTEN loss and KRAS activation cooperate in murine biliary tract malignancies. *Journal of Pathology*, *230*, 165–173.

Martín-Sierra, C., Martins, R., Laranjeira, P., Coucelo, M., Abrantes, A. M., Oliveira, R. C., et al. (2019). Functional and phenotypic characterization of tumor-infiltrating leukocyte subsets and their contribution to the pathogenesis of hepatocellular carcinoma and cholangiocarcinoma. *Translational Oncology*, *12*, 1468–1479.

Massarweh, N. N., & El-Serag, H. B. (2017). Epidemiology of hepatocellular carcinoma and intrahepatic cholangiocarcinoma. *Cancer Control*, *24*, 1–11.

Mavros, M. N., Economopoulos, K. P., Alexiou, V. G., & Pawlik, T. M. (2014). Treatment and prognosis for patients with intrahepatic cholangiocarcinoma: Systematic review and meta-analysis. *JAMA Surgery*, *149*, 565–574.

Melaiu, O., Lucarini, V., Cifaldi, L., & Fruci, D. (2020). Influence of the tumor microenvironment on NK cell function in solid tumors. *Frontiers in Immunology*, *10*, 3038.

Mells, G. F., Kaser, A., & Karlsen, T. H. (2013). Novel insights into autoimmune liver diseases provided by genome-wide association studies. *Journal of Autoimmunity*, *46*, 41–54.

Melum, E., Franke, A., Schramm, C., Weismüller, T. J., Gotthardt, D. N., Offner, F. A., et al. (2011). Genome-wide association analysis in primary sclerosing cholangitis identifies two non-HLA susceptibility loci. *Nature Genetics*, *43*, 17–19.

Meng, Z.-W., Pan, W., Hong, H.-J., Chen, J.-Z., & Chen, Y.-L. (2017). Macroscopic types of intrahepatic cholangiocarcinoma and the eighth edition of AJCC/UICC TNM staging system. *Oncotarget*, *8*, 101165–101174.

Meng, F., Wehbe-Janek, H., Henson, R., Smith, H., & Patel, T. (2008). Epigenetic regulation of microRNA-370 by interleukin-6 in malignant human cholangiocytes. *Oncogene*, *27*, 378–386.

Merdrignac, A., Angenard, G., Allain, C., Petitjean, K., Bergeat, D., Bellaud, P., et al. (2018). A novel transforming growth factor beta-induced long noncoding RNA promotes an inflammatory microenvironment in human intrahepatic cholangiocarcinoma. *Hepatology Communications*, *2*, 254–269.

Mertens, J. C., Fingas, C. D., Christensen, J. D., Smoot, R. L., Bronk, S. F., Werneburg, N. W., et al. (2013). Therapeutic effects of deleting cancer-associated fibroblasts in cholangiocarcinoma. *Cancer Research*, *73*, 897–907.

Mertens, J. C., Rizvi, S., & Gores, G. J. (2018). Targeting cholangiocarcinoma. *Biochimica et Biophysica Acta*, *1864*, 1454–1460.

Michalopoulos, G. K., Barua, L., & Bowen, W. C. (2005). Transdifferentiation of rat hepatocytes into biliary cells after bile duct ligation and toxic biliary injury. *Hepatology*, *41*, 535–544. https://doi.org/10.1002/hep.20600.

Michalopoulos, G. K., & Khan, Z. (2015). Liver stem cells: Experimental findings and implications for human liver disease. *Gastroenterology*, *149*, 876–882.

Mody, K., Antwi, S. O., Hodge, D. O., Ailawadhi, S., Roberts, L., & Bekaii-Saab, T. (2018). A SEER-based multi-ethnic picture of advanced intrahepatic cholangiocarcinoma in the United States pre-and post-the advent of gemcitabine/cisplatin. *Journal of Gastrointestinal Oncology*, *9*, 1063–1073. https://doi.org/10.21037/jgo.2018.07.09.

Moeini, A., Sia, D., Zhang, Z., Camprecios, G., Stueck, A., Dong, H., et al. (2017). Mixed hepatocellular cholangiocarcinoma tumors: Cholangiolocellular carcinoma is a distinct molecular entity. *Journal of Hepatology*, *66*, 952–961.

Monteran, L., & Erez, N. (2019). The dark side of fibroblasts: Cancer-associated fibroblasts as mediators of immunosuppression in the tumor microenvironment. *Frontiers in Immunology*, *10*, 1835.

Mou, H., Yu, L., Liao, Q., Hou, X., Wu, Y., Cui, Q., et al. (2018). Successful response to the combination of immunotherapy and chemotherapy in cholangiocarcinoma with high tumour mutational burden and PD-L1 expression: A case report. *BMC Cancer*, *18*, 1105.

Mu, X., Pradere, J.-P., Affò, S., Dapito, D. H., Friedman, R., Lefkovitch, J. H., et al. (2016). Epithelial transforming growth factor-β signaling does not contribute to liver fibrosis but protects mice from cholangiocarcinoma. *Gastroenterology*, *150*, 720–733.

Murphy-Ullrich, J. E., & Suto, M. J. (2018). Thrombospondin-1 regulation of latent TGF-β activation: A therapeutic target for fibrotic disease. *Matrix Biology*, *68–69*, 28–43.

Nakagawa, H., Hayata, Y., Yamada, T., Kawamura, S., Suzuki, N., & Koike, K. (2018). Peribiliary glands as the cellular origin of biliary tract cancer. *International Journal of Molecular Sciences*, *19*, 1745.

Nakagawa, H., Suzuki, N., Hirata, Y., Hikiba, Y., Hayakawa, Y., Kinoshita, H., et al. (2017). Biliary epithelial injury-induced regenerative response by IL-33 promotes cholangiocarcinogenesis from peribiliary glands. *Proceedings of the National Academy of Sciences of the United States of America*, *114*, E3806–E3815.

Nakagawa, H., Suzuki, N., & Koike, K. (2019). Mouse model for cholangiocarcinoma from peribiliary glands. *Methods in Molecular Biology*, *1905*, 237–245.

Nakamoto, N., Sasaki, N., Aoki, R., Miyamoto, K., Suda, W., Teratani, T., et al. (2019). Gut pathobionts underlie intestinal barrier dysfunction and liver T helper 17 cell immune response in primary sclerosing cholangitis. *Nature Microbiology*, *4*, 492–503.

Nakamura, H., Arai, Y., Totoki, Y., Shirota, T., Elzawahry, A., Kato, M., et al. (2015). Genomic spectra of biliary tract cancer. *Nature Genetics*, *47*, 1003–1010.

Nakano, S., Eso, Y., Okada, H., Takai, A., Takahashi, K., & Seno, H. (2020). Recent advances in immunotherapy for hepatocellular carcinoma. *Cancers*, *12*, 775.

Nakanuma, Y., Kakuda, Y., & Uesaka, K. (2019). Characterization of intraductal papillary neoplasm of the bile duct with respect to the histopathologic similarities to pancreatic intraductal ikosi, mucinous neoplasm. *Gut and Liver*, *13*, 617–627.

Nakanuma, Y., Miyata, T., & Uchida, T. (2016). Latest advances in the pathological understanding of cholangiocarcinomas. *Expert Review of Gastroenterology & Hepatology*, *10*, 113–127.

Nakanuma, Y., & Sato, Y. (2012). Cystic and papillary neoplasm involving peribiliary glands: A biliary counterpart of branch-type intraductal papillary mucinous cystic neoplasm. *Hepatology*, *55*, 2040–2041.

Nakanuma, Y., Sato, Y., Harada, K., Sasaki, M., Xu, J., & Ikeda, H. (2010). Pathological classification of intrahepatic cholangiocarcinoma based on a new concept. *World Journal of Hepatology*, *2*, 419–427.

Nakanuma, Y., Tsutsui, A., Ren, X. S., Harada, K., Sato, Y., & Sasaki, M. (2014). What are the precursor and early lesions of peripheral intrahepatic cholangiocarcinoma? *International Journal of Hepatology*, *2014*, 805973.

Neuzillet, C., Tijeras-Raballand, A., Ragulan, C., Cros, J., Patil, Y., Martinet, M., et al. (2019). Inter- and intra-tumoural heterogeneity in cancer-associated fibroblasts of human pancreatic ductal adenocarcinoma. *Journal of Pathology*, *248*, 51–65.

Nicholás-Boluda, A., Vaquero, J., Laurent, G., Renault, G., Bazzi, R., Donnadieu, E., et al. (2020). Photothermal depletion of cancer-associated fibroblasts normalizes tumor stiffness in desmoplastic cholangiocarcinoma. *ACS Nano*, *14*, 5738–5753.

Nishihara, Y., Aishima, S., Hayashi, A., Iguchi, T., Fujita, N., Taketomi, A., et al. (2009). CD10+ fibroblasts are more involved in the progression of hilar/extrahepatic cholangiocarcinoma than peripheral intrahepatic cholangiocarcinoma. *Histopathology*, *55*, 423–431.

Nissen, N. I., Karsdal, M., & Willumsen, N. (2019). Collagens and cancer associated fibroblasts in the reactive stroma and its relation to cancer biology. *Journal of Experimental & Clinical Cancer Research*, *38*, 115.

Noguchi, S., Saito, A., & Nagase, T. (2018). YAP/TAZ signaling as a molecular link between fibrosis and cancer. *International Journal of Molecular Sciences*, *19*, 3674.

Nomoto, M., Uchikosi, Y., Kajikazawa, N., Tanaka, Y., & Asakura, H. (1992). Appearance of hepatocytelike cells in the interlobular bile ducts of human liver in various liver disease states. *Hepatology*, *16*, 1199–1205.

Nomura, R., Fujii, H., Abe, M., Sugo, H., Ishizaki, Y., Kawasaki, S., et al. (2013). Mesothelin expression is a prognostic factor in cholangiocellular carcinoma. *International Surgery*, *98*, 164–169.

Norris, R. A., Damon, B., Mironov, V., Kasyanov, V., Ramamurthi, A., Moreno-Rodriguez, R., et al. (2012). Periostin regulates collagen fibrillogenesis and the biomechanical properties of connective tissue. *Journal of Cellular Biochemistry*, *101*, 695–711. https://doi.org/10.1002/jcb.21224.

O'Rourke, C. J., Lafuente-Barquero, J., & Andersen, J. B. (2019). Epigenome remodeling in cholangiocarcinoma. *Trends in Cancer*, *5*, 335–349.

Obulkasim, H., Shi, X., Wang, J., Li, J., Dai, B., Wu, P., et al. (2018). Podoplanin is an important stromal prognostic marker in perihilar cholangiocarcinoma. *Oncology Letters*, *15*, 137–146.

Öhlund, D., Handly-Santana, A., Biffi, G., Elyada, E., Almeida, A. S., Ponz-Sarvise, M., et al. (2017). Distinct populations of inflammatory fibroblasts and myofibroblasts in pancreatic cancer. *Journal of Experimental Medicine*, *214*, 579–596.

Okabe, H., Beppu, T., Hayashi, H., Horino, K., Masuda, T., Komori, H., et al. (2009). Hepatic stellate cells may relate to progression of intrahepatic cholangiocarcinoma. *Annals of Surgical Oncology*, *16*, 2555–2564.

Okabe, H., Hayashi, H., Nakagawa, S., Imai, K., Nitta, H., Arima, K., et al. (2016). Inducible factors for cancer-associated fibroblasts in liver cancer versus myofibroblasts in inflammatory liver disease. *Histology and Histopathology*, *31*, 141–148.

Overi, D., Carpino, G., Cardinale, V., Franchitto, A., Safarikia, S., Onori, P., et al. (2018). Contribution of resident stem cells to liver and biliary tree regeneration in human diseases. *International Journal of Molecular Sciences*, *19*, 2917.

Pant, K., Richard, S., Peixoto, E., & Gradilone, S. A. (2020). Role of glucose metabolism reprogramming in the pathogenesis of cholangiocarcinoma. *Frontiers in Medicine*, *7*, 113.

Pellino, A., Loupakis, F., Cadamuro, M., Dadduzio, V., Fassan, M., Guido, M., et al. (2018). Precision medicine in cholangiocarcinoma. *Translational Gastroenterology and Hepatology*, *3*, 40.

Plentz, R. R., & Malek, N. P. (2016). Systemic therapy of cholangiocarcinoma. *Visceral Medicine*, *32*, 427–430.

Plieskatt, J. L., Deenonpoe, R., Mulvenna, J. P., Krause, L., Sripa, B., Bethony, J. M., et al. (2013). Infection with the carcinogenic liver fluke Opisthorchis viverrini modifies intestinal and biliary microbiome. *FASEB Journal*, *27*, 4572–4584.

Primrose, J. N., Fox, R. P., Palmer, D. H., Malik, H. Z., Prasad, R., Mirza, D., et al. (2019). Capecitabine compared with observation in resected biliary tract cancer (BILCAP): A randomized, controlled, multicentre, phase 3 study. *Lancet Oncology*, *20*, 663–673.

Principe, D. R., Park, A., Dorman, M. J., Kumar, S., Viswakarma, N., Rubin, J., et al. (2019). TGFβ blockade augments PD-1 inhibition to promote T-cell mediated regression of pancreatic cancer. *Molecular Cancer Therapeutics*, *18*, 613–620.

Prueksapanich, P., Piyachaturawat, P., Aumpansub, P., Ridtitid, W., Chaiteerakij, R., & Rerknimitr, R. (2018). Liver fluke-associated biliary tract cancer. *Gut and Liver*, *12*, 236–245.

Qian, M. B., Utzinger, J., Keiser, J., & Zhou, X. N. (2016). Clonorchiasis. *Lancet*, *387*, 800–810.

Raggi, C., Correnti, M., Sica, A., Andersen, J. B., Cardinale, V., Alvaro, D., et al. (2017). Cholangiocarcinoma stem-like subset shapes tumor-initiating niche by educating associated macrophages. *Journal of Hepatology*, *66*, 102–115.

Ren, P., Silberg, D. G., & Sirica, A. E. (2000). Expression of an intestine-specific transcription factor (CDX1) in intestinal metaplasia and in subsequently developed intestinal type of cholangiocarcinoma in rat liver. *Americam Journal of Pathology*, *156*, 621–627.

Rhee, H., Ko, J. E., Chung, T., Jee, B. A., Kwon, S. M., Nahm, J. H., et al. (2018). Transcriptomic and histopathological analysis of cholangiolocellular differentiation trait in intrahepatic cholangiocarcinoma. *Liver International*, *38*, 113–124.

Rizvi, S., Khan, S. A., Hallemeier, C. L., Kelley, R. K., & Gores, G. J. (2018). Cholangiocarcinoma-evolving concepts and therapeutic strategies. *Nature Reviews. Clinical Oncology*, *15*, 95–111.

Rizvi, S., Mertens, J. C., Bronk, S. F., Hirsova, P., Dai, H., & Roberts, L. R. (2014). Platelet -derived growth factor primes cancer-associated fibroblasts for apoptosis. *Journal of Biological Chemistry*, *289*, 22835–22849.

Rocha, F. G., Lee, H., Katabi, N., DeMatteo, R. P., Fong, Y., D'Angelica, M. I., et al. (2012). Intraductal papillary neoplasm of the bile duct: A biliary equivalent to intraductal papillary mucinous neoplasm of the pancreas? *Hepatology*, *56*, 1352–1360.

Rőszer, T. (2015). Understanding the mysterious M2 macrophage through activation markers and effector mechanisms. *Mediators of Inflammation*, *2015*, 816460.

Roy, S., Glaser, S., & Chakraborty, S. (2019). Inflammation and progression od cholangiocarcinoma: Role of angiogenic and lymphangiogenic mechanisms. *Frontiers in Medicine*, *6*, 293.

Sabbatino, F., Villani, V., Yearley, J. H., Deshpande, V., Cai, L., Konstantinidis, I. T., et al. (2016). PD-L1 and HLA Class I antigen expression and clinical course of the disease in intrahepatic cholangiocarcinoma. *Clinical Cancer Research*, *22*, 470–478.

Sadri, A.-R., Jeschke, M. G., & Amini-Nik, S. (2016). Advances in liver regeneration: Revisiting hepatic stem/progenitor cells and their origin. *Stem Cells International*, *2016*, 7920897.

Saha, S. K., Parachoniak, C. A., Ghanta, K. S., Fitamant, J., Ross, K. N., Najem, M. S., et al. (2014). Mutant IDH inhibits HNF4α to block hepatocyte differentiation and promote biliary cancer. *Nature*, *513*, 110–114.

Saha, S. K., Zhu, A. X., Fuchs, C. S., & Brooks, G. A. (2016). Forty-year trends in cholangiocarcinoma incidence in the U.S.: Intrahepatic disease on the rise. *The Oncologist*, *21*, 594–599.

Sahai, E., Astsaturov, I., Cukierman, E., De Nardo, D. G., Egeblad, M., Evans, R. M., et al. (2020). A framework for advancing our understanding of cancer-associated fibroblasts. *Nature Reviews Cancer*, *20*, 174–186.

Saito, Y., Nakaoka, T., Muramatsu, T., Ojima, H., Sukeda, A., Sugiyama, Y., et al. (2018). Induction of differentiation of intrahepatic cholangiocarcinoma cells to functional hepatocytes using an organoid culture system. *Scientific Reports*, *8*, 2821.

Saltykova, I. V., Petrov, V. A., & Brindley, P. J. (2018). Opisthorchiasis and the microbiome. *Advances in Parasitology*, *102*, 1–23.

Sasaki, M., Sato, Y., & Nakanuma, Y. (2017). Mutational landscape of combined hepatocellular carcinoma and cholangiocarcinoma, and its clinicopathological significance. *Histopathology*, *70*, 423–434.

Sato, Y., Harada, K., Sasaki, M., & Nakanuma, Y. (2014). Cystic and micropapillary epithelial changes of peribiliary glands might represent a precursor lesion of biliary epithelial neoplasms. *Virchows Archives*, *464*, 157–163.

Sawasdee, N., Thepmalee, C., Sujjitjoon, J., Yongpitakwattana, P., Junking, M., Poungvarin, N., et al. (2020). Gemcitabine enhances cytotoxic activity of effector T-lymphocytes against chemo-resistant cholangiocarcinoma cells. *International Immunopharmacology*, *78*, 106006.

Seehawer, M., Heinzmann, F., D'Artista, L., Harbig, J., Roux, P.-F., Hoenicke, L., et al. (2018). Necroptosis microenvironment directs lineage commitment in liver cancer. *Nature*, *562*, 69–75.

Sha, M., Jeong, S., Qiu, B.-j., Tong, Y., Xia, L., Xu, N., et al. (2018). Isolation of cancer-associated fibroblasts and its promotion to the progression of intrahepatic cholangiocarcinoma. *Cancer Medicine*, *7L*, 4665–4677.

Shimada, K., Sano, T., Sakamoto, Y., Esaki, M., Kosuge, T., & Ojima, H. (2007). Surgical outcomes of the mass-forming plus periductal infiltrating types of intrahepatic cholangiocarcinoma: A comparative study with the typical mass-forming type of intrahepatic cholangiocarcinoma. *World Journal of Surgery*, *31*, 2016–2022.

Shimonishi, T., Sasaki, M., & Nakanuma, Y. (2000). Precancerous lesions of intrahepatic cholangiocarcinoma. *Journal of Hepatobiliary and Pancreatic Surgery*, *7*, 542–550.

Sia, D., Hoshida, Y., Villanueva, A., Roayaie, S., Ferrer, J., Tabak, B., et al. (2013). Integrative molecular analysis of intrahepatic cholangiocarcinoma reveals 2 classes that have different outcomes. *Gastroenterology*, *144*, 829–840.

Sia, D., Losic, B., Moeini, A., Cabellos, L., Hao, K., Revill, K., et al. (2015). Massive parallel sequencing uncovers actionable FGFR2-PPHLN1 fusion and ARAF mutations in intrahepatic cholangiocarcinoma. *Nature Communications*, *6*, 6087.

Sia, D., Villanueva, A., Friedman, S. L., & Llovet, J. M. (2017). Liver cancer cell of origin, molecular class, and effects on patient prognosis. *Gastroenterology*, *152*, 745–761.

Sigel, C. S., Drill, E., Zhou, Y., Basturk, O., Askan, G., Pak, L. M., et al. (2018). Intrahepatic cholangiocarcinomas have histologically and immunophenotypically distinct small and large duct patterns. *American Journal of Surgical Pathology*, *42*, 1334–1345. https://doi.org/10.1097/PAS0000000000001118.

Simile, M. M., Bagella, P., Vidili, G., Spanu, A., Manetti, R., Seddaiu, M. A., et al. (2019). Targeted therapies in cholangiocarcinoma: Emerging evidence from clinical trials. *Medicina*, *55*, 42.

Sirica, A. E. (2012). The role of cancer-associated myofibroblasts in intrahepatic cholangiocarcinoma. *Nature Reviews Gastroenterology & Hepatology*, *9*, 44–54.

Sirica, A. E., Almenara, J. A., & Li, C. (2014). Periostin in intrahepatic cholangiocarcinoma: Pathobiological insights and clinical implications. *Experimental and Molecular Pathology*, *97*, 515–524.

Sirica, A. E., Campbell, D. J., & Dumur, C. I. (2011). Cancer-associated fibroblasts in intrahepatic cholangiocarcinoma. *Current Opinion in Gastroenterology*, *27*, 276–284.

Sirica, A. E., Dumur, C. I., Campbell, D. J. W., Almenara, J. A., Olorunseun, O. O., & Dewitt, J. L. (2009). Intrahepatic cholangiocarcinoma progression: Prognostic factors and basic mechanisms. *Clinical Gastroenterology and Hepatology*, *7*, S68–S78.

Sirica, A. E., Gainey, T. W., & Mumaw, V. R. (1994). Ductular hepatocytes-evidence for a bile ductular cell origin in furan-treated rats. *American Journal of Pathology*, *145*, 375–383.

Sirica, A. E., & Gores, G. J. (2014). Desmoplastic stroma and cholangiocarcinoma: Clinical implications and therapeutic targeting. *Hepatology*, *59*, 2397–2402.
Sirica, A. E., Gores, G. J., Groopman, J. D., Selaru, F. M., Strazzabosco, M., Wang, X. W., et al. (2019). Intrahepatic cholangiocarcinoma: Continuing challenges and translational advances. *Hepatology*, *69*, 1803–1815.
Song, Y., Fu, Y., Xie, Q., Zhu, B., Wang, J., & Zhang, B. (2020). Anti-angiogenic agents in combination with immune checkpoint inhibitors: A promising strategy for cancer treatment. *Frontiers in Immunology*, *11*, 1956.
Sow, H. S., Ren, J., Camps, M., Ossendorp, F., & ten Dijke, P. (2019). Combined inhibition of TGF-β signaling and the PD-L1 immune checkpoint is differentially effective in tumor models. *Cells*, *8*, 320.
Spirlì, C., Fabris, L., Duner, E., Fiorotto, R., Ballardini, G., Roskams, T., et al. (2003). Cytokine-stimulated nitric oxide production inhibits adenylyl cyclase and cAMP-dependent secretion in cholangiocytes. *Gastroenterology*, *124*, 737–753.
Sripa, B., Kaewkes, S., Sithithaworn, P., Mairiang, E., Laha, T., Smout, M., et al. (2007). Liver fluke induces cholangiocarcinoma. *PLoS Medicine*, *4*, e201.
Sripa, B., Tangkawattana, S., & Brindley, P. J. (2018). Update on pathogenesis of opisthorchiasis and cholangiocarcinoma. *Advances in Parasitology*, *102*, 97–113.
Stavraka, C., Rush, H., & Ross, P. (2019). Combined hepatocellular cholangiocarcinoma (cHCC-CC): An update of genetics, molecular biology, and therapeutic interventions. *Journal of Hepatocellular Carcinoma*, *6*, 11–21.
Steiner, P. E., & Higginson, J. (1959). Cholangiolocellular carcinoma of the liver. *Cancer*, *12*, 753–759. https://doi.org/10.1002/1097-0142.
Strauss, J., Heery, C. R., Schlom, J, Madan, R. A., Cao, L., Kang, Z., et al. (2018). Phase I trial of M7824 (MSB0011359C), a bifunctional fusion protein targeting PD-L1 and TGFβ, in advanced solid tumors. *Clinical Cancer Research*, *24*, 1287–1295. https://doi.org/10.1158/1078-0432.CCR-17-2653.
Strauss, O., Phillips, A., Ruggiero, K., Bartlett, A., & Dunbar, P. R. (2017). Immunofluorescence identifies distinct subsets of endothelial cells in the human liver. *Scientific Reports*, *7*, 44356.
Strazzabosco, M., Fiorotto, R., Cadamuro, M., Spirli, C., Mariotti, V., Kaffe, E., et al. (2018). Pathophysiologic implications of innate immunity and autoinflammation in the biliary epithelium. *Biochimica et Biophysica Acta, Molecular Basis of Disease*, *1864*, 1374–1379.
Sun, D., Luo, T., Dong, P., Zhang, N., Chen, J., Zhang, S., et al. (2020). CD86+/CD206+ tumor-associated macrophages predict prognosis of patients with intrahepatic cholangiocarcinoma. *PeerJ*, *8*, e8458.
Takagi, S., Miyagawa, S., Ichikawa, E., Soeda, J., Miwa, S., Miyagawa, Y., et al. (2004). Dendritic cells, T-cell infiltration, and Grp94 expression in cholangiocellular carcinoma. *Human Pathology*, *35*, 881–886.
Tan, D.-W., Fu, Y., Su, Q., Guan, M.-J., Kong, P., Wang, S.-Q., et al. (2016). Prognostic significance of neutrophil to lymphocyte ratio in oncologic outcomes of cholangiocarcinoma: A meta-analysis. *Scientific Reports*, *6*, 33789.
Tang, D., Nagano, H., Yamamoto, H., Wada, H., Nakamura, M., Kondo, M., et al. (2006). Angiogenesis in cholangiocellular carcinoma: Expression of vascular endothelial growth factor, angiopoietin-1/2, thrombospondin-1 and clinicopathological significance. *Oncology Reports*, *15*, 525–532. https://doi.org/10.3892/or.15.3.525.
Tang, Z., Qian, M., & Ho, M. (2013). The role of mesothelin in tumor progression and targeted therapy. *Anti-Cancer Agents in Medicinal Chemistry*, *13*, 276–280.
Taniai, M., Higuchi, H., Burgart, L. J., & Gores, G. J. (2002). p16INK4a promoter mutations are frequent in primary sclerosing cholangitis (PSC) and PSC-associated cholangiocarcinoma. *Gastroenterology*, *123*, 1090–1098.

Tedesco, D., Thapa, M., Chin, C. Y., Ge, Y., Gong, M., Li, J., et al. (2018). Alterations in intestinal microbiota lead to production of interleukin 17 by intrahepatic γδ T-Cell receptor-positive cells and pathogenesis of cholestatic liver disease. *Gastroenterology*, *154*, 2178–2193.

Terada, T. (2013). Combined hepatocellular-cholangiocarcinoma with stem cell features, ductal plate malformation subtype: A case report and proposal of a new subtype. *International Journal of Clinical and Experimental Pathology*, *6*, 737–748.

Thepmalee, C., Panya, A., Junking, M., Chieochansin, T., & Yenchitsomanus, P. T. (2018). Inhibition of IL-10 and TGF-β receptors on dendritic cells enhances activation of effector T-cells to kill cholangiocarcinoma cells. *Human Vaccines & Immunotherapeutics*, *14*, 1423–1431.

Thepmalee, C., Panya, A., Sujjitjoon, J., Sawasdee, N., Poungvarin, N., Junking, M., et al. (2020). Suppression of TGF-β and IL-10 receptors on self-differentiated dendritic cells by short-hairpin RNAs enhanced activation of effector T-cells against cholangiocarcinoma cells. *Human Vaccines & Immunotherapeutics*, *16*, 2318–2327.

Thuwajit, C., Thuwajit, P., Jamjantra, P., Pairojkul, C., Wongkham, S., Bhudhisawasdi, V., et al. (2017). Clustering of patients with intrahepatic cholangiocarcinoma based on serum periostin may be predictive of prognosis. *Oncology Letters*, *14*, 623–634.

Tian, M., Liu, W., Tao, C., Tang, Z., Zhou, Y., Song, S., et al. (2020). Prediction of overall survival in resectable intrahepatic cholangiocarcinoma: IS_{ICC}-applied prediction model. *Cancer Science*, *111*, 1084–1092.

Tian, M.-X., Zhou, Y.-F., Qu, W.-F., Liu, W.-R., Jin, L., Jiang, X.-F., et al. (2019). Histopathology-based immunoscore predicts recurrence for intrahepatic cholangiocarcinoma after hepatectomy. *Cancer Immunology, Immunotherapy*, *68*, 1369–1378.

Trauner, M., Halilbasic, E., Kazemi-Shirazi, L., Kienbacher, C., Staufer, K., Traussnigg, S., et al. (2014). Therapeutic role of bile acids and nuclear receptor agonists in fibrosing cholangiopathies. *Digestive Diseases*, *32*, 631–636.

Tschaharganeh, D. F., Xue, W., Calvisi, D. F., Evert, M., Michurina, T. V., Dow, L. E., et al. (2014). p53 dependent Nestin regulation links tumor suppression to cellular plasticity in liver cancer. *Cell*, *158*, 579–592.

Tsukahara, T., Shimoyama, Y., Ebata, T., Yokoyama, Y., Igami, T., Sugawara, G., et al. (2016). Cholangiocarcinoma with intraductal tubular growth pattern versus intraductal papillary growth pattern. *Modern Pathology*, *29*, 293–301.

Tu, E., Chia, P. Z., & Chen, W. (2014). TGFβ in T cell biology and tumor immunity: Angel or devil? *Cytokine & Growth Factor Reviews*, *25*, 423–435.

Utispan, K., Sonongbua, J., Thuwajit, P., Chau-In, S., Pairojkul, C., Wongkham, S., et al. (2012). Periostin activates integrin α5β1 through a PI3K/AKT-dependent pathway in invasion of cholangiocarcinoma. *International Journal of Oncology*, *41*, 1110–1118.

Utispan, K., Thuwajit, P., Abiko, Y., Charngkaew, K., Paupairoj, A., Chau-in, S., et al. (2010). Gene expression profiling of cholangiocarcinoma derived fibroblasts reveals alterations related to tumor progression and indicates periostin as a poor prognostic marker. *Molecular Cancer*, *9*, 13.

Vale, N., Gouveia, M. J., Gärtner, F., & Brindley, P. J. (2020). Oxysterols of helminth parasites and pathogenesis of foodborne hepatic trematodiasis caused by Opisthorchis and Fasciola species. *Parasitology Research*, *119*, 1443–1453.

van Tong, H., Brindley, P. J., Meyer, C. G., & Velavan, T. P. (2017). Parasite infection, carcinogenesis and human malignancy. *eBioMedicine*, *15*, 12–23.

Vaquero, J., Aoudjehane, L., & Fouassier, L. (2020). Cancer-associated fibroblasts in cholangiocarcinoma. *Current Opinion in Gastroenterology*, *36*, 63–69.

Vesterhus, M., & Karlsen, T. H. (2020). Emerging therapies in primary sclerosing cholangitis: Pathophysiological basis and clinical opportunities. *Journal of Gastroenterology*, *55*, 588–614.

Vicent, S., Lieshout, R., Saborowski, A., Verstegen, M. M. A., Raggi, C., Recalcati, S., et al. (2019). Experimental models to unravel the molecular pathogenesis, cell of origin and stem cell properties of cholangiocarcinoma. *Liver International*, *39*, 79–97.

Vienot, A., & Neuzillet, C. (2019). Cholangiocarcinoma: The quest for a second-line systemic treatment. *Translational Cancer Research*, *8*(Suppl. 3), S275–S288.

Vigano, L., Soldani, C., Franceschini, B., Cimino, M., Lleo, A., Donadon, M., et al. (2019). Tumor-infiltrating lymphocytes and macrophages in intrahepatic cholangiocellular carcinoma. Impact on prognosis after complete surgery. *Journal of Gastrointestinal Surgery*, *23*, 2216–2224.

Vijgen, S., Terris, B., & Rubbia-Brandt. (2017). Pathology of intrahepatic cholangiocarcinoma. *HepatoBiliary Surgery and Nutrition*, *6*, 22–34.

Waisberg, D. R., Pinheiro, R. S., Nacif, L. S., Rocha-Santos, V., Martino, R. B., & Arantes, R. M. (2018). Resection for intrahepatic cholangiocarcinoma: New advances. *Translational Gastroenterology and Hepatology*, *3*, 60.

Wan, X.-S., Xu, Y.-Y., Qian, J.-Y., Yang, X.-B., Wang, A.-Q., He, L., et al. (2013). Intraductal papillary neoplasm of the bile duct. *World Journal of Gastroenterology*, *19*, 8595–8604.

Wang, J., Bo, X., Suo, T., Liu, H., Ni, X., Shen, S., et al. (2018). Tumor-infiltrating neutrophils predict prognosis and adjuvant chemotherapeutic benefit in patients with biliary cancer. *Cancer Science*, *109*, 2266–2274.

Wang, P., Dong, Q., Zhang, C., Kuan, P.-F., Liu, Y., Jeck, W. R., et al. (2013). Mutations in isocitrate dehydrogenase 1 and 2 occur frequently in intrahepatic cholangiocarcinomas and share hypermethylation targets with glioblastomas. *Oncogene*, *32*, 3091–3100.

Wang, J., Dong, M., Xu, Z., Song, X., Zhang, S., Qiao, Y., et al. (2018). Notch2 controls hepatocyte-derived cholangiocarcinoma formation in mice. *Oncogene*, *37*, 3229–3242. https://doi.org/10.1038/s41388-018-0188-1.

Wang, T., Drill, E., Vakiani, E., Pak, L. M., Boerner, T., Askan, G., et al. (2019). Distinct histomorphological features are associated with IDH1 mutation in intrahepatic cholangiocarcinoma. *Human Pathology*, *91*, 19–25.

Wang, J., Li, E., Yang, H., Wu, J., cheng Lu, H., Yi, C., et al. (2019). Combined hepatocellular-cholangiocarcinoma: A population level analysis of incidence and mortality trends. *World Journal of Surgical Oncology*, *17*, 43.

Wang, D., Lin, J., Yang, X., Long, J., Bai, Y., Yang, X., et al. (2019). Combination regimens with PD-1/PD-L1 immune checkpoint inhibitors for gastrointestinal malignancies. *Journal of Hematology & Oncology*, *12*, 42.

Wang, C., Pang, S., Si-Ma, H., Yang, N., Zhang, H., Fu, Y., et al. (2019). Specific risk factors contributing to early and late recurrences of intrahepatic cholangiocarcinoma after curative resection. *World Journal of Surgical Oncology*, *17*, 2.

Williamson, K. D., & Chapman, R. W. (2015). Primary sclerosing cholangitis: A clinical update. *British Medical Bulletin*, *114*, 53–64.

Wójcik, M., Ramadori, P., Blaschke, M., Sultan, S., Khan, S., Malik, I. A., et al. (2012). Immunodetection of cyclooxygenase-2 (COX-2) is restricted to tissue macrophages in normal rat liver and to recruited mononuclear phagocytes in liver injury and cholangiocarcinoma. *Histochemistry and Cell Biology*, *137*, 217–233.

Woo, H. G., Lee, J-H., Yoon, J-H., Kim, C. Y., Lee, H-S., Jang, J. J., et al. (2010). Identification of a cholangiocarcinoma-like gene expression trait in hepatocellular carcinoma. *Cancer Research*, *70*, 3034–3041. https://doi.org/10.1158/0008-5472.CAN-09-2823.

Xu, J., Igarashi, S., Sasaki, M., Matsubara, T., Yoneda, N., Kozaka, K., et al. (2012). Intrahepatic cholangiocarcinomas in cirrhosis are hypervascular in comparison to those in normal liver. *Liver International*, *32*, 1156–1164.

Xu, X., Kobayashi, S., Qiao, W., Li, C., Xiao, C., Radaeva, S., et al. (2006). Induction of intrahepatic cholangiocellular carcinoma by liver-specific disruption of *Smad4* and *Pten* in mice. *The Journal of Clinical Investigation*, *116*, 1843–1852.

Xu, G., Sun, L., Li, Y., Xie, F., Zhou, X., Yang, H., et al. (2019). The clinicopathological and prognostic value of PD-L1 expression in cholangiocarcinoma: A meta-analysis. *Frontiers in Oncology*, *9*, 897.

Xue, R., Chen, L., Zhang, C., Fujita, M., Li, R., Yan, S.-M., et al. (2019). Genomic and transcriptomic profiling of combined hepatocellular and intrahepatic cholangiocarcinoma reveals distinct molecular subtypes. *Cancer Cell*, *35*, 932–947.

Yamada, D., Rizvi, S., Razumilava, N., Bronk, S. F., Davila, J. I., Champion, M. D., et al. (2015). IL-33 facilitates oncogene induced cholangiocarcinoma in mice by an IL-6 sensitive mechanism. *Hepatology*, *61*, 1627–1642.

Yamamoto, M., Oshita, A., Nishisaka, T., Nakahara, H., & Itamoto, T. (2018). Synchronous double primary hepatic cancer consisting of hepatocellular carcinoma and cholangiolocellular carcinoma: A case report. *Journal of Medical Case Reports*, *12*, 224.

Yamanaka, T., Harimoto, N., Yokobori, T., Muranushi, R., Hoshino, K., Hagiwara, K., et al. (2020). Nintedanib inhibits intrahepatic cholangiocarcinoma aggressiveness via suppression of cytokines extracted from cancer-associated fibroblasts. *British Journal of Cancer*, *122*, 986–994.

Yan, C., Koda, S., Wu, J., Zhang, B. B., Yu, Q., Netea, M. G., et al. (2020). Roles of trained immunity in the pathogenesis of cholangiopathies: A novel therapeutic target. *Hepatology*, *72*, 1838–1850.

Yang, X., Li, Y., Zou, L., & Zhu, Z. (2019). Role of exosomes in crosstalk between cancer-associated fibroblasts and cancer cells. *Frontiers in Oncology*, *9*, 356.

Yang, X., Lin, Y., Shi, Y., Li, B., Liu, W., Yin, W., et al. (2016). FAP promotes immunosuppression by cancer-associated fibroblasts in the tumor microenvironment via STAT3-CCL2 signaling. *Cancer Research*, *76*, 4124–4135.

Yang, J., & Yan, L.-N. (2008). Current status of intrahepatic cholangiocarcinoma. *World Journal of Gastroenterology*, *14*, 6289–6297.

Yao, K. J., Jabbour, S., Parekh, N., Lin, Y., & Moss, R. A. (2016). Increasing mortality in the United States from cholangiocarcinoma: An analysis of the National Center for Health Statistics Database. *BMC Gastroenterology*, *16*, 117.

Yao, W., Qian, Y., Campbell, D., Wang, J., & Sirica, A. E. (2018). Su1455-Underexpression of phospho-LKB1, atypical PKCι, and Hes1 associated with disruption of polarized morphogenesis in a 3-dimensional culture model of rat cholangiocarcinoma progression mediated by activated neu receptor. *Gastroenterology*, *154*, S–1152. https://doi.org/10.1016/S00165085(18)33816-2.

Yoo, C., Oh, D.-Y., Choi, H. J., Kudo, M., Ueno, M., Kondo, S., et al. (2020). Phase I study of bintrafusp alfa, a bifunctional fusion protein targeting TGF-β and PD-L1, in patients with pretreated biliary tract cancer. *Journal for Immunotherapy of Cancer*, *8*, e000564.

Yoshida, G. J., Azuma, A., Miura, Y., & Orimo, A. (2019). Activated fibroblast program orchestrates tumor initiation and progression: Molecular mechanisms and the associated therapeutic strategies. *International Journal of Molecular Sciences*, *20*, 2256.

Yu, L., Feng, M., Kim, H., Phung, Y., Kleiner, D. E., Gores, G. J., et al. (2010). Mesothelin as a potential therapeutic target in human cholangiocarcinoma. *Journal of Cancer*, *1*, 141–149.

Zaccari, P., Cardinale, V., Severi, C., Pedica, F., Carpino, G., Gaudio, E., et al. (2019). Common features between neoplastic and preneoplastic lesions of the biliary tract and the pancreas. *World Journal of Gastroenterology*, *25*, 4343–4359.

Zender, S., Nickeleit, I., Wuestefeld, T., Sörensen, I., Dauch, D., Bozko, P., et al. (2013). A critical role for notch signaling in the formation of cholangiocellular carcinoma. *Cancer Cell*, *23*, 784–795.

Zeng, J., Liu, Z., Sun, S., Xie, J., Cao, L., Lv, P., et al. (2018). Tumor-associated macrophages recruited by periostin in intrahepatic cholangiocarcinoma stem cells. *Oncology Letters*, *15*, 8681–8686.
Zhang, X.-F., Beal, E. W., Bagante, F., Chakedis, J., Weiss, M., Popescu, I., et al. (2018). Early *versus* late recurrence of intrahepatic cholangiocarcinoma after resection with curative intent. *British Journal of Surgery*, *105*, 848–856.
Zhang, X, Dong, M, Pan, Y, Chen, J, Huang, X, Jin, Y, et al. (2017). Expression pattern of cancer-associated fibroblast and its clinical relevance in intrahepatic cholangiocarcinoma. *Human Pathology*, *65*, 92–100. https://doi.org/10.1016/j.humpath.2017.04.014.
Zhang, L., Jiang, L., Sun, Q., Peng, T., Lou, K., Liu, N., et al. (2007). Prostaglandin E2 enhances mitogen-activated protein kinase/Erk pathway in human cholangiocarcinoma cells: Involvement of EP1 receptor, calcium and EGF receptors signaling. *Molecular and Cellular Biochemistry*, *305*, 19–26.
Zhang, M., Yang, H., Wan, L., Wang, Z., Wang, H., Ge, C., et al. (2020). Single-cell transcriptomic architecture and intercellular crosstalk of human intrahepatic cholangiocarcinoma. *Journal of Hepatology*, *73*, 1118–1130.
Zheng, B. H., Ma, J. Q., Tian, L. Y., Dong, L. Q., Song, G. H., Pan, J. M., et al. (2020). The distribution of immune cells within combined hepatocellular carcinoma and cholangiocarcinoma predicts clinical outcome. *Clinical and Translational Medicine*, *10*, 45–56.
Zhou, S. L., Dai, Z., Zhou, Z.-J., Chen, Q., Wang, Z., Xiao, Y.-S., et al. (2014). CXCL5 contributes to tumor metastasis and recurrence of intrahepatic cholangiocarcinoma by recruiting infiltrative intratumoral neutrophils. *Carcinogenesis*, *35*, 597–605.
Zhou, G., Sprengers, D., Mancham, S., Erkens, R., Boor, P., van Beek, A. A., et al. (2019). Reduction of immunosuppressive tumor microenvironment in cholangiocarcinoma by ex vivo targeting immune checkpoint molecules. *Journal of Hepatology*, *71*, 753–762.
Zhu, Y., & Kwong, L. N. (2020). Insights into the origin of intrahepatic cholangiocarcinoma from mouse models. *Hepatology*, *72*, 305–314. https://doi.org/10.1002/hep.31200.
Zimmermann, A. (2017). Tumors and tumor-like lesions of peribiliary glands. In *Tumors and tumor-like lesions of the hepatobiliary tract* (pp. 779–788). Cham, Switzerland: Springer. https://doi.org/10.1007/978-3-319-26956-6_41.